CLEMENTINE CHURCHILL

ALSO BY MARY SOAMES

Winston and Clementine: The Personal Letters of the Churchills

A Churchill Family Album

The Profligate Duke: George Spencer Churchill, Fifth Duke of Marlborough, and His Duchess

Winston Churchill: His Life as a Painter

CLEMENTINE CHURCHILL

The Biography of a Marriage

BY HER DAUGHTER

MARY SOAMES

Revised and Updated

A MARINER BOOK

Houghton Mifflin Company

Boston • New York

First Mariner Books edition 2003

Visit our Web site: www.houghtonmifflinbooks.com.

Library of Congress Cataloging-in-Publication Data
Soames, Mary.
Clementine Churchill / Mary Soames
p. cm.
ISBN 0-395-27597-0
ISBN 0-618-26732-8 (pbk.)
1. Churchill, Clementine Ogilvy Hozier, Lady 1885–1977.
2. Churchill, Winston Leonard Spencer, Sir 1874–1965.
3. Great Britain—Politics and government—20th century.
4. Prime minister's wives—Great Britain—Biography.
5. Prime ministers—Great Britain—Biography. I. Title.

Printed in the United States of America

QUM 10 9 8 7 6 5 4 3 2 1

This book is dedicated to my children

NICHOLAS • EMMA • JEREMY
CHARLOTTE • RUPERT
1979

and to my grandchildren, who at the latest count are

HARRY • ISABELLA • CHRISTOPHER
EMILY
GEMMA • FLORA • ARCHIE
CLEMENTINE • ANTONIA
ARTHUR • DAISY • JACK
2002

Contents

Illustrations

Following page 87

Following page 264

Following page 416

Maps

Pictures not credited come from Mary Soames's private collection.

Photographs from the Broadwater Collection are reproduced with permission of Curtis Brown Ltd on behalf of Winston S. Churchill.

The following illustrations are reproduced by permission of the Master and Fellows of Churchill College, Cambridge: 27, 39, 61, 71-5, 81-2 and 98.

The letters reproduced on p. 63 are as follows: WSC to CSC, 1909, CSCT 02/02/08; CSC to WSC, 21 April 1909, CSCT 01/03/12, and WSC to CSC, 31 August 1909, CSCT 02/02/13. These extracts and the letters reproduced on p. 531 are reprinted by permission of the Churchill Archives Centre, Spencer-Churchill Papers.

The maps on pp. 153 and 184 are redrawn from *Winston S. Churchill, Companion Volume III*, by Martin Gilbert.

Every effort has been made to verify sources and copyright-holders, but where photographs bear no clue as to their attribution or have been pasted into albums this has proved impossible. The author apologizes therefore for any inadvertent errors or omissions.

Acknowledgements

TO THE REVISED AND UPDATED EDITION

I WISH TO REITERATE MY INDEBTEDNESS TO ALL THOSE WHOSE ASSISTANCE I acknowledged in 1979 when this book was first published. Sadly many individuals then mentioned are dead – my debt to them remains as great.

May I acknowledge again my gratitude to Her Majesty The Queen for her gracious permission to use material which is in Her Majesty's copyright and to quote from two of her speeches. I wish also to thank Her late Majesty Queen Elizabeth The Queen Mother for graciously allowing me to quote a conversation, and from letters to and from Her late Majesty.

My main sources of information for this book have been the same as for the 1979 edition: namely, my mother Clementine Churchill's own recollections (both written and spoken), and a recent and even more thorough study of the letters between my parents during the course of their life together, and here I have quoted more extensively from their correspondence.

As before, the bedrock of my information concerning the events in my father's life has been the eight-volume official biography started by my late brother Randolph S. Churchill and continued by Sir Martin Gilbert, CBE, D.Litt., whose *Companion Volumes*, and their continuation in the *Churchill War Papers*, are also a mine of information. Furthermore, in his *In Search of Churchill* (1994), Sir Martin has added considerably to the whole picture of my father's life and personality; I am most grateful also to him for his permission to use several of his excellent maps. Again I would like to record my great appreciation to Sir Martin for his ever-ready help and advice.

To members of my own family I once more tender my thanks: firstly, to my nephew Winston S. Churchill for his generosity in granting me use of the great corpus of material which lies within his copyright; for the extra knowledge I have gleaned from his biography of his father: *His Father's Son: The Life of Randolph Churchill* (1996), and his permission to quote therefrom; and for enabling me to see and use my mother's letters to his father. Also for his permission to quote from a number of my father's works as listed in Notes and Sources, and in the Bibliography.

Then my thanks to my niece Edwina Sandys for permission on behalf of her family to quote from her mother, Diana Churchill's letters to her mother. And to my niece Celia Sandys for advice and information on various matters.

I am also most grateful to my cousin Clarissa, Countess of Avon, for allowing me to see and use the most moving letter from my mother to Anthony Eden (the late Earl of Avon) after her husband's death. My thanks also to the University of Birmingham and to Miss C. L. Penney, Head of Special Collections, for sending me copies of letters from Clementine Churchill which are in the Avon Papers.

I am much indebted to my late cousin Peregrine Spencer-Churchill for talking to me at length about his recollections of my parents and his (Jack and Goonie Churchill), and about his memories of Chartwell days. Also for the use of papers in his possession, notably Lady Randolph Churchill's diary for 1882, and some wonderful old family album photographs.

The Churchill Papers are now in the public domain under the aegis of the Sir Winston Churchill Archive Trust, and in the care of the Churchill Archives Centre at Churchill College, Cambridge: I am grateful to the Master and Fellows of Churchill College for their hospitality to myself and my researcher. And my special thanks to Allen Packwood and his staff at the Archives Centre for their unfailing help to us both.

To Ann Hoffmann, my researcher and assistant, I once more offer my heartfelt thanks and appreciation: it was she who first helped me over thirty years ago when I was starting to write the biography of my mother, and she has helped me with my other books as well. It is my extreme good fortune that she has worked with me once again on this revival of *Clementine*: her store of knowledge on our family is by now profound, and I have depended greatly on her wise and expert advice and judgement on many matters relating to the revision and updating of the original work. Ann Hoffmann has researched for me at – among other places – the Bodleian Library; the Churchill Archives Centre; the Fitzwilliam Museum; and the Liddell Hart Centre for Military Archives, and has unearthed some extremely valuable new material. May I thank her particularly for her intense hard work in the last weeks of preparing the book for the publishers, for her eagle-eyed scrutiny of the draft, and her valued criticism. My thanks to her also for compiling the index.

To my Private Secretary, Nonie Chapman, who once again has kept my life and house 'in order', liaised with Ann Hoffmann, done endless copyings and undertaken much correspondence in relation to this book, I give my true thanks. The fact that for the last thirteen years of my mother's life she was her Private Secretary has meant that her memories of that time have been invaluable to me.

I am most grateful to the Trustees of the Bonham Carter Family Papers and the papers of H. H. Asquith at the Bodleian Library, University of Oxford, for their generous permission to use a considerable amount of material from these papers. I would especially like to thank the Hon. Virginia Brand, whose great help in a number of matters I have much appreciated. And I am much indebted to Mrs Priscilla Hodgson for allowing me to quote from her grandmother Margot Asquith's diaries and letters, also at the Bodleian Library. My thanks also to Colin Harris of the Department of Special Collections and Western Manuscripts at the Bodleian Library, who on several occasions has been exceptionally helpful to my research assistant in this connection.

I wish to express my gratitude to Celia Lee, author of *Jean, Lady Hamilton 1861–1941: A Soldier's Wife* (2001), who contacted me about references to both Winston and Clementine in Lady Hamilton's diaries, which she was researching, and who sent me many extracts from them. And my thanks to her husband, John Lee, whose *General Sir Ian Hamilton 1883–1947: A Soldier's Life* (2000) has provided me with interesting information. I gratefully acknowledge the permission of the Trustees of the Liddell Hart Centre for Military Archives to use quotations from Lady Hamilton's diaries, and would like to thank Mrs Patricia Methven and her staff at the centre for help to my researcher, Ann Hoffmann.

Hugo Vickers has been most kind in sending me relevant extracts from Cecil Beaton's unpublished diaries, and a press article; and I thank him for giving me permission to quote from his *Cecil Beaton: The Authorized Biography* (1985); I would also like to thank the Literary Executors of the late Sir Cecil Beaton for permission to quote from *Self-Portrait with Friends: The Selected Diaries of Cecil Beaton 1926–1974*, edited by Richard Buckle (1979). My thanks also to Hugo Vickers for several charming photographs.

The National Trust and Mrs Carol Kenwright, the Property Manager at Chartwell, have been most helpful, particularly in the matter of granting permission to use for the book cover the portrait of Clementine Churchill by Chandor which hangs in the drawing-room there – and I thank them warmly.

I wish to express my great thanks to the following people and publishers for their help:

Mrs Harriet Bowes Lyon for her permission to quote from her late father, Sir John Colville's *Footprints in Time* (1976) and to Hodder & Stoughton for extracts from his *Fringes of Power*; Lady Rose Clowes for putting me in touch with her late aunt, Mrs Borwick, about 'Bay' Middleton; my cousin, the Duchess of Devonshire (Deborah Mitford), for looking through her family papers for any letters from my mother to her parents, the late Lord and Lady Redesdale; Amiral Philippe de Gaulle and Mme Elisabeth de Boissieu for their permission to include two letters from their father, Général de Gaulle, to my mother; Jon Meacham for sending me Margaret Suckley's diaries, *Closest Companion* (1995); Sir Anthony Montague Browne, KCMG, CBE, DFC, for permission to use his letter to *The Times* of 6 June 1966; Viscount Montgomery of Alamein, CMG, CBE, for renewed permission to quote from his father, Field Marshal Viscount Montgomery of Alamein's letters to my mother and myself, and for the use of a photograph taken by 'Monty'; Lord Moran, KCMG, and Constable and Robinson, publishers, for permission to quote from his father's book, *Winston Churchill: The Struggle for Survival 1940–65* (1966); Viscount Norwich, CVO, for permission to quote from his mother, Lady Diana Cooper's *Trumpets from the Steep* (1966); Phil

Reed, Curator of the Cabinet War Rooms, for valuable information concerning my father's occupancy and use; Dr David Reynolds for drawing my attention to Clementine Churchill's letter from Moscow to Averell Harriman in April 1945; Mrs Jeanne Spencer-Churchill for permission to use extracts from her grandfather, Paul Maze's diary; Viscount Thurso, MP, for permission to use two letters from Winston Churchill to his father, Sir Archibald Sinclair, in 1915 and 1916, from the Thurso Papers at the Churchill Archives Centre.

I would also like to thank the following for help given to my researcher, Ann Hoffmann: Colin Clifford, in connection with the Asquith Papers; Mrs Heather Laughton, for finding several out-of-print books; and the staffs of the Bodleian Library; Churchill College Archives Centre; Miss Elizabeth Fielden and staff of the Fitzwilliam Museum Department of Manuscripts; Mrs J. V. Thorpe of the Gloucestershire County Record Office; Hertfordshire Archives and Local Studies; House of Lords Record Office; Imperial War Museum; Liddell Hart Centre for Military Archives; National Register of Archives; Northamptonshire Record Office; Public Record Office; and Dr Lesley Hall of the Wellcome Library for the History and Understanding of Medicine. Dr M. K. Banton of the Research and Editorial Department of the Public Record Office and John Green of the Prime Minister's Office Records, Cabinet Office, were especially helpful with regard to Winston and Clementine's 1943 telegrams.

My agent, Mike Shaw of Curtis Brown, has been a constant support and fount of valuable advice, and I thank him warmly. And I am grateful to Anthea Morton-Saner, also of Curtis Brown, for guiding me through the jungle of copyright.

I am delighted that again my publishers are Doubleday, and it has been a joy to work once more with Sally Gaminara. My thanks also to her team, and especially to Sheila Lee for picture research; to Patsy Irwin, Deputy Publicity Director; and to Rebecca Winfield, Rights and Contracts Director. I also want to express my grateful appreciation to Gillian Bromley for her meticulous copy-editing.

* * *

The author and publishers are grateful to the following authors, publishers, agents, literary executors and others not already mentioned above or where they differ from the acknowledgements made in the first edition: to the Clerk of the Records, House of Lords Record Office, on behalf of the Beaverbrook Foundation, for permission to quote from the Beaverbrook Papers in their custody and a short extract from *Politicians and the War*; to the Syndics of the Fitzwilliam Museum for their kind permission to quote from the Wilfrid Scawen Blunt papers, and for the use of a photograph; to the Controller of Her Majesty's Stationery Office, on behalf of Parliament, for the reproduction of parliamentary copyright material from *Hansard*; to Nigel Nicolson for passages from Harold Nicolson, *Letters and Diaries*

1939–45; to Hertfordshire Archives and Local Studies for the use of a letter from Clementine Churchill to Lady Desborough from the Grenfell Papers; to David Higham Associates for permission to quote from Lady Cynthia Asquith, *Remember and Be Glad*, Lord Hankey, *The Supreme Command 1914–1918*, and Sir Edward Marsh, *A Number of People*; to Weidenfeld and Nicolson for a short extract from *'Chips': The Diary of Sir Henry Channon*, edited by Robert Rhodes James; to Sheil Land Associates Ltd for a short quotation from Robert Rhodes James, *Churchill: A Study in Failure*; to the Carlton Publishing Group (André Deutsch Ltd) and W. W. Norton & Co. for several short passages from Joseph P. Lash, *Eleanor and Franklin*; and to Brandt & Hochman and HarperCollins for an extract from Robert E. Sherwood, *The White House Papers of Harry L. Hopkins*, published in the US as *Roosevelt and Hopkins*.

Every effort has been made to trace the current copyright-holders of all material quoted. In several cases this has proved a difficult and time-consuming task, as since original publication in 1979 some publishers have been amalgamated into other groups, or rights have reverted to an author or an author's estate. We apologize, therefore, for any inadvertent errors or omissions which, if our attention is drawn to them, will gladly be corrected in any reprinted editions of this work.

Acknowledgements
TO THE FIRST EDITION

I WISH FIRST OF ALL TO ACKNOWLEDGE MY GRATITUDE TO HER MAJESTY THE Queen for her gracious permission to use letters which are in Her Majesty's copyright, and to quote from two of her speeches. I also wish to thank Her Majesty Queen Elizabeth The Queen Mother for graciously allowing me to quote from letters to and from Her Majesty.

My main sources of information for this book are my mother's – Clementine Churchill's – own recollections told to me at various times, and her own brief written account of part of her childhood and girlhood, entitled '*My* Early Life'. My parents' relationship to each other, their views on the events in their lives, and their opinions of other people, are for the greater part drawn from their massive correspondence, direct quotations from which, for reasons of space alone, I have had to curtail. I have also drawn on my mother's letters to my sister, Sarah Audley, and on my mother's and my own personal diaries and papers.

For the events in my father's life I have depended in the main on his own published accounts (as listed in the bibliography), and on his official biography started by my brother, Randolph S. Churchill, and continued after his death by Mr Martin Gilbert, M.A., to whom I wish to record my special thanks for his ready help to me and my research assistant in all matters concerning this book on which we have sought his assistance.

Members of my family have given me much help, and I record my true thanks to my late brother, Randolph Churchill, for making copies of my parents' letters to each other which were in his possession available to me. I am very grateful to my sister, Sarah Audley, for permission to use extracts from her letters, and for making available to me our mother's letters to her. Sarah also read the whole script, and I was greatly helped by her comments, and by being able to discuss with her intimate family matters. I wish to thank the children of my late sister Diana, Julian and Edwina Sandys, for permission to quote from her letters, and Celia Walters for searching for photographs of her mother. And I am grateful to my nephew, Winston S. Churchill, for giving me permission to quote from his father's (Randolph Churchill's) letters to his mother, and for allowing me to use the photograph of the sketch portrait of her painted by Winston Churchill, which is in his possession.

I am indebted to my kinsman, the Earl of Airlie, for allowing me to read, and use extracts from, the diaries of his great-grandmother, Blanche, Countess of Airlie, and for allowing me to copy various family photographs. Lord Airlie also kindly read the early chapters of the book, and I am grateful to him for his comments on points of family history. I received valuable help from the late Lady Helen Nutting, who through her own recollections and family letters shed useful light on the relationship between Lady Blanche Hozier and her husband. Also, my thanks are due to Professor Jack Ogilvy of Boulder, Colorado, for writing to me his recollections of my grandmother, Lady Blanche.

I am grateful to my cousin, the late (10th) Duke of Marlborough, for his account of the 'scene' at Blenheim described in Chapter 7 and for allowing me to have copied extracts from the Blenheim Visitors' Book; and I wish to thank his son, the present Duke of Marlborough, for supplying me with a photograph of the Temple of Diana at Blenheim.

I am indebted to my cousin, the late Giles Romilly, who furnished me with copies of Clementine Churchill's letters to her sister Nellie Romilly and to himself. And I wish to thank the Trustees of the Romilly Papers, who allowed me access to them after Giles Romilly's death, and from which I gleaned some valuable background detail for my mother's early years.

I wish to record my gratitude to my late cousin, Judy Montagu (Mrs Milton Gendel), who made available to me in the first instance the extracts concerning my mother in the H. H. Asquith/Venetia Stanley correspondence. And I am most grateful to the Hon. Mark Bonham Carter for assisting me to verify and correct Judy Montagu's transcripts through the good offices of the Warden of Nuffield College, Oxford, Mr Michael G. Brock, M.A., F.R.Hist.S. I also thank Mr Bonham Carter for his permission to quote these extracts, as well as extracts from *Winston Churchill As I Knew Him* by his mother, Lady Violet Bonham Carter, and from her letters to my mother.

I am indebted to Mr Donald Whyte, F.S.A.Scot., L.H.G., who compiled genealogical tables for me which have been invaluable, and who also supplied me with interesting information about the Ogilvy and Hozier families, and other related matters. Also to Mr P. L. Gwynn-Jones, Bluemantle Pursuivant of Arms, for advice and layout of the family pedigree.

I owe a special debt of gratitude to Sir John Colville, C.B., C.V.O., and to Mr Anthony Montague Browne, C.B.E., D.F.C., who have both read much of the manuscript, and whose advice and criticism have been of inestimable value.

The following people have also read and commented on parts of the manuscript, and I am most grateful to them: Miss Nonie Chapman; Lord Clark, O.M., C.H. (re the Sutherland picture); Mr Frank Giles; Miss Grace Hamblin, O.B.E.; the Hon. Mrs Henley, O.B.E.; Sir Robert Mackworth-Young, K.C.V.O. (Librarian at Windsor Castle); Mr Alan Philpotts (Literary Trustee); Sir Frank Roberts, G.C.M.G., G.C.V.O. (Clementine's journey to Russia); the late Mr Michael Wolff.

I would like to record my thanks to the following persons for sending me reminiscences and making information available to me, or for sending me letters and photographs: Sir Richard Barlas, K.C.B., O.B.E. (Clerk of the House of Commons); Miss M. R. Bateman, M.A. (Headmistress of Berkhamsted High School for Girls); the Marquess of Bath; the Dowager Countess of Bessborough; the Countess of Birkenhead; the late Margaret, Countess of Birkenhead; Mr E. H. Cookridge (for valuable information concerning Sir Henry Hozier's writings and military career); Col. R. O. Dennys, Somerset Herald (WSC funeral arrangements); Lord Duncan-Sandys, P.C., C.H.; the Hon. Mrs Averell Harriman; Miss A. M. Johnson; Sir Alan Lascelles; Viscount Montgomery of Alamein, C.B.E.; Mr A. H. Montgomery (Perseverance Trust - Howard House Appeal); Mrs Bryce Nairn; Lord Redesdale; Mr Sydney Reynolds (information concerning the Hoziers' house at Berkhamsted); Mr Kenneth Rose; Mr John Spencer Churchill; Mr Peregrine Spencer-Churchill; the Rev. C. N. White (former Rector of Ellesborough, near Chequers); the late Miss Maryott Whyte; Lord Zuckerman, O.M., K.C.B., F.R.S.

For enabling my research assistant to copy entries in the Chequers Visitors' Book I wish to thank Group Captain J. M. Ayre, C.B.E. (Secretary to the Chequers Trustees) and Vera E. Thomas, O.B.E., Wing Commander W.R.A.F. Rtd. (Curator at Chequers).

For permission to have photographed several family portraits at Chartwell, and for permitting my research assistant to copy entries from the Chartwell Visitors' Book, I wish to thank the National Trust, and Miss Grace Hamblin, O.B.E., and Mrs Jean Broome (the former, and present, Administrators at Chartwell).

I am most grateful for the help afforded me by Mrs Pat Bradford and Miss Clare Stephens, Archivists at Churchill College, Cambridge.

On matters connected with my mother's visits to air-raid shelters during the Second World War I am grateful for information supplied by the late Mrs Olive Prentice and by Mrs J. Fawcett, Archivist of the British Red Cross Society.

For information and reminiscences about both my parents covering the forty years Winston Churchill was the Member of Parliament for the Epping (later Wanstead and Woodford) Division, I wish to thank very particularly: Col. W. H. Barlow-Wheeler, D.S.O., M.B.E.; Mr Donald Forbes, J.P.; Mrs Doris Moss, O.B.E.; Mr A. R. Pittman (Editor of the *Loughton and Woodford Gazette and Guardian*); and Mrs Vera Wilson. In the same connection I also want to thank Mr Gordon (Red) Allen, D.F.C.; Mrs Lilian Henderson; and Mr David A. Thomas.

For information concerning my mother's involvement with the Y.M.C.A. in the First World War (munition workers' canteens) I am indebted to: Mrs G. L. Clapperton; Dame Joan Marsham, D.B.E.; and Mr R. E. Roberts (former General Secretary).

I am very grateful to the following, who have helped to supply information about my mother's work for the Y.W.C.A. in the Second World War:

Miss J. M. Nelson (formerly Personal Assistant to the National General Secretary); Miss Edna Rowe (Archivist); Miss E. Sharples (National General Secretary).

I wish to express my appreciation to Prudence Cuming Associates for photographing the pictures at Chartwell, and those in my possession which appear in the book. And to Marriotts Photo Stores of Tunbridge Wells, who have copied with care and skill many faded and damaged photographs from old albums. Also to the Viking Studios, Dundee, for copying pictures and photographs made available to me by Lord Airlie.

I wish to record my thanks to the staffs of BBC Written Archives; the British Library; Conservative Central Office; Hereford and Worcester Records Office; the House of Lords Record Office; and the London Library, who have helped either myself or my research assistant on various occasions.

I also wish to thank all members of the general public who kindly sent reminiscences in response to my appeals through the Press asking for material – even if their specific contributions do not appear in the text, all facts and reminiscences were valuable in building up the overall picture.

I am truly grateful to all those who, at various times and in diverse ways, have helped me with the preparation of this book.

Specifically I wish to thank Mr Robert Skepper, M.A., who got me started, and under whose direction Mr W. Allen, Mr C. Beauman and Mr de Bunsen did the first arranging and indexing of the material.

I am most grateful to Miss Heather Forbes-Watson, who was my secretary from 1963 to 1967, and who typed all the earlier chapters.

I wish to express my thanks to Mr John Hazelden, who with patience and tenacity recorded on tape all references in *The Times* to my mother; and my thanks too to Mrs Irene Thrower who typed out the tapes.

I am greatly indebted to Miss Ann Hoffmann of Authors' Research Services, who did all my research from 1966 onwards, which included interviewing many people on my behalf. She has housed and organized my archives, and typed the greater part of the manuscript; she has also compiled the index. I am deeply grateful to Ann Hoffmann, not only for her professional skill and hard work, but also for her unfailing enthusiasm and devotion to the task.

I wish to record my thanks to both Mrs Philippa Boland and Mrs Pat Hodgson, who have assisted Miss Hoffmann at various times.

I am very grateful indeed to Mrs Margot Levy, M.A., who has checked my manuscript and advised me on the historical and political background of the biography. When it became clear that the original manuscript was much too long, she gave me invaluable help in rearranging and cutting the material for the final draft.

I wish to record my gratitude to Sir Denis Hamilton, D.S.O., who, both while Editor of the *Sunday Times* and later as Editor-in-Chief and Chief Executive of Times Newspapers Ltd., demonstrated his confidence in me, and whose support and patience I have greatly appreciated.

To Mr Graham Watson of Curtis Brown Ltd. I owe a special debt of gratitude for his sympathetic and ever sensible advice. His constructive criticism and encouragement have throughout been of the greatest help to me.

I wish to express my warm thanks to Miss Anne Carter of Cassell Ltd. and to her assistant, Miss Jannet Grant, who with great patience and meticulous care have seen me through the last stages of preparing the book for publication.

Finally, I want to say a heartfelt 'thank you' to my husband. Although, undeniably, the various events in his distinguished career have mainly served to hinder, rather than prosper, the progress of this book – yet without his belief that I was the person who should write Clementine Churchill's life, and his unwavering confidence in my capacity to do so, I would have lost heart on many occasions.

The author and publishers are grateful to the following authors, publishers, agents, literary executors and others for permission to quote copyright material not already mentioned:

Firstly, to C. & T. Publications Ltd., William Heinemann Ltd. and the Houghton Mifflin Company for permission to quote extensively from the Churchill Papers. Churchill's letters to his wife (the copyright of C. & T. Publications Limited) quoted in this book appear in full in the seven-volume official biography *Winston S. Churchill* by Randolph Churchill and Martin Gilbert (Heinemann: U.K., and Houghton Mifflin: U.S.A.). Next, to Lord Clark, O.M., C.H. and Sir Alan Lascelles for the use of extracts from their own letters to the Author; to Viscount Montgomery of Alamein, M.B.E. for two extracts from his father's letters to Lady Churchill and to the Author; to Lady Madeleine Lytton for two passages from the letters of Neville Lytton to Nellie Hozier; to Lady Diana Cooper for permission to quote from her book, *Trumpets from the Steep*; to the Hon. Mark Bonham Carter for the use of several extracts from the Asquith/Stanley correspondence; to the Beaverbrook Foundation for access to the Beaverbrook Papers, now in the custody of the House of Lords Record Office and reproduced by courtesy of the Clerk of the Records; to the Beaverbrook Foundation and William Collins, Sons & Co. Ltd. for a short extract from Beaverbrook, *Politicians and the War*; to the Director of Publishing at Her Majesty's Stationery Office for the use of Crown Copyright material from *Hansard*; to Times Newspapers Ltd. for permission to quote medical reports on Churchill signed by Lord Moran and Sir Russell Brain and one statement issued by Clementine Churchill also printed in that newspaper; to Mr Anthony Montague Browne, Mr Evan Davies and the executors of Mrs B. E. Moir (deceased) for extracts from letters to *The Times*; to the Hamlyn Group, for passages from Winston S. Churchill, *Painting as a Pastime* and from Lord Riddell, *More Pages from my Diary*; to the Hamlyn Group and Charles Scribner's Sons for *Thoughts and Adventures* and *The World Crisis* by Winston S. Churchill; to Cassell Ltd. and Houghton Mifflin for a number of quotations from Winston S. Churchill, *The Second World War*; to Curtis Brown Ltd. for extracts from Randolph S. Churchill,

Twenty-One Years; to Sarah, Lady Audley and André Deutsch for a quotation from her book, *A Thread in the Tapestry*; to André Deutsch and W. W. Norton & Company Inc. for some short passages from Joseph P. Lash, *Eleanor and Franklin*; to the Hon. Mark Bonham Carter and Thornton Butterworth Ltd., the original publishers of Margot Asquith's *Autobiography*, for two short extracts; to the Hutchinson Publishing Group Ltd. for extracts from Mabell, Countess of Airlie, *Thatched with Gold*, E. Ashmead-Bartlett, *The Uncensored Dardanelles*, Lady Cynthia Asquith, *Diaries 1915–18* and Clementine Churchill, *My Visit to Russia*; to Weidenfeld (Publishers) Ltd. for passages from Harold Nicolson, *Letters and Diaries 1939–45*, and Robert Rhodes James, *'Chips': The Diary of Sir Henry Channon*, and *Churchill: A Study in Failure*; to Eyre & Spottiswoode (Publishers) Ltd. and William Collins, Sons & Co. Ltd. for 8 short extracts from Lady Violet Bonham Carter's *Winston Churchill As I Knew Him*; to George Allen & Unwin (Publishers) Ltd. for permission to quote a short passage from Lord Hankey, *The Supreme Command*; to Collins, Publishers and Curtis Brown Ltd. for one short extract from Tom Harrisson, *Living through the Blitz*; to Brandt & Brandt, Literary Agents Inc. and Harper & Row for permission to quote from Robert E. Sherwood, *The White House Papers of Harry L. Hopkins*; to the Estate of the late Dr Hewlett Johnson for the use of a short extract from *Searching for Light*; to the London School of Economics and Political Science for permission to quote from Beatrice Webb, *Our Partnership*; to the Secretary of the University Court of the University of Glasgow for an extract from the speech of the late Professor John Duncan Mackie; to Miss E. Sharples, National General Secretary of the Y.W.C.A. for the use of letters of former officials of the Association; to Dr John G. Griffith, Public Orator of the University of Oxford, for both verifying the Latin text of the speech of a former Public Orator and clearing up the question of copyright; to the Editor of the *British Medical Journal* for a quotation from that publication; finally, permission to quote 8 lines from the song 'I'm Forever Blowing Bubbles' © 1919 by Kendis Brockman Music Co., sub-published by B. Feldman & Co. Ltd. and reproduced by permission of EMI Music Publishing Ltd.

SELECT FAMILY TREE

Showing Family Members & Connections *as at May 2002*

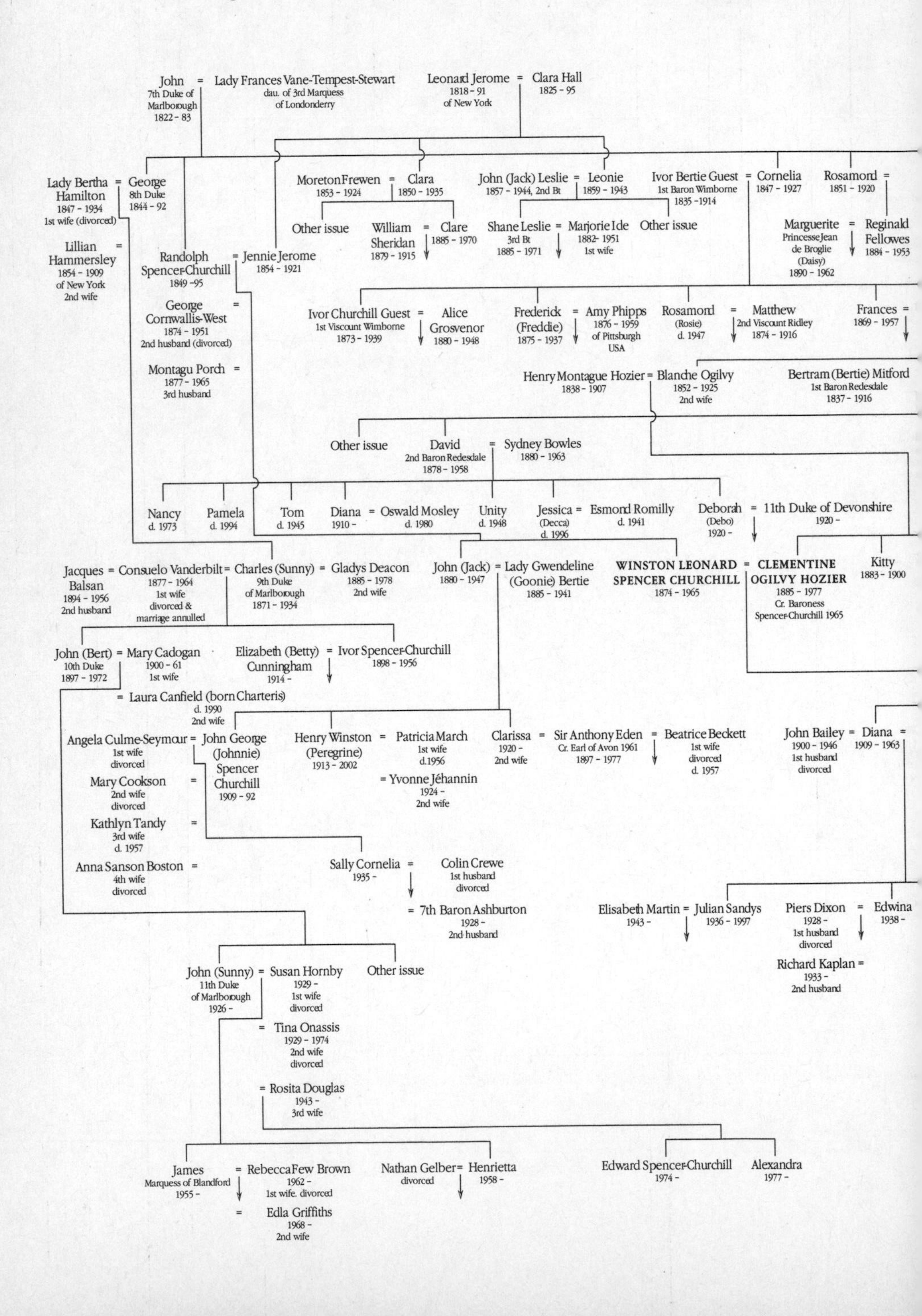

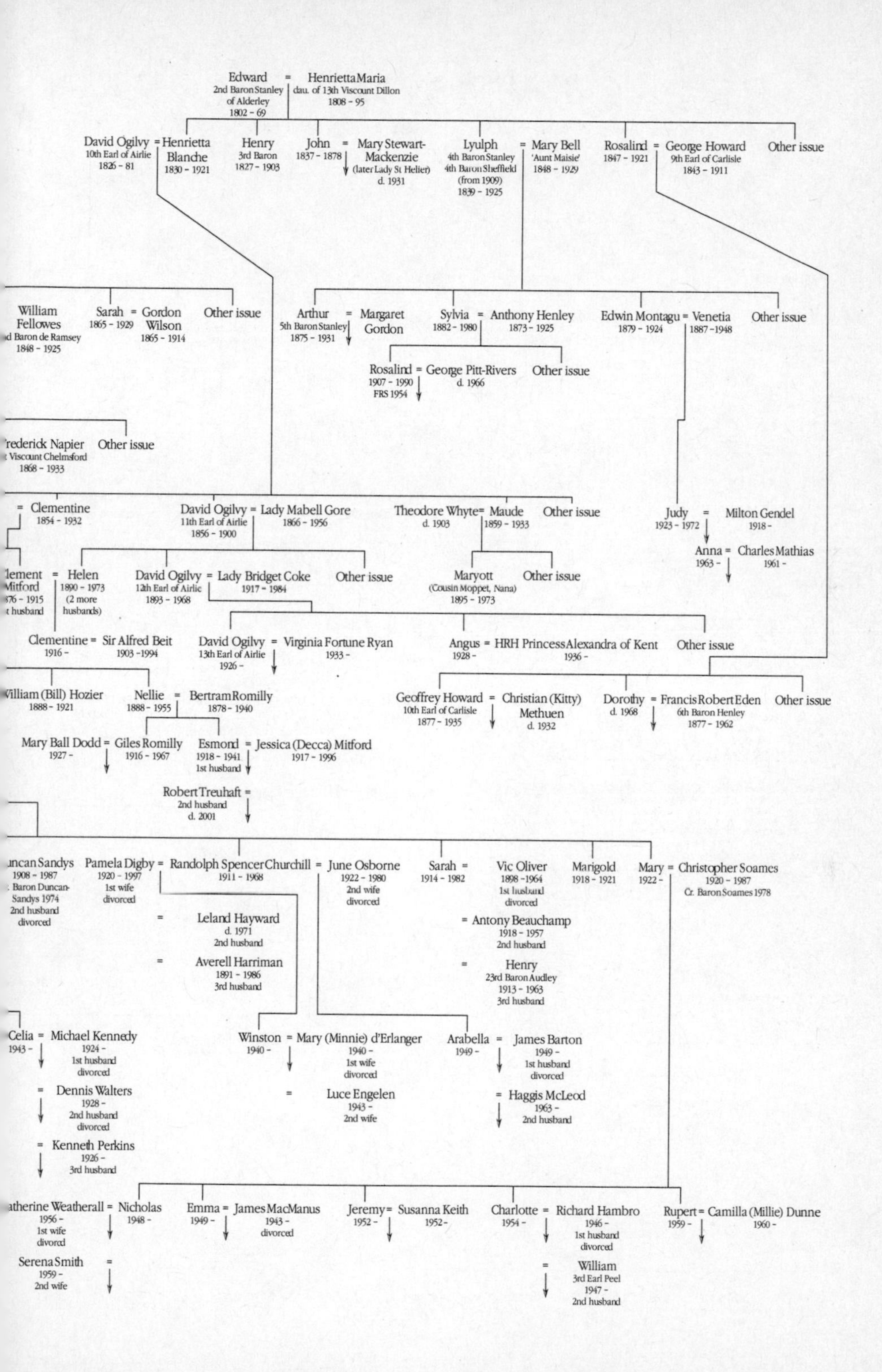

Edward = Henrietta Maria
2nd Baron Stanley of Alderley 1802 - 69
dau. of 13th Viscount Dillon 1808 - 95
David Ogilvy = Henrietta Blanche
10th Earl of Airlie 1826 - 81
1830 - 1921
Henry
3rd Baron 1827 - 1903
John = Mary Stewart-Mackenzie
1837 - 1878
(later Lady St Helier) d. 1931
Lyulph = Mary Bell
4th Baron Stanley 4th Baron Sheffield (from 1909) 1839 - 1925
'Aunt Maisie' 1848 - 1929
Rosalind = George Howard
1847 - 1921
9th Earl of Carlisle 1843 - 1911
Other issue
William Fellowes
d Baron de Ramsey 1848 - 1925
Sarah = Gordon Wilson
1865 - 1929
1865 - 1914
Other issue
Arthur = Margaret Gordon
5th Baron Stanley 1875 - 1931
Sylvia = Anthony Henley
1882 - 1980
1873 - 1925
Edwin Montagu = Venetia
1879 - 1924
1887 - 1948
Other issue
Rosalind = George Pitt-Rivers
1907 - 1990 FRS 1954
d. 1966
Other issue
rederick Napier
Viscount Chelmsford 1868 - 1933
Other issue
= Clementine
1854 - 1932
David Ogilvy = Lady Mabell Gore
11th Earl of Airlie 1856 - 1900
1866 - 1956
Theodore Whyte = Maude
d. 1903
1859 - 1933
Other issue
Judy = Milton Gendel
1923 - 1972
1918 -
Anna = Charles Mathias
1963 -
1961 -
lement Mitford
376 - 1915 t husband
= Helen
1890 - 1973 (2 more husbands)
David Ogilvy = Lady Bridget Coke
12th Earl of Airlie 1893 - 1968
1917 - 1984
Other issue
Maryott
(Cousin Moppet, Nana) 1895 - 1973
Other issue
Clementine = Sir Alfred Beit
1916 -
1903 -1994
David Ogilvy = Virginia Fortune Ryan
13th Earl of Airlie 1926 -
1933 -
Angus = HRH Princess Alexandra of Kent
1928 -
1936 -
Other issue
Villiam (Bill) Hozier
1888 - 1921
Nellie = Bertram Romilly
1888 - 1955
1878 - 1940
Geoffrey Howard = Christian (Kitty) Methuen
10th Earl of Carlisle 1877 - 1935
d. 1932
Dorothy = Francis Robert Eden
d. 1968
6th Baron Henley 1877 - 1962
Other issue
Mary Ball Dodd = Giles Romilly
1927 -
1916 - 1967
Esmond = Jessica (Decca) Mitford
1918 - 1941 1st husband
1917 - 1996
Robert Treuhaft =
2nd husband d. 2001
uncan Sandys
1908 - 1987 Baron Duncan-Sandys 1974 2nd husband divorced
Pamela Digby = Randolph Spencer Churchill = June Osborne
1920 - 1997 1st wife divorced
1911 - 1968
1922 - 1980 2nd wife divorced
Sarah =
1914 - 1982
Vic Oliver
1898 -1964 1st husband divorced
Marigold
1918 - 1921
Mary = Christopher Soames
1922 -
1920 - 1987 Cr. Baron Soames 1978
= Leland Hayward
d. 1971 2nd husband
= Averell Harriman
1891 - 1986 3rd husband
= Antony Beauchamp
1918 - 1957 2nd husband
= Henry
23rd Baron Audley 1913 - 1963 3rd husband
Celia = Michael Kennedy
1943 -
1924 - 1st husband divorced
= Dennis Walters
1928 - 2nd husband divorced
= Kenneth Perkins
1926 - 3rd husband
Winston = Mary (Minnie) d'Erlanger
1940 -
1940 - 1st wife divorced
= Luce Engelen
1943 - 2nd wife
Arabella = James Barton
1949 -
1949 - 1st husband divorced
= Haggis McLeod
1963 - 2nd husband
atherine Weatherall = Nicholas
1956 - 1st wife divorced
1948 -
Serena Smith =
1959 - 2nd wife
Emma = James MacManus
1949 -
1943 - divorced
Jeremy = Susanna Keith
1952 -
1952 -
Charlotte = Richard Hambro
1954 -
1946 - 1st husband divorced
= William
3rd Earl Peel 1947 - 2nd husband
Rupert = Camilla (Millie) Dunne
1959 -
1960 -

Preface

TO THE REVISED AND UPDATED EDITION

IT IS NOW OVER THIRTY YEARS SINCE I STARTED TO WORK ON MY MOTHER'S biography, and over twenty since it was first published; and the book has long been out of print. During this time many new sources of material have become available to me in subsequently published memoirs, letters and diaries, and in collections of papers deposited in libraries and other archive centres, which shed light on Clementine Churchill's life and character, and on her relationships with her husband, children and friends, and which show other people's views of her. I have been able also to try to unravel the tangled skein of her paternity.

Recently, for the purposes of the book *Speaking for Themselves: The Personal Letters of Winston and Clementine Churchill*, edited by myself and published in 1998*, I read again the entire letter-dialogue between my parents during the fifty-seven years of their relationship; and, with the additional material now accessible to me, I pondered anew various aspects of my mother's life and characteristics. With the longer perspective the passing of time has given me, I think I have come to have a better understanding of her at various moments in her life – both before and after her marriage.

For these reasons I welcome the opportunity to revise, update and expand the biography I wrote so many years ago, in the hope that I may do better justice to Clementine Churchill – a remarkable and, in some ways, a heroic woman; and that this new edition of her life may interest a later generation of readers.

Mary Soames
London
April 2002

Preface

TO THE FIRST EDITION

IT WAS MY HUSBAND WHO SUGGESTED TO ME THAT I OUGHT TO WRITE MY mother's life. The idea at first startled me; but presently I broached the subject with my mother. She liked the idea, and with characteristic immediacy sent me, within a week, the tin boxes containing her archives. From the outset it was agreed between us that the book should not appear in her lifetime.

It would always have been a difficult task for me, since I have never before written so much as a pamphlet. A further dilemma was the sheer wealth of the material; and reasons of space alone made it imperative for me to cut severely quotations from my parents' prolific correspondence.

I have not always followed a strictly chronological arrangement: to have done so (without interrupting the flow of the narrative) would have prevented me from dealing in depth with such important aspects of my mother's life as Chartwell, her war work (1939–45), and her personal attitudes and relationships.

It has been a personally moving experience for me to delve back into my mother's long lifetime; and to trace through her spoken recollections, but above all in their letters to each other, the record of her loving and heroic partnership with my father, which lasted over half a century.

One curious effect of writing about my parents is that as time and the book go on, from being an observer in the wings I myself become involved as a witness in the story, and a background, 'incidental' participant in events which have shaped and shaken all our lives in the last forty years.

My mother herself took a great interest in the book, and read, or had read to her, all the chapters up to the early days of the First World War: and she was pleased with them.

Now that the book is finished, as in Parliament I must 'declare my interest': it is a labour of love – but I trust not of blind love.

Mary Soames
Castle Mill House
Hampshire
January 1979

CLEMENTINE CHURCHILL

CHAPTER ONE

Forebears and Early Childhood

IN 1878 LADY BLANCHE OGILVY, THE ELDEST DAUGHTER OF THE 10TH EARL OF Airlie, became engaged to Colonel Henry Hozier. Her family acquiesced with relief rather than enthusiasm in a match which could not have been considered a brilliant, or even a very suitable one for the beautiful daughter of an earl. Henry Hozier was forty, and older by fourteen years than his intended bride. He was a gentleman, but he possessed only a modest fortune, and none of the four Ogilvy girls was well endowed. Moreover, Colonel Hozier had already been married, and only quite recently divorced, a circumstance which would, in those days, certainly have been regarded as a serious impediment. He had met Blanche on one of his visits to her parents' home, Cortachy Castle, Kirriemuir, Angus, for he was a friend of long standing of Lady Airlie – although her husband always thought the Colonel a bit of a bounder.

When Blanche Ogilvy was fifteen, the famous artist G. F. Watts was staying at Cortachy painting the portraits of her parents; he painted also a small and ravishing study of Blanche's head. She had the face of an angel – but this angelic countenance belied a very humanly passionate nature, and, as she grew older, she was to cause her parents grave anxiety by her wayward spirit. In 1878 she was, moreover, twenty-six years old, and Lord and Lady Airlie no doubt thought it high time she was 'settled'. They therefore gave their consent to the engagement, and Henry and Blanche were married in the private chapel at Cortachy Castle, according to the rites of the Episcopal Church of Scotland, on 28 September 1878.

The Ogilvys are of ancient Scottish lineage, and their family has a stormy and romantic history. Their ancestry can be traced back to the middle of the twelfth century when the third son of the Earl of Angus assumed the name of Sir Patrick Ogilvy on being given the land and barony of Ogilvy for services to King William the Lion. His descendant, Sir James Ogilvy, was raised to the peerage in 1491 for services to his country as Ambassador to Denmark. The Lord Ogilvy of the day was one of those who signed the association in defence of Mary, Queen of Scots in 1568. The earldom of Airlie dates from 1639, and was granted to the 8th Lord Ogilvy for supporting Charles I against the Covenanters. This same Lord Airlie paid a high price for his loyalty: by way of punishment Airlie Castle was burned down by the Earl of Argyll. It was after this that Cortachy Castle became the family's home.

Earlier in the seventeenth century, another forebear, John Ogilvie, who was converted from Calvinism and became a Jesuit priest, was martyred for his faith at Glasgow Cross. In 1976 he was canonized – becoming the first Scottish saint since St Margaret, the eleventh-century Queen of Scotland.[1]

Throughout the eighteenth century the Ogilvys were unwavering in their support of the Stuarts, and the Ogilvys of Airlie forfeited their land and titles after the defeat of the Jacobite supporters at the Battle of Culloden in 1745. These were not fully restored until 1826.

The character and career of David Graham, 10th Earl of Airlie, lacked the romantic and dashing characteristics of so many of his ancestors, and during his life he departed from several family traditions. He was, for instance, the first Ogilvy ever to be educated at an English school – Eton College, from where he went up to Oxford. Unlike most of his forebears, soldiering held no attraction for him. As a child his health had been fragile, and his gentle nature, combined with scholarly tastes, led him along different paths. Along with his intellectual predilections, the young Lord Ogilvy (as he then was) developed at university a passion for gambling; and as both a backer and an owner of racehorses he was to lose large sums of money, thereby dissipating the family fortunes. He was to spend much of his later life repairing these losses.

He succeeded to the earldom in 1849 aged twenty-three, and two years later he was the first of his family to seek a bride across the English border. In 1851 he married the Hon. Henrietta Blanche Stanley, one of the daughters of Edward, 2nd Baron Stanley of Alderley. She came from a family whose women in particular were renowned for their intellectual abilities and forceful personalities. Blanche Stanley's mother, Henrietta Maria (a daughter of Henry, 13th Viscount Dillon), was a pioneer of women's education, and was one of the founders of Girton College, Cambridge. Her sister, Rosalind, Countess of Carlisle, was a formidable woman who took an active part in politics, combining radical Liberal views with aristocratic arrogance. Her brothers also demonstrated independence of mind – one of them becoming a Roman Catholic priest, and later Bishop of Emmaus, while another embraced the Mohammedan faith.

When Blanche Airlie first came to Scotland as a bride of twenty-one she was shocked by the crudeness of the manners, the limited conversation and the drinking habits of the new circle in which she found herself. With intrepid firmness, the young Countess ordained that after two glasses of wine the decanters were to be removed from the table; card-playing after dinner was banned, because it interfered with the conversation. Alarming and dictatorial to some, she charmed those who shared her tastes and outlook. Clever men enjoyed her company, and when she was a young woman Thackeray, Matthew Arnold and Thomas Carlyle were her friends.

Gladstone, Disraeli, Lord Rosebery and Ruskin were all guests at Cortachy at various times. A sincerely religious woman, Blanche maintained a friendship and correspondence with Dr Benjamin Jowett, the Master of Balliol, that developed in her over the years a philosophical and catholic outlook on spiritual matters. Unlike her sister, Lady Carlisle, she took no part in public life herself, although, with her family, she held Liberal views. But combined with these elevated qualities of mind Blanche Airlie also possessed a valuable fund of common sense; and to an ancient family she brought the priceless gift of sturdy good health.

Throughout the thirty years of her marriage she governed her rather ineffectual husband, her children and her household with unquestioned authority. Persuading her husband to sell his stud, Lady Airlie sought to divert his interests and energies into less unprofitable and more useful channels. Although he never became prominent in politics (he was far too easy-going), through his wife's influence and encouragement he began to take an interest in political and social problems. It was doubtless due to her influence that when he took his seat in the House of Lords, as one of the Representative Scottish peers, he sat with the Liberals, thus breaking the Ogilvy family's traditional Tory loyalty.

In 1864 Lord Airlie went to North America and travelled widely on that great continent, being much impressed and excited by the golden opportunities it offered. When he returned home to Scotland his mind was full of schemes and new ambitions: selling his remaining bloodstock, Lord Airlie used the proceeds to expand and improve the small herd of Aberdeen Angus cattle which he already owned. Such was the success of this venture that he later invested extensively in ranching in America, settling one of his sons, Lyulph, on land in Colorado. It was on one of his visits of inspection, in 1881, that he was taken ill and died in Denver, Colorado.[2]

The Hoziers had grown from a very different root-stock from that of the Ogilvy family. Their forebears had come into Glasgow from the Stirlingshire countryside at the close of the seventeenth century, and their name was originally McIlhose or McLehose. In the city the family prospered as maltmen or brewers, and William McLehose, the great-great-grandfather of Henry Hozier, became a Burgess and Guildbrother of Glasgow in 1717.

The Hozier family history provides an interesting example of how, from humble and obscure beginnings, in the course of only a few generations the transition was made from the professional, city-dwelling class to that of the landed gentry. As time went on the family bought Partick, Newlands and other properties in the neighbourhood of Glasgow. William McLehose (born in 1760) was admitted Burgess and Guildbrother in 1786; he changed his surname to Hozier, which seems to have been suggested by a pun on the name McLehose – 'Mak' the hose'. His son James (1791–1878), an advocate at the Scottish Bar, purchased in 1850 part of the once extensive barony of Mauldslie, near Carluke, Lanarkshire, including the Adam mansion, Mauldslie Castle. James Hozier married Catherine Feilden, daughter of a Lancashire baronet, and had eight children. Their eldest son, William

Wallace, became the 1st Baron Newlands; and it was their youngest son, Henry Montague Hozier, who married Blanche Ogilvy.

Henry Montague Hozier was born in 1838. He was educated at Rugby and Edinburgh Academy. Going into the Army, he showed early distinction by passing first in, and first out, of the Staff College; during his brilliant military career he held a commission in the Royal Artillery, the Life Guards and the 3rd Dragoon Guards. He served in China and Abyssinia, and was attached to the German army during the Austro-Prussian war of 1866. He was also one of the pioneers of the Intelligence Service. During the Franco-Prussian War Hozier was sent as Assistant Military Attaché to the Prussian forces, being subsequently decorated with the Iron Cross by Emperor Wilhelm I. He was a prolific writer on military subjects and particularly on the campaigns of which he was an observer. In addition to all these activities he also dabbled briefly in politics – and was an unsuccessful Liberal Unionist candidate for Woolwich in 1885.

In 1874 Hozier had left the Army and, aged thirty-six, entered the City as Secretary to the Corporation of Lloyd's of London. During his thirty-two years in this post, this unusually able man was to become the dominating and driving force at Lloyd's. Keenly aware of the possibilities of radio-telegraphy (then in its infancy), he exploited this invention to the great advantage and prestige of the Corporation he served. His name will always be linked with the network of telegraphic signal stations connecting their offices throughout the world, which were his creation. He was appointed a KCB in 1903.

Henry Hozier's private life, however, had not proceeded as smoothly as his career. In 1868, aged thirty, he had married Elizabeth Lyon. Ten years later, in May 1878, she divorced him for adultery and desertion; in September of the same year he married Blanche Ogilvy. In addition to his friendship with Lady Airlie, the distinction of Hozier's military and literary career, and his evident success at Lloyd's, no doubt helped to recommend him to the Airlies as an acceptable husband, offering a secure future for their beautiful but flighty daughter.

With all the advantages of hindsight it is possible to feel the couple's life together was destined for storms. The nearly middle-aged, experienced man, with his forceful and determined character, allied to the wayward, passionate young woman, throbbing with the desire to spread her wings, proved to be too explosive a combination. Moreover, they each possessed the power and – alas – the desire, to tease and taunt the other, and it was not long before their natures were in conflict. An early but profound rift was revealed in their different attitudes to children: he was openly averse to the prospect of a family; Blanche, on the other hand, was deeply maternal, and must have found the five years which elapsed before the birth of their first child a time of grievous frustration.

The Hoziers' first child, a girl, was born in 1883 in London at 18 Queen Anne's Gate: named Katharine, she was always called Kitty. She quickly seized and held pride of place in her mother's warm but uneven affections, and in the seventeen years of her life grew to be a dominating force in her family circle. Two years later, on 1 April 1885, at No. 75 Grosvenor Street (to where the Hoziers had moved in the interval), a second daughter was born – somewhat precipitately, in the drawing-room.

The new baby was called Clementine, an Ogilvy name recalling the family's historic loyalties: for 'Clementina' had been a popular name in Jacobite circles since the marriage of the 'Old Pretender' to Princess Maria Clementina of Poland*. Later, in 1888, the twins, Bill and Nellie, completed the family.

We now know that it is extremely unlikely that Henry Hozier was the father of all – or, indeed, any – of the four children. In her later life Clementine became convinced that she was not his child, and would occasionally speculate in private as to whether Kitty and the twins also were not his. But she had no inkling of this in her youth, and although instinctively frightened of him, as she grew older she became increasingly interested by the brilliant and unusual man who was her (presumed) father. It was a lifelong regret to her that Henry Hozier's death in 1907 precluded his knowing Winston.

But what was her origin? It may be that Henry Hozier believed (at any rate for a time) that his two elder children, Kitty and Clementine, were his progeny – certainly he behaved in an aggressively possessive way towards them, while totally ignoring the existence of the twins. There is no doubt that Blanche Hozier was promiscuous – and known to be so at the time. An entry in Wilfrid Scawen Blunt's† 'Secret Memoirs' for 29 May 1891 reads: 'Lord Napier tells me that the scandal about Lady Blanche Hozier was very public. Hozier found Ashmead-Bartlett‡ in her bedroom and turned her out at once into the street. Lady Gregory who was here yesterday tells the same story and adds there were nine other lovers.'[3] On 2 June he recorded that Lady Carlisle§ had been 'very severe on her niece, Lady Blanche, who she says "goes about parading her adulteries à qui veut l'entendre"' [underlining in original].[4]

Despite the incompatibility of Henry and Blanche and their mutual lack of fidelity, the marriage staggered on until after the scandalous incident recounted above, when, in the autumn of that year, Henry Hozier sued for divorce in Scotland. Blanche's brother, Lord Airlie, helped her to the best of his ability; but Hozier tried to blackmail Airlie (who was endeavouring to obtain better financial support for his sister) by threatening to expose Blanche as the adulteress she was. However, the family lawyers had firm proof of Hozier's own adultery, and he unwillingly agreed to a separation.[5]

One of Blanche's lovers was said to be her own brother-in-law, 'Bertie'

* My mother's name therefore is correctly pronounced 'Clement*een*', not 'Clement*yne*', as in the popular song.
† Wilfrid Scawen Blunt (1840–1922), poet, Arabist and passionate protagonist of Home Rule for Ireland. He himself had a brief affair with Blanche Hozier a year or two before this incident; they remained friends and corresponded on and off all their lives.
‡ Ellis Ashmead-Bartlett (1849–1902), American-born parliamentarian. Knighted 1892.
§ Rosalind Stanley (1847–1921), sister-in-law of 10th Earl of Airlie, m. 9th Earl of Carlisle.

Mitford*, who had married her younger sister, Lady Clementine Ogilvy. Leaving Oxford in 1858, he had joined the Foreign Office, and during his wide travels abroad he became a talented linguist. While in Japan for four years he mastered the Japanese language and was able to conduct negotiations with the Mikado. Back in England, now in his early thirties, he was quite a celebrity, and with his breeding, good looks and vivacity, was popular in high society. But the lure of travel and exotic places had gripped him, and he was soon off again.

Mitford had resigned from the Foreign Office in 1873, and in the following year, on his return to London, Disraeli appointed him Secretary to the Board of Works, where he would remain for twelve years. In that same year he married Lady Clementine Ogilvy, with whom he was to have eight children. After inheriting a considerable fortune from a cousin, Mitford resigned his Government appointment, and he and his family left London to live on the family estate at Batsford Park in Gloucestershire. For some years he was a Member of Parliament, but in 1902 he was created Baron Redesdale, and thereafter was a frequent attender in the House of Lords. In this later part of his life he travelled less, but the lure of the East led him to visit Ceylon, and again Japan. All his journeyings resulted in books and, as he became increasingly deaf, he devoted more and more time to writing – including his *Memoirs*, which occupied the last decade of his life. His eldest son, Clement, was killed in 1915 on the Western Front, in the year before he himself died, aged eighty.

Bertie Mitford was a great character; gregarious and full of life, he was also an outstanding figure of a man, and 'with his fine features, sparkling eyes and elastic figure . . . he was a universal favourite', and doubtless a charming brother-in-law for Blanche, 'Aunt Natty', as she was known in the family; she evoked much sympathy, because of her desperately unhappy marriage. She also had great charm, and no doubt whispers of her 'reputation' added interest: she was a favourite among her younger Mitford relations.[6]

Many social gossipers at the time, and a body of latter-day opinion, believe Lord Redesdale to have been Clementine's father; indeed, Blanche herself is said to have told Lady Londonderry just before Clementine's birth in 1885 that the child she was expecting was 'Lord Redesdale's'.[7]

There is another candidate, however, who must be considered for reasons which cannot be ignored. In his 'Secret Memoirs' Wilfrid Scawen Blunt writes, on 22 June 1892: 'Lunched with Blanche Hozier. She has taken a little house in the "Wild West" [Bayswater!] and lives there with her two younger children, the elder ones having been taken by her husband. She tells me she is happy, . . . also that Bay Middleton was the two elder children's father. He broke his neck at a steeplechase in the Spring.'[8]

William George Middleton came from a prominent Scottish sporting family who owned considerable estates in Ayrshire. His father William had his own pack of foxhounds, which he hunted himself in Dumbartonshire.

* Algernon Bertram Freeman-Mitford (1837–1916), 1st Baron Redesdale (in the second creation).

His mother was a Hamilton (an aunt of General Sir Ian Hamilton, a close friend of Winston Churchill from Boer War days, and commander of the ground forces in the Gallipoli campaign of 1915). Young William (born in 1846) was brought up to ride and hunt, gaining with his background expert knowledge of horses and hounds. His parents also had a house in Prince's Gate, London, where they spent much time 'out of season', and William was educated at a private tutor's establishment in Wimbledon. Joining the Army, he entered the 12th Lancers, who were then stationed in Ireland. He soon made a name for himself as a brilliant rider to hounds and a daring steeplechaser. By now he had acquired the nickname 'Bay', by which he would be known for the rest of his life; it may have derived from the (1836) Derby winner of that name, or from the reddish-brown colour of his hair – called in horses 'bay'. Debonair, and undoubtedly possessing a degree of personal magnetism, he was the best of company to both men and women (the latter falling easy victims to his charms). He was much given to practical jokes which stretched acceptable boundaries even in those days, when such pranks were enjoyed and tolerated. Yet he had another side: he collected Dresden china, and was an interested theatregoer; and his normally ebullient nature was tinged with melancholy, which led him to the conviction that he would die a violent death – which tragically would prove to be true.

In 1870 Bay Middleton was appointed to the Lord Lieutenant (Viceroy) of Ireland, Earl Spencer*, as an extra 'hard-riding ADC' to accompany His Excellency (himself a dashing rider) across country. They became firm friends, and when Spencer's first term of office ended in 1874 and he returned to Althorp, in Northamptonshire, Bay retired from the Army and went with him. His country life absorbed him in the sporting season, and in London he led a fashionable life and was much in demand by hostesses. In 1875 he became engaged to Charlotte Baird, a considerable heiress, whose family's wealth derived from coal mining and iron, and whose property marched with the Middletons'. Bay's future seemed all set for a well-found sporting country life in the superb 'shire' counties where hunting, shooting and steeplechasing found near-perfect conditions, and whose packs of fox-hounds and the sport they showed were world-famous.

In 1876, however, the expected course of Middleton's life was suddenly changed when his friend Lord Spencer asked him to undertake to 'pilot' the Empress Elizabeth of Austria†, who had taken a house, Easton Neston, near Towcester, for the season, and who was looking forward to the thrills of English foxhunting. The Empress Elizabeth, then thirty-eight, was exceedingly beautiful and possessed of magnetic charm when she chose to exert it – and she was also a superb horsewoman.

Earl Spencer, in whose hands lay much of the arranging for this Imperial descent on the shires, was well aware of his responsibility for both the

* 5th Earl Spencer (1835–1910), Viceroy of Ireland 1869–74 and 1882–5.

† Empress Elizabeth of Austria (1837–98). The wife of Francis Joseph I (1848–1916), Emperor of Austria and King of Hungary. She was assassinated by an Italian anarchist in Geneva.

pleasure and the safety of the Empress – a fairly awesome charge. It was clear that, since she would be hunting in the fastest country over large, heavy fences with the Pytchley and Grafton packs, she must have a bold and brilliant pilot. Spencer immediately turned to Bay Middleton and persuaded him to fulfil this role, although Bay was far from enthusiastic at the prospect. But after the first day's hunting, both Elizabeth and her escort were mutually captivated – indeed, within the first few fields Bay discovered his charge was as brave as she was beautiful, and a brilliant rider with an intuitive feel for her horse.

For five seasons (three in England and two in Ireland) the Empress and Bay would be inseparable in the hunting field, and afterwards – both at the dinners and entertainments, and at smaller, intimate parties. Gossip was of course rife, but it is thought unlikely that they ever became lovers in the full physical sense. Elizabeth was shy and frigid in sexual matters, but she liked exerting the power her beauty and magnetism gave her over men, and adulation was a necessity to her. After she left England in February 1882 the Empress never hunted in England or Ireland again; and Bay married Charlotte Baird that autumn. Ten years later he would be killed in a steeplechase.

It is not certain when Blanche Hozier and Bay first met, but there is clear evidence that they saw a good deal of each other in the early spring to summer of 1882, as the appointments diary of Jennie Randolph Churchill clearly shows.[9] Bay had known Jennie from his hunting days in Ireland when he was piloting the Empress: the Duke of Marlborough* was at that time Lord Lieutenant of Ireland, and Lord Randolph was acting as his Private Secretary; the young Churchills, living in Dublin, were keen riders to hounds. The friendship was renewed back in London, where one of Jennie's closest friends at that time was Blanche Hozier, already discontented in her marriage, whom she saw several times a week; Bay's name is often mentioned as one of their party. Blanche's first child, Kitty, was born in April 1883, and Clementine two years later. The twins, Bill and Nellie, born in 1888, may, according to a later view,[10] have been Lord Redesdale's.

Today some people point to 'Mitford' eyes – large and blue – which have appeared in Churchill descendants; but Clementine's eyes were hazel/green, and the painting by G. F. Watts of Blanche Ogilvy, aged fifteen, shows her with large, angelic blue eyes: so this feature could have descended through the maternal line.[11] Wilfrid Scawen Blunt seems to have accepted what Blanche had told him (in 1892); he refers to it again in 1916 in his diary, when, after a gossip with 'valiant' (and now ageing) Blanche Hozier, he wrote: 'Bertie [Lord Redesdale] in later years was one of Blanche's lovers but it was not true . . . that Clementine was his daughter. I doubt if anyone but myself knows who Clementine's father was.'[12]

Given Blanche Hozier's promiscuity, and the conflicting evidence, I find

* John Winston, 7th Duke of Marlborough (1822–83), Lord Lieutenant of Ireland 1876–80.

it difficult to take a dogmatic view about her children's paternity. As so often, the French express it so well: *Je n'y ai pas tenue la chandelle.*

* * *

Memories of early childhood are never continuous, but certain moments and incidents are retained with snapshot clarity. The only recollection Clementine had of the house in Grosvenor Street, where she was born, was of being carried downstairs by her nurse to her parents' bedroom, and of being deposited at the foot of their bed. She recalled: 'I was afraid to move from this position, because although my Mother, whom I remember looking lovely and gay, held out her arms to me, on the other side of the bed was my Father's sleeping form; even then I was frightened of him.'[13] Their parents were rather remote figures for the two girls, for Kitty and Clementine lived the ordered life of upper-class children of that period: in the care of a nanny, their existence was centred on the nursery, with brief daily contacts with their parents when they were at home.

In 1888 the twins, Bill and Nellie, were born, and the family was now complete. The babies had their own domain, the nursery, and their life was made happy or miserable by the nurse or governess of the moment. All her life Clementine remembered that when Kitty was about six their mother took her for a visit to the Netherlands, leaving Clementine at home with the twins in the charge of a cross, unkind nurse, whom she remembered as beating her unmercifully.

But there was another, kindlier reign, the memory of which also endured. In the autumn of 1888 Mlle Elise Aeschimann, a Swiss nursery governess, aged sixteen at the time, came to take charge of the two elder girls. Of Mlle Elise there were only happy memories. She stayed for two years, starting the two elder children on their lessons, and laying the foundations of the good French Clementine was to write and speak all her life.

Both the children loved this charming, kindly young woman, who returned their affection. In particular she showed tenderness to Clementine – 'Rosebud', as her mother called her – who was a somewhat timid child and prone to tears. She had a most sensitive conscience, and suffered untold miseries if the immaculate white of her lace-edged pinafore was marred by spot or stain. Clementine remembered that 'when I cried or was naughty she [Mlle Elise] used to carry me about in her arms, although I was quite a big girl. Mother used to tell her not to spoil me, but she did, just the same.' Perhaps Mlle Elise sensed that Kitty, the more buoyant of the two sisters, was already the star in her mother's firmament, and the perceptive governess sought to comfort and protect the more vulnerable, younger child. Mlle Elise recalled that Clementine was 'more sensitive, more affectionate, more spontaneous' than her elder sister.

Once, Clementine was discovered sobbing her heart out in bed, and the inconsolable child was brought to her mother. The girls' hair was (prior to special occasions) wound up in 'curlpapers' overnight, and between sobs

Clementine managed to tell her mother that she was panic-stricken lest 'Prince Charming' should arrive in the middle of the night and find her in such a plain state. Blanche Hozier had many faults as a mother and was often unjust to Clementine, but she understood romance – the offending 'curl-papers' were laboriously undone, and a sleepy, calmed child returned to her bed.

The Hoziers also had a small house in Scotland, at Alyth, not far from Cortachy, called The Netherton. Clementine wrote of it: 'There was a lawn and a pretty garden just outside the drawing room windows, and beyond this a little wood. I have seen it since, so I know it is a little wood, but to me it seemed like a great forest.' That wood, so forest-like to childish eyes, was perfect for hide-and-seek, and was full of fascination and mystery to the children.

Their social life consisted of visits to and from nearby relations. Their grandmother, Blanche, Countess of Airlie, a widow since 1881, had moved, soon after the marriage in 1886 of her son David (11th Earl of Airlie) to Lady Mabell Gore, to the Cortachy dower house, Airlie Castle, which was within easy driving distance of The Netherton. Cortachy itself was only a little further away, and here a brood of young Ogilvy cousins was starting to make its appearance. Blanche Hozier had a 'tilbury and a cob', which she drove herself, and for nursery outings there was Ginger, the pony, and 'trap', with John, the stable-boy, in attendance.

The family spent regular periods each year in Grosvenor Street. Life in London for the nursery child has changed remarkably little over the years: walks in the park, visits to museums on wet days, and – for treats – occasional expeditions to the Zoo and Madame Tussaud's (and nowadays theme parks) were the main features of a well-found child's daily life. Nor did the young Hoziers lack for companionship; their own close family circle provided them with a large mixed bag of contemporary cousins. Blanche Hozier was herself one of six; her brother, Lord Airlie, had the same number of children, while one of her three sisters, Clementine Mitford (whose husband, 'Bertie', later became the 1st Lord Redesdale), bore eight children; and Blanche's uncle and aunt, Lord and Lady Stanley of Alderley, also had a family of eight. The Mitfords and the Stanleys were often in London, and there were frequent visits between the families.

But town life did not suit Kitty, who was not as strong as Clementine, so there were several seaside visits with Mlle Elise, to recover from chills and colds, and one whole summer was spent by all four children at Westgate-on-Sea. There were other, more glamorous, journeys with their mother: Clementine remembered an Easter visit to Paris, staying at a hotel in the rue de Rivoli. The children spent many happy hours playing in the Jardin des Tuileries, diverted quite as much by a goat-cart in which they went for rides as by *le Théâtre Guignol*.

In 1889, when Kitty was six and Clementine four years old, their portrait was painted by a young and gifted artist, Dampier May. It is an enchanting picture: the two little girls (their hair 'Eton-cropped' in the fashion of the day) in delightful lace pinafores are standing before a round looking-glass; Kitty stands in front, confident and assured, while peeping shyly round her shoulder is her younger sister. Clementine remembered quite clearly the portrait being painted; Kitty had enjoyed the sittings, and was 'cool and dignified', while 'I', Clementine recalled, 'wept and was tiresome'.[14]

Clementine later sadly recognized that her parents probably never knew any true happiness together, and that from the beginning of their marriage their relationship was a stormy one. But for the children, these early years were, on the whole, happy. Safe and secure in their nursery world, they were too young to feel the rising tension between their parents. Flashpoint was reached between Henry Hozier and his wife in 1891, when Blanche was discovered to have been unfaithful (as has been recounted). Her family stood by her loyally, but they were consternated at the prospect of the breakdown of the marriage and deeply shocked by what her brother described as Blanche's 'wrong-doing'.[15] When it was quite evident that reconciliation was out of the question, they sought to ensure that proper provision should be made by Blanche's husband for herself and the children, but they had to settle for what Henry Hozier was prepared to give – which was very little indeed.

In 1891 Blanche and Henry Hozier parted for ever. He must have been an angry, strange man indeed, for his entry in *Who's Who* made no mention of either of his wives, nor of the existence of any children. Despite the fact that he had been divorced by his first wife on grounds of desertion and adultery, and that he himself had been less white than the driven snow since his second marriage, he was to show himself towards Blanche as hard and vengeful.

Blanche therefore became largely dependent upon her family for financial support. The Ogilvys had never been very rich, most of their fortune being tied up in their estates. Lord Airlie was a serving soldier, with a growing family of his own; and of his four sisters only two, Clementine Mitford and Griselda (married in 1897 to James Cheape of St Andrews), had married reasonably well in a worldly sense. Notes at the end of Blanche Airlie's diaries show that she was consistently generous to her more impecunious children. But there is no doubt that, despite help from her family, Blanche Hozier was from now on very badly off financially.

The question of the care and custody of the children was, predictably, the subject of bitter disagreements between their parents, and they became helpless hostages in the ensuing tussle. The first arrangement was that the twins, Bill and Nellie, now aged three, were to remain with their mother, while the two elder girls, aged eight and six, went to live with their father, in the charge of a kind, elderly governess – Miss Rosa Stevenson.

The first memory Clementine had of their new life was when she and Kitty went with their father to Homburg, a fashionable watering-place in Germany. In order to prepare themselves for this journey, the two girls were taken by Miss Stevenson to the Army & Navy Stores, from where, among other things, two large trunks were bought, upon which their names were impressively printed. The trunks were also decorated with gay yellow and red painted bands for easier recognition. The little girls had new best frocks of tartan silk with crochet lace collars. Once in Germany, they did not see a great deal of their father, and indeed stayed in a different hotel, but every morning they escorted him when he went to take the waters, breakfasting afterwards under the trees on deliciously sweet sultana bread and warm goat's milk.

Although Henry Hozier was determined to establish his right to his children, the two small girls were no doubt an encumbrance to him, and on their return home from Homburg, Kitty and Clementine went to stay with Miss Stevenson in her own home at Berkhamsted. For a short time they attended a kindergarten, but for some reason this did not last for more than a week or two. Thereafter, instead of making paper mats or counting beads, the children spent the mornings watching Miss Stevenson do the housework, a labour she performed with the utmost care and attention.

At this time there loomed into the children's lives the formidable figure of 'Aunt Mary': Henry Hozier's unmarried sister, who from time to time visited the children to inspect and supervise. She wore a black, jet-trimmed bonnet and a sombre, though richly embroidered cape and dress of bombazine; the final touch to this imposing presence was given by a *châtelaine* which rattled as she walked. Clementine was terrified of her, but Kitty did not mind her a bit, and was cheerfully impertinent.

The pleasant days at Berkhamsted, however, were numbered: whether or not on the advice of the stern Aunt Mary is not known, but Kitty and Clementine were now sent to a bleak school in Edinburgh kept by Herr Karl Fröbel and his three daughters (reputedly relations of the celebrated German educationalist Friedrich Fröbel). It was situated in Moray Place, a roundel at the west end of Georgian Edinburgh.[16] Clementine remembered lessons in the dining-room, with crumbs on the floor and an all-pervading smell of finnan haddock. Here, in this somewhat grim establishment, Kitty and Clementine knew the first pangs of unhappiness and homesickness.

There is something heartrending about these two doll-like creatures (for their photographs show them as such) contending with a stony world so early in life. The sisters clung together with a fierce loyalty, and developed a strong sense of dependence upon each other. Kitty was the bolder, and the natural leader of the two, but Clementine, although two years younger, could already write a nice clear hand. And so, when the children decided life was unbearable at the Fröbels' school, and Kitty had the idea that an appeal should be made to their father to take them away, she bullied the timorous Clementine into writing the letter. All letters to and from the pupils were scrutinized, so it had to be posted by stealth; this Kitty contrived to do, unperceived by the mistress in charge of the girls, while they were out for their afternoon walk.

The two little girls waited with pent-up anxiety, and four days later Kitty found a letter on her plate – but before she could open it Fräulein Fröbel demanded to have it: she read it on the spot, and immediately burst into loud, hysterical sobs. However, she allowed the children to have the letter, which said:

> My dear Children,
> If you are unhappy, I will take you away at once.
>
> Your affectionate
> Father

Alas for parental honour – he did nothing, and Kitty and Clementine were shunned by the whole school.

Meanwhile, Blanche Hozier had discovered their whereabouts; it seems that they had been sent to this school without her knowledge, and contrary to an agreement she and her husband had entered into as to the management of the children. Indignant, and anxious about the children's well-being, Blanche Hozier behaved with characteristic impetuosity and lack of convention. She hastened at once to Edinburgh, and one afternoon surprised Kitty and Clementine while they were on their weekly expedition to buy sweets: there was great excitement, tears and delight. Their mother then approached the mistress in charge and asked if she could take the children out to luncheon. The mistress demurred, but invited her back to the school to meet the Principal. After some discussion, the children were not permitted to go out with their mother, since Henry Hozier had already briefed Herr Fröbel on his matrimonial difficulties, and had expressly forbidden this. Disappointed, and no doubt angered by this embargo, Blanche Hozier promptly found lodgings immediately opposite the school, from where she used to wave out of the windows to Kitty and Clementine. She also took to attending the children's classes, thereby creating dismay among the staff, but giving great pleasure to the pupils, who all liked her very much and told the now gratified Kitty and Clementine they thought their mother very beautiful. Had the children been older, no doubt they would have been embarrassed by this unusual behaviour, but Clementine remembered they felt only joy and relief at being once more near their mother. Moreover, having been under a cloud of displeasure, they were now the objects of the open envy and interest of their contemporaries. The Fröbels apparently concealed their inner thoughts beneath a diplomatically benevolent aspect.

The Edinburgh episode brought about the negotiation of another arrangement between the elder Hoziers, which was happier from every point of view for the children. Shortly after these happenings, Kitty and Clementine went to live once more with their mother. To mark their joint acceptance of this new policy, Henry and Blanche Hozier were both at the station to meet the children on their return from their northern exile; they all went at once to the house in Bayswater, where Blanche Hozier was now living. The twins, who had been under three years old when they had last seen their elder sisters over a year before, not surprisingly showed no signs of recognizing Kitty and Clementine, who, in their turn, felt acutely embarrassed, and were uncertain as to whether they should kiss them or not. Their father once more took his departure. Thereafter he came once or twice to tea with his wife – the children were sent for on these occasions, which were uneasy and silent – but even these visits soon ceased.

A more settled life now began for Kitty and Clementine – 'settled', that is, in the sense that for the rest of their childhood and adolescent years they were to be in the care and company of their mother; but the family never had a fixed home for long. Even today life is full of problems for a woman living apart from her husband and having to shoulder alone the responsibility for and organization of her family. In those days, on marriage a woman left the protection of her own family and home for that of her husband, and she rarely had any experience of an independent life. Moreover, there was then a considerable degree of social stigma attached to a woman separated from her husband: it was altogether an unenviable situation. Blanche Hozier, with her four children and their governess, lived for much of the time in furnished lodgings. It is possible this way of life may have been dictated in part by her own restless and capricious nature, as well as by financial difficulties, or the stresses of her relationship with her husband.

Clementine remembered very clearly the distinctive imprint which her mother left wherever she lived: although presumably she rented furnished lodgings partly for reasons of economy, this did not deter Lady Blanche from often painting and papering the rooms in which she and her family were established. 'She had very simple, but distinguished taste,' Clementine wrote, 'and you could never mistake a house or room in which she had lived for anyone else's in the world.' Although surrounded by other people's furniture, she always had loose covers of her own made for the sofas and chairs: they were of pure white dimity. 'There were two sets, and these were always being changed. I remember dark green taffeta cushions, and in her bedroom pale blue ones.' The windows were hung with fine book muslin curtains – always fresh and snowy white, so that a friend said, 'I cannot remember the number of Lady Blanche's house, but it does not matter. All I have to do is just walk down the street and look at the windows.'

Nor was Blanche Hozier's fastidious taste confined to interior decoration; she was very much interested in food, and was an excellent cook. At one time (to help her precarious finances) she contributed a series of articles on food to the *Daily Express*. But Clementine remembered rather ruefully that as a housekeeper her mother was distinctly erratic; when she gave her mind to it, the food was delicious, but when she was bored or had other distractions, meals were often somewhat haphazard.

There was nothing haphazard, however, about their educational arrangements. The Hozier children may sometimes have lacked food, but they never went short of learning. They always had a full-time governess, either French or German, and they also attended courses from time to time given by other professors in various subjects, excluding, however, arithmetic, which

Blanche Hozier in her arbitrary way deemed an unnecessary and unseemly study for girls.

When Kitty and Clementine came to live once more with their mother in London, a Frenchwoman, Mlle Gonnard, made her appearance as governess to the four children. Clementine remembered her as 'vivacious, touchy, untruthful and tale-bearing'. But, whatever her shortcomings, the poor woman must have had her hands full with the four children aged between nine and four years old, headed by the saucy and rebellious Kitty. On one of their customary and seemingly interminable walks, while Mlle Gonnard was re-tying her shoe-laces (a frequent occurrence), Kitty, dragging the unwilling Clementine with her, ran away. Mlle Gonnard, anchored to the twins, could not give chase. The two girls ran quite a distance, and then became lost, and were eventually returned to their home by a policeman. Lady Blanche was away, but the outraged Mlle Gonnard threatened to report her daughters' iniquity immediately upon her return – whereupon, Clementine remembered, a terrifying scene ensued: Kitty seized a lighted oil-lamp and, holding it above her head, said to Mlle Gonnard: 'If you don't swear by God and the Holy Virgin that you won't tell, I shall throw the lamp across the room, and then we shall all be burned to death!' Terrorized by Kitty's violent and determined behaviour, Mlle Gonnard promised not to tell – but later she broke her word.

About this time, perhaps because all four children were too much for Mlle Gonnard, Clementine was sent to a course of French lessons with a remarkable woman, Mlle Louise Henri. Clementine regarded this as a special treat, and greatly admired Mlle Henri's powerful intelligence. But Kitty hated lessons and, being much indulged and spoilt by her mother, was not made to attend these classes; instead, she had singing and guitar lessons. She had a lovely little voice, and sang 'A goldfish swam in a big glass bowl' and 'Click, click, I'm a monkey on a stick', and other popular songs.

From 1895 the Hozier family spent many months of the year in furnished apartments at Seaford, then a small resort, on the south coast near Newhaven in Sussex. They lived at Nos 9 and 11 Pelham Place; the landladies were sisters – the Misses Rolls. Miss Caroline was severe, and did not like children, but her sister, Miss Emily, was large and cosy and far more amiable. Fortunately for the children, they and Mlle Gonnard lived under Miss Emily's benevolent rule in No. 11, while their mother occupied the first floor of No. 9 with her Griffons Belges dogs, Fifinne and Gubbins (mother and son), on whom she doted.

Blanche Hozier went away on frequent visits, and in her absence the children were alone with the Rolls sisters, and under the care of Mlle Gonnard. The governess cannot have been over-vigilant, for during their mother's absence the two elder girls indulged in various escapades – in which Kitty was ever the leader and instigator and Clementine her anxious but loyal lieutenant. They were the proud owners of bicycles, and played hectic games of bicycle polo on the rough grass opposite the lodgings, although Mlle Gonnard greatly disapproved of such unrefined pursuits.

Kitty, now in her early teens, was a flirt, and already had several admirers. One of these – more devoted and daring than the others – was in the habit of climbing up the garden wall, and crawling along the narrow parapet of the house, from where he could gossip with Kitty through her bedroom window. Clementine was made to mount guard in the passage outside the bedroom while these clandestine conversations were in progress. Although disapproving intensely, Clementine faithfully paced the passage outside in an agony of apprehension. She asked Kitty one day what she would do if a grown-up came upstairs suddenly, for the young gentleman concerned would hardly have had time to retreat backwards along the narrow parapet in order to avoid discovery. 'Oh,' said Kitty airily, 'I'd just push him off.' She would have, too.

Despite the open preference their mother displayed towards Kitty, the bond of loyalty and affection between the two sisters remained quite unaffected. Clementine accepted the situation philosophically; indeed, she felt her mother's preference to be quite natural, for she too was dazzled by Kitty's prettiness, gaiety and impudence. Kitty adopted a somewhat patronizing air towards her less favoured sister, but was full of affection, and the two girls would often discuss their mother's blatant favouritism. Kitty used to say, 'You mustn't mind. She can't help it.' But both of them felt a contempt for their mother's ungovernable partiality.

When their mother was at home, life was full of interest, uncertainty and amusement. A variety of relations (including their grandmother, Lady Airlie) and friends came and went, staying either at one of the Seaford hotels or in the lodgings in Pelham Place. One of Blanche Hozier's friends, Mrs Mary Paget, who lived not far from Seaford, was a frequent visitor, and, although of an older generation, became Clementine's firm friend. No doubt she observed the preference Blanche Hozier did not seek to hide for Kitty and, feeling sorry for the younger child, showed her kindness and understanding. Mrs Paget took to 'borrowing' Clementine for quite long periods, to stay with her at her farm, West Wantley, near Storrington in Sussex. Kitty was at first included in these invitations, but she was sent home on one occasion for being too naughty and tiresome. Thereafter the visits remained Clementine's personal and particular treat, and the weeks that she passed with Mary Paget and her family became gilded in her memory with the light that shines on remembered happiness and kindness.

When the time came for Clementine to return to the lodgings in Seaford which passed for home, after visiting the Pagets, she felt she was leaving paradise and used to shed many bitter tears. On one occasion her reddened eyes were noticed by her mother, who observed angrily, 'I believe you love Mary Paget more than me,' to which Clementine, with more truth than wisdom, replied, 'Of course I do.' And after that, for some time, there were no more visits to Wantley.

* * *

When Blanche Hozier and her husband parted, The Netherton was given up, and thereafter a firm fixture in the family's life was the annual, month-long visit to their grandmother at Airlie Castle. This enchanting small castle stands on a precipitous rock, and in the gorge beneath flows the river Isla. It is a romantic, beautiful place and as, over the thirty years that she had lived at Cortachy as reigning Countess, Blanche Airlie had improved and beautified it, so in the long years of her widowhood she left the mark of her distinguished taste on the 'dower castle'. The watchful suspicion with which she had for a long time been regarded as a 'foreigner' by the local Scots had long given way to a wry respect, and that affection which a sturdy, independent-minded people slowly but finally accord to real 'characters': in the neighbourhood the Dowager Countess was cordially known as 'Auld Blanche'.

After her husband's death, Lady Airlie led a retired life; her estate and public duties soon devolved on to her daughter-in-law, and her entertaining was necessarily on a smaller scale. She had a small, distinguished circle of friends, and divided her time between Airlie Castle, her house in London, and her villa in Florence. Her tastes were essentially intellectual, and she was extremely well-read. Among her friends she had a reputation for writing, but to the detached eye of her great-granddaughter her essays seem both pretentious and sententious. She also spent a good deal of time covering acres of material with beautiful, elaborate Queen-Anne-style embroidery.

Blanche Airlie was a loving mother and grandmother, and her diary is full of entries recording visits to and from various members of her family, along with their births, illnesses, engagements, marriages and deaths. She had a particularly close relationship with her daughter Blanche Hozier, and it is known that when apart, they wrote to each other nearly every day; unfortunately their correspondence has not survived. Several of her younger relations were devoted to her, but despite her love and concern for her family, old Lady Airlie, now in her sixties, was awe-inspiring, generally arousing feelings of respect rather than of spontaneous affection. Clementine was not one of her grandmother's favourites, and she herself never really came to love or appreciate this undoubtedly remarkable personage. As a child she was frightened of her, and as she grew older she dared to survey this matriarchal figure with a not wholly uncritical gaze.

It can hardly, therefore, be a cause for surprise that the annual holiday at Airlie was for the Hozier children a period they regarded with mixed feelings. The dates were settled months ahead, and any attempt to alter them was met with tight-lipped disapproval. Clementine's chief memories of these visits were of the devices whereby the children sought to circumvent the many restrictions placed on their activities by the wide range of their grandmother's prohibitions. Blanche Hozier, although strict in some ways, wanted her children to enjoy the liberty and opportunities for fun that life at Airlie offered after the more confined life they led at home. So she allowed her children to bring their bicycles with them, but they had to be kept out of sight, and used surreptitiously – for Lady Airlie (like Mlle Gonnard) disapproved of bicycles for young ladies. They used sometimes to bicycle to Lintrathen Loch, which is about three miles from Airlie Castle; here they would fish happily for hours with a kindly fisherman. On their return their grandmother would remark approvingly what excellent walkers the children were.

One year, Kitty and Clementine were given a croquet set, but their grandmother said it would spoil the appearance of her lawn in front of the castle, and ordered that it should be sent back immediately to Seaford. However, their mother persuaded the gardener's wife, Margaret Milne, to have the hoops put up on the small green patch at the back of her cottage, and here, out of sight of their grandmother's eagle eye, the children played croquet; although they were somewhat hampered by the length of the grass, and by having to dodge under kind Mrs Milne's washing when she hung it out to dry.

In November 1898, aged thirteen, Clementine was confirmed in Kirriemuir Church, near Airlie, by the Bishop of St Andrews, who had also prepared her for confirmation. Present with her at this serious moment, as well as her mother and grandmother, was her beloved friend Mary Paget. Clementine went through an emotionally religious phase at this time and, on one occasion, being deeply moved by a sermon on sacrificial giving, snatched off the pendant she was wearing and placed it in the offertory bag. After the service, however, she had some qualms, for the pendant was a precious one and had been given to her by a relation. When her quandary became known to the grown-ups, she was admonished for excess of zeal, and the pendant was retrieved – a more reasonable offering being substituted.

Kitty and Clementine also often visited their Stanley cousins, usually in their London house in Mansfield Street, but also at Easthope, their country home in Yorkshire. Great-aunt Maisie Stanley was kindness itself, and there was always a warm welcome for the Hozier children. Here they were thrown into the enjoyable hurly-burly of a large family, and here began friendships which were to endure throughout a lifetime. Venetia and Sylvia Stanley were the immediate contemporaries of Kitty and Clementine; both became and remained Clementine's friends. But it was with Sylvia that she had most in common, and Sylvia's loyalty, love and powers of companionship never wavered; in her Clementine was to find a tower of strength and a true friend to the end of her life.

In the schoolroom days there were, of course, the special friendships, quarrels and larks, into which Kitty entered with zest; but Clementine, who was rather a prig and always full of anxious forebodings, came in for a lot of teasing. On the brink of adolescence, she showed little sign of beauty; she suffered from red eyes, and was prone to be timorous and tearful. In all things she was overshadowed by the fascinating, boisterous Kitty.

CHAPTER TWO

Dieppe and Afterwards

THE SPRING AND SUMMER OF 1899 PROVED A HARASSING AND DIFFICULT TIME for Blanche Hozier. In the first place she was in debt, and had to ask her brother for help. Henry Hozier had shown over the last six or seven years little concern for the children's welfare, and was at this time defaulting on the meagre allowance he had undertaken to pay his wife. Kitty was now sixteen, Clementine fourteen, and Bill and Nellie eleven years old. In addition to paying the wages of a governess and other teachers for the girls, the fees for Bill's education at Summerfields Preparatory School had to be found. Blanche Airlie helped her daughter with money on a considerable scale this year, and Lord Airlie, as well as paying some of his sister's bills, caused his lawyer to press Henry Hozier for the money he owed. In addition to these worries, Blanche was taken quite seriously ill with influenza and quinsy towards the end of March.

In the last week of July, with the unpredictable suddenness that marked so many events in their lives, the Hozier family moved at one day's notice (as far as the children were concerned) from Seaford to Dieppe, on the French coast. Blanche Hozier never discussed her personal life or difficulties with her children, but it seems that she had always thought her husband might some day try to regain the custody of the two elder children, who were the only ones in whom he ever evinced any interest. And it may be that when she picked on Seaford as a place to live, she had this easy 'escape route' in mind. Dieppe was only four or five hours away, and the daily steamer from Newhaven was a familiar sight to Seaford residents. When, therefore, during this summer of 1899, Blanche Hozier had reason to believe that her husband was about to take action of some sort, she quickly made her plans, and removed herself and her family from England by the quickest and nearest route.

The announcement by their mother that they were all to leave Seaford at once came as a complete surprise to the children, who greeted the news with shouts of glee. There must have been a flurry of packing, and hurried good-byes to the Misses Rolls, and also to Mlle Gonnard, whose mediocre services were now dispensed with. Family legend has it that on her arrival in Dieppe with her four children (and, of course, Fifinne and Gubbins), Blanche Hozier spun a coin to decide whether they should seek lodgings in Puys or Pourville, two villages in the neighbourhood of Dieppe. The coin declared for Puys and, hailing a cab, Lady Blanche told the coachman to drive there. Once in the village, she smote the driver on the back with her umbrella as they were passing a charming-looking farm – La Ferme des Colombiers; leaving her family in the cab, she disappeared into the farm, emerging triumphantly a short while later accompanied by a stout lady, Mme Balle, with whom she had concluded an agreement by which the Hozier family would stay there for two months.

At Puys, which is by the sea, the family spent an idyllic summer: bathing, canoeing, picnics and blackberrying filled the happy days. With the coming of the autumn, schooling had once more to be thought of, and Lady Blanche took a furnished house in Dieppe itself – No. 49, rue du Faubourg de la Barre. Outside it is not an attractive house, the front door opening straight on to the narrow pavement, but it was most conveniently situated in the same street as the Convent of the Immaculate Conception, which the three girls would attend. As the children already spoke good French, they quickly adapted themselves to their new life, and to the convent uniform of black overalls with brilliant green belts.

One of the most historic towns on the French Channel coast, Dieppe at the end of the nineteenth century was a popular summer resort, being the nearest seaside town to Paris. It offered the lure of a casino, and was one of the few places at that time to have a golf course. In the early 1900s the town also was second only to Vichy as a musical centre: during the season, and before Deauville and Le Touquet exerted their rival and more powerful attractions, Dieppe was full of chic Parisian society, which crowded the sea-front hotels and paraded on the promenade in elaborate *toilettes*.

Apart from these seasonal visitors, there was a resident English community, such as was to be found in many French towns up to the Second World War. *La Colonie*, as the English inhabitants were collectively known, consisted largely of retired and, for the most part, impoverished military men and their families, diversified by the presence of various writers and artists who either lived there, or were passing visitors to the picturesque town and the lovely countryside around, and whose presence lent a bohemian and intellectual flavour to Dieppe society. The famous French painter Jacques-Emile Blanche had a house and studio there, where the intellectual and artistic *monde* met; and at various times Aubrey Beardsley, Oscar Wilde, George Moore, Arthur Symons, Max Beerbohm and Walter Sickert all frequented Dieppe.

France and French people were much to Blanche Hozier's taste, and she soon made friends (and not a few foes) among both the locals and *La Colonie*. Artists and writers, such as Blanche, Sickert and George du Maurier, found her company agreeable, and she quickly established an elegant manner of life. She also began to divert herself with frequent visits to the Casino, becoming an inveterate gambler – thus further compromising her precarious finances.

Lady Blanche was soon recognized as *une vraie caractère*; her circumstances also no doubt aroused curiosity and speculation. At forty-seven she was still very beautiful, dressing elegantly, but quite independently of changing fashions. Instead of a hat, she often wore a lace mantilla, and Clementine remembered being embarrassed by the way her mother would sometimes wear her hair in a long thick plait down her back, instead of piled conventionally on her head, or in a bun. It appears also that her children's clothes were the cause of comment: people expected to see the granddaughters of an earl attired otherwise than in check gingham.[1]

Kitty and Clementine were a noticeable pair as they went about Dieppe. Their looks and natures formed a striking contrast: Kitty, with her long dark hair in plaits and deep blue eyes, was ever in the fore, compared with Clementine, who was so much fairer and much less obviously pretty, and whose nature was shy and retiring. But if Kitty was already a beauty, to the discerning eye there were signs of dawning loveliness in her younger sister. George du Maurier, visiting Blanche Hozier at about this time, on seeing Clementine declared: 'She's my Trilby – she's come to life!' – a comment which Blanche Hozier did not allow him to repeat in Clementine's presence, it being considered inappropriate in those days for young people to be 'noticed' or praised to their faces.[2]

Living at this time in Neuville-les-Dieppe was Walter Sickert, the painter. Although he had many friends in the fashionable, conventional part of the town, Sickert found it more congenial to live on the far side of the harbour, where most of the seaport and fishing community lived. He had lodgings in the house of Mme Augustine Eugénie Villain, the acknowledged queen of the Dieppe fish market. *La Belle Rousse*, as she was known, was locally famous for her red hair and her fiery temper.[3] In Sickert's life she had the triple role of landlady, model and mistress.

Mr Sickert soon became a frequent visitor of Blanche Hozier and her children. No doubt he enjoyed both her unusual, fascinating company, and the delicious food which was always to be found in her house. In 1899 he was about forty years old; he was then clean-shaven and had thick honey-coloured hair; his eyes were a piercing sea green. His mode of dress was generally eccentric, his most usual form of headgear being a workman's *casquette*. Clementine was deeply struck by him, and thought he was the most handsome and compelling man she had ever seen.

Her mother often sent her to do the shopping: armed with a large basket, Clementine frequently came upon Sickert sketching at the corner of one of the small, narrow streets leading to the church of St Jacques. One day, when she had stopped to look at his painting, he asked, 'Do you like my work?' Clementine, after a pause, replied, 'Yes.' 'What is it you don't like?' he asked, noting her hesitation. Clementine struggled between truth and politeness, but finally said, 'Well, Mr Sickert, you seem to see everything through dirty eyes.'

Evidently this frank criticism did not offend him, for he presently asked her mother if she could come to tea with him. Clementine wrote her own vivid account of this occasion. After toiling up the hill to Neuville-les-Dieppe, she presently found the right house.

> The door was open, no bell, no knocker; so I rapped with my knuckles. After a long pause a magnificent commanding woman appeared, and I recognised Madame Villain. She was very big and tall, dressed in a black dress with a blue and white check apron tied tightly round her shapely waist, her hair stood out round her head like a golden red halo, her sleeves were rolled up, and she stood akimbo displaying beautiful white arms and hands like those in a Vandyck picture . . . 'Que voulez-vous?' she asked abruptly. I was terrified, but determined to keep my end up. 'Je viens prendre le thé avec Monsieur Sickert.'
>
> 'Eh bien, il n'est pas là.'
>
> I was dumbfounded. She then said, 'Mais si vous voulez vous pouvez l'attendrc.' She then flung open the door leading into a very dirty bedroom. The bed was unmade and on a rickety deal table were the remains of a herring. I was profoundly shocked and thought, 'Perhaps Mother can find Mr Sickert a better house-keeper.' I waited – no Mr Sickert – so then I decided to clean up his room. I made the bed, found an old broom and swept the floor, and finally seizing the skeleton of the herring by the tail flung it out of the window upon a rubbish heap just below. I then washed the plate and put it away. Just then I heard loud melodious singing and there he was, with a large bag of hot brioches under his arm.
>
> 'What have you been doing to my room?' he asked suspiciously.
>
> 'It was very untidy, so while I was waiting for you I cleaned it up; which I think your landlady might have done.'
>
> 'Where's my herring?'
>
> 'I threw it away.'
>
> 'You little interfering wretch, I was just going to paint it. And where, pray, is the handsome plate it was sitting on?'
>
> 'I have washed it and put it on the shelf.'
>
> We then consumed the brioches, accompanied by tepid cider, after which Mr Sickert took me home.

That winter Sickert engraved Clementine's portrait with a sharp, red-hot poker on her hockey stick: she had just returned from playing in a match at school, and he disapproved of such violent and aggressive sports for young women. 'I must show you what you look like,' he said, and burnt her silhouette along the length of her hockey stick. It was the very image of her, but Clementine was mortified, and covered her poker-work portrait with sticking-plaster. Shortly afterwards the hockey stick was anyhow broken, during a fierce bout with the opposing centre-forward.

During the winter of 1899 an extraordinary and dramatic incident took place, which shows that Blanche Hozier was right in thinking her husband might try to get hold of the children. An entry in her mother's diary reveals that Lady Airlie still had some contact with her son-in-law, for on

16 December 1899 she recorded: 'Anxious about Blanche's children. HH wants to see them. Goes to Dieppe today.' She obviously had no time to warn her daughter, and the sequel is taken up in Dieppe.

It was a snowy evening, and Blanche Hozier, Kitty and Clementine, and Mr Sickert (who often came to supper) were all in the sitting-room, the window of which gave onto the street. The curtains were not yet drawn, and the room was lit only by the flickering flames of the log fire. Suddenly Clementine saw a man looking in at the window; her mother noticed him at the same moment, and instantly dragged both the children down onto the floor. The man continued to stare into the room for a few minutes; then he turned away, and they heard the front door bell tinkle. Kitty got up to answer it, but she was immediately dragged down again by her mother. The bell rang several times more; no one moved, and they heard something being thrust into the letter-box. The man reappeared at the window, peered again into the darkened room and then walked away. Clementine, besides being extremely frightened, was also acutely embarrassed by the presence of Mr Sickert, who was the astonished witness of this scene. However, Blanche Hozier remained quite calm, and said simply, 'That was my husband; I expect he brought a letter. Kitty, go and see.'

Kitty returned with two letters, one for herself and one for Clementine. Both letters contained similarly worded invitations to dine and lunch on succeeding days with their father at the Hôtel Royal. Clementine was much alarmed by this prospect, but Kitty was thrilled, and rushed out the next morning to buy a new pale blue sash to adorn her white net dress for the occasion.

Although Blanche Hozier must have been anxious and apprehensive, she made no demur about the girls accepting their father's invitations; but she sent with them on each occasion their stalwart French maid Justine, to accompany them to and from the Hôtel Royal. Kitty's dinner with her father was a great success: he had invited her to stay with him, and had promised to give her a dog and a pony. The next day it was Clementine's turn; again Justine acted as escort, with strict instructions to tell 'Monsieur le Colonel' that she would collect 'Mlle Clémentine' promptly at two o'clock. Henry Hozier replied that Justine need not return, as he would bring Mademoiselle home himself. Clementine, hearing this, cast imploring looks at Justine, who responded with a reassuring wink. During luncheon (an omelette followed by larks *en brochette*), Clementine could hardly speak from shyness and anxiety. Henry Hozier asked numerous questions about the children's life and upbringing, and expressed strong disapproval when he learnt that they were being educated at a convent. Presently he said, 'I suppose your mother has told you that in future you will be living with me, or rather with your Aunt Mary.' Clementine was aghast at this information, and replied that she had heard of no such plan; plucking up courage, she added that she did not think her mother would let her and Kitty go. All this time her eyes were on the clock, watching the hands creeping on towards the hour of deliverance. At five minutes to two, she started surreptitiously pulling on her gloves; at two o'clock exactly, Justine reappeared. Clementine rose and bade her father goodbye, but he looked angry, and, striding towards Justine, he thrust a gold coin in her hand and pushed her out of the door. Clementine felt trapped and wondered desperately what to do – but when her father went across the room to get a cigar, she saw her chance. Quick as lightning, she rushed to the door, dashed down the passage and out of the hotel – and there was the faithful Justine waiting anxiously for her in the snowy street. They started to run as fast as they could on the icy pavements, Henry Hozier pursuing them at a fast walk, swearing angrily – but he soon gave up the chase – nor did he reappear at their house. Later, Lady Blanche heard from the bank manager, and from the captain of the Dieppe–Newhaven steamer, that her husband had intended to kidnap Clementine, and take her back to England with him that very afternoon.

The Anglo-Boer War, which had begun in October 1899, was now raging fiercely, and although personal friendships were maintained, England's role was viewed with hostility by the French. Kitty and Clementine were aware of the thinly veiled disapproval of their companions at school; nor did the nuns maintain too saintly a detachment from these worldly events. In November the British forces in the town of Ladysmith were besieged by the Boers; the siege continued for four months, but rumours that the town had been relieved were circulating in January, causing rejoicings among *La Colonie*, which were not shared by their French neighbours.

One morning the Mother Superior swept into the classroom, interrupting the lesson. 'I wish,' she said, 'to make an announcement. Despite many rumours to the contrary, Ladysmith has not been relieved.' On her withdrawal from the room, all the French children stood up and cheered. Kitty and Clementine burned with indignation and resentment: after a brief and whispered consultation they rose and, heads high, left the room. Once in the passage, a difficulty arose; all the doors were glazed halfway down to permit easy observation, so the two girls (determined to carry through their patriotic gesture of defiance) crawled on all fours along the passage to the entrance, where they also evaded the notice of the concierge, and managed to reach the street unhindered.

But then what? Their courage had now rather evaporated, and uncertainty had set in as to their reception at home. They wandered down to the beach, and whiled away the time by playing ducks and drakes. At luncheon time they were of course missed at home, and Blanche Hozier sent her maid to enquire as to her elder daughters' whereabouts; her messenger was somewhat coldly informed that 'les Mesdemoiselles d'Hozier' had removed themselves earlier in the day. When, eventually, the two (by now rather rueful, and very hungry) heroines returned home, they were received with a mixture of relief and reproaches. However, all in all, they were thought to have done rather well; and no more was said at school about this episode.

The two elder girls were now in their middle teens and, as when they were children, were inseparable companions. Kitty was sixteen, and the naughty, fascinating, fearless child was fast maturing into a beautiful and alluring young woman. She used often to sit in the open window of their house playing her guitar, the gaily coloured ribbons which were attached to it waving gently as she played and sang in her charming natural voice; she caused quite a sensation among old and young alike.

But the pleasant life in Dieppe to which the Hozier family was so quickly becoming accustomed was soon to be darkened by tragedy. In the middle of February 1900 Kitty fell ill, and typhoid fever was diagnosed: it soon became clear that her condition was desperate. She was subjected to the most drastic and primitive treatments (Mr Sickert was constantly in the house carrying great jugs of icy water upstairs). Lady Airlie sent a nurse out from England, following herself a short time later. Meanwhile Blanche Hozier decided to send Clementine and Nellie (Bill was already back at boarding school) to stay with their aunt, Lady Griselda Cheape, at St Andrews, so that she could devote every ounce of her strength to the now grim battle for her adored Kitty's life. Hurriedly the girls' trunks were packed; they were to travel by themselves, but Sickert offered to take them to the boat, which left late at night, and to see them safely on board. When they were all ready to go, Clementine and Nellie went upstairs to say goodbye: their mother met them at the bedroom door, and peremptorily told them to take off their hats, so that Kitty should not guess they were leaving. The sisters said goodnight to the strangely still and silent figure in the bed – and did not know it must be for ever. Mr Sickert saw the two rather subdued girls off on the first stage of their long bleak journey to St Andrews.

Kitty died on 5 March, a little more than a month before her seventeenth birthday. The telegram bearing the tragic news to her sisters in Scotland was somehow delayed or lost, and a second telegram, telling them of the plans for the funeral, was the only one they received. It read: 'Am taking Kitty to Batsford. Come at once.' Hope and joy were rekindled in the hearts of the family at Strathtyrum – surely this meant that Kitty must be better? She must be convalescent – and well enough to travel to Batsford, their Aunt Clementine Mitford's home? The children were discussing this seemingly hopeful news with their Aunt Griselda, when suddenly Mr Black, Lord Airlie's factor from Cortachy, was announced: his aspect was respectful and grave. 'I have come, m'lady, to offer my condolences upon the death of your niece . . .' The hope that had soared like a bright star was extinguished. Kitty – the bold, beautiful Kitty – was indeed gone for ever. Once more, the children packed up, and set out, with sad and heavy hearts, to bury her at Batsford.

* * *

Kitty's death marked the end of a whole period in Clementine's life. Kitty, it is true, had caused her younger sister moments of embarrassment and anxiety, but the sisters loved each other dearly, and together they had faced the sorrows and dramas of their childhood, and shared pleasures and secrets. Kitty had ever tempered her mother's injustice to Clementine, and had been her leader and champion; the younger sister, in return, gave Kitty unwavering love, loyalty and admiration. And now that Kitty was gone, she was a lonely figure in the family scene; for the twins, at twelve, were still children, and bound up in each other's lives.

Blanche Hozier was paralysed with grief: there were bonds of affection and duty between her and Clementine, but not much understanding, and little ease. Kitty and her mother had been kindred spirits, but Clementine's nature and her own were too dissimilar. The unspoken knowledge that Kitty had been so much her favourite child may have added remorse to Blanche Hozier's bitter grief, and it made a shyness in Clementine's approaches to her mother. They both sorrowed deeply and separately.

After Kitty died, Blanche Hozier and her three remaining children returned to England, and settled for the next four years at Berkhamsted in Hertfordshire. The decision to live at Berkhamsted was governed by educational considerations – the town offered excellent schools both for the two girls and for Bill. Clementine and Nellie went to the Berkhamsted High School for Girls, while Bill went to the Berkhamsted Grammar School for Boys. By the middle of the summer Blanche Hozier had found a small but charming eighteenth-century house, No. 85 (now renumbered 107) in the High Street, with a long narrow garden running uphill at the back. She soon made the house pretty, and imported a cook from Dieppe. And so once more, despite adverse circumstances and abiding sorrow, Blanche Hozier created around herself a charming and civilized atmosphere.

As far as the girls were concerned, the family and economic circumstances which combined to make them grammar-school pupils certainly meant that they received a superior and wider education than they would have had had they been taught by governesses at home, which was still the predominating pattern of education for girls of their class. Already accustomed to changes in their place and way of life, Clementine and Nellie became quickly acclimatized to their new school and companions. But for Bill this change was one that caused him bitterness of heart: at Summerfields he had been with boys who, almost without exception, were going on to one or other of the great public schools, and by right and custom he should have done so too; but his mother was simply unable to afford the fees, and Henry Hozier showed not the slightest intention of contributing more than he already grudgingly did towards his family's maintenance.

Clementine (now fifteen) and Nellie (twelve) started at the High School in the autumn, and Clementine's studious nature now came into its own. Despite so many years of governess teaching, she relished the competitive atmosphere of work, and the wider opportunities this larger community offered her. Starting in the Upper Fifth Form, Clementine was soon promoted to the Sixth for French, but suffered humbling demotion to a much lower form for mathematics (which was hardly surprising, since her mother had always prevented the girls from learning this subject). Nellie also settled in well and was happy at school; she had a much more easy-going and carefree nature, and did not overburden herself with too much scholastic endeavour.

It was not only the work that Clementine enjoyed; she flung herself into school life with zest, and enjoyed it to the full. Many years later (in 1947), in a speech on Commemoration Day at the school, she would speak of her time there:

> I began my school life in the old building in the High Street, and although it has been described as bare and gloomy, I loved it. I felt I was out in the world, and on my own, away from the fiddling little tasks of arranging the flowers, folding the newspapers and plumping up the cushions, and the tender (though I sometimes ungratefully thought) exaggerated solicitude of my Mother over wet feet, colds in the head, and difficult homework.

In 1901 Clementine distinguished herself by winning, in open competition with students from all over the country, a handsome solid silver medal for French, presented by the Société des Professeurs Français en Angleterre. She received it from the hands of M. Jules Cambon, the French Ambassador. The following year, partly as a reward, Blanche Hozier gave Clementine a special treat: she sent her to Paris for a fortnight, in the care of Mlle Louise Henri, whose classes Clementine had so greatly enjoyed even when she was a child of nine or ten. They stayed at a modest hotel, and the days flew by, each one full of interest and pleasure. Museums, lectures and art galleries were interspersed with thrilling glimpses of other worlds and ways. Mlle Henri had procured tickets for a *vernissage*, where Clementine's attention was distracted from the pictures by a most handsome and fantastic-looking woman: she had a wasp-waist, and her hour-glass form was encased in a white cloth dress; both her gown and her large hat were trimmed lavishly with ermine tails. Mlle Henri, seeing that Clementine was fascinated by this astonishing person, drew her aside and whispered to her that this vision was none other than 'Polaire', the most famous *demi-mondaine* and music-hall artiste in Paris. No wonder Mlle Henri was such an interesting and agreeable companion, for her knowledge and interest were not confined merely to academic subjects.

Another delightful event in this magical fortnight was the sudden arrival, out of the blue, of Mr Sickert, who turned up one evening and announced his intention of taking Clementine out with him the following day. Mlle Henri was somewhat dubious as to the wisdom or propriety of this suggestion: however, she was soon won over by the Sickert charm and by Clementine's assurances that he was an old friend of her mother.

Clementine had a wonderful day: they started by breakfasting in a little café (she noticed Mr Sickert did not pay the bill, but that it was chalked up on a blackboard). He then took her to the Luxembourg Gallery, where he asked her to point out to him her favourite picture. Clementine chose John Sargent's *La Carmencita* (a sparkling Spanish dancer in yellow satin); Mr Sickert was shocked by the vulgarity of her taste, and directed her attention to the gentle, dreamy painting, *Le Pauvre Pêcheur*, by Puvis de Chavannes. They lunched in a bistro in a rather poor neighbourhood, and once again the bill was chalked up. During luncheon Clementine asked him, 'Who is the greatest living painter?' He looked astonished at the question, and answered quite simply, 'I am, of course.'

In the afternoon they called upon Camille Pissarro, who lived in a garret overlooking the Madeleine: 'He was a magnificent old man, with a great silver beard and a black sombrero hat. There were at least ten people sitting in a small attic, and they were all drinking tepid beer, which we shared with them.'[4] To wind up the day, Mr Sickert took Clementine to dinner with M. Jacques-Emile Blanche in his elegant house on the outskirts of Paris, a great contrast to the bohemian flavour of their day so far. 'Everything was exquisite,' Clementine recalled, 'the furniture, the food and Monsieur Blanche's manners, though I did overhear him enquiring from Mr Sickert how he came to be entrusted with the care of a *jeune fille bien élevée*'. After a most pleasant evening, Mr Sickert took Clementine – exhausted but happy – back to her hotel. She did not see him again for twenty years.

This wonderful fortnight came all too quickly to its end, and Clementine had to come down to earth, and back to school.

The Headmistress of Berkhamsted High School at this time was Miss Beatrice Harris, whom Clementine greatly admired and described as an 'inspired and inspiring personality'. Miss Harris held advanced views on women's suffrage, and the school was imbued with the spirit of feminine advance and independence: influenced by her, Clementine nursed dreams of going to university. This was an unusual thing for a girl to do then, and required a strenuous, single-minded dedication. Miss Harris, who wished to encourage Clementine in her ambition, offered to have her to stay during the holidays, so that she could continue her studies.

This helpful invitation was never taken up, for despite the fact that Blanche Hozier believed in women being well educated, the idea of her daughter becoming an earnest 'bluestocking' never gained any favour with her. Indeed, when she saw how much engrossed by school life and studies Clementine was becoming, she deliberately set out in trivial and annoying ways to hinder her from doing her homework, and to distract her into more feminine interests and pursuits.

Clementine minded very much her mother's open or covert attempts to sabotage her educational progress and thwart her ambitions, and the conflict between their outlooks and wills increasingly produced abrasive encounters. Clementine was often reduced to tears by her mother, and now she had no champion – although Nellie was ever a loyal confidante and peace-maker. Their mother's curious and arbitrary notions forced the girls into subterfuges, arithmetic often being studied in secret on the flat tombstones in the graveyard which adjoined the top of their garden.

Clementine's character, so long eclipsed by Kitty's more dominating nature, now began to show its strength and fortitude. She found an unexpected ally in her great-aunt Maude Stanley (Blanche Airlie's sister), a benevolent old spinster who lived in Smith Square and of whom Clementine was very fond. Sharing the reforming zeal of so many of her family, Miss

Stanley had started the first Working Girls' Clubs in London. She was a great supporter of women's education and emancipation, and was delighted to find in one of her younger relations a somewhat kindred spirit.

But Blanche Hozier was not to be so easily circumvented; she was quite determined that Clementine should not pursue her aim of going to university. Totally socially unambitious for herself, Blanche Hozier now pocketed her pride and approached an aunt of hers by marriage, Lady St Helier, and sought her help in launching Clementine into the world to which by birth and upbringing she properly belonged. It was not easy for Blanche to take this step, for not only was she averse from asking favours, she did not like Lady St Helier personally, thinking her worldly and snobbish. It is difficult now to understand the degree to which Blanche Hozier was cut off from the normal social life of the day by the fact of her separation from her husband, and by her straitened financial circumstances. Moreover, as we now know, she had acquired a scandalous reputation.

Lady St Helier, a Miss Mary Mackenzie by birth, had first married Colonel John Stanley, one of Blanche Hozier's uncles. After his death, she had married Sir Francis Jeune, a distinguished judge, who became the first Lord St Helier; he died in 1905. She was a remarkable personality, combining energetic and intelligent service with the London County Council and work for numerous good causes with a flair and taste for large-scale entertaining. A prominent London hostess, she pursued the lions of society, and always liked to have several of them roaring at her dinner parties.

Although Lady St Helier greatly relished the company of the successful and the famous, her warm heart led her to many kindnesses to young people not yet arrived in any conspicuous way upon the world's stage. Only a few years earlier she had helped Winston, the adventure-seeking son of her friend Lady Randolph Churchill, to become attached to the British Army in Egypt by intervening on his behalf with Sir Evelyn Wood, the Adjutant-General. Now, when she was asked to help a beautiful but obscure and poor young relation, with her usual warm-heartedness she agreed to do so.

Blanche Hozier's financial circumstances were such that a whole 'season' was out of the question; moreover, Clementine was still at school. However, from time to time she would go to stay with her Great-Aunt Mary, who would take her to a ball or invite her to one of her many dinner parties; indeed, she gave Clementine as a present her gown for her first ball.

During the summer of 1903, Clementine passed her Higher School Certificate in French, German and biology. There was no more talk now of her desire to go to university; her mother's opposition, reinforced by the pleasurable and beguiling glimpses she had had from time to time of the London scene, where she was already making a mark, had settled the issue. And so, after taking her examinations, Clementine left the High School. All her life she was to remember with happiness and gratitude her time there; she kept in touch over the years with Miss Harris, and visited the school on several occasions. That summer, Bill too left school and, deciding on a naval career, passed in to HMS *Britannia*.

Perhaps to tide over the break from school and the fading of Clementine's university aspirations, Lady Blanche took a house for that winter in Paris, in the rue Oudinot in the *7me arrondissement*. Clementine attended lectures at the Sorbonne, which thrilled and absorbed her; she used to return home for luncheon between her lectures, but when the windows were being cleaned (a frequent happening in Blanche Hozier's house) or when her mother was bored with ordering meals, Clementine was given a franc and sent to have her luncheon at the *laiterie*. Here she sat up at the counter, her skirts and feet tucked up off the floor, which seemed always wet from frequent washings. Her one franc bought her a lightly boiled pullet's egg, a glass of milk, a fresh roll and a pat of butter: marvellous value for money, but hardly a winter meal for a hungry girl of eighteen.

Despite the straitness of the family finances, Lady Blanche was not parsimonious, and she had the gift of providing unexpected treats. One day she took Clementine and Nellie to Voisin (an elegant, fashionable and extremely expensive restaurant), where they were regaled on *tournedos* and *pommes soufflées*, with excellent wine. Clementine was delighted, and enjoyed it all immensely, but none the less asked her mother if this were not a gross extravagance, considering that they were passing through a particularly lean time financially. 'My dear child,' replied her mother, 'to know how to order and enjoy good food is part of a civilized education.'

The following spring Clementine was sent for three months to stay with the Von Siemens family, who had a beautiful house on the Wannsee, near Berlin. Here she had an agreeable time taking part in fashionable Berlin life; she also learnt to ride – and, inevitably, fell in love with the riding master.

The house at Berkhamsted having now fulfilled its purpose, Lady Blanche thought it better for them all to live in London, and more 'in the world'. She accordingly took a house in Kensington – No. 20 Upper Phillimore Place.

When Clementine returned from her stay in Germany, in order to supplement her small allowance she started to give French lessons at 2s 6d an hour. Her moment of triumph came when she assured the success of a young boy sitting for an entrance examination by running him through various technical terms and data in French. The fact that he knew the correct word for 'ball-bearing' apparently sealed the successful result, and called forth praise for his tutor – Miss Hozier – from the child's gratified parents.

The Hoziers had lived in Upper Phillimore Place for only about a year when they moved to a charming cream-painted stucco house, No. 51 Abingdon Villas, just behind Kensington High Street, where they were to remain for several years. In addition to giving French lessons, Clementine now started to work for a cousin, Lena Whyte, who had a small dressmaking business in North Audley Street. Clementine was not gifted with her hands, but she toiled away with perseverance, and learnt enough to make some of her own clothes and to do the endless refurbishing of her very limited wardrobe.

Although there were difficulties in Clementine's life, and her activities were considerably curtailed by being short of money, she had all the resilience and hopefulness of youth, and she flung herself with zest into the pleasures and ploys that came her way. Admiration adds much to a woman's morale – and Clementine lacked neither admirers nor serious proposals of marriage. Nellie, three years younger, watched all this with admiration and often shared amusement. Indeed, she saucily suggested at one moment that a file of 'Proposals to Clementine' should be kept, with the sub-divisions: 'Discussed', 'Answered' and 'Pending Decision'.

But Clementine had much to endure from her mother's unreasonable nature and capricious temper. Even when she was nineteen, her mother used to insist that she should be home from balls by midnight; this rule was patently ridiculous, and, needless to tell, Clementine was often late, returning to find her mother awaiting her on the doorstep, and was subjected to furious upbraidings. One morning, after a particularly stormy scene, Lady Blanche, beside herself with temper, boxed her daughter's ears. What is endurable for a child is a mortification not to be borne by a young woman of nearly twenty. It was certainly too much for Clementine's proud spirit: she left the house and, not waiting to put on her coat or hat, ran all the way across the park to Harley Street, where Mrs Paget, her kind friend of Seaford days, lived: here she poured out her woe and indignation. Clementine stayed the night there, and before she returned home, Mary Paget visited her old friend Blanche Hozier, and no doubt delivered herself of some words of warning and moderation.

When she was eighteen Clementine had met at one of her great-aunt's parties a man whose love and friendship were to colour her life for the next few years – Sidney Peel. He was thirty-three at the time; the younger son of the 1st Viscount Peel, he was a man of brilliance and distinction, a Fellow of Trinity College, Oxford, and sometime lecturer there. In the South African War he had served with the Imperial Yeomanry and had been decorated, and he afterwards wrote a book about his campaigning experiences; now, in 1903, he was a barrister and banker. Sidney Peel became deeply and devotedly in love with Clementine, whose serious nature and intelligence appealed to him as much as her great beauty. He courted her with a tenacity and grace which seems strange to a generation used to more casual, hurried ways. Clementine never left London but he was there to escort her to the train; he always met her on her return. He lent her books, took her to the theatre, and (fired by the example of the hero of a contemporary novel) sent her a bunch of white violets every single day – except during the month of August, when even his devotion and resource could not produce one white violet. During the winter the Hozier family spent in Paris, Sidney Peel would make the long journey from London every weekend in order to spend a day or two in her company.

Clementine was deeply fond of him; she valued his love, and rejoiced in his companionship. These were difficult, lonely years for her: she longed at times to escape from her circumscribed home life with its tensions to the spacious security that marriage to such a man could offer. Moreover, Sidney Peel loved her truly, and she might so easily have persuaded herself that esteem and affection are substitutes for love. Perhaps she nearly did – for twice she was secretly engaged to him. Those closest to her hoped she would marry Sidney; her mother would have been happy to see Clementine settled, and was herself very fond of him. But although it would have been a good marriage, Blanche Hozier's own romantic nature prevented her putting any undue pressure on Clementine.

Not so Lady St Helier – who strongly recommended Clementine to accept this admirable suitor, and subjected her great-niece to searching enquiries as to why she held back from so satisfactory a match with a man who loved her dearly, and who could offer her so much. Lady St Helier was not impressed by Clementine's explanation that she did not feel herself to be 'in love'. But despite all these pressures, Clementine remained true to her own innermost instincts and feelings.

It was about this time that Clementine formed a friendship that was to prove lifelong: Horatia Seymour, beautiful, brilliant and poor herself, and a few years older than Clementine, always showed her warmth and kindness; and to a mutual liking was added the understanding each had of the other's situation. Clementine had many ardent admirers and true friends, but there were some sharp tongues and jealous, watchful eyes. She was greatly touched by Horatia Seymour's friendliness to a newcomer on the scene, and valued her helping hand.

Now was the heyday of Edwardian times, and entertaining was conducted on a scale of luxury, formality and elaboration unknown to the Victorians. Society had never been so brilliant or so mixed, and money set the pace in the smart world. Leisure, for the upper classes, was a common commodity; servants were plentiful, and large hospitable houses abounded both in London and in the country. Spirits and confidence were undimmed by any premonition of the holocausts to come; nor had values in modes and manners been changed by the social revolution which would transform the face and form of society beyond recognition. Country-house visits were a great feature of life: weekends were from Friday to Monday, but many house parties lasted for longer, centred on a coming-of-age, shooting or amateur theatricals (which Clementine loved), when the entire cast would stay for weeks at a time.

Dressing for women was time-consuming and intricate; coiffures, hats and gowns were beautiful and complicated, and needed hours of care and maintenance. Armies of personal maids (of varying degrees of grandeur, strictly related to their employers' status) washed, starched, goffered, ironed and stitched.

The clothes needed for a mere weekend visit usually occupied several sizeable trunks, with at least one large, unwieldy hat-box, for it was quite usual to change into three different ensembles during the course of the day. Ball gowns were beautiful, lavish creations, with yards of frilling, flouncing or lace, and it took hours to repair the ravages of dancing. A great extra expense felt very much by poorer girls was the absolute necessity for long, spotless white kid gloves for all evening occasions. It is not surprising, therefore, that

Clementine's social life was strictly rationed by sheer lack of money. But she was then, as later, immensely fastidious, and had simple good taste; and so, despite the not inconsiderable difficulties, Clementine soon became remarked upon for her beauty, charm and natural elegance by her own, as well as the older generation.

Various contemporary memoirs contain references to Clementine's outstanding beauty, to which photographs (either then or later) rarely did justice. Lady Cynthia Asquith was to write of her: 'Clementine Hozier; classical, statuesque; yet full of animation. A Queen she should have been; her superbly sculptured features would have looked so splendid on a coin. "There's a face that will LAST," said everyone. How right they were!'[5] Apart from her beauty and elegance, Clementine was lively and enthusiastic, and one of her great charms was that, despite being an acknowledged 'beauty', she was completely lacking in self-consciousness. Sir Alan Lascelles recalled most charmingly the first, indelible impression Clementine made upon him:

> My first glimpse of her was in the Xmas holidays, 1903, when, as a 16-year-old schoolboy, I was staying with old Judge Ridley & his wife at Crabbet. There was a largish party, for some private theatricals.
>
> One day at luncheon, Lady Ridley announced, 'The beautiful Miss Hozier is coming tonight. She will be with us at tea-time.'
>
> About 5 o'clock we were all sitting in the hall having tea. Suddenly the butler threw open the door, & shouted (as if he were on the stage at Covent Garden), 'Miss Clementine Hozier'.
>
> There was an instant (& for her, poor girl, most shy-making) silence, for there in the doorway was a vision so radiant that even now, after 61 years, my always roving, always fastidious, eye has never seen another vision to beat it.[6]

CHAPTER THREE

To Thine Own Self Be True

DURING THE SUMMER SEASON OF 1904, WHEN CLEMENTINE WAS NINETEEN, she met Winston Churchill at a ball given by Lord and Lady Crewe at their house in Curzon Street.

Winston Churchill was now thirty. He had been born in 1874 at his grandparents' home – Blenheim Palace – the elder of the two sons of Lord Randolph Churchill and the American beauty, Jennie Jerome. After schooldays which showed no surface sign of ability to come, he had opted for the Army, and passed (at his second attempt) into the Royal Military College at Sandhurst. His father died in January 1895 at the early age of forty-six, his tragic illness* and death cutting short a political career of exceptional and meteoric brilliance. Shortly afterwards, Winston was commissioned as an officer in the 4th Hussars, a cavalry regiment. He yearned for excitement and action, and during his first prolonged period of leave he and a brother officer sought both in Cuba, where the Spaniards were fighting a guerrilla uprising. He received his baptism of fire on his twenty-first birthday.

The following year the 4th Hussars were posted to India, where Winston discovered the beauty of butterflies; was much occupied with the cultivation of the roses in the garden of his bungalow; and became a keen and effective polo player. It was during this time that a great desire for knowledge seized him, and he embarked on a prodigious programme of serious reading, his mother finding it difficult to keep him adequately supplied with all the books he demanded from home. Ever on the look-out for action and for a chance to distinguish himself, Winston took part (first as a war correspondent, and then as an officer) in the campaign of 1897 on the North-West Frontier, afterwards publishing his first book, *The Story of the Malakand Field Force*, which attracted considerable attention and some favourable comment.

Soon a new field of action caught Winston's impatiently roving eye – Egypt, where Lord Kitchener was fighting the Mahdi in the Sudan. Partly

* The diagnosed cause of Lord Randolph's death was a condition known as general paralysis of the insane (GPI): a long-term consequence of syphilis. Some recent medical opinion holds that his illness and death were caused by a brain tumour.

through the good offices of the benevolent Lady St Helier, a friend of his mother, he succeeded in joining the Army of the Nile, and in September 1898 took part in the charge of the 21st Lancers at the battle of Omdurman. The *Morning Post* published his letters describing the battle, and on his return to India he wrote the greater part of his two-volume account of the Egyptian campaign, *The River War*.

In 1899 Winston resigned his commission in the Army and made his first venture into politics, fighting a by-election as a Conservative in Oldham, Lancashire, in which he was narrowly defeated. In October of that year the Anglo-Boer War broke out, and he sailed for Cape Town as correspondent for the *Morning Post*. He secured a commission in the Lancashire Hussars, and then made for Natal, where it seemed the serious fighting would be. Then followed a series of novel-worthy episodes, starting on 15 November 1899 with the 'Armoured Train' when Winston, exposed to fierce enemy fire, according to all reports acquitted himself with gallantry. During this foray he was captured by the Boers and was incarcerated in a prisoner-of-war camp in the State Model School in Pretoria. He made his escape from prison camp on the evening of 12 December 1899. After a series of adventures, he arrived in Durban on 23 December to find himself a popular hero – his capture, imprisonment and escape, and subsequent speculation as to his fate, having been widely reported in the press.

He returned home in July 1900, and in the General Election that autumn stood once more for Oldham; this time he was elected, and in the New Year of 1901 he made his maiden speech as a Conservative Unionist Member of Parliament. The House listened with sympathy and interest to the daring and enterprising son of the brilliant, ephemeral Lord Randolph.

When, therefore, some three years later, Winston Churchill met Clementine Hozier for the first time, he had already packed a great deal into three decades and was a colourful public personality. Furthermore, he had added to his dashing career as a soldier and war correspondent that of a successful author, with six books already published. Now, in this spring of 1904, he had also become politically controversial: he bitterly opposed the Conservative Party's drift towards Tariff Reform, and in defence of Free Trade he abandoned his father's party, crossing the floor of the House to sit on the Liberal benches.

The Liberals were, of course, enchanted by this new acquisition to their ranks, but in the eyes of his own party he was regarded as a renegade and a turn-coat. Moreover, success (especially if it is judged to be premature) makes many enemies. Although, at thirty, Winston Churchill had his friends and champions, he was also a controversial figure and, especially among his own class and contemporaries, was judged by many to be bumptious, opportunist and generally insufferable. Politics were more vehement in those days, and permeated social life to a marked degree. Since Winston's defection the doors of nearly all Conservative houses were closed to him. But he was welcome at Crewe House, since Lord Crewe was a prominent member of the Liberal Party.

Arriving at the ball with his mother, Winston saw, standing alone in a doorway, the girl of whom it was later to be written, 'her beauty needed no adornment. She looked like a lily.'[1] He asked his mother who she was – Lady Randolph* did not know, but said she would find out. Presently she returned with the news that the admired being was Clementine Hozier, the daughter of an old friend of hers – Lady Blanche Hozier – whom she had not seen for some years. In addition, Lady Randolph's brother-in-law, Sir Jack Leslie, was Clementine's godfather. These links made it easy and natural for Lady Randolph to introduce her son to Clementine.

Their first meeting was, however, brief and unpropitious – for, upon his mother's introducing him, instead of asking Clementine to dance, Winston stood rooted to the spot, staring at the vision which had so powerfully beguiled him. Clementine felt embarrassed by his scrutiny and, far from thinking that this strange, intent young man had asked to be introduced to her, imagined that Lady Randolph had noticed that she was without a partner and, taking pity on her plight, had out of kindness introduced her son. Moreover, Clementine had heard about Winston Churchill; she had heard him discussed on several occasions, and she was not at all favourably disposed towards him. Now his gauche, singular behaviour settled the matter. Luckily, help was at hand: a beau of hers was standing nearby and, obeying a discreet signal, he came up and asked her to dance with him. While dancing, he asked Clementine what she had been doing talking to 'that frightful fellow Winston Churchill'.

This brief encounter has significance only in retrospect – it left no profound impression at the time on either Winston or Clementine, whose paths were not destined to cross again for nearly four years. If she thought of the incident again, it was with amused disdain. As for Winston Churchill, on one of his rare excursions into purely social life he had been struck by a girl's beauty and grace. But his life was not lived in ballrooms, nor even in the fashionable country-house parties where they might have met again far sooner, for already the world of politics was claiming him for its own.

During these last years he had had friendships with several beautiful and clever girls. Pamela Plowden, the daughter of the Resident at Hyderabad (whom he had first met in India as a young subaltern), was his first great love. She married the Earl of Lytton in 1902; but she and Winston were to remain friends always. A little later, Muriel Wilson, an heiress and a great beauty, whose family were Hull shipowners, held his attention. He thought of marrying her, but she rejected him (it is said she did not think he had much of a future); they too always remained on the friendliest terms. Winston also in his time proposed to the famous actress Ethel Barrymore, who later said she had been much attracted to him, but that 'she felt she would not be able to cope with the great world of politics'.[2]

* Winston's mother was now Mrs George Cornwallis-West, having in 1900 married a man twenty years her junior. She divorced him in 1913. In 1918 she married again – once more a man little more than half her age – Montagu Porch. For convenience she will be referred to throughout as Jennie, or Lady Randolph. Her daughters-in-law called her 'Belle Maman' or 'BM'.

But although he relished the company of beautiful, lively women, Winston was never – then or later – a 'ladies' man'; somewhat gauche and un-handy in his manner, he neglected those charming but trivial small attentions which so often pave the way in gallant relationships. Politics were always to fill and dominate his thoughts – and now his star was in the ascendant.

The Conservatives had been in power for twenty years, but A. J. Balfour's Government was by now seriously weakened by the split over Protection. In December 1905 Balfour resigned, and the King sent for the Liberal leader, Sir Henry Campbell-Bannerman. The new Prime Minister offered the post of Under-Secretary at the Colonial Office to Winston Churchill. Parliament was dissolved, and early in 1906 a General Election took place. In a dramatic landslide, the Liberals were swept into power with a large majority over all other parties.

Winston at once began to play a prominent part in affairs, for although he was only a junior minister, his chief, Lord Elgin, the Colonial Secretary, sat in the House of Lords; Winston, as Under-Secretary, was therefore the chief government spokesman on colonial affairs in the House of Commons.

* * *

In April 1906 Clementine's cousin and great friend, Sylvia Stanley, married Colonel Anthony Henley, and she was one of the bridesmaids. Shortly before the wedding Sylvia had broken her arm, but despite the necessity for a sling, and although still suffering considerable pain, she insisted that the plans should not be altered. Her fortitude made a deep impression on Clementine, who was at this time secretly engaged to Sidney Peel; she felt that were such an accident to befall her on the eve of her marriage, she would not have the tenacious devotion to go through with it. This reflection made her realize that her feelings for Sidney Peel fell short of true love; and so, soon after Sylvia's wedding, she broke off her own engagement.

Clementine must, all this time, have subconsciously been looking for an avenue of escape to a freer and more fulfilled life, for during the summer of this same year she suddenly became engaged to Lionel Earle – a man nearly twice her age – whom she had met for the first time some six weeks before at a house party. He was forty, a distinguished civil servant, with intellectual tastes; he was very well off, and had a somewhat stiff nature and pompous manner.

They met for the second time at Castle Ashby in Northamptonshire during the brilliant party held at the end of July to celebrate the coming-of-age of Lord Compton, heir to the Marquess of Northampton. During the festivities, which lasted for five days, Lionel Earle, who knew very well how to make himself agreeable to a beautiful woman, laid siege to Clementine's affections, and she fell in love with him: no doubt life and personalities assumed a rosy hue in the general atmosphere of gaiety and pleasure. Before the party broke up, Clementine had accepted his proposal of marriage.

Lady Blanche was not enthusiastic about the engagement; she may have felt that in the greyer light of everyday life Lionel Earle's sophistication and the somewhat formal side to his character would come to jar on Clementine's more spontaneous nature. Moreover, she may well have felt uneasy that Clementine should be plunging so hastily into such a serious relationship, when only a short time before she had been engaged to Sidney Peel. But Blanche Hozier had learnt to reckon with her daughter's determination, and, after all, it was a good solid match for a penniless girl.

The engagement was announced on 15 August 1906. A few days after the announcement Lady Blanche and Clementine left for a pre-arranged holiday with old friends, Mr and Mrs Charles Labouchère, at Bantam in the Netherlands; the hospitable Labouchères also invited Lionel Earle. During the fortnight that ensued, when she was constantly in the company of her betrothed, Clementine began to have serious doubts. Later that year Blanche Hozier wrote to her brother Lyulph Ogilvy (who ranched in Colorado, in the United States): 'I v. soon saw that Clementine was not happy – and I asked her what was amiss – told her that if she had made a mistake she must say so. She said she was frightened, would not dare – I told her she need not be frightened if I stood by her – so – the engagement was broken off.'[3] Blanche Hozier was in many ways difficult and unpredictable, but in matters of the heart her sense of priorities was clear. It was of course most tiresome and awkward: the families had to be told; the dis-engagement announced; and many wedding presents packed up and sent back to their givers. But through it all Blanche consistently supported Clementine.

Now was the time for the Hozier family's annual visit to Airlie. But this year, Clementine's heart failed her – she did not feel she could face her formidable grandmother's cold disapproval, for old Lady Airlie took a far from indulgent view of her granddaughter's conduct in this whole affair. In her estimation Clementine had behaved in a foolish and hysterical fashion. Fortunately, just at this time Lena Whyte (the kind dressmaking cousin) was convalescing from an operation for appendicitis, and she begged that Clementine, instead of going to Scotland, might keep her company; this was a happy solution. The whole episode affected Clementine deeply: her equilibrium and self-confidence were shaken, and she felt a sense of humiliation. For a month or two she was quite ill – no doubt a direct result of her mental and emotional distress.

At the end of February 1907 the news came of Henry Hozier's death in Panama, at the age of sixty-nine. He had resigned from his position as Secretary at Lloyd's after thirty-two years in 1906. All his considerable talents and force of character had been devoted to this job; in 1903 he had been knighted. Even after his resignation, Sir Henry had still busied himself with the Corporation's affairs, and was visiting one of its signal stations when he became unwell and, after a short illness, died.

With Henry Hozier's death, an element of tension was removed from his family's life; and yet Clementine was deeply saddened. Although she had feared her father, she had always cherished the hope that in time she might come to know and to understand this strange, violent, yet most compelling man. After his attempt to take Kitty and herself from their mother in Dieppe in the winter of 1899, Clementine had seen her father on only one other occasion. It was in the summer of 1900, when the family had returned to live in England following Kitty's death. Henry Hozier had expressed a wish to see Clementine; this was duly arranged, and Blanche Hozier accompanied her fifteen-year-old daughter to his house in North Audley Street. She was taking leave of her on the doorstep when the door suddenly opened, and out came her husband. On seeing his wife, Henry Hozier flew into a rage, and upbraided her for daring to appear at his house. Blanche Hozier remonstrated that she had merely come to leave Clementine for the appointed meeting, but he was not to be pacified, and an angry scene took place between them in the street, of which Clementine was the unwilling and mortified witness. Presently her mother left, and Clementine followed her father into his house; the interview that ensued, not surprisingly, was not a source of pleasure or comfort to either of them, and saddest of all, though neither could know, it was to be their last encounter.

* * *

One evening in March 1908, Clementine arrived home rather bent and weary after giving French lessons to find a message from Lady St Helier asking her to dine that very evening, as one of her guests had fallen out, and she was in danger of being an unlucky thirteen. Clementine was reluctant to go; she was tired, she had no suitable gown ready, and no clean gloves. But her mother was adamant that she must brace up and go to the dinner party, reminding Clementine of the many kindnesses she had received at Aunt Mary St Helier's hands. And so, without any relish for the evening ahead, Clementine went upstairs to change.

Lady St Helier, in fact, was lucky not to have another defection from her dinner: one of her other guests, Winston Churchill, decided the whole evening was going to be a great bore, and told his Private Secretary, Eddie Marsh* (who found him dawdling in his bath), that he did not intend to go to the party. Eddie was shocked, and reminded Winston of Lady St Helier's efforts on his behalf when he was having such difficulty in getting out to Egypt in 1898. So, with none too good a grace, Winston hurriedly dressed and set out, already late.

At Lady St Helier's house, No. 52 Portland Place, the assembled company, besides Clementine, included F. E. Smith† and his wife Margaret; Lord Tweedmouth, the First Lord of the Admiralty, and a kinsman of Winston; Sir Henry Lucy, a political journalist, who wrote for *Punch* under the name

* Later Sir Edward Marsh (1872–1953). Private Secretary to Churchill, 1905–29. As well as being a career civil servant, he was a patron of the arts, classical scholar and *littérateur*.

† Frederick Edwin Smith (1872–1930), later 1st Earl of Birkenhead. Brilliant advocate and orator. Conservative politician, successively Solicitor-General, Attorney-General, Lord Chancellor and Secretary of State for India. His close friendship with Winston was never affected by their political differences.

of 'Toby M.P.'; Miss Ruth Moore, an American heiress who later married Lord Lee of Fareham; and Sir Frederick Lugard, the great West African explorer and colonial administrator, and his wife, Flora Shaw, herself an expert on colonial affairs. Winston being very late, they started dinner without him. Clementine in consequence had a gap on her right; on her left was Sir Henry Lucy. Halfway through dinner Winston arrived, full of explanations and excuses, and took his place between Lady Lugard and Clementine.

Clementine and Winston's first meeting at the ball at Crewe House four years before had not been a signal success, and this second encounter seemed to have got off to a bad start as well. But things went very differently now. Since Winston was at the time Under-Secretary for the Colonies, he might well have been expected to devote a good deal of his attention to the distinguished and knowledgeable Lady Lugard; however, he turned almost at once to the ravishing young woman on his left, and monopolized her for the rest of dinner. Afterwards, when the gentlemen joined the ladies, Winston, who usually liked to linger behind, was first into the drawing-room, making straight for Clementine, in whose company he remained for the rest of the evening. His attentions were so noticeable that Clementine had to endure a little gentle teasing from the other ladies, when they all went to get their cloaks.

The romantic in Winston Churchill's nature sought in women beauty, distinction and nobility of character: Clementine Hozier possessed all three. On their first brief encounter he had recognized the first two qualities; now, after an evening spent in her company, he realized that here was a girl of lively intelligence and great character. Clementine, for her part, was enthralled by the brilliance, warmth and charm of his personality, which filled his whole being and made his attraction so irresistible, despite his forgetful, unpunctual and even at times un-gallant ways. During their conversation he asked Clementine if she had read the life he had written of his father, which had been published in 1906. She said she had not read it, but would like to. 'I will send it to you tomorrow,' he said. But he forgot!

Events were moving in Winston's political life as well. In early April 1908 Sir Henry Campbell-Bannerman, the Prime Minister, mortally stricken by an illness from which he died only a few weeks later, resigned and was succeeded by H. H. Asquith. The new Prime Minister made few changes in the Government, but one of them was to move Churchill from the Colonial Office and to make him President of the Board of Trade, with a seat in the Cabinet. At that time, on being appointed a minister, a Member of Parliament had to seek re-election. Winston was therefore faced with a by-election in his seat at North West Manchester, which he had won with a respectable majority for the Liberals at the General Election two years earlier.

Although there was so much afoot politically for Winston, he was now in love – and he had no intention of allowing even weeks, let alone years, to slip by this time before he saw Clementine again. He therefore asked his mother to invite her, with Lady Blanche, to spend the weekend of 11–12 April at her house, Salisbury Hall, near St Albans in Hertfordshire. To her lasting credit, Lady Randolph from the first encouraged Winston's attachment to Clementine. It would clearly have been advantageous for him to have married a girl with some money, and Lady Randolph was a worldly woman; but she instantly recognized in Clementine those sterling qualities which were to make her the perfect wife for her brilliant son.

It was during this weekend that Asquith announced his Cabinet, so that spring Sunday at Salisbury Hall must have been full of excitement at the present, as well as of unspoken hopes for the future.

The next day Clementine (most reluctantly now) left England with her mother, on a long-arranged journey to Germany. During the previous summer Nellie had been diagnosed with tuberculosis; she had therefore been packed off to a clinic at Nordrach to undergo a rigorous cure. The treatment (which would be complete and lasting) was now in its final weeks, and Lady Blanche and Clementine were going out to collect her. On their way home they planned to visit old Lady Airlie in Florence, where she invariably spent the summer months.

At this juncture of her life it must have been agony for Clementine to contemplate such a long and distant separation from Winston, who now occupied her thoughts and, increasingly, her heart. As for Lady St Helier, she expressed herself strongly to Blanche Hozier to the effect that she must be out of her mind to remove her daughter from the country for over a month just when it was clear that Winston Churchill was seriously interested in her.

As Clementine rested in Paris between trains on Monday 13 April, she wrote to Winston's mother:

> Dear Mrs West,
>
> I want to thank you very much for making me so happy during my visit to you.
>
> At this moment your whole mind must be filled with joy & triumph for Mr Churchill, but you were so kind to me that you make me feel as if I had known you always. I feel no one can know him, even as little as I do, without being dominated by his charm and brilliancy. I wish I could be with you in this stirring fortnight.
>
> This is a poor scribble after a rough crossing, but I feel I want to write to you at once.
>
> Yours affectionately,
> Clementine Hozier

Three days later Winston wrote Clementine a long letter from his house, No. 12 Bolton Street (which he shared with his brother Jack). He had returned from the political battlefront to 'kiss hands' on his new appointment. He gave a lively account of the campaign in Manchester, and was full of confidence that he would have a 'substantial success'.

A considerable part of his letter was taken up by his account of Lady Dorothy Howard's electioneering activities and talents. Dorothy Howard was the daughter of Lord and Lady Carlisle, and was Clementine's cousin. She was beautiful, gifted and intensely politically minded. Clementine had distinctly mixed feelings on reading Winston's enthusiastic recital of how Dorothy had arrived 'of her own accord, alone and independent' to help him in his campaign. However, she need not have worried – earnest women were never really to Winston's liking. Nor can Lady Dorothy's passionate advocacy of 'Votes for Women' and her support for the 'Anti-drink' lobby have aroused in him much enthusiasm. Indeed, Winston told Clementine how, when he had teased her 'by refusing to give a decided answer about women's votes, she left at once for the North in a most obstinate temper. However on reading my answers given in public, back she came, and is fighting away like Diana for the Greeks.'

But his letter was not wholly about the election or Lady Dorothy. Of his meeting with Clementine he wrote:

> what a comfort & pleasure it was to me to meet a girl with so much intellectual quality & such strong reserves of noble sentiment. I hope we shall meet again and come to know each other better and like each other more . . . Meanwhile I will let you know from time to time how I am getting on here in the storm; and we may lay the foundations of a frank & clear-eyed friendship which I certainly should value and cherish with many serious feelings of respect.[4]

On 23 April Clementine replied to him from Nordrach:

> Your letter found me here only yesterday – Seemingly, our maid at home thought there was no hurry in forwarding letters. If it were not for the excitement of reading about Manchester every day in the belated newspapers I should feel as if I were living in another world than the delightful one we inhabited together for a day at Salisbury Hall . . .
>
> I feel so envious of Dorothy Howard. It must be very exciting to feel one has the power of influencing people, ever so little. One more day & we shall know the result of the Election. I feel as much excited as if I were a candidate . . .
>
> I don't know if wishing & hoping can influence human affairs – if so – poor Joynson-Hicks!
>
> Yours very sincerely,
> Clementine Hozier

Before Winston wrote his next letter on 27 April the battle in North West Manchester was over. Joynson-Hicks, the Conservative candidate, had won by 429 votes. The Tory press was beside itself with exultation. 'Winston Churchill is out, OUT, OUT!' wrote, or rather shrieked, the *Daily Telegraph*. Winston took his defeat philosophically, and wrote to Clementine:

> If I had won Manchester now, I should probably have lost it at the general election. Losing it now I shall I hope get a seat wh will make me secure for many years. Still I don't pretend not to be vexed. Defeat however consoled explained or discounted is odious.

There was more – somewhat annoyingly – about Lady Dorothy (who had changed identity, style and century since his previous letter):

> [she] fought like Joan of Arc before Orleans. The dirtiest slum, the roughest crowd, the ugliest street corner. She is a wonderful woman – tireless, fearless, convinced, inflexible – yet preserving all her womanliness.

Is it possible that Winston could have been gently tantalizing his true love? But the letter ends with a very special message for her:

> How I should have liked you to have been there. You would have enjoyed it I think. We had a jolly party and it was a whirling week. Life for all its incompleteness is rather fun sometimes.
>
> Write to me again. I am a solitary creature in the midst of crowds. Be kind to me.
>
> Yours vy sincerely,
> W

Towards the end of May Clementine, with her mother and sister, returned home. In the interval, Winston had fought and won Dundee at a by-election. He was to represent this constituency for twelve years. During the months of June and July he and Clementine saw each other several times, but as unmarried girls did not in those days lunch or dine alone with men, they met in the main only on social occasions. Clementine was by now deeply in love, and in an agony lest their growing friendship should be remarked upon. Such was her anxiety on this count that when Winston invited her to a garden party he was giving in the gardens of Gwdyr House in Whitehall (then the Board of Trade) she declined to go. Presumably their closest family must have been aware that something was brewing, but their discretion was complete; more than most girls, Clementine could keep her own counsel, so now at this crucial moment of her life even her few close friends were unaware of her feelings.

She spent part of these, for her, most trying summer months staying with her great-aunt Maude Stanley, who once more proved to be a sympathetic refuge for her great-niece during a time which, for Clementine, was filled with alternating hopes and fears.

The parliamentary summer recess was soon imminent. Both Winston and Clementine were committed to pre-arranged visits, but they planned to meet at Salisbury Hall in the middle of August. In the interval, Clementine went to stay with Mrs Godfrey Baring at Nubia House in Cowes. She took part in the round of balls and entertainments, but she was a somewhat distracted guest, as her thoughts and heart were elsewhere.

Winston was staying with his cousin, Freddie Guest, and his American wife, Amy Phipps, at Burley Hall, near Oakham, in Leicestershire. During the early hours of 6 August, soon after everyone had gone to bed, a fire broke out, and a whole wing of the house was burned to the ground by daybreak. Rumours flew fast, and Clementine's heart turned over when she heard a garbled account of the disaster. She knew Winston to be one of the house party, and first reports were alarming about loss and injury. Fortunately *The Times* put her out of her misery: the damage had been great, but no loss of life or injury had occurred. Unable to contain her relief and joy, she rushed to the local post office and sent a telegram to Winston.

On 7 August he wrote her a long letter from Nuneham Park (Sir Lewis and Lady Harcourt's house, where some of the Burley Hall party had taken refuge), in which a lively account of the fire took third place to an alteration in plans for the visit to Salisbury Hall and an account of his brother Jack's marriage to Lady Gwendeline Bertie (daughter of the Earl of Abingdon) at Abingdon on 4 August:

> Jack has been married to-day – civilly [the bride was a Roman Catholic]. The service is tomorrow at Oxford but we all swooped down in motor-cars upon the little town of Abingdon and did the deed before the Registrar – for all the world as if it was an elopement – with irate parents panting on the path . . . & then back go bride & bridegroom to their respective houses until tomorrow. Both were 'entirely composed' & the business was despatched with a celerity and ease that was almost appalling.

The alteration in plans consisted of interposing a two-day visit to Blenheim Palace before going on to Salisbury Hall. Winston had arranged everything with his cousin Sunny Marlborough*, who invited Clementine at Winston's urgent behest. Clementine jibbed at this, to her, somewhat alarming suggestion, but allowed herself to be persuaded by Winston's assurance that

> Sunny wants us all to come, & my Mother will look after you – & so will I. I want so much to show you that beautiful place & in its gardens we shall find lots of places to talk in, & lots of things to talk about.

Of the fire at Burley Hall, he wrote:

> I was delighted to get your telegram this morning & to find that you had not forgotten me. The fire was great fun, & we all enjoyed it thoroughly. It is a pity such jolly entertainments are so costly. Alas for the archives. They soared to glory in about ten minutes. The pictures were of small value, & many, with all the tapestries & about ½ the good furniture were saved. I must tell you

* Charles Spencer-Churchill, 9th Duke of Marlborough (1871–1934).

> all about it when we meet. My eyes smart still & writing is tiring.
>
> It is a strange thing to be locked in deadly grapple with that cruel element. I had no conception – except from reading – of the power & majesty of a great conflagration . . . Poor Eddie Marsh lost everything (including many of my papers) through not packing up when I told him to. I saved all my things by making Reynolds [his manservant] throw them out of the window. It was lucky that the fire was discovered before we had all gone to sleep – or more life might have been lost – than one canary bird; & even as it was there were moments of danger for some.

Winston ended his letter by asking her what she had been thinking of during her days at Cowes,

> & whether you would have thought of me at all – if the newspapers had not jogged your memory! You know the answer that I want to this.

He omitted in his letter to mention that he had taken a leading part in salvaging pictures and other valuables from the conflagration, and had escaped possible death or serious injury when, just as he left one part of the house carrying two marble busts, the roof, which had been blazing furiously, collapsed.

The following evening, 8 August, Clementine wrote from Cowes.

> I was so glad to get your delightful letter this morning. I retired with it into the garden, but for a long time before opening it I amused myself by wondering what would be inside. I have been able to think of nothing but the fire & the terrible danger you have been in – The first news I heard was a rumour that the house was burnt down – That was all – My dear my heart stood still with terror. All the same I did not need that horrible emotion to 'jog my memory'.
>
> We all went to a ball the next night which I hated. I was extremely odious to several young partners not on purpose, but because they would interrupt my train of thought with irrelevant patter about yachts, racing, the weather, Cowes gossip, etc. So I was obliged to feign deafness –
>
> Please do not think there is any real reason for my not at first wanting to go to Blenheim. It was only a sudden access of shyness.

Another reason for Clementine's hesitation in accepting the invitation was that she was down to her last laundered and starched dress and, having no personal maid, she would have to make this one last the unexpected visit.

Winston's next letter, also written on 8 August, must have crossed with Clementine's, and in it he reproaches her for being an indifferent correspondent. However, he had tidings of her from his aunt, Lady Leslie, who had arrived at Nuneham Park from Cowes to attend Jack Churchill's wedding, and who, he wrote,

> brings me news from Cowes of a young lady who made a great impression at a dance four nights ago on all beholders. I wonder who it could have been!

But the letter started with a description of Jack's church wedding:

> My dear – I have just come back from throwing an old slipper into Jack's departing motor-car. It was a vy pretty wedding. No swarms of London fly-catchers. No one came who did not really care, & the only spectators were tenants & countryfolk. Only children for bridesmaids & Yeomanry with crossed swords for pomp. The bride looked lovely & her father & mother were sad indeed to lose her. But the triumphant Jack bore her off amid showers of rice & pursuing cheers – let us pray – to happiness & honour all her life.

Clementine had evidently telegraphed Winston that same Saturday morning to acquiesce in the Blenheim plans. Winston hastened to reassure her that there would not be a large party, and to introduce her, as it were ahead of the event, to his first cousin, the Duke.

> I hope you will like my friend, & fascinate him with those strange mysterious eyes of yours, whose secret I have been trying so hard to learn . . .
>
> He is quite different from me, understanding women thoroughly, getting into touch with them at once . . . Whereas I am stupid and clumsy in that relation, and naturally quite self-reliant and self-contained. Yet by such vy different paths we both arrive at loneliness!

His letter ended:

> Till Monday then & may the Fates play fair.
>
> Yours always
>
> W

His cousin's loneliness was due to the fact that since 1906 he had been living apart from his beautiful Duchess. She had been born Consuelo Vanderbilt of New York City, and her marriage to the Duke of Marlborough had been arranged by her domineering mother in the manner of a bygone century, and in the face of her daughter's declared attachment to another man. The inner wishes of the Duke's heart also lay elsewhere, but the duty and desire to perpetuate the historical splendours of Blenheim were paramount to him. His well-dowered and beautiful bride came to the altar in tears, but she stayed at Blenheim for eleven years, fulfilling her duties as Duchess with grace and distinction, and providing her husband with two sons. But the differences between their natures were deep and irreconcilable, and finally they separated. Later the Duke married again, as also did she; but now, at this moment, he was a solitary figure in a great setting.

Winston had a strong sense of consanguinity, and between him and his first cousin Sunny were not only bonds of blood, but also true liking and affection. Winston's radical views were not those of the Duke; but neither then nor later, despite some brisk verbal passages, did political differences interfere for long with their friendship.

Throughout his life Winston was a frequent visitor to Blenheim. It was in a special sense his home: there he was born; there he became engaged; there he spent the first few days of his honeymoon. Later, when Winston wrote his splendid life of the great 1st Duke of Marlborough, Blenheim epitomized for him the triumphs and glories of England and of his hero. And so it seemed most right and fitting that in the end he should have wished to return there – to Bladon, the little churchyard just outside the park – to lie close to his parents and kinsfolk.

It seems certain from his letters that Winston had determined to ask Clementine to be his wife. He had taken his cousin, the Duke, into his confidence, and the Blenheim visit was undoubtedly arranged so that he could propose to this beautiful girl with whom he was so deeply in love, in a setting which combined the romantic with the heroic, and where he felt so strongly the ties of family and friendship.

In the train travelling between Cowes and Blenheim, Clementine wrote to her mother, who was in London.

Darling mother,

Thank you so much for your letter. I shall get to Oxford at 5.20 where I shall be met by motor.

I feel dreadfully shy & rather tired . . . I will write to you tomorrow morning from Blenheim.

I shall see 2 of Winston's greatest friends there – the Duke and F. E. Smith. On Wednesday we go to Salisbury Hall . . .

I cannot write a good letter – I feel like in a dream & can do nothing more intelligent than count the telegraph posts as they flash by or the pattern on the railway cushions.

Much love from
Clem

The somewhat hastily arranged house party gathered at Blenheim Palace on Monday, 10 August. It was, as promised, a small one. Winston came with his mother, and a Private Secretary from the Board of Trade (for affairs of state must not be neglected, even at such an hour). The F. E. Smiths were the only other guests.

Before retiring to bed on the Monday evening Winston made an assignation with Clementine to walk in the rose garden the following morning after breakfast. Appearing promptly at breakfast was never at any time in his life one of Winston's more pronounced qualities – even on this day of days he was late!

Clementine came downstairs with characteristic punctuality, and was much discomfited by Winston's non-appearance. While she was eating breakfast, she seriously turned over in her mind the possibility of returning immediately to London. The Duke observed that Clementine was much put out, and took charge of the situation. He despatched a sharp, cousinly note upstairs to Winston and, deploying his utmost charm, suggested to Clementine that he should take her for a drive in his buggy. He whirled her round the estate for about half an hour, and upon their return there was the dilatory Winston anxiously scanning the horizon.

During the course of the late afternoon Winston and Clementine went for a walk in the garden; overtaken by a torrential rainstorm, they took refuge in a little Greek temple, which looks out over the great lake. Here Winston declared his love, and asked Clementine to marry him. When in due course, the rain shower being over, they emerged from the temple – they were betrothed.

Clementine had enjoined absolute secrecy upon Winston until she had been to London to tell her mother their great news, and to seek her formal consent. But as they drew near the house they met the Duke and his other guests gathered on the lawn, whereupon Winston's triumph and joy were too much for him to contain – he started running across the grass waving his arms with excitement, and, despite his promise to the contrary, told his assembled friends and relations his glorious tidings.

It was arranged that Clementine should go to London the next morning to see her mother. Early on that morning Winston and Clementine exchanged notes. He wrote:

> My dearest,
>
> How are you? I send you my best love to salute you: & I am getting up at once in order if you like to walk to the rose garden after breakfast & pick a bunch before you start. You will have to leave here about 10.30 & I will come with you to Oxford.
>
> Shall I not give you a letter for your mother?
>
> Always,
> W

She replied:

> My dearest,
>
> I am very well – Yes please give me a letter to take to Mother. I should love to go to the rose garden.
>
> Yours always
> Clementine

A few hours later Clementine left Blenheim with Winston, who was to accompany her to the station. She bore with her the following letter:

Blenheim Palace
12th Aug. 1908.

My dear Lady Blanche Hozier,

Clementine will be my ambassador to you today. I have asked her to marry me & we both ask you to give your consent & your blessing.

You have known my family so many years that there is no need to say vy much in this letter. I am not rich nor powerfully established, but your daughter loves me & with that love I feel strong enough to assume this great & sacred responsibility; & I think I can make her happy & give her a station & career worthy of her beauty and her virtues.

Marlborough is vy much in hopes that you will be able to come down here today & he is telegraphing to you this morning. That would indeed be vy charming & I am sure Clementine will persuade you.

With sincere affection
Yours ever,
Winston S. Churchill

But in the event his letter was hardly necessary, as we see from the account Clementine gave of that whirlwind twenty-four hours in a somewhat breathless and disjointed letter a day or two later to Nellie who, throughout these emotional and exciting days for her sister, was staying away with friends in the country:

Well it happened Tuesday evening – & on Wednesday morning I telegraphed mother to say I was coming to see her from Blenheim. The Duke telegraphed asking her to come & stay – Winston was by way of coming in the motor to the station only, as he had some work to do with his secretary at Blenheim, but when we got there we hopped into the train together. When we got to Paddington I went off alone to Abingdon [Villas] & told him to follow in ½ an hour – I wanted to see the Min [the nickname the children used for their mother among themselves] alone first & also I was not sure if she would be dressed! However, my dear, she was! with her hair beautifully done by Miss Blount looking quite beautiful – much nicer than me – I have become very ugly & thin & I don't know what to do about it.

Well W arrived presently & I left them alone & everything was all right. Gubbins [Lady Blanche's dog] looked much disgusted. I tried to remove him from the room as I thought he was a disgrace to the establishment, but was prevented – When I am married I am going to have a dog. You must come and help to choose it – a Pekinese.

My dear, I have the most lovely ring – a fat ruby with 2 diamonds – I must tell you about it – When Lord Randolph married he gave her [Lady Randolph] 3 lovely rings – one all diamond, one sapphire & diamond & one ruby & diamond – she wears the diamond one, & Goonie [Gwendeline, Jack's wife] has the sapphire & diamond & I have the twin ruby one – Lord Randolph said they were for his sons to give their wives.

Well to return to the narrative – Min, Winston & I returned to Blenheim in the afternoon – We were all dog-tired, but no matter – Min was most delightful at dinner & looked lovely in her white lace arrangement . . .

Blanche Hozier must have expected the news Clementine brought her. Perhaps she had hoped Clementine would marry a man who could offer her, as well as personal happiness, solid financial security. Winston Churchill was a poor man by the standards of the world in which he lived; his talents were his only prospects; and although he was undeniably brilliant, he was ill-regarded by many on both public and personal grounds. But in many ways, as we have seen, Blanche Hozier was unworldly, and, intuitive in all things concerning the heart, she probably recognized that now, for the first time, Clementine was truly in love.

Moreover, Lady Blanche had herself been captivated by Winston's personality and charm. Clementine wrote to Nellie: 'I am so afraid you will fall in love with Winston – Mother has already & he with her. He thinks she is the most delightful person almost in the world!'[5] Blanche Hozier herself wrote, in a letter to her sister-in-law, Mabell Airlie: 'Clementine is engaged to be married to Winston Churchill. I do not know which of the two is the more in love. I think that to know him is to like him. His brilliant brain the world knows, but he is so charming and affectionate in his own home life.'[6] And to an old friend – the poet and author Wilfrid Scawen Blunt: 'He [Winston] is so like Lord Randolph, he has some of his faults, and all his qualities. He is gentle and tender, and affectionate to those he loves, much hated by those who have not come under his personal charm.'[7]

During the following day or two, notes flew along the corridors of the Palace. No more formality now, no veiled words – both could give rein to the love and spontaneous joy each felt.

From Winston:

My dearest,

I hope you have slept like a stone. I did not get to bed till 1 o'clock; for Sunny kept me long in discussion about his affairs wh go less prosperously than ours. But from 1 onwards I slept the sleep of the just, & this morning am fresh and fit. Tell me how you feel & whether you mean to get up for breakfast. The purpose of this letter is also to send you heaps of love and four kisses XXXX

from your always devoted
Winston[8]

Clementine replied:

My darling,

I never slept so well and I had the most heavenly dreams.

I am coming down presently – Mother is quite worn out as we have been talking for the last two hours.

Je t'aime passionément [*sic*] – I feel less shy in French.

Clementine[9]

From Blenheim, Winston and his mother took Clementine with them to Salisbury Hall for the weekend, and also Jack and his bride, Gwendeline (always known as Goonie); and they were all there when the engagement was officially announced on Saturday, 15 August. Close friends and relations were hurriedly told the great news – to nearly everybody it came as a surprise, and was the cause of much excited comment. If Winston's name had been linked romantically, it had certainly not been with the beautiful, but obscure and penniless, Clementine Hozier.

To one person at least, the tidings brought bitter pain: Sidney Peel had always told Clementine that he would wait for her for ever, but in the event of her marrying someone else it would be the end of everything between them. While she was free he could endure anything, but once hope (however frail) was dead he could not transmute his deep devotion into terms of pallid friendship. So when Clementine wrote to tell him of her engagement she knew she was severing for ever a relationship she greatly valued, and in which she had found much comfort and companionship in bleaker days. Many years later – in 1921, soon after Bill's tragic death – Clementine met Sidney Peel (now DSO, and since 1914 married) walking towards her down Conduit Street in the West End. They passed – then both stopped and turned. He said, 'Oh, Clementine – poor Bill – I am so sorry.' And then, 'Are you happy?' She said, 'Yes – I'm happy.' He turned on his heel; she never saw him again.*

The news of Clementine's engagement, with various inaccurate embellishments, travelled to Dieppe, and caused much local interest. Mme Villain, still the reigning queen of the fish market, mounted a herring box and, clapping her hands for silence, announced to the assembled company: 'Voilà des nouvelles! . . . Vous vous rappelez de notre petite Clementine? Eh bien, elle s'est fiancée . . . avec un ministre anglais . . . millionaire et décoré!'

Clementine's grandmother, Lady Airlie, noted in her diary for 13 August 1908: 'I get news of Clem Hozier's engagement to W. Churchill sitting in the garden seeing the hedges clipt.' She later wrote to her daughter-in-law, Mabell Airlie: 'Winston is his father over again, with the American driving force added. His mother and he are devoted to one another, and I think a good son makes a good husband. Clementine is wise. She will follow him and, I hope, say little.'[10] It is evident that old Lady Airlie had not fully estimated her granddaughter Clementine's strength of character, nor her capacity to express strongly held opinions.

Winston would not brook a long engagement. Then as later, he saw no reason for delay once a plan was decided upon; moreover, the dictates of the parliamentary year set certain time limits.

In a letter to her Aunt Mabell (young Lady Airlie) Clementine wrote of their plans:

* Sidney Peel, later 1st Baronet Peel, had married Lady Delia Spencer, daughter of the 6th Earl Spencer, in 1914.

I cannot describe my happiness to you. I can hardly believe it is true. I have been engaged nine days now. I cared very much for him when he asked me to marry him, but every day since has been more heavenly. The only crumpled roseleaf is the scrimmage getting ready in time. September 12th has now been settled for our wedding. We shall have only seventeen days away as Winston has to be back in London on October 3rd.[11]

Even today, when an elaborate trousseau is no longer a necessity, a month is scant time for a bride to organize herself. Clementine was soon swept up in a tidal wave of preparations and correspondence, which at moments seemed daunting. Writing from Abingdon Villas (in an undated letter), she told Winston:

My darling,

Thinking about you has been the only pleasant thing today.

I have tried on so many garments (all of which I am told are indispensable) . . .

My tailor told me he approved of you & had paid 10/6d. to hear you make a speech about the war at Birmingham – After that I felt I could not bargain with him any more . . .

Goodbye my darling I feel there is no room for anyone but you in my heart – you fill every corner.

Clementine

It is not an uncommon experience in the interim between betrothal and marriage for the sunshine of joy and confidence to be overcast by a fleeting cloud of doubt. Clementine had her moment of trepidation. Perhaps in part it was reaction to the excitement, the public interest and the exhausting bustle of the preparations. Maybe she even wondered if Winston truly loved her. Even during these few weeks of their engagement, public life laid constant claim to both his time and his interest. Already she saw the face of the only real rival she was to know in all the fifty-seven years of marriage that lay ahead – and for a brief moment she quailed. She confided her doubts and fears to Bill, who showed himself in a firmly steadying role. He wrote to her saying that although he was younger than her, he was now the head of the family, and that as such, he must remind her that she had already broken off at least two engagements – one of them publicly announced; he told her that she could not make an exhibition of, and humiliate, a public personage such as Winston. But more than Bill's brotherly admonition, it was the warmth of Winston's swift reassurance, and the force of his own supreme confidence in their future together, which swept away the doubts that had beset her.

Their letters to each other during those short weeks of their engagement breathe love and hope and joy – long pent up – at last finding release and expression. In one of her letters to Winston, Clementine wonders 'how I have lived 23 years without you. Everything that happened before about

5 months ago seems unreal.'[12] Winston's underlying mood was one of wonder and thankfulness. 'There are no words,' he wrote to her, 'to convey to you the feelings of love & joy by wh my being is possessed. May God who has given me so much more than I ever knew how to ask keep you safe and sound.'[13]

CHAPTER FOUR

Early Days Together

THERE WAS MUCH PUBLIC INTEREST IN THE MARRIAGE. IT WAS AN INDICATION of the extent to which Winston Churchill had already captured people's imagination, by his exploits and his writings as much as through the name he bore. Moreover, the British public is incurably romantic, and in this match they found all their favourite ingredients: youth, courage, beauty, birth – and, above all, true love.

As the day of the wedding, 12 September, drew near, many of the daily papers carried running commentaries on the progress the bride was making with her preparations, and published long lists of wedding presents. The bridegroom's status as a national figure received a measure of significant recognition on the day of his wedding, when his life-size model went on display at Madame Tussaud's.

Kind and hospitable Lady St Helier offered her house in Portland Place for the reception, and since Abingdon Villas was small and would be full of relations and bridesmaids, it was arranged that Clementine should spend the night before her wedding at her great-aunt's house, and that she would leave from there for the church.

On the eve of her wedding, therefore, Clementine went to stay with Lady St Helier. The next morning – her wedding day – she awoke very early indeed. The bedroom seemed chill and vast, and in the grey dawn light her wedding dress glimmered remote and unreal. A sudden wave of panic and homesickness swept through her, and she wished with all her heart that she was back in the familiar bustle of Abingdon Villas.

On the spur of the moment she decided to go home for breakfast; but this was easier wished than done, as all her everyday clothes had been swept away the night before by her great-aunt's lady's maid. In her dressing-gown, Clementine crept downstairs: the great rooms were still in darkness, but in the morning room a young under-housemaid was busily blacking the grate. Clementine confided to her the great longing she had to go home; the maid understood at once – probably she too found the house in Portland Place stiff and grand. Clementine seemed near enough the same size; the friendly girl ran upstairs and returned with a complete set of her own best 'going out' clothes. Shortly afterwards Clementine was let out of the back door, attired in a pretty print dress, dark close-fitting coat and black-buttoned boots, with a neat bonnet and kid gloves. The early buses were running, and she caught one on the route she often used. As he gave her a ticket, the conductor eyed her narrowly. 'Ain't you going the wrong way, Miss?' he said.

Great was the astonishment and merriment when she appeared at home, and dressed so strangely. After a cheerful breakfast with her family she was packed off back to Portland Place, her spirits restored. The bride-in-disguise managed to slip in amid the general confusion of preparations before her absence had been reported to Lady St Helier.

Although September was a time of the year when many people were away from London, none the less St Margaret's, Westminster, was packed to the doors with wedding guests. For a day a truce was called between radical and Tory, and indeed, one of the first guests to arrive was the Tory member for North West Manchester, William Joynson-Hicks, who had so narrowly defeated Winston earlier in that same year. A rather late arrival, Wilfrid Scawen Blunt, had room made for him in the bride's family pew by Lady Blanche, and found himself next to 'Hugo Wemyss who is I think like me a former lover. Redesdale a third was with her in the front.'[1] No doubt the piquancy of these seating arrangements was not lost on various members of the congregation.

Great crowds had assembled in Parliament Square, and both the bridegroom, with his best man (Lord Hugh Cecil), and the bride were enthusiastically cheered as they arrived. Clementine was – from all accounts – a glorious sight, as she came up the aisle on the arm of Bill, a handsome and devoted brother, now a sub-lieutenant in the Royal Navy. Her gown was of shimmering white satin, and her voluminous veil of tulle was held in place by a coronet of fresh orange blossoms; she wore the diamond earrings given her by Winston. Her bouquet was of white tuberoses, and she also carried the white parchment prayer book which was the gift of her godfather, Sir Jack Leslie. Nellie was of course a bridesmaid, along with Clementine's cousins Venetia Stanley and Madeline Whyte, Clare Frewen (Winston's cousin) and one dear friend – the exquisite Horatia Seymour. The bridesmaids wore gowns of biscuit-coloured satin with large, romantic black hats wreathed in roses and camellias; they carried bouquets of pink roses. St Margaret's was a bower of palms and ferns, with white lilies everywhere.

The Bishop of St Asaph conducted the service. Wilfrid Scawen Blunt recorded in his diary: 'Winston's responses were clearly made in a pleasant voice, Clementine's inaudible.'[2] Bishop Welldon (now Dean of Manchester), Winston's former Headmaster at Harrow, gave the address.

While the bridal party was in the vestry, the bridesmaids, carrying wickerwork baskets, distributed 'favours' of orange blossom to the guests. Mr Lloyd George, the Chancellor of the Exchequer, signed the register, and Winston immediately set to talking politics with him.

As the bride and bridegroom left the church they were cheered vociferously, and another crowd greeted them at No. 52 Portland Place.

After being received by Lady Blanche and greeting the bride and bridegroom, the guests were able to view the wedding presents, which were many and splendid. These were still spacious times; the day had not yet come when tea towels, dishwashers and ovenware are welcome presents. Clementine's present list was sprinkled with pendants, muff chains and a variety of ornamental boxes. Winston received several silver candelabra, wine coasters, pepper pots, and no fewer than seventeen silver inkstands. King Edward sent him a gold-topped walking stick. Dundee Liberal Association weighed in with a large and beautiful Georgian silver dinner service; and his colleagues in the Government gave him a large scallop-edged silver tray, with all their signatures engraved on the back. Prominently displayed among these rich trophies was a glistening turbot with a lemon in its mouth – the gift of Mme Villain from Dieppe.

While the fine company admired the presents and toasted the bride and bridegroom, jollity reigned outside in Portland Place. The crowd of admirers and unknown friends included a contingent of 'Pearly Kings and Queens', their clothes resplendent with pearl buttons sewn in intricate patterns. As President of the Board of Trade, Winston had helped to protect the costermongers' rights to trade in the streets; so, on this his wedding day, many of them gathered in their splendid traditional finery to wish him well. Late in the afternoon the bride and bridegroom left Lady St Helier's house to start their journey to Blenheim, where they were to spend a day or two before going to Italy. With her grey cloth ensemble Clementine wore a large black satin hat adorned with one long, sweeping ostrich feather. The Pearly Queens, whose own large hats are always decorated with waving ostrich plumes, were delighted. 'Ooh, wot a luv'ly fevver!' they cried, as Winston and Clementine got into their carriage.

The poor bridegroom, however, did not draw such favourable comments on his apparel – at least from the *Tailor and Cutter*, which described his wedding outfit as '"neither fish, flesh, nor fowl" . . . one of the greatest failures as a wedding garment we have ever seen, giving the wearer a sort of glorified coachman appearance'.

The bridal couple travelled to Woodstock by train, and were greeted at the station by an enthusiastic crowd, their cheers mingling with the bells of St Mary Magdalene. It had been a long and glorious day, and presently, amid the splendours of the past, and where their own present happiness had begun – Winston and Clementine were alone together.

* * *

After a few days spent at Blenheim recovering from the emotions and excitements of their wedding day, the honeymooners travelled to Italy, visiting Lake Lugano and Venice. From there they went to stay with Baron ('Tuty') de Forest and his English wife at Eichorn in Austria. Born in 1879 of Austrian extraction, the Baron had been educated and brought up in England, and both he and his wife had been friends of Winston from childhood. As his guest, Winston enjoyed some excellent shooting, but Clementine found the large house party rather grand, stiff and social, and was glad to begin the homeward journey. Winston wrote to his mother on 20 September 1908: 'We have only loitered & loved, a good & serious occupation for which the histories furnish respectable precedents.'

One of their first visits on their return to England was to old Lady Airlie at Airlie Castle, where they stayed on 8 October on their way to Dundee, in order that Winston might present his bride to his constituents.

* * *

There had been no time for house-hunting in the weeks before their wedding, and so they had decided to start their married life in Winston's bachelor establishment in Bolton Street. Winston had entrusted the necessary rearrangements to his mother, who had set about the house with characteristic energy and perhaps rather too much zeal, and on their arrival home in early October Clementine was greatly surprised to find her bedroom completely redecorated. To her simple and rather austere taste, the sateen and muslin covers trimmed with bows, which decked the chairs, dressing-table and bed, appeared vulgar and tawdry. She did her best, however, to conceal her lack of appreciation of her mother-in-law's well-meant efforts.

Clementine was never to have a really close relationship with Lady Randolph. This once supremely beautiful woman was still, in her fifties, remarkably handsome; but, accustomed all her life to attention and adulation, Lady Randolph had now to accept a long, cool look from her newest daughter-in-law. Although as time went on Clementine came to salute her courage and her unfailing zest for life, now in these early years of her marriage she passed (in her heart) fairly severe judgement on her celebrated mother-in-law: she thought her vain and frivolous and, in her marriage to George Cornwallis-West – a man so much younger than herself – somewhat ridiculous.

As Jennie Jerome, she had been brought up as one of the three beautiful daughters of a generous father – Leonard Jerome of New York City – who made, lost and recouped fortunes with surprising speed. Jennie was open-handed and a great spender, and Clementine was on many occasions to feel resentment (fully shared by Jack's wife, Goonie) at her mother-in-law's extravagances, which often weighed heavily upon her two sons.

But Jennie was a celebrated character, and we see her through more indulgent eyes in Eddie Marsh's vivid sketch of her:

> She was an incredible and most delightful compound of flagrant worldliness and eternal childhood, in thrall to fashion and luxury (life didn't begin for her on a basis of less than forty pairs of shoes) yet never sacrificing one human quality of warm-heartedness, humour, loyalty, sincerity, or steadfast and pugnacious courage.[3]

Clementine knew how dearly Winston loved his mother, and she kept (for the most part) her opinions and reservations about Lady Randolph to herself. And mutual good manners, surface affection and a shared loyalty to Winston maintained a civilized relationship between these two very different women.

On their return from their honeymoon, while Winston plunged once more into politics, Clementine proceeded to arrange their domestic affairs. She started with the kitchen: needless to say, Winston's bachelor establishment had not been run on economical lines. Clementine set to with energy to improve upon this state of affairs. Accustomed all her life to managing on very little money, she had a horror of extravagance, and from her mother she had learnt that delicious food is more a matter of taking trouble than lavish spending. To assist her in these laudable endeavours, she was armed with a little booklet (given to her as a wedding present) entitled *House Books on 12/- a week*, which was considered to be a generous allowance per head in those days. Clementine also took up the cudgels for economy on the wardrobe front. Violet Asquith (the Prime Minister's daughter with his first wife) recorded that soon after her marriage Clementine confided to her that 'Winston was most extravagant about his underclothes. These were made of very finely woven silk (pale pink) and came from the Army and Navy Stores and cost the eyes out of the head.'[4] Whether Clementine's campaign for cheaper chemises was successful is not recorded, but Winston protested that he had a most delicate skin (which was true) which necessitated such luxurious lingerie. The author's memory is that he always wore very fine silk underclothes.

But Clementine was not entirely engrossed by domestic affairs. Political life thrilled her, and she now flung herself with passion into the radical cause. The noble, puritan element in her nature responded naturally to the great reforms in which Churchill, working closely with Lloyd George, was deeply involved. Old age pensions had been introduced earlier in that year, and Winston, who had taken up the cause of social reform with ardour and determination, was now pushing forward the schemes for trade boards, which would attack the exploitation of sweated labour, and of labour exchanges to combat unemployment.

Clementine was proud to be the wife of one of the leaders of such noble measures. As Winston's wife, she moved naturally into the orbit of the young reformers such as Lloyd George and C. F. G. Masterman (then Parliamentary Secretary to the Local Government Board) and the higher echelons of the Liberal leadership, including the Asquiths; the Foreign Secretary, Sir Edward Grey; and the Secretary of State for War, Richard Haldane. Bitterness in politics had spread into social life, and some die-hard Tory acquaintances would even cross the road rather than meet her in the street now that she was Mrs Winston Churchill. This only served to fan her crusading ardour, and was no impoverishment, for the young Churchills' friends included parliamentary colleagues, eminent civil servants, radical politicians, and writers and journalists drawn from the galaxy of talent surrounding the Government in these years before the Great War. And Winston maintained his links across the 'barricades' with such Tories as

F. E. Smith, Lord Hugh Cecil and A. J. Balfour, with whom friendship never faltered.

Wherever Winston took his bride, Clementine's beauty, charm and enthusiasm gained her friends and praise. She even earned the (albeit somewhat patronizing) approval of that 'high priestess' of the left Beatrice Webb, who wrote in her diary: 'On Sunday we lunched with Winston C. and his bride – a charming lady, well bred and pretty, and earnest withal – but not rich, by no means a good match, which is to Winston's credit.'[5]

Clementine soon had a preoccupation of her own, more personal than but just as exciting as radical politics – she was expecting her first child. The lease of No. 12 Bolton Street was due to expire in February 1909, and in any case it was of doll's-house proportions, so in the latter part of 1908 she was busy house-hunting. In the New Year they found, and acquired on a long lease, No. 33 Eccleston Square, a substantial family house in one of the large squares with tree-filled gardens behind Victoria Station, most conveniently situated for the House of Commons. It needed, however, considerable alterations and complete redecoration, and to bridge the time between leaving Bolton Street and moving into Eccleston Square, Freddie Guest, Winston's cousin, lent them his house in Carlton House Terrace.

The baby was expected in July, and towards the end of April Clementine was invited to Blenheim, where she stayed for nearly a fortnight. It was the first time in their married life that she and Winston had been apart, for he had to remain in London, as the House was sitting, and much was afoot politically. Also staying at Blenheim was Clementine's sister-in-law, Goonie Churchill; she too was expecting her first child, and it was during this visit that the two sisters-in-law, born in the same year, cemented a liking that had been immediate. Although they were quite different in their style of beauty and their natures, their mutual affection and loyalty were to be lifelong. Lady Cynthia Asquith described Goonie: 'Her alluring mermaid beauty has a strange translucent quality; her wide-opened eyes are like blue flowers. "Queen Queer" we called her, in homage to an enigmatic loveliness that cast so subtle and so lasting a spell.'[6]

But there had been a piquant prologue to the long and happy relationship which would exist between the two brothers and their wives. Jack Churchill, twenty-six and just beginning his career as a stockbroker, had met twenty-two-year-old Lady Gwendeline Bertie, daughter of the Earl of Abingdon, in 1907, and both fell romantically (and somewhat impractically) in love. A little later, on a visit to his mother at Salisbury Hall, Winston met Goonie, who was also staying there with Jack. Winston and Goonie took to each other, and a flirtatious 'jokey' relationship sparked between them in which – to judge from the three light-heartedly teasing letters extant which she wrote him in late August – Goonie set the pace. Jack would tell Winston, in a letter of November 1907, that he and Goonie had declared their love for each other earlier that summer, and that they were secretly engaged. Meanwhile no one was to know of their situation – and certainly her August letters to Winston contain no hint at all of how things stood between her and Jack.

It had been planned that Winston, as Under-Secretary of State for the Colonies, would make a journey of several months' duration to East Africa that autumn: the Royal Navy ship which would convey him was to be at Malta ready for departure in the first days of October. Before embarking on this long official journey, Winston decided to take a holiday on the Continent, combining it with a visit to watch the French army on their manoeuvres. It has been suggested that this European jaunt was arranged in a great hurry because Winston wished to distance himself from a situation which seemed to be developing between him and Gwendeline Bertie.[7] Certainly her letter to him of 27 August 1907 seems to show surprise and disappointment that there would be no opportunity for farewells:

Wytham Abbey,
Oxford

Dear Mr Winston,

It is positively cruel of fate to determine we should not say goodbye, not to allow us the opportunity of bidding each other a friendly farewell, for it is a long time to lose sight of someone, it is five months.[8]

Be that as it may, Winston left London for Paris on 28 August, and after the manoeuvres drove down in a leisurely way through France, Italy and Austria. Agreeable and stimulating companions for him en route were his new Tory friend, the brilliant attorney F. E. Smith, and – joining them in Italy – his kinsman Sunny Marlborough.

In Malta, where Winston arrived on 2 October, Eddie Marsh and the rest of the official party were awaiting him. After some days of business and official visiting there, they embarked in HMS *Venus*; sailing via Cyprus, the Red Sea and Aden, they reached Mombasa at the end of October, where Churchill's East African journeyings would begin.

It was during the course of these African travels that Winston received a long letter from Jack, written on 14 November, telling him that he and Goonie were secretly engaged, and swearing him to absolute silence, only their mother and her husband, George Cornwallis-West, being in the know at this point. Jack poured out his heart to Winston about his love and happiness, but also confessed: 'I am driven to despair at the thought of all the difficulties in the way: There is money – there is Religion [Goonie was a Roman Catholic] – there are her parents – who will be very angry.'

A week later Lady Randolph wrote to Winston with all the home news:

In the first place Jack will have told you <u>his</u> news – this will probably surprise you. I sometimes thought you had designs in that quarter – but not serious ones – & Goonie has always cared for Jack. They are both much in love but will have to wait a long time I'm afraid – ways & means are not brilliant . . . Goonie is a good girl besides being charming, & she is determined to wait for Jack.[9]

In a further letter, on 5 December, Lady Randolph, in updating Winston on Jack's news, made a prophecy: 'There is nothing fresh about the situation which has to be kept dark until he can go to Lord A[bingdon] with your financial plan. I suppose you will be the next to "pop off"; it is always so in a family.'[10]

* * *

It was now that the truly remarkable 'letter-dialogue' began between Winston and Clementine which would last for over fifty years. Fortunately for posterity, neither of them was a habitual 'telephoner': even in later times when that instrument had become the chief means of communication for most people, they would always prefer writing to ringing.

In the Churchill Archives (which include the Baroness Spencer-Churchill Papers) there exist over 1,700 letters, telegrams and memoranda between Winston and Clementine*. Even an absence of two days would find two or three letters or telegrams flying between them, and when they were under the same roof they quite often sent each other notes; these are referred to as 'House-post'. (Clementine adopted this habit especially in later years, when she wanted to make a particular point – perhaps a vexatious one – to Winston.) This long, revealing correspondence makes compelling, and at times deeply moving, reading, and it forms the principal source of this book.†

Now, in these early days of their marriage, although Winston was held in London, deeply embroiled in political affairs, he never failed to give Clementine a detailed report of the work being carried out in the Eccleston Square house. On 21 April he wrote to her:

> The marble basin has arrived. Your window is up – a great improvement. All the bookcases are in position. (I have ordered two more for the side windows of the alcove.) The dining room gleams in creamy white. The Big room is papered. The bath room well advanced. Altogether there will be a fine show for you on Monday.

After her visit to London to inspect progress, Clementine wrote to Winston on 27 April 1909:

> I had a long afternoon with Baxter & carpets. The green carpet is lovely & will do beautifully for library. It looks like soft green moss . . . I tried hard to make the red stair carpet do for the dining room, but it is really too shabby – The edges of the stairs have made ridges along it – & there are awful stains (not my dog this time!) . . . Green sickly looking carpet out of Jack's bedroom

* The entire correspondence forms part of the Churchill Papers housed at the Churchill Archives Centre, Churchill College, Cambridge, under the care and control of the Sir Winston Churchill Archives Trust.
† See also Soames (ed.), *Speaking for Themselves: The Personal Letters of Winston and Clementine Churchill.*

> in Bolton Street does Puppy Kitten rooms. Another bit does the Cook's room – A big rug (from dining room in Bolton Street?) does my little sitting room . . .
>
> The whole house is now carpeted except one big servants room (which can be done with cheap linoleum for about £2) and the Dining room! for which new carpet will have to be got. I have written to the people who are making the blue stair carpet to ask what it will cost to cover dining room entirely with the blue – (4/6 a yard).

These thrifty details show how much Winston and Clementine needed to economize and husband their capital resources, and how conscientiously Clementine tried to make do with leftovers from Bolton Street. She must have been relieved to know her carpeting decisions met with Winston's approval, when he wrote in reply: 'You certainly have made a judicious selection of carpets & I entirely approve it.'[11]

With the first appearance in Clementine's letter of the term 'Puppy Kitten' – the name by which she and Winston referred to their expected baby – it would seem to be the moment to describe the various soubriquets which they employed, and also the emblems and devices with which they adorned their letters and signatures.

From the earliest days of their marriage, Winston and Clementine conferred on each other honorary titles of affection. Clementine was always the 'Cat' or 'Kat', and her letters were invariably adorned with cats in various moods and poses. But although resembling the drawings on a child's slate, they are full of character and meaning.

Winston was at first the 'Pug', then the 'Amber Pug'; and pugdogs rampant, joyful, fatigued or frivolous always accompanied his signature. In one letter Clementine wrote, 'I must have lessons in Kat-drawing as your pugs are so much better than my Kats.'[12] Gradually the Pug motif was replaced by the Pig symbol, and many are the pigs, expressive of every varying mood, which adorn Winston's letters to her throughout their correspondence; but Clementine's cypher always remained a cat. In due course their children, unborn or born, acquired 'pet' names, some of which stuck to them in the family for all time.

Meanwhile, events were moving swiftly at Westminster. The Liberals had determined upon the policy which was to plunge the country into one of the most dramatic periods of Britain's political history – the Budget of 1909, which provoked the conflict between the Liberal Government and the House of Lords. Winston had been much occupied with the passage through the House of his Trade Boards Act; he wrote to Clementine daily, and his letter of 27 April 1909 ends: 'Tomorrow – Sweated Trades! Thursday – the deluge!!! [the Budget] Thus the world wags – good, bad, & indifferent intermingled or alternating, & only my sweet Pussy cat remains a constant darling.' The following day he could report that his Bill had been 'beautifully received & will be passed without division'. But there was no doubt of the storms that lay ahead: 'Tomorrow is the day of wrath! I feel this

must stop now as
...ketch is on my ...
& finish it to-night
will write again ...
telegraphing you from ...
...you looked so lovely yester...
...d. do concentrate ...
Your ever loving husband
W.

at Carlton. I telegraphed
to Edith to find
some curtains ...
your windows last
night. I hope she
did—
Goodbye my Darling
Winston
Your loving
Clemmie

The Wow The Woo

than make mad."
Please telegraph early,
& whether I can bring
anything down for you.
& with best love
I remain
Your loving & devoted
W

Early examples of letters between Clementine and Winston.

Budget will be kill or cure. Either we shall secure ample funds for great reforms next year, or the Lords will force a Dissolution in September.' Writing from Blenheim the same day, Clementine mentions that 'Sunny is much preoccupied about the Budget. It will make politics vy bitter for a long time.' It would indeed.

Lloyd George introduced his Budget on 29 April 1909. A considerable increase in revenue was needed to sustain the Government's reforming legislation; and the 'People's Budget' – as it came to be called – aimed to secure this by increasing taxation, and introducing for the first time super-tax, a tax on motor-cars and petrol, and land value duties. The last of these, necessitating the valuation of all land by the Government, was luridly depicted as the thin end of a very sinister and odious wedge. The generation of the Welfare State, attuned to Budgets of astronomical figures, and accustomed to bearing taxation of penal proportions, may find it difficult to understand the passions aroused by these measures.

But deep and bitter was the gulf between those who believed they stood on the threshold of a brighter day, when the wealth of the nation could be mobilized for the elimination of poverty, and those who felt their whole way of life was threatened by wild men with wild tongues, who showered wrath and contempt on property, place and privilege. And these latter determined to resist the Budget to the end.

* * *

Early in May 1909 Winston and Clementine moved into their new home, with some weeks to settle down before the arrival of their first baby. At the end of the month Winston took a few days' respite from political turmoil to take part in the annual camp of the Oxfordshire Yeomanry, in which he was an officer. He greatly enjoyed these military gallivantings, which often (as in this year) took place in the Park at Blenheim. Among the officers were also his brother Jack, F. E. Smith and various other agreeable cronies. While the menfolk endured the rigours of camp life, their wives were entertained in the Palace – so a good time was had by all. But on this occasion Clementine did not feel up to a large house party, and chose to spend part of these last, and tedious weeks of waiting with her cousin, Dorothy Allhusen (the daughter of Lady St Helier by her first marriage) and her family at their house at Stoke Poges Court in Buckinghamshire.

In a letter dated 30 May 1909*, Winston wrote to Clementine from 'Camp Goring' (the Yeomanry Camp); he was in buoyant form after a strenuous Field Day: 'My poor face was roasted like a chestnut . . . We had an amusing day. There were lots of soldiers & pseudo soldiers galloping about.'

On 31 May, Goonie Churchill, Jack's wife, gave birth to her first child, John George. Winston wrote (in the same letter quoted above) to

* WSC must have made a mistake, as the latter part of his letter makes it clear that he was writing on 31 May.

Clementine to give her the news, and taking pains to describe how expeditiously the event had been accomplished. 'It seems to have been a most smooth and successful affair.' He went on: 'My dear Bird – this happy event will be a great help to you & will encourage you. I rather shrink from it – because I dont like your having to bear pain & face this ordeal. But we are in the grip of circumstances, and out of pain joy will spring & from passing weakness new strengths arise.'

Clementine replied (on 1 June) that his letter found her 'in the schoolroom here, playing games with my small cousins . . . It is wonderful Goonie having her Baby so easily – I feel it is nothing now & only wish there was not another month to wait.'

Winston and Clementine's first child was born on 11 July 1909 at No. 33 Eccleston Square. The baby was a girl, and they named her Diana. In a letter later that summer Winston was to write: 'Kiss especially the beautiful P.K. [Puppy Kitten] for me. I wonder what she will grow into, & whether she will be lucky or unlucky to have been dragged out of chaos. She ought to have some rare qualities both of mind & body. But these do not always mean happiness or peace. Still I think a bright star shines for her.'[13] These were strangely prescient words. Diana grew up to have an ardent, sensitive nature, and the capacity to give and to love largely, but life brought her much unhappiness and disillusion.

When Diana was about two weeks old, Clementine went to recuperate at Carpenters, a small house at Southwater, near Brighton, on the property of the poet Wilfrid Scawen Blunt, who was a great friend of Lady Blanche and her daughters (he was particularly fond of Nellie) and used to lend them Carpenters from time to time. Clementine joined her mother and sister in this charming place, from where she wrote to Winston on 29 August that she found the countryside 'quite wild, savage and altogether delightful . . . The butter is yellower, the cream thicker & the honey sweeter . . . than anywhere else in the world.'

The baby remained in London with her father, who kept a vigilant eye on nursery matters. 'The P.K. is vy well,' he wrote on 31 August, 'but the nurse is rather inclined to glower at me as if I was a tiresome interloper. I missed seeing her (the P.K.) take her bath this morning. But tomorrow I propose to officiate!' On 6 September he further reported: 'I have just seen the P.K. She is flourishing and weighs 10 tons!'

During the ten days Clementine was recuperating at Carpenters, Winston managed two fleeting visits to her, in the intervals making a speech at Leicester and attending Army manoeuvres at Swindon. During this time he wrote her four long letters; her health and strength were (now and always) a preoccupation to him: 'Dearest Clemmie do try to gather your strength,' he wrote on 6 September, in the train on his way back from Swindon:

> Don't spend it as it comes. Let it accumulate. Remember my two rules – No walk of more than ½ a mile; no risk of catching cold. There will be so much to do in the autumn & if there is an Election – you will have to play a great

> part . . . My darling I do so want your life to be a full & sweet one, I want it to be worthy of all the beauties of your nature. I am so much centred in my politics, that I often feel I must be a dull companion, to anyone who is not in the trade too. It gives me so much joy to make you happy – & often wish I were more various in my topics. Still the best is to be true to oneself – unless you happen to have a vy tiresome self! Good night my sweet Clemmie, give my love to your Mamma and Nellinita – and keep yourself the fondest wishes of my heart – now & always.

Although Clementine's confinements were generally straightforward and relatively speedy, the births of her children were always to leave her exhausted and low. The visit to Carpenters did her a great deal of good, but to complete her convalescence she went almost at once to stay with her Aunt Maisie Stanley (now Lady Sheffield, since her husband's succession to the barony) at Alderley Park, Cheshire, this time taking Diana and her nurse with her.

Clementine loved her visits to her Stanley cousins, who had from her earliest years always received her with warmth and kindness. This time she joined an already large family party. 'There are six babies here,' she wrote to Winston on 11 September; 'The eldest is 4 years & the youngest 4 days old! None of them are fit to hold a candle to our P.K. or even to unloose the latchet of her shoe.' A day later she had more grist for the mill of parental pride: 'Our Nurse is purring with joy because of the compliments which are showered on our P.K. who is universally proclaimed (below stairs) to be the finest specimen ever brought within these walls.'[14]

Despite her summer visits, Clementine's health was not fully restored, and in order to consolidate her strength, she spent a few weeks in the early autumn at the Crest Hotel at Crowborough, Sussex, again taking with her Diana and her nurse. Here she played some golf, and took up her riding again; this gave her much enjoyment, and was greatly encouraged by Winston. When he came down at the weekends they rode together, and on her return to London she kept up her riding in Rotten Row, sometimes taking lessons in the Riding School of Knightsbridge Barracks.

The joys and triumphs of her maternal state did not lessen the attention Clementine gave to public events; throughout her life, whether at home or away, she was an assiduous reader of the daily newspapers. One of her first letters to Winston from Alderley shows how she was following politics and making her own assessments with a keen eye and penetrating judgement. She noted that Lord Rosebery's alternative to the Budget proposals – cutting down the Civil Service, spending less on Ireland, and encouraging self-reliance and thrift in the working classes – 'would seem hardly to provide the necessary cash'; and his remark that 'being a Peer . . . I do not regard myself as a financier' drew from Clementine (in view of his marriage to Hannah Rothschild) the somewhat tart observation that 'His delicate & refined nature has been kept aloof since early youth (by a thrifty marriage) from the sordid consideration of how to make both ends meet.'[15]

The first anniversary of their wedding found Clementine still staying at Alderley, while Winston was on his way to attend the German army manoeuvres at the personal invitation of the Kaiser. But their thoughts were of each other on 12 September, and from Strasbourg Winston wrote:

> My darling Clemmie,
>
> A year to-day my lovely white pussy-cat came to me, & I hope & pray she may find on this September morning no cause – however vague or secret – for regrets. The bells of this old city are ringing now & they recall to my mind the chimes which saluted our wedding & the crowds of cheering people. A year has gone – and if it has not brought you all the glowing & perfect joy which fancy paints, still it has brought a clear bright light of happiness & some great things.

And from Alderley, Clementine wrote:

> The year I have lived with you has been far the happiest in my life & even if it had not been it would have been well worth living.[16]

CHAPTER FIVE

At the Home Office

IN THE SUMMER OF 1909 WINSTON HAD WARNED CLEMENTINE THAT SHE would need all her strength and energy later in the year, and he was proved right. Political tension was rising, and public meetings, which in those days drew large audiences, were held throughout the country to plead the rights or wrongs of the Budget. While staying at Alderley in September, Clementine travelled to Birmingham with her great-aunt Lady Sheffield to attend a large meeting addressed by Mr Asquith. These great political meetings never failed to thrill her, and she gloried in the popularity and affection engendered by Winston's reputation as one of the most energetic members of the reforming radical Government. 'The P.M. made a splendid speech,' she wrote the following day: 'The meeting was in a state of wild excitement – We came out by a side door. A steward said to the crowd "There's Mrs Churchill" and they all cheered The Pug – Two boys leant into the carriage and said "Give him our love." Those poor people love and trust you absolutely. I felt so proud.'[1]

After vehement debate, reflecting the passion which Lloyd George's proposals had aroused in the country, the Commons finally passed the Finance Bill on 4 November. The Conservatives were powerless against the enormous Liberal majority in the elected Chamber, but they had determined to use their built-in majority in the House of Lords against the Government. The Opposition benches in the Lords were swelled by the presence of 'backwoods peers', who arrived in Westminster in the last days before the crucial debate to take the Oath so that they could join in defeating what they viewed as the infamous 'People's Budget'. On 28 November the Lords rejected the Finance Bill by a large majority, thus shattering the centuries-old tradition that control over the raising and spending of revenue was the exclusive prerogative of the Lower House. The Government thereupon dissolved Parliament and appealed to the country – the Liberals' rallying cry being 'The Peers against the People'.

Winston at once set off, accompanied by Clementine, for a speaking tour in Lancashire and Cheshire. He addressed thronged meetings on the burning issues of the hour in eight towns. At nearly all of them he was constantly interrupted by shrill voices screaming 'Votes for Women!' At Bolton, a young woman hurled a heavy piece of iron, round which was wrapped a political message, at Winston's car. She chose to go to prison for seven days rather than pay the forty-shilling fine.[2]

From 1905 until the outbreak of the First World War, prominent politicians were subject not only to disturbances and interruptions at public meetings, but also to personal physical attacks of quite astonishing violence at the hands of the militant suffragettes. Churchill never publicly opposed the principle of giving women the right to vote (unlike Asquith, who was unwavering in his antipathy to the idea). Clementine strongly supported the suffragists, who sought by legal and constitutional methods to press for the right for women to vote, but were opposed to the militant, law-breaking activities of the suffragettes. Although the suffragists commanded wider and more influential support from both men and women, the suffragettes caused the greater stir. Clementine publicly declared that she was 'ardently in favour of votes for women'[3] and privately lobbied Winston, but she had ruefully to acknowledge that she was never able to make him support the cause, except in carefully qualified terms. Her disapproval of the violent aspects of the suffragettes' campaign was, not surprisingly, strengthened by the quite frequent attacks on Winston himself.

An alarming incident had taken place when Winston and Clementine arrived at Bristol railway station in November 1909.[4] As they descended from the train and were being greeted by local Liberal notables, a young woman advanced on Winston and attempted to strike him in the face with a riding whip; in self-defence he seized her wrists, whereupon she started to manoeuvre him to the very edge of the platform. The train on which he had travelled was beginning to draw slowly out of the station. Clementine realized the danger instantly and, scrambling over a pile of luggage which was between her and the struggling Winston, seized him by the coat-tails and pulled him back with all her might. The suffragette was then laid hold of and arrested, but there had been a moment of very real danger, when Winston could easily have been toppled under the train. Such incidents were the frequent lot of all prominent members of the Government, and were difficult to guard against because of the fanaticism of the women involved, who courted arrest and were indifferent to personal pain or injury.

Winston and Clementine (with baby Diana) spent the Christmas of 1909, as was their custom in the early years of their marriage, amid a convivial family gathering at Blenheim. In the New Year the General Election campaign launched into full swing. It was Clementine's initiation into the excitements and fatigues of electioneering, and the first of fifteen electoral battles in which she was to take part with Winston over the years. Dundee must have been a somewhat bleak and bitter battleground at that season; but she was not Scottish for nothing, and the fight evoked all her enthusiasm and energy. But although Winston scored a handsome victory in Dundee, the Government overall lost 104 seats – the great Liberal majority of 1906 had vanished.

Asquith now had to rely on the votes of the 82 Irish Nationalists and 40 Labour Members of Parliament in the hard struggle which lay ahead to secure the Budget, and the Government's programme as a whole.

The Prime Minister immediately reconstructed his Government, appointing Churchill as Home Secretary. At thirty-five, he was the youngest holder of that office since Sir Robert Peel.

* * *

When Churchill went to the Home Office in February 1910, he took with him all his reforming zeal, and an essentially humane attitude towards those who lie in gaol. It had been his own experience to suffer imprisonment, if only for a short period, and to his ardent temperament the loss of personal freedom seemed the blackest of all curses.

He was less than two years at the Home Office, but in that time he did much to forward the then comparatively novel idea that the reform of the criminal was more important than mere punishment. He sought to humanize conditions in the overcrowded gaols, and to improve the treatment of young prisoners. Although he was totally out of sympathy with the violent law-breaking of the suffragettes, he considered that they should be treated more as prisoners-of-war than criminals; he therefore instituted a new category of 'political prisoner', which carried with it certain amenities and privileges. But since 1909 the suffragettes had adopted the new tactic of going on hunger strike as soon as they were arrested, and as Home Secretary Winston had to follow his predecessor in authorizing forcible feeding (a brutal and disgusting process) to prevent women who were bent on martyrdom from dying of starvation.

One of the most onerous duties of the Home Secretary was the grim responsibility for the signing of death sentences. Clementine remembered how heavily this burden weighed upon Winston, and she recalled with what scrupulous attention he sifted every case.

The Home Secretary at that time held a wide brief: he was responsible for the well-being of seven million people in factories and workshops, and a million more in the mines. Churchill was able to effect major improvements in working conditions and safety in the mines; he tried (but failed) to limit the appallingly long hours then worked by shop assistants; and he shared the task of seeing through the House of Commons Lloyd George's National Insurance Act of 1911 – the apogee of the Liberal Government's social reforms.

But the prime duty of the Home Secretary is the maintenance of law and order, and these last years of peace on the international scene were marked in Britain by domestic strife. In the summer of 1910 the country entered upon two years of mounting industrial unrest, increasingly accompanied by violence – circumstances that required the Home Secretary to take critical decisions to maintain order and to prevent loss of life.

It was against this background of swiftly moving events, dramatic incidents and political argument that Clementine led the first few years of her married life. At a time when most women are preoccupied with arranging their first home and brooding over their first baby, her interest and energy had other pressing claims as well. Right from the start of her married life she learnt to live with crisis and controversy. Looking back, one can see that this period was a fitting training for the more sombre, deadly days that lay further ahead in their lives.

The hazards and excitements of political life, and the joys of marriage and motherhood, were not, moreover, unalloyed by anxieties. By the prevailing standards of the times, Winston and Clementine were by no means well off: although when they moved into Eccleston Square, which was to be their home for nearly four years, they had a household of five, this was not, by any contemporary yardstick, a lavish establishment for their particular station in life; and they lived in a world of people who were, in the main, much better off than themselves.

Neither Winston nor Clementine possessed any significant personal fortune. Winston earned somewhat less than their keep by his ministerial salary and by his writings. He could already command quite large sums as a writer, but politics and the demands of high office severely curtailed his literary activities. In his bachelor days, he had constantly been short of money; but it is one thing for a young Cavalry officer to have debts, quite another for a Cabinet minister, with a wife and family to support and a position to keep up, to be encumbered financially.

Clementine had always been accustomed to manage on very little money for herself, but she soon discovered that it was a teasing problem housekeeping for an extravagant and – although loving – most inconsiderate husband. Often at very short notice Winston would announce that he was bringing home several cronies to dinner. Few wives (and fewer cooks) will endure such behaviour. It made housekeeping not only exasperating but also needlessly expensive, and very soon in their married life they found themselves beset with financial difficulties.

Clementine had a proud and independent spirit and was meticulous in all money matters; above all, she dreaded accepting favours from the very rich. Her mother, although an inveterate gambler and always short of money, had brought her children up to have an unmercenary attitude towards wealth. She despised money-grubbing ways and the asking of favours, and badly off though she was, she would never allow anxieties about money to dominate either her thoughts or her conversation. 'Never talk rich; never talk poor; never talk money,' she used to say.

Within a few months of her marriage, Clementine found herself preoccupied with money worries and struggling to make ends meet. Unpaid bills were a nightmare anxiety to her, but it was one which was to haunt her continually, and constant financial troubles had a wearing effect on her health and spirits. Winston, of course, felt the burden of his financial difficulties too; but he was so deeply engrossed by other and larger problems that they preoccupied him only at intervals. Moreover, his confidence in his ability to provide by his pen, and his generally optimistic outlook, let him off more lightly.

Clementine had an inborn tendency to worry about many things, causing her anguish which a more carefree, happy-go-lucky spirit would have been spared, although one cannot but be thankful, on looking back, for her sensitivity and watchfulness. The British public has always expected high standards of probity from its public figures, and, had Winston married a feckless woman who could shrug off such inconveniences as unpaid bills, or who would easily have accepted favours from her husband's rich friends or admirers – who knows whether some worldly catastrophe might not have left a tarnish, which might have for ever hampered, if not ruined, Winston's political career?

Soon, too, Clementine discovered in Winston a trait which was to be a cause of deep anxiety to her throughout their life together: he loved gambling. Clementine was a good spender – when she had the money to spend – but, although she might have an occasional 'flutter' at the tables, basically she loathed gambling. To her it was a senseless, almost wicked waste of money, and yet it was her fate in life to be surrounded by those who were drawn to it: her mother, her brother, her sister and, now, her husband were all, to varying degrees, gamblers. Over the years, she schooled herself to display a tolerance of gambling which she did not feel, and she suffered much anxiety and unhappiness on this score.

Clementine was by nature shy and reserved, and all the circumstances of her childhood and youth had combined to accentuate these characteristics. The shyness she soon conquered, and the civilized education she had received, and her lively intelligence, amply equipped her for the political and social world into which her marriage brought her. She could quickly break the ice of any conversation, and people warmed to her spontaneity, liveliness and charm. But although Clementine's manner was generally warm and friendly, her inner reserve never left her, and, throughout her life, she fended off attempts to make intimate relationships. She never felt the need, as so many women do, for confidential gossipings; and she could keep her own counsel even with close friends. In these early years of marriage, as in later life, Clementine focused all her energy and interest upon Winston, his career and their home.

She had two staunch friends from before her marriage – Sylvia Henley (Stanley) and Horatia Seymour, both of whom were greatly liked by Winston – and it was happy and fortunate that from the start, as we have seen, Clementine and Goonie should have made friends, for the brothers were devoted to each other. Cultivated and well-read, Goonie had a puckish sense of humour which was devoid of any touch of malice, and she distilled a sense of enchantment; she had a rare power of sympathy, and in her Clementine found an affectionate, loyal and companionable sister-in-law.

At the start of their married life Winston and Clementine had few friends in common. This was not surprising, since their worlds had not touched, and Winston was ten years older than Clementine. She approached his close circle with understandable shyness. A great friend of Winston's was Violet Asquith (now twenty-one years old), the Prime Minister's daughter. Winston and she had first met in 1906, and they had established a close and confidential friendship. On his many visits to No. 10, Winston used to 'drop in' to see her in her sitting-room on his way to or from Cabinet meetings; he greatly enjoyed her company, for she had a brilliant mind, and was as passionately interested as he was in the high politics of the day. When Winston married, Violet ostensibly extended her friendship to Clementine.

But now, in these early days, Clementine felt an instinctive reserve towards this well-ensconced friend of Winston's. Moreover, Violet's bosom companion was Venetia Stanley (Clementine's cousin and Sylvia Henley's younger sister). Venetia had a strong personality, and her standards and outlook were very different from Clementine's. In the Violet–Venetia combination Clementine sensed an indefinable threat; and she felt that in any difficulties between her and Winston, she would not find support from that quarter. Her wariness was indeed justified, as we can now see in the letters exchanged between Violet and Venetia at the time of Winston and Clementine's engagement. On 14 August 1908 Violet wrote to Venetia from Slains Castle in Aberdeenshire, which the Asquiths had rented:

> The news of the clinching of Winston's engagement to the Hozier has just reached me from him, I must say I am gladder for her sake than I am sorry for his. His wife could never be more to him than an ornamental sideboard as I have often said & she is unexacting enough not to mind not being more. Whether he will ultimately mind her being as stupid as an owl I don't know – it is a danger no doubt – but for the moment she will have rest at least from making her own clothes & I think he must be a little in love. Father thinks it spells disaster for them both* (unberufen†). I don't know that it does that. He did not wish for – though he needs it badly – a critical, reformatory wife who would stop up the lacunas in his taste etc. & hold him back from blunders. But as in Arnold's case no one who saw these things would have the nervous system to cure them. I have wired begging them both to come here (as he was going to) on the 17th – won't it be amusing if they do? Father is a little chilly about it – & W. generally, & Margot has an odd theory that Clementine is mad! which she clings to with tenacity in spite of my assurances that she is sane to the point of dreariness.[5]

Venetia wrote on the 16th (the letters must have crossed) from Alderley Park, Cheshire, her family's home:

> My darling Aren't you thrilled about Winston. How I wonder whether Clementine will become as much of a Cabinet bore as Pamela [McKenna]. I

* In his introduction to *Lantern Slides*, p. xxvi, Roy Jenkins comments that this '. . . was not one of Asquith's more prescient remarks'.

† The German equivalent of 'touch wood' [*Lantern Slides*, p. 162].

> don't expect she will as she is too humble. Poor Pamela I'm afraid will be awfully bored at no longer being the only young liberal matron. I had a very ecstatic letter from Clemmy saying all the suitable things. I wonder how stupid Winston thinks her . . . Also what do you think of Clementine's accomplishments as set forth in the Manchester Guardian – six languages, a good musician, a brilliant conversationalist. Surely that influences Margot's opinion a little. Father has a theory that her mother is mad, but I've never heard any doubt cast as to her sanity. I think he must be a good deal in love with her to face such a mother in law.[6]

Of course, these hurtful, uncharitable and dismissive opinions did not surface – outwardly, friendliness and affectionate good manners prevailed. It would be fair to think that on Violet's side there was almost certainly an element of jealousy; but she soon came to appreciate that Clementine was, indeed, just the woman for Winston. Only a few years on, she would be writing: 'I love her so much now – more than W.'[7] Although Violet and Clementine came to have a genuinely affectionate friendship (especially later on in the Second World War, and afterwards in the years of Winston's sad decline), yet one detects from Violet's spontaneous and revealing diaries that there always remained a ghost of the patronizing attitude of the earlier years.

Venetia and Clementine were first cousins; they had known each other from early childhood, and the cousinly bond remained strong. Moreover, Winston was very fond of Venetia – she was excellent and stimulating company, and would be a frequent weekend guest at Chartwell in years to come. But one cannot help thinking that her cousinly loyalty was somewhat lacking at this point.

It would seem that Winston may have had an inkling that Violet and her *côterie* had their reservations about Clementine. A few months after the marriage Violet wrote to Venetia: 'I lunched with Clemmie & Winston on Tuesday & drove him to the House in a taximetre afterwards during which time he confided to me that Clemmie had more in her than met the eye Padlock I thought I got out rather well from a point of view of combined truth & emollient as I cloyingly reiterated: "But so much meets the eye" '[8] One must feel that Clementine's 'antennae' were remarkably well-tuned – and that subconsciously she sensed more than she could possibly have known.

* * *

Clementine must have found it difficult, in the early days with Winston, to accept that his established work patterns and pace of life made frequent absences – short though they were – inevitable. Luckily, she was thrilled by the world of high politics into which her marriage had plunged her; all through her life she would accept the demands and priorities set by public life. And, fortunately for them both (and for posterity, as it turned out), Winston and Clementine were excellent communicators.

But occasionally – and mostly attributable to an absence, or a rare failure to be in touch – there were misunderstandings. No clue exists to the cause or circumstances of the tiff which drew from Winston the following vigorous and moving protestation – but it shows that Clementine could nurse jealous fears:

> Dearest it worries me vy much that you should seem to nurse such absolutely wild suspicions wh are so dishonouring to all the love & loyalty I bear you & will please god bear you while I breathe. They are unworthy of you & me. And they fill my mind with feelings of embarrassment to wh I have been a stranger since I was a schoolboy. I know that they originate in the fond love you have for me, and therefore they make me feel tenderly towards you & anxious always to deserve that most precious possession of my life. But at the same time they depress me & vex me – & without reason.
>
> We do not live in a world of small intrigues, but of serious & important affairs. I could not conceive myself forming any other attachment than that to which I have fastened the happiness of my life here below. And it offends my best nature that you should – against your true instinct – indulge small emotions & wounding doubts. You ought to trust me for I do not love & will never love any woman in the world but you and my chief desire is to link myself to you week by week by bonds which shall ever become more intimate & profound.
>
> Beloved I kiss your memory – your sweetness & beauty have cast a glory upon my life.
>
> You will find me always
> Your loving & devoted husband
> W[9]

* * *

As for Winston's male friends, Clementine viewed them too with shyness and caution. His closest political colleague was Lloyd George, whom she admired yet, even now, vaguely mistrusted. But Winston's greatest personal friend was F. E. Smith: their friendship had begun and prospered despite the party issues which divided them deeply, and it was to endure, unwavering, until F.E.'s early death in 1930, at the age of fifty-eight. Clementine never took to this brilliant and unusual man, and from the first she disapproved of his influence upon Winston politically. Moreover, she thought F.E. encouraged Winston to gamble. They were fellow officers in the Oxfordshire Yeomanry, and Clementine learnt to dread the camps, for they afforded opportunities for gambling at which Winston often lost a good deal of money. F.E., an outstanding advocate, could always quickly recoup his losses by some brilliant brief, but no such means of replenishment was open to Winston. Conversely, however, Clementine greatly liked and admired F.E.'s wife, Margaret.

Winston's friendship with F. E. Smith was particularly remarkable in view of the bitterness of party strife at that time; and it was much disapproved of in some quarters. Both Clementine and Margaret Smith were approached on several occasions by Liberals or Conservatives and each asked to use her wifely influence to wean her husband away from a friendship which was considered politically harmful.

Clementine had often to suffer the company of people she did not really like, whether friends of Winston's or some of his relations to whom she never warmed. Into this latter category came Lord Wimborne (who had married Winston's aunt, Lady Cornelia Spencer-Churchill) and his sons Ivor and Freddie Guest, who were not only Winston's first cousins, but also his great friends – particularly Freddie Guest, who was his exact contemporary. Unfortunately, Clementine could never abide them; and yet in the first years of their married life she was often in their company, for both Lord and Lady Wimborne and their sons invariably showed kindness and hospitality to both Winston and Clementine. She generally managed to conceal her feelings, but despite good intentions, her antipathy sooner or later revealed itself.

There was a scene on one occasion when Winston and Clementine were staying with the Wimbornes at Canford Manor in Dorset. During bridge after dinner, Ivor Guest lost his temper, and threw his cards at Clementine's head; not surprisingly, she left the card table and went upstairs to bed. The next day, despite profuse apologies from the contrite Ivor, she insisted on returning to London, taking Winston (one suspects rather reluctantly) with her. Clementine undoubtedly sometimes took things too much to heart, and when she was aroused her reflex actions were swift and determined.

The difficulties which Clementine experienced with some of Winston's friends in these early days were to be a recurring feature all through her life. Her highly developed critical faculty made it hard for her to accept and enjoy people at their face value, and this trait in her character, combined with her inborn reserve, was to be the cause of much loneliness in a life which, to the casual observer, would appear to be teeming with friends.

It was fortunate for Clementine (and indeed for Winston too) that purely social life held few charms for her. Immersed as Winston was in politics, a rift might well have developed had she yearned for the bright lights of the ballrooms, or had wanted to swim with the tide of the social season. Nevertheless, from time to time they would for a few days enjoy the life of the great country houses, and the sports and pleasures of the social world into which he (particularly) had been born, and from which his ardent pursuit of politics made him a cheerful exile. When he hunted, he enjoyed hunting; when he shot, he enjoyed shooting; and he did both remarkably well, for one who had so little practice.

One such occasion was in December 1910, when Winston stayed for a few days at Warter Priory in Yorkshire (the home at that time of the Nunburnholme family). Clementine had been kept at home by a heavy cold. Winston wrote to her of his visit:

A nice party – puissant, presentable, radical in preponderance – a rare combination. I wish you were here . . . Nearly all men. Tweedmouth, Elcho, Lovatt, Cowley, daughters of the House, Granards. Tomorrow pheasants in thousands – the vy best wot ever was seen. Tonight Poker – I lost a little – but the play was low.

On the whole survey, how much more power and great business are to me, than this kind of thing, pleasant tho it seems by contrast to our humble modes of entertainment.

I expect I will have a headache tomorrow night after firing so many cartridges. All the glitter of the world appeals to me: but not thank God in comparison with serious things.

How naughty of the Kat to be enrhumed. Tell Beauchamp [the doctor] to telegraph fully and take the greatest care. I will be back on Weds.

The letter was signed, 'Your pug in clover, W.'[10]

Clementine wrote back on 20 December:

My Own Darling Pug,

Now that I feel better, I do so wish I was at Warter with you enjoying the Flesh Pots of Egypt! It sounds a delightful party and your frivolous Kat would have purred with pleasure . . . Dearest you work so hard and have so little fun in your life. I wish you had more of this sort of thing. But I would not have liked to marry a Cowley or a Granard, or Even a Lovat (tho' I should *then* have had porridge every morning and not only when I go to Dundee).

One feels in these letters that although both Winston and Clementine could enjoy these fleeting distractions, they were more like spectators looking in on a brilliantly lighted scene; they gladly returned to the blustery weather outside.

* * *

It was during his time at the Home Office that Winston (by his own account) endured prolonged bouts of depression – 'Black Dog', as he called them. Many years later, in August 1944, he spoke about this draining and undermining experience to his doctor, Lord Moran*:

When I was young . . . for two or three years the light faded out of the picture. I did my work. I sat in the House of Commons, but black depression settled on me. It helped me to talk to Clemmie about it. I don't like standing near the edge of a platform when an express train is passing through. I like to stand right back and if possible to get a pillar between me and the train. I don't like to stand by the side of a ship and look down into the water.

* Sir Charles Wilson, MC (1882–1977), created 1st Baron Moran, 1943. Consulting Physician and Dean of Medical School, St Mary's Hospital, Paddington. President of Royal College of Physicians, 1941–50.

> A second's action would end everything . . . It helps to write down half a dozen things which are worrying me. Two of them, say, disappear; about two nothing can be done, so it's no use worrying, and two perhaps can be settled.[11]

Again, many years later, he would tell Diana that he thought his near-breakdown was caused by the anguish he suffered in carrying out his duty as Home Secretary to weigh up death sentences and take the awesome decisions. And there were so many cases to consider: in his twenty-one months' tenure, forty-three cases came before him; in twenty-one cases he recommended clemency.

Lord Moran would tell Winston: 'Your trouble – I mean the Black Dog business – you got from your forebears. You have fought against it all your life.'[12] Winston on his father's side was descended from several generations of unions between ancient and interconnected families. The great gift of 'fresh blood' was twice over brought to the Churchills, first by the beautiful, robust Yankee, Jennie Jerome – Winston's mother – and in the same generation by the exquisite Consuelo Vanderbilt of New York, who married Sunny, the 9th Duke of Marlborough.

In all the conversations I had with my mother when I was beginning to write this book, I find no mention in any notes I kept that we ever discussed 'Black Dog' in general – or its presence at this particular period. But looking back as far as my own memory and understanding go – and drawing on the witness of his contemporaries, coupled with the hindsight of later commentators – my considered opinion is that 'Black Dog' was increasingly effectively kennelled by two circumstances or events in my father's life. The first was the build-up of the confidence and strength he increasingly found in his relationship with my mother. Writing at this time, 11 July 1911, he told her:

> Very nice dinner last night with Ivor & Alice [Guest] . . . Alice interested me a good deal in her talk about her doctor in Germany, who completely cured her depression. I think this man might be useful to me – *if my black dog returns* [emphasis added]. He seems quite away from me now – It is such a relief. All the colours come back into the picture. Brightest of all your dear face – my Darling.[13]

And in March 1925, when she was staying with the Balsans* at Lou Sueil on the Riviera, Winston wrote:

> When do you think you will return my dear one. Do not abridge yr holiday if it is doing you good – But of course I feel far safer from worry and depression when you are with me & when I can confide in yr sweet soul . . . You are a rock & I depend on you & rest on you. Come back to me therefore as soon as you can.[14]

* Consuelo had previously been married to the 9th Duke of Marlborough (Sunny); they had divorced in 1920, and in 1921 she had married Jacques Balsan, a lieutenant-colonel in the French air force.

Secondly, I am convinced of the 'therapeutic' aspect of his discovery of painting: and we shall see presently how this ploy or hobby came fortuitously into Winston's life in 1915, when he was forty, at a moment of crisis and deep humiliation and frustration – and from then on played a fundamentally important role in his life. I believe this engrossing and compelling occupation played a real part in renewing the source of the great inner strength that was his – enabling him to confront storms, ride out depressions, and rise above the rough passages of his political life.

* * *

In April 1910 the House of Lords finally passed the bitterly contested 'People's Budget'. However, this brought no end to political strife, because the Liberals were determined to break the power of the Lords to veto legislation passed by the Commons; they had already introduced the Parliament Bill designed to accomplish this purpose. King Edward VII died in May, but national mourning, which was deep and real, brought only a brief respite to the parliamentary struggle.

The atmosphere of clash and crisis was not confined to Westminster. The late summer saw mounting unrest in the country, and although Winston and Clementine were able to get away in late August for a cruise in Baron de Forest's yacht, they returned home early, and Winston was able to take part in the Army manoeuvres with the Yeomanry. Clementine wrote to him from London: 'I am thinking of you galloping about on Salisbury Plain in this delicious crisp weather & wishing you were Commander in Chief on the eve of war instead of Home Secretary on the brink of an Autumn Session. I hope your horse is good & I am sure my sweet Pug looks more amber than ever in his lovely new russet clothes.'[15]

Winston had then to grapple with the mounting industrial strife. Thirty thousand miners went on strike in the Rhondda and Aberdare valleys. The strikes were accompanied by rioting and often looting, and proved beyond the power of the local police forces to control. The General Commanding Southern Command on his own initiative ordered troops into the area. Anxious to avoid the possibility of armed troops clashing with civilians, Churchill sent in a strong force of Metropolitan Police, who restored order at Tonypandy by wielding their rolled-up mackintoshes among the rioters. Nevertheless, this aroused the fury of the left wing, and the whole incident became grossly distorted, 'Tonypandy' throughout his career being used as a stick with which to belabour Winston. The right, on the other hand, took him severely to task for doing too little – so he was roundly abused by both sides.

Drama and action seemed to attend Winston for most of his life. Sometimes they were of his own making; and a good example of this is his own participation in an extraordinary incident which took place in early January 1911 and became known as the 'Battle of Sidney Street'. Some notorious anarchists had barricaded themselves into a house in the East End of London, fired on the police, and were successfully resisting all attempts to dislodge them. Authority was sought of the Home Secretary for troops to be sent. Churchill gave the necessary permission, and characteristically hurried himself to the scene of battle to share the excitements and the responsibilities. After a fierce resistance by the cornered men, the house – No. 100 Sidney Street – caught fire; Churchill used his authority to prevent the Fire Brigade intervening (three policemen having already been killed), and two bodies were later found in the smoking ruins. He was sharply criticized in Parliament and trounced by the press for his presence and intervention during the affray. Clementine was early on receiving some sharp lessons in the vagaries and hazards of political life.

The All-Party Constitutional Conference called by King George V in June 1910 had failed to find a workable compromise. Asquith therefore decided to hold a General Election specifically on the issue of the reform of the House of Lords. He had wrung from the uneasy King a secret pledge that if the Liberal Party won the forthcoming election, and the House of Lords obstructed the passage of the Parliament Bill, the King would create sufficient new peers to ensure the passing of the measure in the Upper House.

On 28 November 1910 Parliament was dissolved, and the country was faced with a second General Election within a year. Although Clementine was expecting her second child, she played her part in the campaign. Winston held his seat at Dundee with a slightly smaller majority than in the previous February. The overall result of the General Election was to leave the balance of power between the parties practically unchanged.

After the election, Winston went for a few days' rest to Alderley, staying with Clementine's Stanley relations. But she herself had to remain in London, nursing a heavy cold which threatened to settle on her chest. Winston, although absent, was not forgetful, as Clementine's letter shows:

> My Own Darling,
>
> After lunching with your Mamma today and doing a little Christmas shopping I again returned to bed – I dozed off – When I woke up, imagine my surprise and delight to see my room transformed into a Paradise of exquisite flowers and a lovely melon sitting near the bed. You are a sweet Darling Lamb Bird!![16]

Winston and Clementine's tender devotion for each other was plain for all to see, and we have a charming glimpse of them lunching with Wilfrid Scawen Blunt in the late summer of 1909:

> We lunched in the garden, and sat on there after it till four o'clock, discussing nearly all the great questions of the day . . . I like him much. He is *aux plus petits soins* with his wife, taking all possible care of her. They are a very happy married pair. Clementine was afraid of wasps, and one settled on her sleeve, and Winston gallantly took the wasp by the wings and thrust it into the ashes of the fire.[17]

We get another peep into the young Churchills' domestic happiness in the diary of Lord Esher:

> Yesterday I dined with W.C. at his home in Eccleston Square . . . Only 6 people . . . And crackers. He sat all the evening with a paper cap, from a cracker, on his head. A queer sight, if all the thousands who go to his meetings could have seen him.
>
> He and she sit on the same sofa, and he holds her hand. I never saw two people more in love.[18]

During the spring of 1911, in the last few weeks before the birth of their second child, Winston and Clementine spent Easter at Blenheim together. Later that April Clementine went by herself to stay with Lord and Lady Sheffield at Penrhos, Holyhead. She was always happy staying with Aunt Maisie and her family. 'I am counting the days till May 15th when the Chumbolly* is due,' she wrote on 18 April, adding, 'I hope he will not have inherited the Pug's unpunctual habits.' In the event, the baby kept her waiting a fortnight. But then all was rejoicing, for on 28 May their second child and only son, Randolph, was born at their home in Eccleston Square.

A few days after this exciting event in their family life, Winston had to go into camp with his Yeomanry at Blenheim, but his thoughts were constantly with Clementine. Writing on 2 June from Blenheim Camp, he described the beauty of the park in high summer, and the agreeable and spirited company assembled. He went on:

> Many congratulations are offered me upon the son. With that lack of jealousy wh ennobles my nature, I lay them all at your feet.
>
> My precious pussy cat, I do trust & hope that you are being good, & not sitting up or fussing yourself. Just get well & strong & enjoy the richness wh this new event will I know have brought into your life. The Chumbolly must do his duty and help you with your milk, you are to tell him so from me. At his age greediness & even swinishness at table are virtues.

Clementine wrote back the next day:

> Blenheim sounds gay & entrancing, & I long to be there sharing all the fun & glitter. But I am very happy here, contemplating the beautiful Chumbolly who grows more darling & handsome every hour, & puts on weight with every meal; so that soon he will be a little round ball of fat.
>
> Just now I was kissing him, when catching sight of my nose he suddenly fastened upon it & began to suck it, no doubt thinking it was another part of my person![19]

* Randolph's in-the-womb nickname, possibly derived from a beautiful flower that grows in north-west India, or from the Persian language (Farsi) meaning a healthy, chubby new-born baby. See Winston S. Churchill, *His Father's Son: The Life of Randolph S. Churchill*, p. 3.

Winston hated to feel out of touch with family affairs. 'A telephone message was received this morning that all was well with you: but no telegram from you. Tell the nurse to send one every day without fail,' he directed in his letter on 4 June.

Clementine's letters during this period breathe happiness and contentment.

33 Eccleston Square, S.W.
Whit Sunday Evening [4 June 1911]

My Sweet Beloved Winston,

The Chumbolly & I are both very well. It is a lovely calm evening after a most sultry airless day & I have been thinking of you with pleasure among the green trees & cool waters . . .

I am so happy with you my Dear. You have so transformed my life that I can hardly remember what it felt like three years ago before I knew you. Goodbye Dearest One.

Your own loving
Clemmie

Please be a good Pug & not destroy the good of your little open air holiday by smoking too many fat cigars. This recommendation comes rather late!

The Coronation of King George V was to take place on 22 June, just over three weeks after the birth of her baby. Clementine had resigned herself to missing the ceremony in Westminster Abbey, as she would not have been strong enough to endure the hours of waiting and standing which such an occasion entails; nor could she have left her precious new baby so long unsustained. However, during an audience some weeks previously the King had enquired of Winston how his wife was; and, on hearing that to her great disappointment Clementine would be unable to attend the Coronation, with charming thoughtfulness and solicitude, he made special and exceptional arrangements, which enabled Clementine after all to be present at what is one of the most moving, grandiose and beautiful ceremonies – the crowning of the Sovereign. A royal brougham called for her at her house and drove her to Westminster Abbey, so that she arrived at the latest possible moment; she watched the ceremony until after the actual crowning of the King, then slipped discreetly away and was driven home to the hungering Randolph. This was the first of three Coronations at which both Winston and Clementine were destined to be present.

The day after the Coronation Clementine was also able to go with Winston when he accompanied the King and Queen on the first of their Coronation State Drives through London. A day or two later she went with the children and their nurse to Seaford, where they all spent a happy sunshine month. As she had spent part of her childhood at Seaford, the neighbourhood was well known to her. Her mother was there too, staying with an old friend, Mrs Jack, of whom Clementine also was very fond; and there were other friends besides, and old acquaintances it gave her pleasure to see again. It was a restful, idyllic interlude.

Winston kept her regularly informed about his doings. He had accompanied the King and Queen on another of the Coronation 'royal progresses', and he gave Clementine a diverting account of his drive, his two carriage companions being the Duchess of Devonshire and the Countess of Minto: 'Of course all the whole route I was cheered and in places booed vigorously. It was rather embarrassing for these two Tory dames. They got awfully depressed when the cheering was vy loud, but bucked up a little around the Mansion House where there were hostile demonstrations.'[20] The King, Winston told Clementine, had been most friendly, and the Queen had asked solicitously after her and had 'urged the importance of a full rest before beginning gaieties'.

Winston had also given dinner to Lloyd George at the Café Royal; they had had a very good talk, and had 'renewed treaties of alliance for another seven years'. Lloyd George had been 'full of your praises – said you were my "salvation" and that your beauty was the least thing about you'.[21]

This summer of 1911 saw a period of political and constitutional crisis. Despite the verdict of two General Elections, the Conservatives struggled on bitterly through the long sweltering summer to defeat the Parliament Bill. There were scenes of unparalleled hysteria in both Houses before and after the Government made public the King's pledge to create enough Liberal peers to ensure the passage of the Bill: only then did the Lords surrender.

Meanwhile, an ominous warning note was sounded from abroad. London society was in a whirl of Coronation celebrations when, in June, Germany galvanized Europe by sending a gunboat to Agadir, as a challenge to French claims to Morocco. On 21 July, in a speech at the Mansion House, Lloyd George in measured terms warned Germany that if it intended war it would find England against it. During those sultry weeks the threat of war crystallized as a reality: the moment of danger passed, but to those who could read the omens aright, this incident meant only one thing – that Germany was bent on trouble, and that its expanding economic vitality and growing physical power made it a grave threat to the peace of Europe.

Winston's mind ever roved over a wide field, and his interest and superabundant energy were never confined to matters relating solely to his office of the moment. Even in his earliest days as a minister, at first concerned with colonial affairs, and then with various aspects of social reform, he had always shown a keen interest in military and naval matters. Ever since the days of his childhood, when he passed hours at a time manoeuvring his armies of tin soldiers on the schoolroom floor, military tactics, movements and problems had always held an intense fascination for him. And later, the days he spent in camp with the Yeomanry always brought this inborn preoccupation to the surface. In a letter to Clementine from Camp Goring on 31 May 1909, Winston had confided to her:

> Do you know I would greatly like to have some practice in the handling of large forces. I have much confidence in my judgment on things, when I see clearly, but on nothing do I seem to feel the truth more than in tactical combinations. It is a vain and foolish thing to say – but you will not laugh at it. I am sure I have the root of the matter in me – but never I fear in this state of existence will it have a chance of flowering – in bright red blossom.

But if his pulses raced at the excitement of these exercises and the epic aspects of war, he saw clearly behind the wheeling columns the savagery and tragedy. When later that same summer he had attended the German manoeuvres, and received marked attention from the Kaiser, he had written to Clementine of the deep impression this display of military might had made upon him (15 September 1909): 'This army is a terrible engine. It marches sometimes 35 miles in a day. It is in number as the sands of the sea . . . Much as war attracts me & fascinates my mind with its tremendous situations – I feel more deeply every year – & can measure the feeling here in the midst of arms – what vile and wicked folly and barbarism it all is.'

In the winter of 1908–9, Winston had joined Lloyd George in heated denunciations of the greatly increased Naval Estimates, which provided for the building of extra Dreadnoughts (the most powerful battleships hitherto known), believing that Germany at that time did not constitute a threat to Britain, and that the country could not afford the extra battleships as well as the great social reforms they had at heart. Since then the Agadir crisis had brought into sharp focus the need to consider in a different light all matters concerning the country's defence; and Winston's sense of priorities continued to change in direct relation to the altered international situation. The Navy and the organization of the Admiralty caused him increasing disquiet; and his preoccupations over this were shared by the Prime Minister.

But his immediate concern and responsibility as Home Secretary was the preservation of law and order, which all through this summer were threatened as industrial disruption increased. In June there was a seamen's strike, and a dock strike which spread from Southampton to London and other ports. These events prevented Winston from visiting Clementine and the babies at Seaford as much as he would have wished. He missed them very much: 'This house is vy silent without you; & I am reverting to my bachelor type with melancholy rapidity,' he wrote on 29 June.

The dialogue of their letters continued with great regularity: his full of public happenings, hers on a quieter theme. Plans were made for the Chumbolly christening in the autumn. Winston suggested that they should invite Sir Edward Grey, the Foreign Secretary, to be a godfather; this shy, withdrawn and, since his wife's tragic death a few years earlier, sad man was touchingly fond of the young Churchills. In accepting their invitation he wrote on 28 June 1911: 'My best wishes will always be with him & I hope he will never cease to contribute to the happiness of your home which it always rejoices me to see.' The other godparents were F. E. Smith and the Wimbornes' daughter, Rosie (Lady Ridley). The Chumbolly was duly christened in the Crypt of the House of Commons on 26 October 1911, and given the names of Randolph Frederick Edward.

As Diana's second birthday approached, Winston was charged with the important task of procuring a suitable present for her. On 11 July he wrote to report progress:

> My dearest one,
>
> I bought some toys for the P.K. last night – but she is so little that it is vy difficult to know what will amuse her. Be careful not to let her suck the paint off the Noah's Ark Animals. I hovered long on the verge of buying plain white wood animals but decided at last to risk the coloured ones. They are so much more interesting. The Shopman expressed himself hopefully about the nourishing qualities of the paint & of the numbers sold – and presumably sucked without misadventure. But do not trust to this.

It is interesting, and touching, to see from these letters how much Winston participated in the lives of his small children: perhaps it shows how deeply he had felt the great distance which separated him as a child from his parents.

After nearly a month at Seaford with her nursery, Clementine returned home to London, where she stayed a fortnight before going off with Goonie to Garmisch, a resort high in the Bavarian Alps, to complete her convalescence after Randolph's birth. Various enjoyable projects for Winston and Clementine to go away together after the rising of the House feature in their letters to each other just now, but in the end none of these agreeable plans came to fruition, for events on the home front demanded his presence and attention. The strikes in the docks had been quickly succeeded by partial strikes by railwaymen, which in turn escalated into a general railway strike in August; there were stoppages in the London markets, and there was a serious threat that food supplies generally might be jeopardized. These grave troubles were the source of several violent incidents.

Clementine would certainly have accepted the scrapping of their holiday plans with a good grace; she must have shared the general anxiety over events and, indeed, would have preferred to be with Winston at a time when he bore such a burden of responsibility. This, however, must have been one of her early experiences of the uncertainties and inconveniences of political life in general, and of life with Winston in particular. In the course of years, over and over again some agreeable project, or carefully planned holiday, had to be abandoned or drastically altered.

Fortunately Clementine had a highly developed sense of duty and order of priorities in all matters concerning public affairs or Winston's political obligations, and she invariably accepted last-minute cancellations and changes of plan philosophically. But she was to suffer many real disappointments, particularly when a little holiday *à deux* had been planned, or a cosy spell with the family at the seaside. Sometimes, however, her feelings of disappointment would have a tinge of bitterness, for although often it was a question of crisis or public duty, there were occasions when Winston simply changed his mind, or could not bear to tear himself away from the hub of events, even when he was not directly involved: this was harder to accept.

* * *

Late in September 1911, Winston and Clementine went to Scotland. It was customary, then as now, for senior Cabinet ministers to be invited for a day or two to Balmoral Castle in Aberdeenshire, and Winston planned to combine his formal ministerial visit to the King with a descent upon Dundee to attend to constituency affairs. He and Clementine had also been invited by the Prime Minister and Mrs Asquith to spend a few days with them at Archerfield, on the East Lothian coast. While Winston was at Balmoral (it not being the custom for ministers' wives to be invited), Clementine paid a visit to her grandmother at Airlie Castle.

Clementine could by this time view her formidable grandmother (now eighty-one), once a source of awe to her, with detached amusement. On 25 September she wrote to Winston:

> I hope you are happy my sweet Pug and that you are being properly petted, & that you will secure a huge stag. I am very happy here – Granny is become much kinder with age tho' she points out all my defects for my 'good'. She, like you, finds my handwriting detestable and observes while looking at me severely and stroking her lace lappets, that 'a gentlewoman of consequence should not write like a housemaid'. She is going to give me samples of the handwritings she most admires so that I may try & copy them – Lord Melbourne and Lord Palmerston are her favourite guides in this.
>
> She sends you her love & is looking forward to seeing you on Wednesday for luncheon which is at 1.30 to the second by Greenwich time. Afterwards we fly away to Archerfield in the new motor.

The 'new motor' was a great event in the family: a red Napier, it had cost £610, and in Winston's letters to Clementine while she had been in Austria he had expatiated upon its wonders and comforts. It was due to be delivered to him at Balmoral, whence he would drive to Airlie Castle to pick up Clementine.

During their stay at Archerfield with the Asquiths, there hung over Winston and Clementine the tantalizing question of the designation of the new First Lord of the Admiralty. The need for a change at the Admiralty had been clear since the Agadir crisis. There were two obvious candidates for this glittering post: Lord Haldane, at present Secretary of State for War, whose reorganization of that ministry had been acclaimed – and Winston himself. In addition to other weighty recommendations, Haldane was a close and long-time friend of the Prime Minister. Piquancy was lent to the situation when Lord Haldane (who lived in the neighbourhood) paid two long visits to Archerfield, when he was closeted with Mr Asquith.

Winston was far from confident that he would be offered the post, and Haldane's visits made him even more uncertain. One evening he confided his fears to Clementine. On a sudden impulse she opened the Bible which, as in most Scottish houses of the day, was in their bedroom and, after a brief pause, she said to Winston, 'I know it's all right about the Admiralty,' and read him the passage she had lit upon at random. It was the verse from the 107th Psalm – 'They that go down to the sea in ships, that do business in great waters . . .'

Before the end of their visit, the Prime Minister offered Winston the post he so earnestly desired – First Lord of the Admiralty. Haldane remained for the time being at the War Office, becoming Lord Chancellor in June 1912. It is agreeable to relate that Winston and he worked together at their vital posts closely and harmoniously.

Churchill's appointment was announced on 24 October 1911. He flung himself into this new and coveted job with ardour and confidence. Here was the work for which all his talents fitted him, the task for which he was destined – the defence of Britain.

CHAPTER SIX

Those that Go Down to the Sea in Ships

A NEW MINISTRY BRINGS FRESH CONTACTS AND A DIFFERENT SPHERE OF DUTY and activity to the minister's wife as well as to himself. Agreeables must be exchanged between herself and the wives of her husband's new colleagues (and in those days more time and importance were devoted to such amenities); there may be an official house to inspect and inhabit; and, apart from all these obligations, a minister's wife is always in demand to play her part in the good works, social functions and entertaining which automatically pertain to senior Government departments.

Clementine had always thrown herself with energy and enthusiasm into whatever duties of this sort came her way. Whether it was visiting labour exchanges or prisons, attending socials or giving away prizes at police sports, it was all part and parcel of her day's work as Winston's wife. Her natural liveliness and the genuine interest she felt in all these different tasks made lighter work for her, and evoked warmth from the people she met.

But however interesting the work of civilian departments may be, there is no doubt that in the service ministries a dash of panache is added. Not only is there the dignity of a Government department to be upheld in a decorous manner, but the prestige and public image of the particular service is involved, and each looks after its own with meticulous care – none more so than the Navy which, as someone once remarked, 'always travels first class'. And indeed, one of Clementine's first engagements as wife of the new First Lord of the Admiralty was a glamorous and enjoyable one: she launched the battleship *Centurion* at Devonport on 18 November 1911.

Until the Great War the Lords Commissioners of the Admiralty, and in particular the First Lord, had at their disposal a steam yacht, the *Enchantress*, in which they could visit the British Navy, which in those days was large, prestigious and far-flung. Winston made the fullest use of the Admiralty yacht for his naval business, and on many occasions business and pleasure could be combined. Writing of it himself, he described how 'the Admiralty Yacht *Enchantress* was now to become largely my office, almost my home; and my work my sole occupation and amusement. In all, I spent eight months afloat in the three years before the war. I visited every dockyard, shipyard and naval establishment in the British Isles and in the Mediterranean and every important ship.'[1] When Winston was on the *Enchantress* he was, and felt himself to be, in close touch with the men and ships whose destiny was his special charge.

Another agreeable adjunct to the historic and then still powerful office of First Lord was Admiralty House. Tucked between the Admiralty itself and the Horse Guards buildings, it faces onto both Horse Guards Parade and Whitehall, on which side a courtyard bounded by a stone screen (designed by the brothers Adam) provides a shield against the hurly-burly of the Whitehall traffic. Now this beautiful eighteenth-century house has been converted into convenient flats for ministers, some of the main rooms being used for Government entertaining, but in 1911 it was still the residence of the First Lord, and must surely have been one of the most desirable and grandly elegant 'tied' houses of all time.

It was entirely to Winston's taste (and indeed to Clementine's), but since they were considerably stretched financially at this time even by the outgoings of the much smaller house in Eccleston Square, Clementine used all her powers of persuasion – on this occasion with success – to prevail upon Winston to remain in their more humdrum civilian house, which could be managed with five servants, rather than to move into the agreeable splendours of Admiralty House, which would require a staff of eleven or twelve. She won a delaying action, for it was not until the spring of 1913 that they made the move; and, during their tenure, apart from the ground floor, where the reception rooms were kept for official entertainments, the rest of the house was used as offices.

With Winston immersed in his new tasks, and spending much of his time on the *Enchantress*, Clementine took advantage of the opportunity which now presented itself to discover a new (to her) sporting activity – fox-hunting. Lithe and lissom, but blessedly without a trace of heartiness, she loved sports and games. We have a romantic and vivid description of her bathing at the seaside from the pen of the artist Neville Lytton. Visiting his brother and sister-in-law, Lord and Lady Lytton (she was Pamela Plowden – Winston's early love) at Broadstairs in September 1911, he had found Winston and Clementine also staying there. Through his wife, Judith Blunt (Wilfrid Scawen Blunt's daughter), he already knew the Hozier family well, and he described beach life in a letter to Nellie Hozier:

> Winston went off to dig castles in the sands and the rest of us bathed. It was a broiling day and the water was heavenly. Clemmie came forth like the reincarnation of Venus re-entering the sea. Her form is most beautiful. I had no idea she had such a splendid body. She joined in a game of water-polo with Victor [Lytton] and me and then she and I swam out half way across the Channel, and had an animated conversation bobbing up and down between the waves.[2]

Perhaps in different circumstances Clementine would have become a noted sportswoman. She played both golf and tennis, and of these she shone at the latter; and with her good eye and rock-like steadiness, she was much in demand as a partner.

Now, she leapt at the chance to hunt, which was offered her by several of Winston's friends, and her descriptions of her hunting activities bubble over with the excitement and fun of it all. Her first sporting report was from Blenheim. She wrote to Winston on 3 January 1912:

> My Darling,
> I have just come back from a long day's hunting . . . It was the greatest fun – Sunny took charge of me and gave me a lead over the stone walls. We found at once, and had a lovely run over the vale . . . Sunny was most kind, he thinks the little bay mare is a splendid jumper. We went over some quite big places . . . I took all the fences after Sunny. No one was in at the death, but Captain Daly gave me the brush.

Below the signature to this letter are four rough drawings – a fox's brush, a couple of fences, and an excited-looking feline labelled 'Kat about to leap'.

Later that winter she paid several visits to Freddie and Amy Guest at Burley-on-the-Hill (the scene of the great fire). Here she had some highly enjoyable hunting, of which she wrote long and entertaining descriptions to Winston. But she missed him and wanted his news: 'Do write to me my darling – I feel a long way off from my Pug down here with the Fox hunting Squires.'[3] Although appreciating their kindness and hospitality, Clementine was never quite on the same wavelength as her hosts, and one of her letters contains the pithy comment, 'Amy is kind, but more Suffragetty, Christian Sciency and Yankee Doodle than ever. Poor Freddie is a sheep in Lion's clothing.'[4]

Winston was delighted that she was having so much fun, but it caused him some anxiety, and he expected daily reports, writing on 23 January 1912: 'I hope you have had a good day – no telegram! bad Cat – but no news is good news. Do not do foolish things.' Clementine replied on the 24th, assuring Winston that she had sent two telegrams the previous day, and adding: 'Lovely hunting yesterday with the Belvoir – lots of fences & a long run. Today I was too tired. But I hope tomorrow to hunt with Mr Ferney's. But Alack it looks like freezing.'

Clementine's hunting activities became so important to her that Sir Edward Grey, with whom she used to bicycle energetically whenever Winston and she visited him at his home at Fallodon in Northumberland, wrote to her at about this time: 'I hear such accounts of your prowess in hunting that I fear you will despise a bicycle, & I shall never see you on one any more.'[5] In fact, although she had taken so enthusiastically to this sport, and was to hunt again the following season and on isolated occasions from time to time in the future, these two brief hunting seasons were not to be repeated; hunting is an expensive business, and sterner days were imminent.

Although Churchill's prime preoccupation was now with the Navy, his attention and interest roamed over the whole field of domestic and foreign affairs. National security was his overriding concern, and the prospect of violence and civil war in Ireland now loomed ever closer. Home Rule for Ireland had been an explosive ingredient in British politics ever since Gladstone had espoused the cause in 1886. Lord Randolph Churchill had rallied the Protestant North with the cry 'Ulster will fight and Ulster will be right'. In 1912 the Liberals introduced the third Home Rule Bill; because of the Parliament Act, the House of Lords could now delay, but not prevent, its passage, and the militant Ulstermen thereupon determined that they would resist it by force.

Winston was one of the Government's chief spokesmen on the Irish Question, and early in 1912 plans were made for him to address a rally in Belfast, the very heart and centre of the Unionist North, where was focused all the passionate determination of the Protestants never to submit to rule by a Catholic parliament in Dublin. The Ulster Hall, the only building suitable for a large gathering, had been booked by the Orangemen* for the night before Winston was due to speak; it was known that they intended to remain in the Hall overnight and, if necessary, to hold it by force. The Ulster Liberal Association retaliated by hiring the Celtic Football Ground on the outskirts of Belfast, where they proposed to erect a vast tent to accommodate the meeting.

The situation was so tense that the civil and military authorities tried to dissuade Winston from going, but he was determined upon it: '"coute que coute",' he wrote to Clementine on 23 January 1912, 'I shall begin punctually at 8 o'clock on 8th February to speak on Home Rule in Belfast.' Clementine approved his determination, and decided to accompany him, even though the organizers of the meeting were doubtful of the wisdom of her coming, and several people tried to dissuade her. Winston himself warned her on 24 January: 'Do not be too venturesome hunting. Keep some of your luck unused.' But her courage was equalled by shrewd good sense, for she believed that her presence at the meeting would probably discourage violence.

Winston and Clementine set out for this momentous occasion on the evening of 7 February 1912, crossing from Stranraer on the night ferry. They hardly had a wink of sleep because some suffragettes who were on board ran round and round the decks all night shrieking 'Votes for Women'. When they berthed at Larne at half-past seven on the morning of 8 February, they found the walls surrounding the quay placarded with anti-Home Rule slogans and a large hostile crowd who hooted at them as they disembarked and sang 'Rule Britannia' and the National Anthem. Another unfriendly reception from a great crowd awaited them at the station in Belfast.

They spent the several hours before the meeting at the Grand Central Hotel, and by the evening the crowd outside – Unionist to a man – had swelled to some ten thousand people.

* Orangemen are members of a Protestant political organization originating in the eighteenth century. They were and are noted for their vigorous and intolerant Protestantism.

As they started on their long drive to the Celtic Football Ground, the crowd surged round the car and, despite the strong escort of police, very nearly overturned it. As the cars in the cortège slowly made their way through the milling mass, men thrust their heads through the windows, uttering menaces.[6]

This was Clementine's first experience of mass hostility and must have been extremely frightening to her. But despite her initial natural alarm, she remained steadfastly at Winston's side throughout the day. She later told the diarist and newspaper proprietor George Riddell (chairman of the *News of the World* and later 1st Baron Riddell) that when they were in the car with the crowd pressing round them, 'she was not afraid of being killed, but feared she might be disfigured for life by the glass of the motor being broken or by some other means.'[7] She also observed Winston's calm, remarking that 'the opposition and threats seemed to "ginger him up".'[8]

It must have been a bewildering sensation to experience the dramatic change in the emotional climate as the cars crossed the boundary line into the Falls (the Nationalist Roman Catholic area of the city), where the yells of angry resentment turned to cheers and cries of greeting, as their car was suddenly surrounded by ardent supporters.[9]

After these dramatic preliminaries, the meeting itself passed off surprisingly peacefully. Despite deluging rain, five thousand people crowded the stands of the Football Ground and the huge marquee; they gave an ecstatic welcome to Winston and Clementine, shouting and waving sticks, hats and handkerchiefs. Winston then addressed the meeting for over an hour. Having well and truly carried out his advertised intent to speak on Home Rule in Belfast, he accepted the advice of the city authorities to return by a different route. And so, after leaving the Nationalist quarter, where they were once more cheered enthusiastically, he and Clementine were driven back to the railway station through quiet back streets.

Some months afterwards, a deputation from the Liberals of Ulster called at the Admiralty and presented blackthorns (hefty walking sticks) to Winston and Clementine as souvenirs of their stormy visit to Belfast. The gentleman making the presentation spoke 'enthusiastically of the pluck shown by Mrs Churchill in accompanying her husband to Ireland'.

* * *

Whether it was due to hunting or to their Belfast visit, Clementine suffered a miscarriage at the end of February 1912. She appeared to make a swift recovery, and towards the middle of March went to Brighton for a short visit to recuperate. But when she returned to Eccleston Square, she was still obviously far from well. She wrote to Winston, who was away on the *Enchantress*:

My Own Darling,

Yesterday was such a soft spring day, that I longed to be out in the sunshine instead of lying flat on my back looking at the tops of the sooty trees in the square.

It is so strange to have all the same sensations that one has after a real Baby, but with no result. I hope I shall never have such another accident again.[10]

After a short while she again seemed recovered, and certainly she was 'taking notice', for in a letter to Winston dated 1 April 1912, Mr Asquith wrote: 'Much the best thing that I have read for a long time on the Woman Question is a short but very pointed letter in The Times today signed C.S.C. Have you any clue to the identity of the writer?'11 The letter referred to was in reply to one over three thousand words long, and of exceptional pomposity and tedium, written by the eminent physician Sir Almroth Wright, which had appeared in The Times on 28 March, opposing the granting of votes to women on psychological and physiological grounds. The reply which had attracted Asquith's attention appeared under the heading: 'Ought not women to be abolished?' and read as follows:

To the Editor of *The Times*. March 30th

Sir,

After reading Sir Almroth Wright's able and weighty exposition of women as he knows them the question seems no longer to be 'Should women have votes?' but 'Ought women not to be abolished altogether?'

I have been so much impressed by Sir Almroth Wright's disquisition . . . that I have come to the conclusion that women should be put a stop to.

We learn from him that in their youth they are unbalanced, that from time to time they suffer from unreasonableness and hypersensitiveness . . . Later on in life they are subject to grave and long-continued mental disorders, and, if not quite insane, many of them have to be shut up.

Now this being so, how much happier and better would the world not be if only it could be purged of women? . . . Is the case really hopeless? . . . Cannot science give us some assurance, or at least some ground of hope, that we are on the eve of the greatest discovery of all – i.e., how to maintain a race of males by purely scientific means?

And may we not look to Sir Almroth Wright to crown his many achievements by delivering mankind from the parasitic, demented, and immoral species which has infested the world for so long?

Yours obediently,
C.S.C.
('One of the Doomed')

Two days later, Margot Asquith sent her a congratulatory note:

Rows of ladies may write on Almroth Wrights supreme letter but none will touch yours!

Bravo my Darling Clemmy
Margot[12]

About a fortnight later, Clementine seemed well enough to go for a few days to Paris, where she stayed at the Hôtel Bristol with Mrs Beatty (the American wife of the Rear-Admiral, later of Jutland fame), and Lord Ridley and his wife Rosie (Guest, Randolph's godmother). But after a few days Clementine felt so tired and ill that the kind and watchful Rosie Ridley sent for a French gynaecologist. His report was disconcerting, as it cast grave doubts on the treatment of her miscarriage by her English doctor – who, in the French specialist's view, had allowed her to get up far too soon – in consequence of which she must now lead an invalid's life for a month, if she was to avoid becoming really ill and having to have an operation. He counselled her to return home at once. Poor Clementine arrived home therefore a bare four days after she had set out on this potentially enjoyable jaunt. She found that Winston had had to go away in the *Enchantress* for several days of meetings and inspections, but flowers and anxious enquiries awaited her.

Throughout the summer of 1912, Clementine was in poor health and undergoing treatment on and off from her doctors. She wrote Winston debonair letters, but she had spells of great depression, for she hated to be out of things and longed to take up her active life again.

In May, however, despite her indifferent health, she went with Winston in the *Enchantress* for a cruise in the Mediterranean. The party included the Prime Minister and Violet Asquith, Goonie Churchill and Nellie Hozier, as well as the Second Sea Lord (Prince Louis of Battenberg) and Winston's Private Secretaries, Eddie Marsh and James Masterton-Smith. They embarked at Genoa on 22 May and visited Naples, Syracuse, Malta, Bizerta and Gibraltar, before returning to Portsmouth.

Violet Asquith wrote a lively account of this holiday in *Winston Churchill As I Knew Him**, describing how the voyage provided diversions and interest for all members of the party – from her father, whose bent was classical and intellectual, to Winston, who, apart from his naval inspections and conferences at every port of call, tended to see the ancient scenes of bygone conflicts in modern strategic terms, while their companions pursued various pleasures – sightseeing, bathing, parties ashore and on board, according to their various temperaments. In the evenings there was bridge, to which game the Prime Minister was an addict, though (in the words of his own daughter) 'an eager and execrable player'; Clementine played a socially sufficient game, and 'Winston was even more dangerous, for he played a romantic game untrammelled by conventions, codes or rules.'[13]

Violet Asquith's more recently published letters, written at the time, give

* Published in 1965 under her married name, Bonham Carter.

some additional details. On the journey out to join the *Enchantress* she wrote to Venetia Stanley that 'poor Clemmie was in tears of nerves & exhaustion (padlock) she is not at all strong yet & excitement keeps her going much too much.' Nor could she – of course – resist a sharply pointed dart or two when describing her hosts and fellow guests:

> So far as the personnel is concerned everyone has been at their most typical. Clemmie [evidently happily recovered enough to enjoy herself] spending almost her entire time at Naples in hat & glove shops – being very nice – vital – frolicsome – soignée & kind – Goonie 'drimmy' [*sic*] & vague but full of aromatic charm – Nellie the especial prey of Father & his questions that rain in a steady shower . . . she is very disarming & has an amusing little face – but is thoroughly light metal don't you think? always borrowing, thanking, apologising, fagging, telling one of her country house floaters with attractive frankness.
>
> Eddie – most characteristic, longing not to miss anything & by an evil fate always being left behind . . . Winston is passionately absorbed in naval arrangements – his enthusiasms become obsessions – a delightful trait in the abstract but which might a little spoil one's balance as a companion.[14]

Although generally she throve on sea air, Clementine was dogged by ill-health on this otherwise perfect holiday, and on their return home her condition deteriorated further – so much so that she began to be really alarmed lest her health was permanently undermined. Winston also was greatly worried by her continuing illness, and himself took matters in hand, seeking further medical advice. During his tours on the *Enchantress* he demanded not only reports from Clementine, but daily telegrams from her doctors.

She seemed to be responding to the new treatment, and Winston wrote encouragingly and tenderly on 30 June 1912:

> Aarons' [the doctor] telegram yesterday was vy satisfactory. I really believe you are going to have a new lease of health & vigour. I do trust he did not hurt you at all this morning.
>
> I love you so much & thought so much about you last night & all your courage and sweetness.

But the better news which was evidently signalled by the doctors was again followed by a relapse. From the *Enchantress*, he wrote on 9 July:

> My poor darling one,
>
> I am most grieved to find that the rosy views of Parkinson's [another doctor] earlier telegrams have not been confirmed, & that you have had a return of the pain & only 'a fair night' . . .
>
> My dear darling I am so distressed to think of you in pain & in discouragement. Hope deferred makyth the heart sick. Always to be expecting

that you are going to get well & then one thing after another. It is a cruel trial for you. But you will bear it bravely – my own darling & emerge triumphantly into the sunlight.

While she was ill, Clementine seems to have relied chiefly on telegrams as her means of communication with Winston – fortunately at that time they cost little to send. But although she was so unwell at this time, she would not allow her thoughts to be dominated by her ill-health. She was much enthralled and excited when the National Insurance Act, which Winston had helped to draft when he was at the Home Office, came into operation on 12 July; and she wrote him a long letter:

My own Darling,
. . . it [the Insurance Act] is a real triumph. I suppose it is almost the first time in the world that every single person in the country will have to perform a definite action as the result of a single act of Parliament. Every single person must be affected by it as everybody must be employed or else employ someone.[15]

Clementine's excitement over the implementation of this measure may well seem exaggerated to a later generation, who have long become accustomed to their everyday actions being governed by Acts of Parliament. But her enthusiasm is a reminder of how those who viewed the Liberal reforms as the dawn of a new day felt at that time. Although the great wave of radical legislation had now lost much of its early impetus, and although Winston's whole interest had shifted to the field of national defence, Clementine had lost none of her reforming zeal.

During these politically charged years Clementine had worked up a healthy hatred of the Tories, and her radical enthusiasm made her deeply distrustful of Winston's Tory friends. In the same letter in which she saluted the implementation of the National Insurance Act, she made a passionate outburst, prompted no doubt by her fears that the wicked Tories might cast their toils around her shining, progressive Winston:

My dear Darling Amber Pug – Do not let the glamour of elegance & refinement & the return of old associations blind you. The charming people you are meeting to-day – they do not represent Toryism, they are just the cream on the top. Below, they are ignorant, vulgar, prejudiced. They can't bear the idea of the lower classes being independent & free. They want them to sweat for them when they are well & to accept flannel & skilly as a dole if they fall ill, & to touch their caps & drop curtsies when the great people go by – Goodbye my Darling. I love you very much.

Your Radical Bristling
[diagram of cat]

Many years later, when asked what she regarded as the happiest time of her life, Clementine was to point without hesitation to these early years of her life with Winston. She had embraced not only the man, but also his cause, with wholehearted fervour. When in the early twenties Winston left the Liberal Party, and by stages eventually rejoined the Conservatives, Clementine too shifted her public allegiance; but while she followed his transition with her reason, she never made a good Tory, and from time to time her natural radicalism would burst through the layers of reasonable compliance like a volcanic eruption, often to people's astonished bewilderment.

* * *

By late summer 1912, the new course of medical treatment recommended by her doctors was at last producing satisfactory results, and Clementine's health was really on the mend. To establish the cure, she spent a few seaside weeks with the children at Sandwich, staying at Rest Harrow, a charming house lent to them by Waldorf and Nancy Astor*. The House was sitting, and Winston was therefore only an intermittent visitor; and, moreover, he was deeply engaged in introducing his Supplementary Estimates to Parliament. These had become essential in order to maintain the superiority in strength of the British Navy over the German fleet; but they were strongly contested.

During the debate, which lasted several days, Winston spoke a number of times. Clementine was disappointed she could not be in the Gallery to hear his main speech on 22 July, and her first letter from Sandwich starts a little wistfully: 'All this afternoon I was thinking of you making your speech. I am so jealous of Goonie and Nellie hearing you & not your Kat.'[16] However, she read avidly the reports of the debate in the newspapers and commented: 'I must own that I should like a handful more dreadnoughts – 23 to 29 [British to German ships] is certainly a close shave, especially as we have come off so badly in the Olympic Games!!'[17]

Meanwhile she had established a happy holiday life with the children:

> I have had a delicious blowy afternoon on the Beach with the Babies. You would have laughed if you had seen Randolph eating his pap for supper. If ever the spoon contained too much milk & not enough solid he roared with rage & insisted on solid food being scooped up. Diana says her prayers so sweetly & with great dignity. God keep Papa & Mama & little Randolph & make Diana a good little girl for Jesus Christ's sake, Amen. Randolph goes to sleep holding his Nurse's hand & gurgling & cooing like a Ring Dove. Tomorrow comes George†. Thursday Goonie & Saturday Oh Joy my Pig. Come early not late dear one.[18]

* Waldorf Astor, later 2nd Viscount Astor, and his American wife Nancy (Langhorne) were both prominent in British politics; in 1918 she became the first woman to take her seat in the House of Commons.
† John George, later always called Johnny, Jack and Goonie's elder son; the same age as Diana.

The following day she wrote again:

> My Darling,
> Another speech! I cannot follow you in my thoughts more swiftly than you fly from one scene to another thro' space, like a glorious star shedding comet. And all the time your lazy Kat sits purring & lapping cream & stroking her kittens &, after these exhausting operations, sleeps soundly.[19]

Life was not all halcyon calm, however: Clementine had to report to Winston that the kitchen-maid had become demented, threatened to commit suicide and tried to murder the cook. The poor girl had been returned to her own home in the custody of the children's nurse.

Usually such a faithful letter-writer, Winston had told Clementine in a letter from the yacht on 25 July, 'I have been too busy and too tired to write these last two days. But I hope my telegrams have kept you apprised of my violent movements.' But despite being constantly on the move between London and the *Enchantress*, he managed to spend the weekend after the Estimates debate with his family at Rest Harrow.

After he had returned to his toil once more, Clementine continued her account of her seaside life:

> Two windy days since you departed, but tonight I think the weather is abating. Neville Lytton & Nellie turned up yesterday evening late, just before dinner & we had a merry evening – singing songs, dancing one-steps & flute playing by Neville. This morning Neville, Bill & Nellie bathed, but your Kat was good & sat looking on with her tail just dipping enviously in the water. This afternoon we motored to Ramsgate & sat on the sands . . . Come down soon my dear one. I think it is going to be lovely & hot & I am longing to talk with you, walk with you, bathe with you, sing with you.
>
> Your own
> Clemmie[20]

But there were quiet spaces for Winston in his hectic life, and he looked forward to them. On 1 August he told Clementine, 'I have made no plans for Sunday next except to come down quietly and see you. We ought to find a really good sandy beach where I can cut the sand into a nicely bevelled fortress – or best of all with a little stream running down – You might explore and report.'

They stayed at Sandwich until the middle of August, and Winston was able to make several visits to them during that time. However, even ministers' holiday retreats were not free from disturbance by the ubiquitous suffragettes. Returning to Rest Harrow on the afternoon of 13 August, Winston found the road about a hundred yards from the house blocked by two women, who had wheeled their bicycles in front of the car; the driver was, however, able to circumvent them. And a day or two later, several parties gathered on the high road, hoping to intercept his car; but Winston had already left his lair before they arrived.[21]

In mid-August Winston and Clementine, with a family party, went for a cruise in Scottish waters in the *Enchantress*. This was a perfect plan, combining pleasure with Winston's visits to naval establishments, and it added an ideal postscript to Clementine's convalescence. The *Enchantress* was comfortable and beautifully run, involving the hostess in the minimum amount of planning and domestic preoccupation – an aspect of holiday life which greatly appeals to all women.

The party, which apart from other members of the Board of Admiralty included Lady Blanche and Nellie, embarked at Chatham Dockyard on 19 August and sailed up the east coast, stopping at various ports and harbours for Winston to make his inspections, to the Tyne and Aberdeen Bay, from where Winston and Clementine were bidden to dine at Balmoral. They then cruised round the wild and beautiful Scottish coastline to the west coast.

In mid-September they broke their holiday to visit Winston's constituency in Dundee, where he held several meetings. These were constantly subject to noisy interruptions by suffragettes; at one meeting at least there was considerable disorder, and several women were forcibly ejected. Clementine herself in a short speech, referring to the interruptions, said: 'All women must feel a sense of humiliation and degradation at the scenes we have just witnessed.'[22]

Although supporting their cause, Clementine was appalled by the outrageous methods employed by the militant suffragettes, whose campaign of violence was reaching its zenith. Their attacks were becoming increasingly dangerous; they included setting fire to railway stations and planting bombs (one of which exploded) in a house being built for Lloyd George. At the Derby in 1913 Miss Emily Davison flung herself in front of the horses at Tattenham Corner, dying later from her injuries, and becoming a martyr to the cause.

Dining with the Asquiths in November 1913, Clementine heard from them of an unnerving attack on the Prime Minister, of which she wrote to Winston:

> The suffragette affair on the road to Stirling was horrible & upset poor Violet a good deal. One of the women slashed Mr Asquith 4 times with a crop over the head, before she was seized. Luckily his face was saved (like yours at Bristol) by the brim of his stiff hat. He never seems to defend himself on these occasions but remains calm & stolid & unshrinking! About two pints of pepper were thrown into the car, but luckily no one's eyes were injured.[23]

The measures which individuals in the public eye were forced to take will have an all too familiar ring to the modern reader; Winston had to warn Clementine not to open 'suspicious parcels arriving by post without precautions . . . These harpies are quite capable of trying to burn us out.'[24] Following letters threatening to kidnap the children, Diana and Randolph had for a while to be escorted by plain-clothes detectives on their walks in the parks.

It was not until 1918 that women over thirty were given the vote, and it was the general view that this was in recognition of their unstinted contribution to the war effort, rather than the campaign of disruption and violence. But Clementine's opinion in later years was that, although most rational opinion was affronted by the excesses of the suffragettes, none the less the day would not have been won when it was, had not their cause been championed by women with a passion which exceeded constitutional and legal bounds.

1. Blanche, Countess of Airlie, Clementine's formidable grandmother.

2. Lady Blanche Hozier, in 1884 (aged thirty-two), a year before her daughter Clementine's birth.

3. Colonel (later Sir) Henry Hozier, Lady Blanche's husband, in his fifties.

4. Algernon Bertram Freeman-Mitford ('Bertie'), later 1st Baron Redesdale, in 1983 (aged fifty-six).

5. William George Middleton ('Bay'), in 1883 (aged thirty-seven).

6. Clementine (aged about four), with Mlle Elise at The Netherton.

7. Kitty (left) and Clementine, with Sir Henry Hozier at The Netherton, 1888.

8. Kitty (left) and Clementine, at one of their 'lodgings'.

9. Lady Blanche Hozier, 1893.

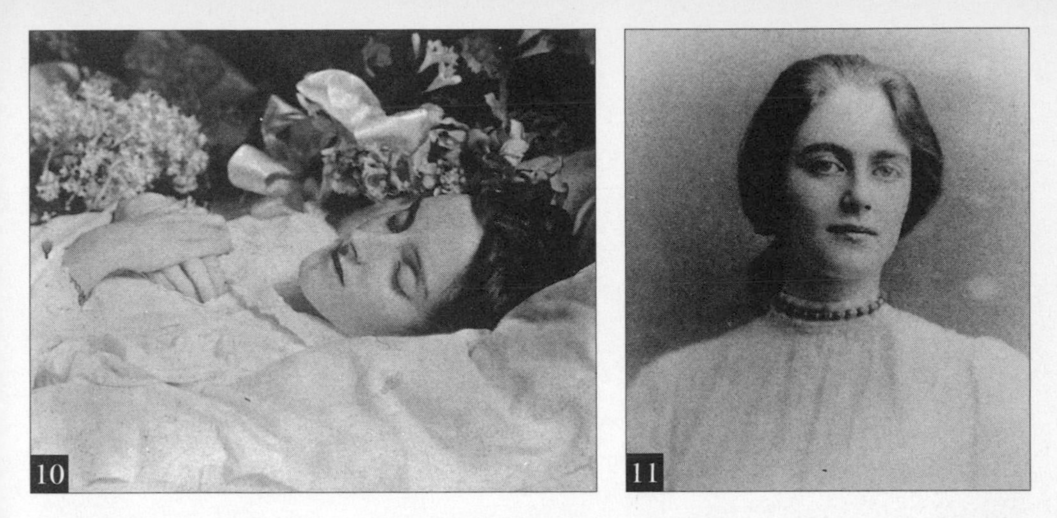

10. Kitty on her deathbed, 5 March 1900, aged nearly seventeen. 'Cover her face; mine eyes dazzle; she died young.' (John Webster, *c.*1580 – *c.*1625, *The Duchess of Malfi*)

11. Clementine, 1901, aged sixteen.

12. Clementine at twenty, 1905.

13. 51 Abingdon Villas, Kensington (white door). The home of Lady Blanche and her family at the time of Clementine's marriage to Winston Churchill in 1908.

14. Crewe House, in Curzon Street, where Winston and Clementine first met at a ball in 1904.

15. Horatia Seymour: a friend for life.

16. Clementine (standing left), bridesmaid to her cousin Sylvia Stanley, who married Anthony Henley in 1906. Sitting left is Venetia Stanley (later Montagu).

17. Winston and Clementine at the time of their engagement, August 1908. Taken by Margaret (Mrs F.E.) Smith.

18. The bridegroom arrives at St Margaret's, Westminster, 12 September 1908. At right, his best man, Lord Hugh Cecil.

19. The bride arrives. With his back to us is Bill Hozier, who gave her away.

20. Picture drawn for the *Daily Graphic* of Clementine's wedding gown. At left, a bridesmaid's dress.

21. Clementine, now Mrs Winston Churchill, leaves her wedding reception at 52 Portland Place.

22. Goonie and Jack Churchill, 1908.

23. Nellie Hozier, Clementine's younger sister, in 1914.

24. Lady Randolph Churchill ('Jennie'), around 1915.

25. 'Sunny', 9th Duke of Marlborough, Winston's cousin: they were great friends.

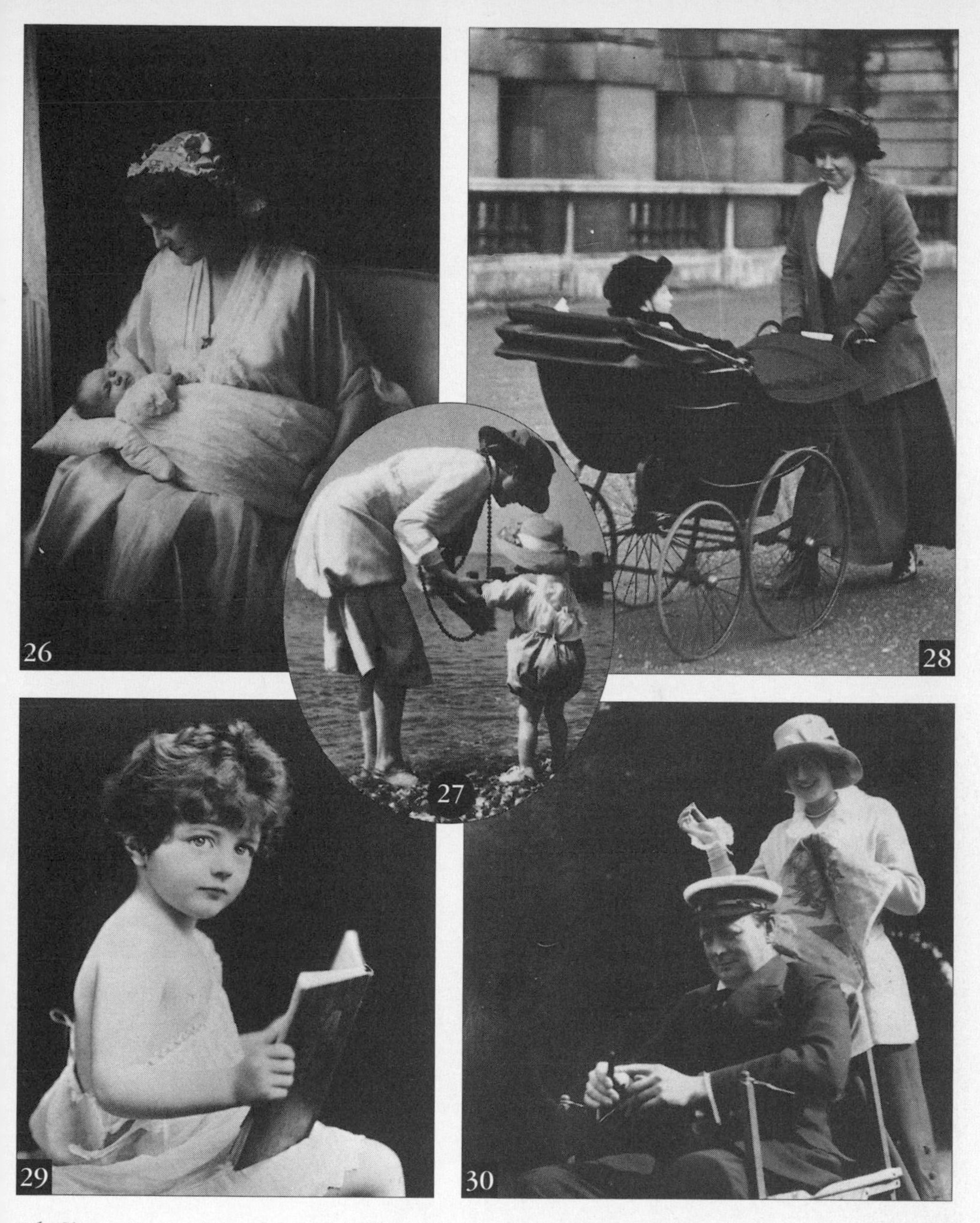

26. Clementine with her first child, Diana, born in July 1909.

27. Clementine with Diana, aged two years, at Seaford, June/July 1911.

28. Nanny Higgs wheels Randolph in his pram on Horse Guards' Parade, 1913/14.

29. Sarah learning to read – quite early on!

30. Winston and Clementine at Penrhos (the Stanleys' home near Holyhead, Wales), 1913.
(I know of no other photograph of my mother sewing!)

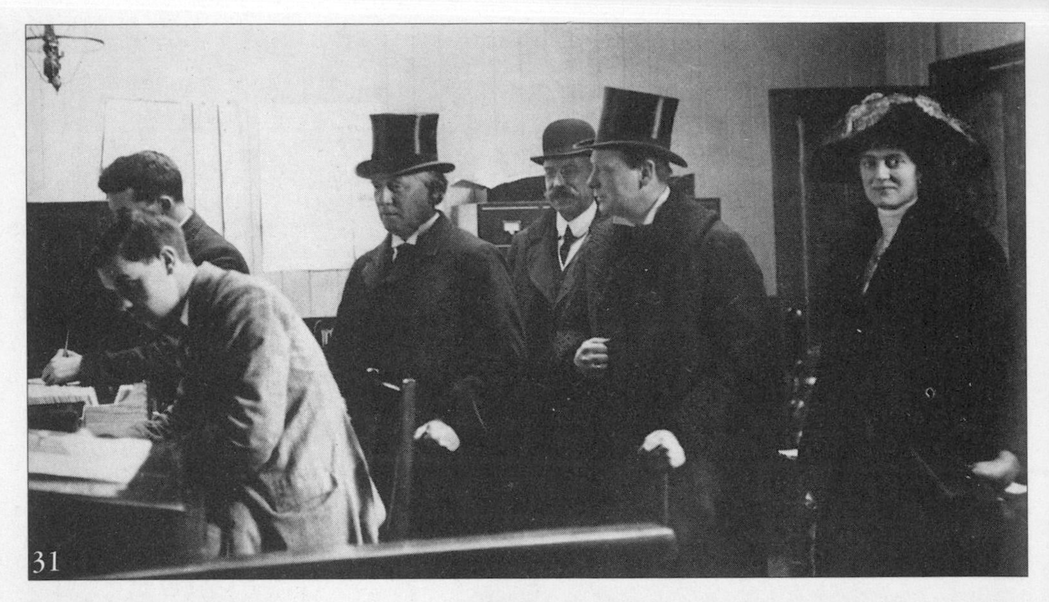

31. In 1910 Winston (now Home Secretary) and Clementine escort the Prime Minister, H. H. Asquith, on a tour of labour exchanges, which were pioneered by Churchill while President of the Board of Trade.

32. Margot Asquith, the Prime Minister's wife. (She got as good as she gave from Clementine.)

33. H. H. Asquith and David Lloyd George at The Wharf, in 1916. By the end of the year Lloyd George would have engineered Asquith from power.

34. Violet Bonham Carter, aged twenty-eight, at about the time of her marriage in November 1915.

35. Clementine playing in the Ladies Parliamentary Golf Tournament at Ranelagh, 1913.

36. Venetia Stanley, Clementine's cousin, *c.*1914. A close friend of H. H. Asquith, she would marry Edwin Montagu the following year.

37. Clementine playing tennis, 1918.

38. Bill Hozier, Lieutenant R.N., 1914, aged twenty-six.

39. Nellie in uniform, 1914, aged twenty-six.

40. Clementine speaking at the opening of the first YMCA hut canteen for women munition workers, in Edmonton, London, 30 August 1915.

41. Hoe Farm, near Godalming, Surrey, which the Churchills rented for the summer of 1915. Here Winston first tried his hand at painting.

42

43

44

45

42. Nellie married Colonel Bertram Romilly at the Guards' Chapel, London, on 4 December 1915. Their attendants were (left to right): Diana (six), Randolph (four), and John George (Jack and Goonie Churchill's son, six).

43. Lt.-Col. Churchill (wearing his *poilu*'s steel helmet) with Sir Archibald Sinclair, Bt., at the Front, early 1916.

44. Clementine with Lloyd George (Minister for Munitions), when he opened a canteen for munition workers at Ponders End, London, 3 February 1916. On the left is Violet Bonham Carter.

45. Lullenden, East Grinstead. Bought by the Churchills, and their country home from 1917 to 1919.

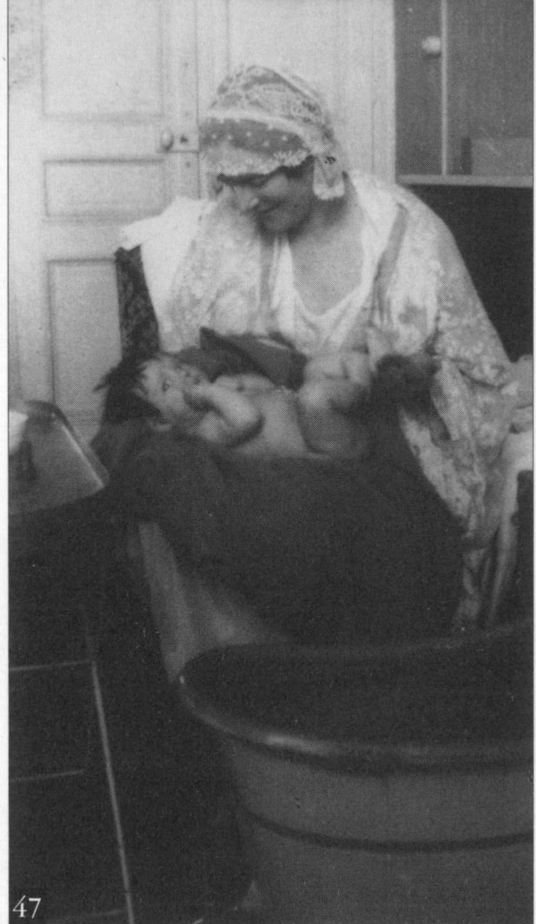

46. Marigold – the beloved 'Duckadilly'. This snapshot was taken not long before she died on 23 August 1921, aged two years and nine months.

47 Bath time for Mary, born 15 September 1922, Winston and Clementine's fifth – and last – child.

48. Circa summer 1923. Sarah and Randolph with their first cousins, Giles and Esmond Romilly, at Hosey Rigge, near Westerham, where the Churchills lived while waiting to move in to Chartwell.

49. Chartwell Manor as it was when Winston bought it in 1922.

50. Chartwell in the late 1920s.

51. The dining room at Chartwell.

52. Winston and his two bricklayer's mates, Sarah and Mary, at Chartwell, in the late 1920s.

53. Lady Blanche in her old age, 1922.

54. 'Bendor', 2nd Duke of Westminster, Winston's great friend since Boer War days.

55. Clementine with 'Sunny' Marlborough at Aintree, for the 1930 Grand National.

56. Winston and Mrs F. E. (Margaret) Smith (later Countess of Birkenhead), at Blenheim Palace, in the summer of 1923.

57. Winston in earnest conversation with his great Tory friend, F. E. Smith (later 1st Earl of Birkenhead), the brilliant advocate and orator, in 1929, the year before he died.

58. In August 1929 Winston embarked on a three-month tour of Canada and the United States. Here is the 'Churchill troupe' (WSC's expression): Winston, his brother Jack, and their sons Randolph (left front) and Johnnie.

59. In Epping, General Election 1924. Winston would represent this constituency until his retirement from Parliament in 1964.

60. A family party goes to hear the Chancellor of the Exchequer's first Budget, April 1925: Clementine, with Randolph (nearly fourteen) and Sarah (ten).

CHAPTER SEVEN

The Sands Run Out

THE DELAYING ACTION FOUGHT AND WON BY CLEMENTINE AT THE TIME OF Winston's appointment as First Lord, to defer their move to Admiralty House on grounds of economy, now ended in her reluctant, but graceful, defeat; they moved from Eccleston Square to Admiralty House in early April 1913. Clementine, however, still succeeded in winning a compromise victory: in order to diminish the number of staff they would need, they decided to seal off the main first floor with its large suite of grand and beautiful rooms. In this way she could run the house with nine servants instead of the twelve needed if four floors were used.

The main rooms were already furnished by the Government, but the wife of the First Lord could exercise her personal choice when it came to the sitting-rooms, bedrooms and nurseries. 'I have spent yesterday & today morning in grapple with the Admiralty installation & the Office of Works officials,' wrote Clementine on 31 January 1913;

> I am afraid altho' we are allowed to 'choose' our bedroom furniture it will be difficult to get anything attractive as the 'choosing' is to be done out of a grim catalogue; so far I have avoided but not chosen anything.
>
> I really think they ought to have a woman at the head of the Office of Works – someone like your Mama.[1]

Winston had been champing to move from the start, and was looking forward to moving into Admiralty House with undisguised relish. House-moving being an essentially female occupation, Clementine was no doubt relieved that he had a programme of visits and inspections over the days of the move, leaving her free to conduct operations.

From the *Enchantress* in Portsmouth he wrote to her on 6 April:

> It will be nice coming back to the Admiralty. I like the idea of those spacious rooms. I am sure you will take to it when you get there. I am afraid it all means vy hard work for you – Poor lamb. But remember I am going to turn over a new leaf! That I promise – the only mystery is 'What is written on the other side'. It may be only 'ditto ditto'!

Although even now the move was against her better judgement, Clementine was in good spirits, and not weighed down by the rigours of house-moving. From a forlorn and dismantled No. 33, she penned a hurried note to Winston on Saturday, 5 April:

> My darling,
> House moving is going on & there is no resting place for the soles of my feet. I am now sitting in a desolated library (ink-pots & pens gone) . . .

Clementine spent the weekend with the Asquiths at their riverside house, The Wharf, at Sutton Courtenay, and wrote to Winston on the Sunday:

> My darling Winston,
> A very pleasant Sunday here. The Prime & Violet picked me up yesterday morning at Eccleston Square & as I was whisked away, I felt it was the end of a chapter in my life. Leaving a house where one has lived nearly 4 years is as much of an event in a 'Kat's' life, as changing from Home Office to Admiralty for a statesman!
> We stopped on our way to Huntercombe [golf course] & lunched at 'Skindles' where we had 'all the delicacies of the season' – Plover's eggs (I hope you will not have had a surfeit of them on the yacht, as they are to smile at you on your return), salmon, lamb, asparagus, rhubarb. Then we proceeded on our way & played very badly. We said it was the wind, but I suspect it was Skindles.
> This morning we played again, a 3 ball match – I halved with Violet & was badly beaten by the Prime, but defeated him in the bye . . . Tomorrow I fly back by the early morning train as I am longing to take possession of our Mansion.[2]

Any anxieties Winston and (especially) Clementine might have had about their treasured first home in Eccleston Square must have been allayed by their good fortune in finding an ideal tenant – their friend and Winston's colleague, Sir Edward Grey.

An amusing story arises from his tenancy of their house. One day he entertained the French Ambassador, and during the course of his visit showed him over his new abode. The Ambassador was greatly startled by the decor in Sir Edward Grey's bedroom (formerly Clementine's room). In the main, Clementine had lovely, unaffected taste and particularly disliked decorators' effects, but while she was planning her first house she had fallen under the influence of the 'art nouveau' movement, and although the main rooms at No. 33 Eccleston Square excited no undue comment, she had thoroughly let herself go in her own bedroom (perhaps in revulsion from the 'sugary' muslin-and-bows decoration which her mother-in-law had so kindly, but indiscreetly, prepared for her as a surprise in their first home in Bolton Street). At any rate, Clementine's bedroom in the Eccleston Square house was decorated in shades of green, brown and orange, with a large orange-tree laden with oranges appliquéd all over the walls. The Ambassador was transfixed with undisguised amazement and horror. When, shortly afterwards, he lunched at Admiralty House, he remarked to his hostess that he had been astonished by what he considered 'un drôle de décor pour la chambre à coucher d'une ravissante jeune femme'.

* * *

Clementine soon established her own domestic life at Admiralty House. She spent much time and thought on her housekeeping; and then, as always, the food in her house was delicious. Winston was exacting in standards of comfort and hospitality: 'I am easily satisfied,' he would say disarmingly, 'with the very best.' Clementine herself had perfectionist standards and tussled willingly, but sometimes wearily, to produce a perfectly run house and elegant and plenteous hospitality on too little money.

The menu book was never 'sent up' to her for approval, as was the custom in many similar households; she always saw the cook herself, usually at about half-past eight, when a conference would take place between them in her bedroom, every detail being gone over with care and interest. These morning interviews were always a feature of the day – Clementine sitting up in bed, looking ravishing in a lace cap and bedjacket; the cook, clad in a dazzling white apron, bearing the menu book; and probably a child sprawled on the bed or rampaging round the room. Sometimes their deliberations would be interrupted by Winston, who would come stumping in in his dressing-gown to say 'Good morning' or to draw Clementine's attention to something in the newspapers. 'Let's have Irish stew,' he would say, 'with lots of onions.'

From the beginning of their life together, they always had separate bedrooms. This habit was not exceptional; indeed, it was customary among upper-class couples (sharing bedrooms being considered very middle-class). The husband always had a dressing room, where his valet would lay out his clothes; his wife's lady's maid performed the same ritual in her bedroom, morning and evening. In fact this tribal custom suited Winston and Clementine very well, for he nearly always went to bed very late, while she much preferred an early bedtime and was always 'up with the lark'.

They invariably breakfasted in their respective rooms; he had usually worked into the small hours and woke up later than Clementine. She was a habitual early bird, and was ravenous by eight; indeed, by that time she had probably written several letters, curled her hair and read several chapters, for it was not unusual for her to wake between five and six o'clock. In some ways this early morning pattern was a handicap to her, for by the evening her energy had waned, and, especially as she grew older, she yearned for an early bedtime. At this time, she often went up to bed hours before Winston, sometimes leaving little 'Kat'-signed notes for him. One found among her papers reads:

My darling,

When your business is finished do come to bed as early as you can. You look weary from want of sleep & tomorrow you will need all your cool brain & judgement.

[sketch of a cat]

Another note reads:

Mrs Grimalkin [an old she-cat] presents her compliments & would appreciate a little visit with the Speech.

[sketch of a cat] Miaow, but sitting tight

One of Clementine's most attractive characteristics was her vitality – but it was not accompanied by stamina, and she was dogged all her life by feeling strain and fatigue to a marked degree. Even as a young woman, it was a necessity for her to rest during the afternoon if she was to be equal to her evening's activities.

Since Winston and Clementine rarely partook of 'fashionable' social life, those with whom they lunched and dined were for the most part either relations or Government and ministerial colleagues, and their close friends were all in the political world.

The Prime Minister found pleasure and relaxation to a very great extent among the younger members of his Government such as the Churchills, preferring small dinner parties, after which he liked to play bridge. Asquith's affectionate regard for Winston had grown with his appreciation of the younger man's abilities, although he regarded his rampageous enthusiasm and loquacity with ironic amusement. Since Winston's marriage, Clementine's beauty, good manners and sufficiency at bridge had made them, as a couple, welcome in the fairly small circle of the Asquiths' private social life; Winston's friendship with his daughter Violet also was a further link. Winston and Clementine were frequent guests both at Downing Street and, at weekends, at The Wharf. Included also in many of these invitations were Goonie Churchill (whom the Prime Minister greatly admired) and Nellie Hozier (known as 'the Bud' or 'Mlle Beauxyeux').

Margot Asquith, a Tennant before her marriage to Mr Asquith as his second wife, was a most eccentric woman, warm-hearted, but famed and feared for her whiplash tongue. She was fond of both Winston and Clementine, although she frequently criticized the former, very often to his wife; Clementine, however, would not allow herself to be bullied, and quite often there were sharp exchanges. But Margot had rather a touching admiration for her, writing once to Winston, 'I have a great feeling for Clemmy, she is so rare not to be vain of her marvellous beauty.'[3] This was generous praise from an older woman, among whose attributes it was impossible to include beauty.

For Winston's part, he had strong reservations about Margot Asquith: for one thing, ugly women rarely appealed to him. But he was probably more put off by her aggressive manner and overbearing personality. His attempts to avoid her did not go unobserved; in a rather naïve postscript to a note to Clementine, she wrote: 'Whenever I meet Winston he sits out of range – such a bore!'[4]

Although treated with kindness by the Asquiths, Clementine was not really in her element with them; their world was one of high intellectual peaks (as well as not very good bridge), and although Clementine was intelligent and well-educated, she was not an intellectual, and never had any pretensions as such. The austere side of her nature made her dislike Mr Asquith's predilection for peering down 'Pennsylvania Avenue' (the contemporary expression for a lady's cleavage) whenever he was seated next to a pretty woman. Moreover, she was uncomfortable in the company of Violet Asquith and Venetia Stanley, in whose close friendship, as we have seen, she had always (rightly) sensed a latent threat, despite an outward show of smiling affection.

The lack of self-confidence which had been a feature of Clementine's character right through her childhood and adolescence had not yet been exorcised by the depth and abiding strength of Winston's love for her and by discovery of her own capacities. Although kindly received, Clementine felt in a sense an 'outsider' in this small, brilliant circle, with its criss-cross currents and watchful eyes, to which she was a late-comer.

Nor was her discomfiture without foundation. In January 1912, the Prime Minister and Edwin Montagu (a junior minister in the Government) were in Sicily, and were joined there by Violet and Venetia. The friendship between Asquith and Venetia, which had already existed for several years, in these holiday weeks began to become a very close and important one. He was then fifty-nine years old, and she was twenty-four. Venetia was handsome, clever and worldly, and Asquith became infatuated by her. Their relationship developed virtually unremarked, camouflaged as it was by the already existing friendship between Violet and Venetia.

Asquith and Venetia Stanley wrote constantly to each other – he often doing so during the course of Cabinet meetings. Only a few of her letters to him survive, but she kept the greater part of his letters to her; and it is from them that one learns details of state affairs, many of them of a highly secret nature at the time. In addition, Asquith expressed his private opinions of people, including those in the close circle of his friendship.

There are a number of references to Clementine in these letters, but one sentence sums up his opinion of her: 'Clemmie, of whom I am quite fond, is <u>au fond</u> a thundering bore!'[5] His view is comprehensible, for their characters and outlook could not have been more different: Asquith was supremely sophisticated, erudite and worldly; and – unlike Winston, who lived and breathed politics in all his waking hours – he, having dealt with affairs of state, preferred to put them aside, and to distract himself with golf, long motor-car drives, bridge and intellectual pursuits – preferably in the company of beautiful, easy-going, clever women. Clementine was earnest, enthusiastic, highly moral and, although he never actually used the word, obviously – in Mr Asquith's view – a prig.

His descriptions of Clementine in his letters to Venetia were generally critical. As a motoring companion, for instance, she appears to have been defective: after a wet weekend in March 1914 he complained to Venetia, 'The weather was vile, and we could not golf – only trundle about at snail's pace (in deference to Clemmie's fears) in a shut up motor.'[6]

During another weekend party, both Clementine and Goonie came under criticism: 'The critical among us don't think either Clemmie or Goonie quite in their best looks. In spite of her excellent manners & many good qualities, I think the former would rather try one as 'yoke-fellow' & perpetual companion. It is by no means a deep well.'[7] Soon afterwards, dining with the Wimbornes, Asquith sat next to Goonie; he reported to Venetia: 'Goonie & I had a delightful talk about you. She is a loyal & devoted & appreciating friend of yours – worth (I think) 100 Clemmies – not that I am the least anti-Clemmie.'[8]

Of the two sisters, Clementine and Nellie, the latter, with her gayer and more easy-going temperament, appealed (although not without evaluation) more to Mr Asquith: 'The Bud herself is rather a curious study,' he wrote to Venetia; 'I like her (not vehemently) and she is really much cleverer and more original than Clemmie. But her pose of a sort of simple fatuity is apt to take one in and sometimes one is not quite sure whether it is a pose or the real person.'[9]

Mr Asquith's geniality and good manners, however, must have concealed his reservations about Clementine, for she felt quite able to confide in him, while walking in the grounds of Walmer Castle, an incident that had occurred a few days previously when Mrs Keppel (the beautiful and amiable friend of King Edward VII) had warmheartedly offered to give her a dress from some fashionable couturier, which Clementine had declined to accept. Her scrupulous rejection of a lovely present, however, earned her no marks at all: 'Clemmie tells me,' Asquith wrote to Venetia, 'you were present at a curious scene . . . when Mrs Keppel offered her a dress, wh. she was too proud & "well brought up" to accept, notwithstanding your advice to the contrary. Clemmie is very "particular", isn't she?'[10] One cannot but feel a little sorry for Clementine – surrounded as she was by such critical eyes, and at the mercy of such sharp pens.

Given the puritan streak in her nature, Clementine's own highly developed critical faculty, and the shyness which made her so reserved and withdrawn at times, it is not surprising that worldly sophisticates should not have found her totally sympathetic. And some people, at first introduction anyway, preferred the ebullient and light-hearted Nellie, though they might warm to Clementine on further acquaintance. Such a one was the painter Neville Lytton (who later that year would be so much struck by Clementine's 'Venus'-like appearance at the seaside), who was to become both an admirer and a friend. He had written to Nellie on 8 February 1911:

> It is only right and just that I shld write and tell you how immensely I liked

> your sister yesterday. I have always been the one to say disagreeable things about her, but I am quite wrong . . . there is something so natural and impulsively light-hearted about your sister as well as such far greater claims to beauty than the stars [fashionable women] I have been seeing lately.[11]

Throughout her life Clementine was not easy to know well; she rarely lowered the barriers of her reserve, leaving people baffled, and often with a false impression of her nature. Sylvia Henley, her cousin, whom she liked and trusted, once told me that she found Clementine 'very un-even' – one day she would be completely approachable and ready to confide her thoughts, and the next, her defences would be up, and they might have been mere acquaintances. (This would be particularly marked during the Second World War, when Clementine was the repository of so much secret knowledge.) But if people found it difficult to establish real and intimate friendship with her, she was not shy in terms of daily social intercourse, or in the contacts she had with Winston's colleagues.

During 1913 Clementine led an increasingly active social life, quite frequently attending dinner parties and other functions on her own, since Winston was so often away on naval business. Her letters to him during that summer are full of her goings-about in London, and plans for their own entertaining at home.

While Winston was on manoeuvres with the Fleet towards the end of July, she wrote to tell him that she had met Lord Kitchener (at this time a member of the Committee of Imperial Defence) at Lady Crewe's, and that he had expressed a desire to see Winston; she suggested that they should invite him to luncheon. 'By all means ask K [Kitchener] to lunch,' he answered. '. . . Let us be just à trois. I have some things to talk to him about.'[12] This is an interesting indication that Winston had already acquired the habit he practised throughout his life of talking 'shop' at meals. For this reason, when he was in office he very much deplored merely 'social' meals, as he then naturally felt inhibited from discussing many confidential subjects. From this letter we also see that Clementine is accepted into his full confidence, and that he had no secrets from her. She never in fifty-seven years betrayed that trust by deed, or sign, or word.

Winston, in marrying Clementine, had found an equal in spirit and temperament; but in her character were also a super-sensitivity and a proneness to worry which were to cause her moments of great anxiety and unhappiness throughout her life. 'It may never happen,' Winston would say consolingly, when she was fretting about some vexing possibility. 'Cast care aside,' he would enjoin when worries were threatening to spoil some enjoyable plan. One must observe that these wise and kindly meant injunctions did not always have the soothing effect he desired. Clementine was not quarrelsome, but she could easily become vehement and agitated. And it was hardly strange that the union of these two highly charged natures sometimes generated flashes of lightning.

In retrospect one sees that they shared a very important characteristic (a vital ingredient in a long-enduring marriage) – namely, the ardent wish to make up any quarrel without delay. 'Let not the sun go down upon your wrath' was a biblical precept which Winston often quoted, and which he practised in all his relationships. Clementine, too, was always swift and ready to make up any quarrel, but she tended to suffer inner 'bruising' for longer than Winston.

There is an exchange of letters at the end of January 1913 which is illustrative. They had evidently had a sharp disagreement just before Winston left home to go aboard the *Enchantress*, which was in Scottish waters, for on 30 January he wrote to her about their prospective move into Admiralty House, continuing:

> I was stupid last night – but you know what a prey I am to nerves and prepossessions. It is a great comfort to me to feel absolute confidence in your love & cherishment for your poor P.D. [Pugdog].

In a postscript he added:

> Don't be disloyal to me in thought. I have no one but you to break the loneliness of a bustling and bustled existence.
>
> Write to me at Queensferry on receipt of this. It will reach me Satdy morning, and tell me how much you care about your W. XXX Here are three kisses one for each of you. Don't waste them. They are good ones.

This letter, with its moving avowal of how much in five years of marriage Winston had come to depend upon Clementine as his sure refuge, received a swift reply.

'My Sweet Darling Winston,' she wrote on 31 January,

> I love you so much & what I want & enjoy is that you should feel quite comfortable and at home with me – you know I never have any arrière pensée that does not immediately come to the top and boil over; so that when I get excited & cross, I always say more than I feel & mean instead of less. There are never any dregs left behind.
>
> The only time I feel a little low is when the breaks in the 'bustling & bustled existence' are few & far between. I suppose they are not really few, but I am a very greedy Kat and I like a great deal of cream.

One of the causes of worry and tension which they both shared, but which Winston cast off more easily than Clementine, was money – or rather, their lack of it. Admiralty House was proving the burden Clementine had always feared it would be. In April 1914 she was in Paris for a few days on her way home from a holiday in Madrid with Winston, who had returned ahead of her. In a letter which also gave her good news of 'the Kittens', he reviewed their financial situation and found that 'Our finances are in a condition wh requires serious & prompt attention. The expense of the 1st quarter of 1914 with our holiday trip is astonishing. Money seems to flow away.'[13] A few days later he wrote again, and returned to their money situation: 'I . . . am preparing a scheme which will enable us to clear off our debts & bills & start on a clean ready money basis. We shall have to pull in our horns. The money simply drains away.' But even when oppressed by these concerns, Winston was not stingy: 'If you have anything left out of the £40, spend it on some little thing you like in Paris.'[14]

It may have been after this financial crisis that the sad and touching episode of the ruby necklace occurred. When Winston and Clementine were married, his aunt Cornelia Wimborne had given her a muff chain: it was very pretty, and at first Clementine had supposed the gems composing it to be moonstones – however, they turned out to be diamonds! Winston had given her as a wedding present a beautiful ruby and diamond-cluster necklace, which she proceeded to have combined with the 'moonstones' to make an even more ravishing possession. But, sad to relate, during a moment of financial panic Clementine sold this exquisite necklace to pay the housebooks. On discovering her precipitate action, Winston rushed to the jeweller to try to retrieve her treasured possession, but – alas – he was too late.

In the last eighteen months before the war engulfed personal life and enjoyment in its overwhelming tide, Winston and Clementine had a series of delightful holidays. In February 1913 they went to Cannes for a short break, and in May there was yet another sunlit cruise on *Enchantress* in the Mediterranean, when Winston inspected the naval establishments in Malta. The Prime Minister, this time accompanied by Margot as well as Violet, was among the party, which also included Lady Randolph (Mrs Cornwallis-West) and James Masterton-Smith (Winston's Private Secretary).

They embarked at Venice, and on the way to Malta they visited Athens, seeing the Acropolis by moonlight. We are again indebted to Violet Asquith's well-pointed pen for some entertaining *vignettes*, written to a friend:

> the personnel have also been harmonious & sympathetic though two rather explosive elements (both singly and conjointly) no doubt exist in M[argot] and Mrs Cornwallis-West. They both illustrate the difficulties of growing-old-well – in very different ways. Both have rather thrown 'dignity' overboard and are clinging onto the skirts of youth by the teeth . . . I have formed the firm resolve to go into a lace-cap & mittens at 35 [Violet was then twenty-six].
>
> The rest of us are divinely happy day in day out – Clemmie is most smooth & serene & delicious to live with – & looking more beautiful than I have ever seen her.[15]

On this voyage Violet's letters are not our only source of comment on people and places: Margot Asquith kept a voluminous diary which concentrated chiefly on dissecting her fellow travellers.[16] It is also quite clear that physically she found the programme of sightseeing very hard going (she was forty-nine):

> We sight-saw all day & every day – temples museums churches palaces in boiling sun. Hundreds of steps going up, hundreds of steps going down: I stood on my legs till I felt numbed . . . To make up for these personal miseries I saw many beautiful things & made one or 2 nice friends. I also made a complete study of Winston & Clemmy.

Enormously critical of Winston, and stressing again and again his self-centredness, Margot also was at times acutely perceptive about his character. Masterton-Smith confided in her that he had greatly preferred working under Reginald McKenna, whom Winston had succeeded as First Lord, and ended a comparison of the two men with: 'take away that something wh. men call genius & brain for brain McKenna's is as good as the 1st Lord's [Winston] I personally think.' Margot shrewdly observed: 'That something is however just what has made Winston 1st Lord: & has also made him unchangeable – you can alter talent but never genius.'

Moving on to Clementine, she starts dismissively, but continues to make a profound observation:

> neither of the females are Xactly interesting to me; Clemmy or Ly Randolph, but Clemmy is the greatest study. She is not at all what the little cats in society think she is. She is always loving & delicious to me but I see she has little or no maternal feeling a great deal of flimsiness self indulgence & considerable temper . . . Socially & intellectually she is a slight bore. She has <u>much</u> more self-control than one wd. suspect – She has got Winston tighter than a cleverer woman wd. have got him. It was really pretty to see them together & amusing to see what she cd. do with him. The moment he gets back from some futile fishing shooting or inspecting Xpedition, his first question is always 'Where is Clemmy?' & he is quite gloomy if after hunting about he can't find her.

On 18 May, after dining at the British Legation in Athens, the whole *Enchantress* party visited the Acropolis by brilliant moonlight, and during this expedition Margot was the witness of an intimate and touching scene between Winston and Clementine. Margot was daunted by the high marble steps and was somewhat left behind,

> but Winston & Clemmy were with me Winston helped me up the steps . . . luckily for me Winston & C. stopped now & then. In one of these stops Clemmy took exception to Winston's hat & put it on at another angle – this rather irritated him & he flung his head back & with slight – very slight roughness pushed her hand away – Clemmy instantly started walking steadily but very <u>very</u> quickly away from him up the beautiful high steps – Winston not quite liking to desert me shouted to her to stop but she never halted or turned. We cd. see her sprightly almost dancing figure on the plateau at the top white in the light of the blue night. I told Winston to leave me wh. he promptly did, racing up the steps with a last shout 'Clemmy!!!' he caught her & I felt almost ashamed of being a spectator to their embrace.

It was on one of these voyages that Clementine, paying a visit to the galley to talk to the cook, found a large and beautiful turtle confined in a tin bath; it had, she said, 'such beautiful eyes', and with sinking heart she enquired what it was proposed to do with the turtle. The answer was obvious: 'Which evening would Madam prefer turtle soup?' Clementine fled aloft and besought the Captain for a dinghy and a party of men sufficient to grapple with a bathful of live turtle. She herself accompanied the expedition which, after rowing a little distance from the yacht, carefully decanted the turtle into the crystalline Mediterranean. Winston, who, although very fond of turtle soup, was always tender-hearted towards animals, fully supported her deed of mercy. The cook's reaction is not recorded.

Once they arrived at Malta on 22 May, private pleasures and romantic sightseeing were succeeded by official receptions, inspections and conferences on shore. At sea there were firing exercises, and a mock submarine attack. Only too soon these dress rehearsals were to become real-life drama, and the sunlit holiday interludes acquired a strange unreality – faint echoes from a life and a world that was to vanish in the smoke and thunder of a thousand guns.

In September 1913, as in the previous year, there was a cruise up the west coast of Scotland. Among other calls, they visited the Lloyd Georges at their home at Criccieth Castle, overlooking Cardigan Bay, and then sailed on up the coast to Greenock. Clementine stayed on the *Enchantress*, with Goonie and Nellie for company, while Winston made his annual ministerial visit to Balmoral.

The cruise wound up with a ministerial conference, which assembled at Castle Brodick on the Isle of Arran (the home of Mr Illingworth, the Government Chief Whip). Several other ministers' wives were there, and the gathering, which lasted two days, had a social as well as a political flavour. On the last evening the entire party dined on board *Enchantress*. It made a fitting climax to this autumn cruise, destined to be the last of those glamorous enjoyable holiday voyages, which would always glow in Clementine's memory of those Admiralty days.

* * *

Although Winston and Clementine were part of the small but brilliant world at the heart of events, and were totally preoccupied by their crowded and demanding life, there is an interesting passage in one of Winston's letters which shows he was aware of the social isolation in which they found themselves. He was always particularly anxious to maintain the links which bound him to his family; but it was not always easy for Clementine to meet his wishes, for, despite her reserve, dignity and graceful ways, she found it difficult to overcome or conceal her feelings about people. From the *Enchantress* at Newcastle, Winston wrote to her at Alderley on 19 October 1913: 'Here is a letter from Cornelia [Wimborne]. I want you to write & accept. I have a great regard for her – & we have not too many friends. If however you don't want to go – I will go alone. Don't come with all your hackles up & your fur brushed the wrong way – you naughty.'

Clementine's dislike of the Wimbornes was centred on Ivor and Freddie Guest, rather than on their generous and amiable mother, Winston's aunt, and she answered dutifully: 'I will write tomorrow to Aunt Cornelia – I would like to go, & I will be very good I promise you, especially if you stroke my silky tail.'[17]

There were no political differences with the Wimbornes, who were Liberals; nor had politics ever interfered with Winston's friendship with his cousin, Sunny Marlborough, and despite real differences in point of view, they had maintained a cordial relationship. Since their marriage Winston and Clementine had often stayed (together and separately) at Blenheim. But about now politics were the cause of an incident between Clementine and Sunny which resulted in a distinct chill between the cousins for nearly two years.

It is hardly surprising that the controversies surrounding the People's Budget and the Parliament Bill should have caused tension between the ducal and radical branches of the family. Lloyd George had pilloried the dukes in his famous Limehouse speech in July 1909; and Winston too had had fiery things to say about peers and primogeniture which did not pass unnoticed, and must have been exceedingly irritating to Sunny Marlborough, despite his tolerant and hospitable ways.

In the autumn of 1913, Clementine, Goonie and Nellie were all staying at Blenheim. Also among the guests was Margaret Smith (Mrs F. E. Smith) who, like Clementine and Goonie, was there without her husband. Apart from the Duke himself, Lady Norah Churchill, one of his sisters (thirty-eight, and still unmarried), and his elder son, Blandford, a boy of sixteen, made up the family side of the party.

On 22 October Lloyd George made an important speech at Swindon, in which he outlined the Government's proposals for Land Reform, which, as can be imagined, held no joy for the land-owning classes who, in addition, were roundly castigated and held up to ridicule in the more inflammatory passages of the Chancellor of the Exchequer's oration. The following morning, in the Great Library, Clemmie and Nellie were observed to be reading reports of the speech, not only with interest, but with loud expressions of admiration and approval, which cannot have failed to have caused irritation to the Duke, already ruffled by his own perusal of the morning's papers. That same day Clementine wrote to Winston:

> How splendid Lloyd George's Swindon Speech is! I think it is the greatest speech I have ever read. Nellie read it aloud to me this morning & we were both moved to tears. Sunny is ('au fond') broad-minded about the land, but he does not let it appear on the surface & indulges in the sourness & bitterness of a crab apple! But he works it all off at tennis.[18]

The Duke also relieved his feelings at mealtimes by teasingly referring to the Prime Minister's drinking habits, which were the subject of comment at the time; this greatly offended Clementine's sense of loyalty to Winston's chief, and she begged Sunny to refrain from these unseemly jokes (particularly while the servants were in the room) – whereupon he repeated them in a clearer tone. Tempers were clearly rising, and flashpoint was reached when a telegram was handed to Clementine while they were all at luncheon; it was from Lloyd George and concerned some arrangements in which Winston and she were involved. Declaring that the message was most important, and that she must answer it at once, Clementine got up, and proceeded to write a reply at one of the writing tables in the Green Room, where the family lunched when they were few in numbers. Sunny said to her, 'Please, Clemmie, would you mind not writing to that horrible little man on Blenheim writing-paper' – upon which Clementine left the room, went straight up to her bedroom and rang for the maid, whom she instructed to pack her cases; she then ordered a 'fly' from Woodstock to take her to the station.

Presently Goonie came upstairs with Norah Churchill. 'You're not really going?' she asked. Upon Clementine assuring her that she most certainly was, Goonie loyally said she would go too, but Clementine dissuaded her from doing this; Lady Norah watched with consternation (possibly not unmixed with admiration). When the packing was done, Clementine descended to the front hall, where she was met by Sunny Marlborough, who begged her not to leave and apologized for upsetting her. But Clementine, once roused, was not easily calmed, and despite this proffered twig of olive, she left the house and took a train from Woodstock to London.

When Winston heard of it, he thought the whole affair a storm in a teacup, and took Clementine's part only rather half-heartedly, which mortified her, for she felt she had nobly defended his colleagues against Tory malice. The incident upset her very much, and not long after the quarrel, when she was alone with Margaret Smith, and they were talking about it, she became very agitated and burst into floods of tears. Looking back in after years Clementine said she thought she had been wrong not to have accepted Sunny's apology.[19]

It may have been this scene at Blenheim, or perhaps some other passing cause of friction, which accounts for a letter she wrote to Winston on 2 November 1913:

> My sweet and Dear Pig, when I am a withered old woman how miserable I shall be if I have disturbed your life & troubled your spirit by my temper. Do not cease to love me. I could not do without it. If no one loves me, instead of being a Cat with teeth & Claws, but you will admit soft fur, I shall become like the prickly porcupine outside, & inside so raw & unhappy.

In his reply Winston made a touching avowal of his sense of his own inadequacies:

> I loved much to read the words of your dear letter. You know so much about me, & with your intuition have measured the good & bad in my nature. Alas I have no good opinion of myself. At times, I think I cd conquer everything – & then again I know I am only a weak vain fool. But your love for me is the greatest glory & recognition that has or will ever befall me: the attachment wh I feel towards you is not capable of being altered by the sort of things that happen in this world. I only wish I were more worthy of you, & more able to meet the inner needs of your soul.[20]

'Thank you my darling for your dear letter,' she replied the next day, 'I feel so good now, as if I could never be naughty again!'

* * *

As the year 1914 opened, the sands of peace were fast running out; but even to those who saw clearly the growing menace of a European conflagration, the day and the hour, and the immediate cause, were all hidden. The dominating issue for Great Britain was the Irish Question. Passions were fanned white hot when the Home Rule Bill had its third reading in the House of Commons, while gun-running and military drilling in Ulster emphasized the danger of a descent into civil war.

Despite these convulsions, Winston and Clementine were able to take a short holiday abroad together at Easter; it was to be the last for a long time. They went to Madrid as the guests of Winston's old friend and patron, Sir Ernest Cassel; also in the same party were Mrs Keppel and her daughter Violet (later Trefusis). After Winston had to return home, Clementine stayed on to visit Seville and Granada. She was once more expecting a child, and, since the mishap she had sustained in 1912, was observing all sensible precautions; so she did not join the party at the Great Bull Fight in Seville, from which she recounted that Mrs Keppel and Violet returned 'ashy grey & dreadfully upset'.[21]

Sir Ernest also had to return home ahead of the party, and he generously suggested that they should, on their homeward journey, all spend a few days in Paris as his guests; this they gratefully did.

It was probably during these pleasant holiday weeks that Mrs Keppel made to Clementine the startling, if well meant, suggestion that she could greatly assist Winston in his career by finding herself a rich and well-placed lover! Mrs Keppel indicated her willingness to assist in the search for a suitable candidate, and implied that it would be positively selfish of Clementine to neglect to help Winston forward in this way. It is a measure of her affection for Mrs Keppel that Clementine was much amused by this quaint suggestion.

But although Clementine enjoyed her time in Spain and, as was usual with her, was in good health during her pregnancy, there had been for many months a constant fear which gnawed away at her: Winston had taken up flying.

Churchill was the chief inaugurator of the Royal Naval Air Service, the purpose of which was for the Navy to have its own independent means of defending Britain's harbours and other vital naval installations. He believed passionately in the role of this new arm for the Navy, and took a keen and close interest in its development. And, of course, quite soon he was not content to watch from the ground. Flying was still in its infancy; it was only in 1909 that Blériot had flown the Channel. Now both the Army and the Navy were exploring the immense possibilities the aeroplane presented as a war machine. But aircraft were in a rudimentary state of development, and these early machines took their toll of the daring men who experimented with them.

During 1912 and 1913 Winston made many flights, and this new activity was a cause of deep anxiety to Clementine. She felt he took unnecessary risks and that flying was for him 'above & beyond the call of duty'. Her view was shared by Sunny Marlborough, who on 12 March 1913 wrote Winston a letter starting: 'I do not suppose I shall get the chance of writing you many more letters if you continue your journeys in the air,' and going on to tell Winston that he owed it to his 'wife, family and friends' to desist from this hazardous occupation.

But flying thrilled and fascinated Winston. In October 1913, while he was in the *Enchantress*, he visited the airfield at Eastchurch, near Sheerness, and flew in a seaplane, a bi-plane, and finally in a naval airship. On 23 October he wrote Clementine a high-spirited account:

> Darling, We have had a vy jolly day in the air . . . It has been as good as one of those old days in the S. African war, & I have lived entirely in the moment, with no care for all those tiresome party politics & searching newspapers, and awkward by-elections . . .
>
> For good luck before I started I put your locket on. It has been lying in my desk since it got bent – & as usual worked like a charm.

Clementine was much alarmed by this casual account of Winston's exploits, and at once sent him an imploring telegram, which she followed up with a letter: 'I hope my telegram will not have vexed you, but please be kind & don't fly any more just now.'[22] Her entreaties fell on deaf ears, for on 25 October (by which time he would have certainly received her telegram) he flew again – and he continued to do so whenever the opportunity presented itself.

Winston was aware of the immense urgency for Great Britain to develop this new science, with all its far-reaching implications for both war and peace. Not only did flying grip him, but also he came to know and admire many of the men who were developing and testing these new machines; he loved to share some of the risks taken by them; and he always found in physical danger a release from mental stress. That winter, no longer content to be merely a passenger, he started receiving instruction as a pilot. Of course, this served only to increase Clementine's fears for him, and he wrote somewhat apologetically on 29 November: 'I have been naughty today about flying. Down here with twenty machines in the air at once and thousands of flights made without mishap, it is not possible to look upon it as a vy serious risk. Do not be vexed with me.'

Three days later Captain Wildman-Lushington, with whom Winston had spent such a happy day on 29 November, was killed in his plane. This tragedy further alarmed Clementine, and fanned the anxiety already felt by many of his friends. 'Why do you do such a foolish thing as fly repeatedly?' wrote F. E. Smith on 6 December 1913. 'Surely it is unfair to your family, your career & your friends.'[23] Winston, however, continued to be deaf to all reasonings and entreaties.

Even while she was enjoying her Easter holiday in Spain in 1914, Clementine's fears were not asleep, and she wrote: 'I have just been seized with a dreadful anxiety that you are making use of my absence to fly even more often than you do when I am there – I beg of you not to do it at all, at any rate till I can be there.'[24] Her anxieties were not ill-founded, for on the very day she wrote to him, Winston not only made a flight in a seaplane, but was involved in a minor mishap, when engine failure forced the pilot to make an emergency descent on the water close to Clacton jetty. But Winston was completely undeterred, and made a further flight a day or two later.

At the end of May, Clementine took Diana and Randolph (now aged five and three) to stay with her mother in Dieppe. Despite the tragedy of Kitty's death, Lady Blanche had retained agreeable memories of Dieppe life, and she had acquired a few years previously a charming house, St Antoine, 16 rue des Fontaines; across the small garden was an old converted coachhouse, which was known as the Petit St Antoine and served as an annexe to the main house.

Winston's first letter to Clementine at Dieppe on 29 May must have caused her an agony of worry: 'I have been at the Central Flying School for a couple of days – flying a little in good & careful hands & under perfect conditions. So I did not write you from there as I knew you would be vexed.' By the time she replied, the papers had been full of accounts of the loss over the Channel of Gustave Hamel, a famous aviator, known to both Winston and Clementine. All this naturally added to Clementine's anxiety, and her first letter from Dieppe on 30 May has a note of despair in it:

St Antoine
16 rue des Fontaines
Dieppe
30th May [1914]

My Darling,

I began writing to you, but was nipped in the bud by reading of your exploits & of your determination to repeat them the next day! . . .

I felt what you were doing before I read about it, but I felt too weak & tired to struggle against it. It is like beating one's head against a stone wall . . .

The babies are well & very happy & I think are a pleasure to mother. She

> & they inhabit the large house while Nellie & I conduct our establishment in the dower house [the coachhouse] opposite! Everything is minute but very comfortable. We go out every morning with a large basket and do our marketing which is very engrossing, altho' not on a large scale. Goodbye my dear love. Perhaps if I saw you, I could love and pet you, but you have been so naughty that I can't do it on paper. I must be 'brought round' first.
>
> Your loving
> Clemmie
> [Sketch] ears down.

Two more letters in swift succession from Winston kept Clementine in touch with his plans; but in neither of them did he mention the sore subject of flying. Perhaps he felt no explanations could remove her fears, which were undoubtedly beginning to prey on her spirits, and she was now nearly five months with child. On 1 June 1914 she wrote again:

> Life here is so tame and uneventful that the days slip by without being noticed – almost too easily, for one might grow old without having felt any acute sensation of either pain or pleasure. The Daily Mail keeps me in touch with the outside world where things happen such as ghastly shipwrecks, & pigs flying & other events; when the Times arrives at 6 o'clock I am too lazy to open it . . .

Clementine was obviously feeling a little neglected; the letter goes on:

> Meanwhile I have had much leisure to play with & observe the 'kittens'. You will be surprised to hear that they are getting quite fond of me. I am finding out a lot of things about them. They ask occasionally with solicitude & respect about you . . . When shall I see you again!!! Mother & Nellie have gadded off to the Casino to have a little flutter at Chemin de Fer, but I feel apathetic about that too.
>
> I shall go quite to sleep unless someone takes a lot of trouble soon. I am making <u>such</u> good salads & coffee & even roasted a chicken tho' the stove is so tiny here it would hardly go inside.
>
> I shall walk out & post this letter & then I think I shall have done very well for one day.
>
> Goodbye absent & wandering Pig!
> Your loving
> Clemmie
> [sketch of cat with two kittens]

Winston was able to spend one day with them all on 3 June, the *Enchantress* dropping him off at a point near Dieppe early in the morning, and picking him up that same evening, to continue his naval peregrinations. The children were enchanted by this sudden appearance, but, as Clementine wrote the next day, 'The Babies were sad when they found the *Enchantress* had sailed away in the night & there was no "Papa" in the Bed in the Day Nursery.'[25]

They must have talked about the flying, and Clementine seemed to have resigned herself to the fact that Winston meant to continue his exploits, for she wrote in the same letter: 'I cannot help knowing that you are going to fly as you go to Sheerness & it fills me with anxiety. I know nothing will stop you from doing it so I will not weary you with tedious entreaties, but don't forget that I am thinking about it all the time & so, do it as little, & as moderately as you can, & only with the very best Pilot. I feel very "ears down" about it.'

A very long letter she wrote to Winston the next day shows how much it preyed on her mind, and how overwrought her nerves had become. Nellie and she had made an expedition to Puys, the little village about three miles along the coast where the Hozier family had spent the summer of 1899, before they moved into Dieppe itself. They had a delightful day visiting old haunts, and prospecting possible seaside houses. 'We got home very late,' wrote Clementine,

> But after this pleasant day I have had a miserable night haunted by hideous dreams, so this morning I am sad & worn out.
>
> I dreamt that I had my Baby, but the Doctor & Nurse wouldn't show it to me & hid it away. Finally after all my entreaties had been refused I jumped out of bed & ran all over the house searching for it. At last I found it in a darkened room. It looked all right & I feverishly undressed it & counted its fingers & toes. It seemed quite normal & I ran out of the room with it in my arms. And then in the Daylight I saw it was a gaping idiot. And then the worst thing of all happened – I wanted the Doctor to kill it – but he was shocked & took it away & I was mad too. And then I woke up & went to sleep again and dreamt it a second time. I feel very nervous and unhappy & the little thing has been fluttering all the morning.
>
> Your telegram arrived last night, after we were in bed. Every time I see a telegram now I think it is to announce that you have been killed flying. I had a fright but went to sleep relieved & reassured; but this morning after the nightmare I looked at it again for consolation & found to my horror it was from Sheerness & not from Dover where I thought you were going first – so you are probably at it again at this very moment.
>
> Goodbye Dear but Cruel One
> Your loving
> Clemmie[26]

Every woman who has borne a child will surely recognize in Clementine's vivid description of her ghastly nightmare the sudden panic-fear that no logic can dispel. Winston may have talked and reasoned down her pleas and arguments and discounted her fears up to now – but his reaction to this letter was instantaneous. 'My darling one, I will not fly any more until at any rate you have recovered from your kitten,' he wrote to her on 6 June. He went on:

> This is a wrench, because I was on the verge of taking my pilot's certificate. It only needed a couple of calm mornings; & I am confident of my ability to achieve it vy respectably. I shd greatly have liked to reach this point wh wd have made a suitable moment for breaking off. But I must admit that the numerous fatalities of this year wd justify you in complaining if I continued to share the risks – as I am proud to do – of these good fellows. So I give it up decidedly for many months & perhaps for ever. This is a gift – so stupidly am I made – wh costs me more than anything wh cd be bought with money. So I am vy glad to lay it at your feet, because I know it will rejoice & relieve your heart . . .
>
> You will give me some kisses and forgive me for past distresses – I am sure. Though I had no need & perhaps no right to do it – it was an important part of my life during the last 7 months, & I am sure my nerve, my spirits & my virtue were all improved by it. But at your expense my poor pussy cat! I am so sorry.

Winston honoured his word: he abandoned flying at once.

With this gnawing anxiety removed from her mind, Clementine was able to relax and enjoy the summer. The sisters-in-law had arranged to spend the best part of July and August by the sea at Overstrand, near Cromer, with their respective nurseries. Clementine, Diana, Randolph and their nanny installed themselves in a little house called Pear Tree Cottage, while five minutes away Goonie Churchill and her two sons, Johnny and Peregrine, aged five and one, had taken Beehive Cottage. Here Clementine passed some happy sunshine weeks, garnering her strength for her new child, and enjoying the last carefree days she, and many others, were to know for four long dark years.

The Home Rule Bill had its third reading and was passed by the House of Commons on 26 May 1914, and during the ensuing weeks tension over the Irish Question seemed to be reaching breaking point. In a last-minute attempt to avert an explosion, the King called a Home Rule Conference at Buckingham Palace on 21 July. The next day Winston, under a heading Secretissime, reported to Clementine that 'the conference is in extremis'; two days later it broke up, having failed to reach agreement.

While the country's attention was engrossed by these bitter domestic struggles, an event occurred in Central Europe that passed almost unnoticed by the mass of the British people: on 28 June the Archduke Franz Ferdinand of Austria (heir to the Emperor Francis Joseph I) and his wife were assassinated at Sarajevo, in Bosnia, setting in train a sequence of events which was to prove fatal to Europe's peace.

But despite troubles at home and menacing news from abroad, Winston managed to spend several Sundays with his family at Pear Tree Cottage, where Clementine declared to him in a letter, 'I am very happy perched among its branches.'[27] After his second visit to them on 12 July, Winston left by sea (being taken off by *Enchantress*), writing the next day to Clementine:

> It was quite forlorn leaving you last night. I don't know why a departure to the sea seems so much more significant, than going off by train. We watched your figures slowly climbing up the zigzag & slowly fading in the dusk; and I felt as if I were going to the other end of the world.
>
> The kittens were vy dear & caressing. They get more lovable every day. Altogether Pear Tree is a vy happy, sunlit picture in my mind's eye. Tender love my dearest – I must try to get you a little country house 'for always'.[28]

Early in 1914 it had been decided by the Cabinet that, as an economy measure, the customary Grand Manoeuvres of the Fleet (involving the wide deployment of many ships) would be replaced by a test mobilization of part of the Fleet. This took place in July, and was followed by a grand review of the ships of every Fleet, passing in front of the King in the Royal Yacht at Spithead. 'It constituted', Churchill later wrote, 'incomparably the greatest assemblage of naval power ever witnessed in the history of the world.'[29] The ships of the Third Fleet, having completed their test mobilization, were about to disperse to their various ports, when it was learnt that Austria had delivered an ultimatum to Serbia.

After attending a Cabinet meeting on Friday, 24 July, Winston wrote to Clementine: 'Europe is trembling on the verge of a general war. The Austrian ultimatum to Serbia being the most insolent document of its kind ever devised. Side by side with this the Provincial govt. in Ulster wh is now imminent appears comparatively a humdrum affair.' But despite the gravity of the situation, there was no positive feeling in Government circles that England would be involved, even if there were to be a conflict on the European Continent, and Winston was able to spend the weekend at Overstrand.

On Sunday, 26 July, between telephone calls to and from the Admiralty, Winston played with the children on the beach: 'We dammed the little rivulets which trickled down to the sea as the tide went out. It was a very beautiful day. The North Sea sparkled to a far horizon.'[30] However, this idyllic and domestic day was to be cut short. After speaking on the telephone to Prince Louis of Battenberg (the First Sea Lord) at midday, Winston decided that events demanded his presence and returned to London, where Prince Louis had already given orders to halt the dispersal of the Fleet.

During the ensuing days Winston kept Clementine in touch with events by both letter and telephone. Pear Tree Cottage had no telephone, but the kind neighbours from whom the Churchills had rented their holiday house, Sir Edgar Speyer and his wife*, offered Clementine the use of their telephone. From there at pre-set times, and talking in veiled language, she was able to keep track of the increasingly menacing course of events. On

* Sir Edgar Speyer (1862–1932), a German Jewish financier, philanthropist, art collector and Liberal Party supporter; naturalized, 1892; created baronet, 1906; Privy Counsellor, 1909. He and his (Christian) wife Leonora were well-known in social and musical circles, and entertained lavishly in both London and Overstrand. Following allegations of disloyalty during the First World War, his citizenship and Privy Counsellorship were revoked in 1921, when he and his family left England to settle in the United States.

returning to the cottage after talking to Winston one night, she wrote: 'Goodnight my Dearest one. I trust the news may be better tomorrow. Surely every hour of delay must make the forces of peace more powerful. It would be a wicked war.'[31]

Austria rejected Serbia's conciliatory reply to its ultimatum, and the prospect of Russian intervention on Serbia's side, which would draw in France and Britain, grew more likely every hour. At midnight on 28 July Winston wrote to Clementine from the Admiralty: 'Everything tends towards catastrophe & collapse. I am interested, geared up & happy. Is it not horrible to be built like that? The preparations have a hideous fascination for me. I pray to God to forgive me for such fearful moods of levity. Yet I wd do my best for peace.' Despite the imminence of catastrophe, he could still turn his attention to small and peaceful things: 'The two black swans on St James's Park lake have a darling cygnet – grey, fluffy, precious and unique. I watched them this evening for some time as a relief from all the plans & schemes.'[32]

The next day, Clementine replied:

> My Darling,
>
> I have just returned from telephoning you & find your letter. I am glad that you are 'cap à pied' but how I do pray that the perfection of your plans may not be brought to the test in this foreign quarrel. I am sad that I shall not see you this week (last Sunday was so delicious) but I fear it is far for you in these critical times, in spite of the Speyers' splendid telephone.
>
> While I am waiting for the connection in Sir Edgar's 'study' I contemplate a lovely Madonna over the mantelpiece who gazes with melancholy eyes at the piles of Business Books on the writing table.[33]

Two days later, on 31 July, Winston told her: 'There is still hope although the clouds are blacker and blacker.'

Happy though she was with her seaside life and in the company of her babies Clementine longed to be with Winston: 'I much wish I were with you during these anxious thrilling days & know how you are feeling – tingling with life to the tips of your fingers.' [34]

On Sunday, 2 August the Cabinet sat almost continuously; there were deep divisions among its members at this agonizing hour of decision. In the early hours of that morning Winston had written a hurried note to Clementine to tell her the latest, fateful piece of news.

> Admiralty,
> Whitehall.
> 2nd August 1914
> 1 a.m.
>
> Cat – dear,
>
> It is all up. Germany has quenched the last hopes of peace by declaring war on Russia.

Events now took charge of men, and rolled them forward relentlessly towards the terrible abyss. Following on Germany's ultimatum, the King of the Belgians appealed to France and Great Britain to protect his country's neutrality. On 3 August Germany declared war on France and invaded Belgium. At eleven o'clock on the night of 4 August the British ultimatum to Germany expired, and Great Britain was at war. As the first strokes of Big Ben sounded the fatal hour, there flashed from the Admiralty to every British ship and naval establishment throughout the world the signal: 'Commence hostilities against Germany.'

CHAPTER EIGHT

All Over by Christmas

BRITAIN'S DECLARATION OF WAR AGAINST GERMANY WAS GREETED BY A GALE of patriotic hysteria and flag-waving, in strong contrast to the country's quiet, dogged acceptance of war twenty-five years later. The spectre of civil war with Ireland, which had loomed so large, was, for the time being, exorcized; and, despite bitter divisions between the parties, the Conservatives gave the Government their wholehearted support. The outbreak of war unified all classes and shades of opinion throughout the country. Conscription was not to come till much later, and in this autumn of 1914 men flocked to the recruiting stations – among them many Irish volunteers. Guarded by the Navy, the British Expeditionary Force was transported safe and sound across the Channel during the first weeks of August. The general feeling throughout the country was one of elation and excitement, and a confident certainty (shared even in some well-informed quarters) that it would 'all be over by Christmas'.

All Churchill's efforts since his appointment to the Admiralty to put the Navy on a war footing had culminated in the swift and efficient mobilization of the Fleet. The dramatic speed at which the final crisis had developed meant that many decisions concerning the High Command had to be taken without delay. One of these, which Winston found painful, was to relieve Admiral Sir George Callaghan of the Supreme Command of the Home Fleets, replacing him by Sir John Jellicoe. Callaghan was due to retire on 1 October 1914 but Churchill was doubtful whether his health and physical strength were equal to the immense strain that would now be cast upon him, and this was no time to consider individuals. Winston told Clementine of the decision on 31 July, when he wrote to her: 'I have resolved to remove Callaghan & place Jellicoe in supreme command as soon as it becomes certain that war will be declared.'

Clementine's reply provides an example of the sage and perceptive advice she was to give Winston many times during their long life together. Often he did not take it, and sometimes he regretted he had not. In this case she did not doubt the wisdom of the decision, but she expressed concern about the manner in which Sir George Callaghan should be relieved of his command. On 4 August she wrote to Winston:

> I have been cogitating for an hour or two over the 'Callico Jellatine' crisis, which, Thank God, is over as far as essentials are concerned.
>
> There only remains the deep wound in an old man's heart.
>
> If you put the wrong sort of poultice on, it will fester – (Do not be vexed, when you are so occupied with vital things my writing to you about this, which, in its way, is important too). An interview with the Sovereign and a decoration to my mind is the wrong practice – To a proud sensitive man at this moment a decoration must be an insult.
>
> Please see him yourself & take him by the hand and (additional) offer him a seat on the Board, or if this is impossible give him some advisory position at the Admiralty. It does not matter if he 'cannot say Bo to a Goose' – His lips will then be sealed and his wife's too. Don't think this is a trivial matter. At this moment you want everyone's heart & soul – you don't want even a small clique of retired people to feel bitter & to cackle. If you give him a position of honour and confidence, the whole service will feel that he has been as well treated as possible under the circumstances, & that he has not been humiliated.
>
> This will prevent people now at the top of the Tree feeling 'In a few years I shall be cast off like an old shoe' – Jellicoe & Beatty & Warrender & Bailey, now the flower of the service, are only a few years younger than Callaghan.
>
> Then, don't underrate the power of women to do mischief. I don't want Lady Callaghan & Lady Bridgeman* to form a league of retired Officers' Cats, to abuse you. Poor old Lady Callaghan's grief will be intense but if you are good to him now it will be softened; if he is still employed she is bound to be comparatively silent.
>
> Anyhow I beg of you to see him.
>
> If I were doing it – if he refused the appointment I would earnestly urge him again & again . . . to accept it, saying you need his services – then he will believe it & come round & there will not be any disagreeableness left from this difficult business.
>
> [sketch of a cat]

In the crash of war, molehills retain their proper proportions; but Clementine's perception and awareness of people's feelings were over and over again to be a help to Winston, who in striding towards the main goal was often totally oblivious of the side-effects of his actions, or of other people's reactions to them. No immediate appointment was announced for Sir George Callaghan, but in December 1914 he was appointed Commander-in-Chief at the Nore.

Another example of Clementine's moderating and sage advice is to be found in an undated letter of hers which must belong to this period.

* The wife of Admiral Sir Francis Bridgeman, who had prematurely retired as First Sea Lord in 1912 to make way for Prince Louis of Battenberg.

Winston formed the habit of making frequent short visits to see Sir John French, the Commander-in-Chief of the British Expeditionary Force, at his Headquarters in France. From these visits he gained much useful information, and was able to form a wide and general view of the military situation; but they were criticized by some of his colleagues. One of these expeditions was in the offing towards the end of September, when Clementine wrote to Winston:

> Now please don't think me tiresome; but I want you to tell the PM. of your projected visit to Sir John French. It would be very bad manners if you do not & he will be displeased and hurt.
>
> Of course I know you will consult K [Lord Kitchener, now Secretary of State for War]. Otherwise the journey will savour of a week-end escapade & not of a mission. You would be surprised and incensed if K skipped off to visit Jellicoe on his own. I wish my Darling you didn't crave to go. It makes me grieve to see you gloomy & dissatisfied with the unique position you have reached thro' years of ceaseless industry & foresight. The P.M. leans on you & listens to you more & more. You are the only young vital person in the Cabinet. It is really wicked of you not to be swelling with pride at being 1st Lord of the Admiralty during the greatest War since the beginning of the World. And there is still much to be done & only you can do it . . .
>
> Be a good one & rejoice & don't hanker. Great & glorious as have been the achievements of our army, it is only a small one, ⅛ of the allied forces. Whereas you rule this gigantic Navy which will in the end decide the War.[1]

Winston paid heed to Clementine's advice: he wrote on 26 September to Lord Kitchener, who made no objection. 'How right you were about telling K.,' he acknowledged, in a letter written aboard the ship taking him to France.[2] However, during the next three months Winston (although always 'asking permission') made a number of visits to Sir John's Headquarters, and also corresponded with him. Kitchener became suspicious and annoyed, and complained to the Prime Minister, who, in a letter of 18 December 1914, effectively vetoed any further visits.

Although she must have longed to be at his side and at the centre of events, Clementine, now in the seventh month of pregnancy, wisely stayed on at Pear Tree Cottage with the children. She gleaned what news she could from Winston's letters and the veiled conversations she was able to conduct over the Speyers' telephone. Winston had left their car with her; in the back of his mind lurked the fear that the east coast might be the objective of a German raid, and he wished her and the children to have a ready means of flight if necessary. However, the car gave trouble: 'The motor is a lame duck till Wednesday at least,' Clementine wrote on Saturday, 8 August, 'so I hope the Germans won't arrive before that date. I have taken I hope efficient measures to have it seen to.'

But in any case the future of the 'motor', it seems, was uncertain, for her next letter on the following day puts forward a delightful suggestion: 'I have had such a good idea. Later on, in the spring . . . let us instead [of the motor] have 2 chestnut horses "with nice long tails" & ride for an hour every morning. It would be far more value to you than a motor & I should enjoy it so much. It would not cost so much surely?'[3] It is interesting that even Clementine, living so much in touch with affairs, had not grasped, at that moment, the severity of the ordeal that lay ahead.

Most of the summer visitors to Cromer were packing up and leaving for their homes, to the great distress of the local landladies and shopkeepers. Clementine described to Winston in the same letter how

> Yesterday the local authorities in a frantic effort to stem the ebbing tide of tourists had the following pathetic appeal flashed on the screen of the local cinema show. (I am not sure of the words, but this is the gist.)
>
> 'Visitors! Why are you leaving Cromer? Mrs Winston Churchill and her children are in residence in the neighbourhood. If it's safe enough for her, surely it's safe enough for you!'

In his letter of 9 August, Winston reverted to his fears for her safety:

> It makes me a little anxious that you should be on the coast. It is a 100 to one against a raid – but still there is the chance, and Cromer has a good landing place near.
>
> I wish you would get the motor repaired and keep it so that you can whisk away at the first sign of trouble.

His anxiety over a possible raid was not merely fanciful, for that very December the east coast towns of Scarborough, Whitby and Hartlepool were bombarded by German ships, with considerable loss of life.

Clementine thirsted for news; Winston must have alerted her (on the telephone) to the arrival of a letter from him, for in her letter of 9 August she goes on:

> I am longing to get your letter with the secret news. It shall be destroyed at once. I hope that in it, you tell me about the expeditionary force. Do I guess right that some have gone already? Be a good one and write again & feed me with tit-bits. I am being so wise & good & sitting on the Beach & playing with my kittens, & doing my little housekeeping but how I long to dash up & be near you and the pulse of things.

The same day Winston sent her a document to put her in the picture, with strict instructions to burn it at once. Replying on 10 August, she assured him:

> The Secret document has been read & burnt before my eyes in the kitchen fire! It was most interesting but I was disappointed because I hoped you were going to tell me about the Expeditionary Force. Do send me news of it. When

it is going, where it will land, which regiments are in the first batch, etc. I long for it to arrive in time to save the Liège citizens from being massacred in their houses.

On 11 August Winston wrote a very short note, which revealed how tired he was; it contained also a very gentle rebuke regarding her undeniably indiscreet questions about the Expeditionary Force.

My dear one,

This is only a line from a vy tired Winston. The Expedy Force about wh you are so inquisitive is on its road & will be all on the spot in time. I wish I cd whisk down to you & dig a little on the beach. My work here is vy heavy & so interesting that I cannot leave it.

Now I am really going to knock off.
Ever your loving
W

The note of fatigue in Winston's letter produced a reply the next day, full of practical suggestions as well as sympathy:

Now are you doing everything you can not to be too tired?
1) Never missing your morning ride.
2) Going to bed well before midnight & sleeping well and not allowing yourself to be woken up every time a Belgian kills a German. (You must have 8 hours sleep every night to be your best self.)
3) Not smoking too much & not having indigestion.

Now shall I come up for a day or two next Monday & tease you partly into doing all these things? Or are you being a good one all by yourself? . . .

The golden sands are too delicious & long to have fortifications made on them, & there is a little stream that really ought to be dammed. But Alas it is too far. Goodbye my sweet one. I am so proud of you & love you very much.[4]

Although removed from the hub of affairs, life at Pear Tree and Beehive Cottages was not without its diversions and excitements. Out of the blue a Mr Robert Houston, a wealthy shipowner and Conservative MP, although unacquainted with either Winston or Clementine, sent her, enclosed in a letter of praise of Winston, a huge emerald and diamond ring. Clementine of course returned this generous but unsuitable present with an appreciative letter. But during the brief time she had this lovely jewel in her possession it caused a sensation. 'Goonie was struck all of a heap,' she wrote to Winston, and 'Lady Speyer who is a "connoisseur" says she thinks it is an antique. She is covered with emerald rings . . . none of them come anywhere near it!' Clementine went on to philosophize on the unfairness of things: 'What an illogical world it is. Becos' I have a Pig who is a Genius and a "Poodle-Ching" I am given a costly jewel, whereas if I had an inferior husband & needed consolation I should get nothing! for nothing.'

Winston wrote back on 13 August, full of praise for his high-minded 'Kat': 'How right you were not to hesitate to send back the ring, & what a good thing it is to have high & inflexible principles! You are a dear & splendid Kat. I am vy proud of you. What you have done for me no one can measure.'

Both Clementine and Nellie thought it unwise for their mother to go on living in Dieppe, and Nellie was therefore despatched to bring her home; Clementine invited her to join them at Pear Tree Cottage, where she arrived around the middle of August. Her arrival coincided with Overstrand's local attack of 'spy-mania' – an epidemic of highly coloured (and generally not well substantiated) stories of spies and their alleged activities, which swept through the country just then.

Clementine reported a local incident to Winston on 13 August 1914:

> Yesterday one of the cottager's wives saw two men walking along the cliff. Their coats were bulging & she thought they were carrying some puppies. It was rather a lonely part of the cliff & they looked annoyed when they saw her, & spoke to each other in a foreign tongue. When they had passed, she hid herself & then followed them at a distance. She then saw them open their jackets & let fly 4 carrier pigeons! She kept them in sight for some time but eventually they disappeared into a lane. Luckily on her way home she met 2 officers in a little car. She told them about it, & they pursued the men & caught them . . .*
>
> Mother is much alarmed at the 'carrier pigeon story' & insists that the message carried by the pigeons was that 'the wife of the First Lord is at Overstrand, & that the Germans are to send an aeroplane to kidnap me & that then I am to be ransomed by the handing over of several of our handsomest ships.'
>
> If I <u>am</u> kidnapped I beg of you not to sacrifice the smallest or cheapest submarine or even the oldest ship . . . I could not face the subsequent unpopularity whereas I should be quite a heroine & you a Spartan if I died bravely & unransomed.

But although Clementine recounted the 'spy' incident with such verve and gaiety, she was in fact in a highly agitated, almost hysterical state of mind, engendered by what she considered to be Nellie's inconsiderate and wrong-headed behaviour. She poured out the whole story in great detail to Winston, in a letter on 14 August which covered pages and pages of writing paper.

Nellie had escorted Lady Blanche back to England, and Clementine had expected that she would deliver their mother herself to Pear Tree Cottage. However, in the event Lady Blanche arrived alone, in a feeble and exhausted

*Local knowledge that the Speyers were of German origin led to accusations that they were signalling to submarines. See footnote p. 120.

state, bearing a letter from Nellie telling Clementine she had gone to Cliveden (the Astors' house in Buckinghamshire) to help in its conversion into a convalescent home.

Clementine was displeased by this news, as she had arranged with Nellie that she would come with their mother to Overstrand, and remain to look after her, so that Clementine could go up to London to spend some time with Winston. She was made even more angry when another letter arrived from her sister announcing that she was going immediately to Belgium as part of a nursing unit headed by her friend Angela Manners, whose parents, Lord and Lady Manners, and their friends were financing the team, which was to consist of a surgeon, six nurses and Nellie, who was to act as secretary and interpreter.*

Clementine was outraged – at this distance of time, it seems, rather unreasonably so. She wrote to Winston:

> It is all cheap emotion. Nellie is not trained, she will be one more useless mouth to feed in that poor little country which in a few days will be the scene of horrible grim happenings.
>
> Nellie's obvious and natural duty is to look after Mother whom she has brought over from Dieppe and 'dumped' down thus leaving her responsibilities without a thought to others.
>
> I feel quite ill this morning, as I have had a very bad night & this on top of it has really upset me. I long to see you & put my arms round your neck. You are always so sane and sensible my darling one, & you would calm my hurt and angry feelings . . .
>
> But my Dear, in the midst of your work it is wrong of me to bubble over like this, but my heart is full & I can't help it.
>
> Goonie is sweet & placid & I have confided my indignation to her.[5]

In a long sequel to this stormy letter, written a few hours later, Clementine, now somewhat calmed, attempted to be 'reasonable and calm & to analyse why I was so upset'. She had to admit it was because she felt she had been left in the lurch by Nellie: 'just now when I am not feeling very well & would appreciate comfort & support she has left me to cope with Mother alone, which she knows, even when I am feeling very brisk & well is a tax on the nerves!'[6]

It must be remembered that Clementine was now within six weeks of her confinement and was, of course, feeling her long separation from Winston. Having got it all off her chest in this tempestuous outburst, however, she avowed:

* This is an interesting example of what scope there was in the First World War for 'private enterprise' schemes. There were many similar teams and organizations.

I feel better since writing to you my Darling. Do not misjudge me for being so garrulous.

Your loving very tired
Clemmie

Despite her elder sister's disapproval, the impetuous and intrepid Nellie lost no time in carrying out her intentions, and left for Brussels on 14 August with the nursing unit as she had intended. She and her friends were in Brussels when it was occupied by the Germans on 20 August, but managed to get to Mons a few days after the battle, and in the confusion of the retreat which followed their unit was taken prisoner by the Germans. Nellie's spirits were undaunted, however, and her natural impudence unquenched. While locked in the waiting-room of a railway station with her companions, she scrawled upon the walls the following jingle:

Our good King George is both
Greater and wiser,
Than all other monarchs,
Including 'der Kaiser'.

The nursing unit, however, was eventually allowed to continue its work with the British wounded, and several fascinating letters from Nellie survive, giving a lively and moving account of daily life in the little hospital, and the horrors and ordeals to which the Belgian civilian population were subjected by their conquerors. Clementine and Lady Blanche were able to communicate with her occasionally, but their anxiety was intense, not only because of Nellie's proximity to the fighting line, but also on account of her rash tongue and plucky spirit, which could easily have got her and her fellow prisoners into trouble with their captors.

At the end of the summer Clementine left Pear Tree Cottage and returned with the children to Winston in London, where she spent the last month before the birth of her child. This joyous natural event was deeply overcast by the news of the terrible events which were sweeping over Europe like relentless waves. As the weeks went by, more and more people bore the burden of bereavement as the armies came to grips, and the remorseless slaughter of a whole generation began on the battlefields of Flanders and France.

The early sunlit seaside weeks of the war must now have seemed like a dream to Clementine, who, in addition to feeling national events and the sorrows of her friends, had personal and family preoccupations. Nellie was still a source of anxiety; in addition, Bill was now commanding HMS *Thorn*, a torpedo destroyer, and Jack Churchill was in France with the Oxfordshire Hussars. And these worries were much increased by Winston's sudden departure on 3 October to take part in the defence of Antwerp.

The Germans' plan had been to take the Allies by surprise – to sweep through Belgium and then destroy the French army in a single mighty onslaught. This plan was foiled by the Allies' desperate resistance on the River Marne. The opposing armies became locked in bitter conflict, and on both flanks a race for the sea began, to gain possession of the strategic ports on the Belgian coast. Of these, one of the most vital was Antwerp, which was not only important to the strategy and safety of the British Army but was made also the heart and stronghold of Belgian resistance by the presence there of the King and Queen of the Belgians and their Government.

On 2 October a telegram had been received in London from Antwerp announcing the intention of the Belgian Government and the King and Queen to leave Antwerp the following day. This was grave news indeed; and at a midnight conference Sir Edward Grey, Lord Kitchener and Churchill agreed that someone of standing and authority should be sent immediately to the beleaguered fortress city. This task was confided to Churchill, who, needless to say, was an eager volunteer. Very early on the morning of 3 October he was on his way to Antwerp. It was a situation which appealed to his every instinct: a desperate struggle; action in which he himself could play a decisive role; and, above all, the chance to share dangers with brave and patriotic men.

On learning that reinforcements were to be sent and that Winston Churchill – a Cabinet minister – was on the way to join them as the emissary of the British Government, the Belgian Council of War suspended the order to evacuate Antwerp.

During the few hectic and dangerous days Churchill remained in the now closely besieged city, he became deeply and emotionally identified with its fate, and on the morning of 5 October he telegraphed to the Prime Minister asking that he might be permitted to resign as First Lord of the Admiralty and take formal charge of the British forces in Antwerp. This extraordinary request, which Asquith described as being received by his colleagues with a 'Homeric laugh',[7] was firmly refused. The Cabinet decided to send out Lieutenant-General Sir Henry Rawlinson, and Churchill was asked to hold the fort as best he might until Sir Henry's arrival. Churchill handed over to him on the evening of 6 October, and travelled back to London overnight. It was clear that Antwerp could not hold out much longer, and the city in fact capitulated on 10 October.

Clementine had been startled by the suddenness of Winston's departure for Antwerp, and during the ensuing days her fears for his safety were mingled with deep doubts as to the wisdom of his action. In her heart, she thought his impetuous intervention – albeit as the official representative of the Government – unwise. She was even more disconcerted when she learnt of his wild telegram: to her it smacked of an almost frivolous approach to this affair. Both to his wife and to his colleagues it seemed crystal clear that Winston's sense of proportion had deserted him. The siege of Antwerp was a dramatic incident; the prolongation of the city's resistance was certainly of vital importance at that moment, but in the whole scale of events it could not compare in significance with Winston's own ministerial responsibility for the overall direction of the maritime power of Great Britain. Politically astute, Clementine realized also that Winston's rash and ill-conceived request would harm him both in the opinion of his colleagues and in the eyes of the general public.

It was towards the end of these days of doubt and sharp anxiety that, early on the morning of Wednesday, 7 October, Clementine gave birth at Admiralty House to their third child – a red-headed daughter. During the course of the same day Winston arrived home, and the Court Circular records that he was received at Buckingham Palace by the King. He also was received 'in audience' by his Clemmie and the new Sarah (as the baby was to be called) – and great were the relief and the rejoicings.

In after years, Clementine was always to speak of the Antwerp incident with a tinge of resentment. She never really accepted that Winston's inescapable public duty required him to be there, and with her anxieties for his safety and her apprehensions over the wisdom of his actions was mingled the very understandable wish for her husband to be near her for the birth of their child.

Although Churchill's impetuous intervention at Antwerp brought him discredit in many quarters, and was attacked by most of the press, he was given a warm reception by his colleagues on his return. Whatever reservations they may have felt, they appreciated the heroic efforts he had made to prolong Antwerp's resistance. One of them, Sir Edward Grey, wrote Clementine this warm and charming letter:

Committee of Imperial Defence,
2, Whitehall Gardens, S.W
October 7, 1914

Dear Mrs Churchill

I am so glad to hear that you & the new baby are well.

I am sitting next Winston at this Committee, having just welcomed his return from Antwerp.

And I feel a glow imported by the thought that I am sitting next a Hero. I cant tell you how much I admire his courage & gallant spirit & genius for war. It inspires us all.

Yours very sincerely
Edward Grey

Later this incident, or 'escapade', as many regarded it, was used to brand Winston as an irresponsible adventurer; but in retrospect two things emerge quite clearly: Churchill's presence in Antwerp undeniably stiffened the resistance of the Belgians; and, by delaying the capitulation of the city, a signal service was rendered to the Allies. But it would be Violet Bonham Carter's measured opinion that 'No event in his whole career, with the one exception of Gallipoli, did him greater and more undeserved damage.'[8]

* * *

In order to recuperate after the birth of Sarah, Clementine spent a few weeks in November at Belcaire, Lympne, in Kent, the house of the millionaire politician and art collector Sir Philip Sassoon, where she and Winston were often to be guests in the years to come. Winston was not able to visit her much during this time, the pressure and gravity of events being too great. Although Clementine longed to see him, she completely understood the reason for his absence. 'I am sure you ought not to leave London just now even for a few hours,' she wrote on 18 November 1914, 'so don't even think of coming next Sunday.'

Happily she was not without company, as Goonie was able to be with her for part of the time. 'Poor Goonie is in a great state of anxiety about Jack,' she wrote to Winston on 19 November, 'as the 2nd Division was in action on Tuesday, and she calculates from his last letter that Jack would that day have been in the trenches.' Many, many women must have been making these agonizing calculations. But before she had finished this letter, Winston had telephoned that Jack (quite coincidentally) had been posted to Sir John French's staff, and so Clementine was able to finish her letter on a brighter note: 'Goonie is so happy about Jack being on the staff. She has been so sad all day, but now she is overjoyed. We have both been sitting completely buried by papers all the afternoon paying our bills. Outside the snow is falling thick and fast.'

Later that same month, to her family's relief and joy, Nellie arrived home; the Germans had apparently decided to repatriate the whole nursing unit, after they had gone on strike and refused to nurse German wounded. Evidently the authorities felt they had enough troubles on their hands without being encumbered by a small party of non-combatants, imbued with a saucy and rebellious spirit; Nellie and her companions were therefore bundled home. Routed through Norway, they arrived safe and sound, but shivering, as they were still in the thin summer uniforms in which they had set out the previous August.

In the first months of the war, the Navy had sustained some severe blows. But in early December cheering news of a splendid victory at sea warmed British hearts when Admiral Sturdee's ships sank four out of five of a German squadron near the Falkland Islands. The British Navy was now supreme, and the country's supply lines across the world were (for the time being) free from the fear of surface raiders.

As 1914 drew to its close, such good news was indeed welcome, for all hope of a quick victory had faded from even the most sanguine heart. From the Channel to the Swiss frontier the armies faced each other, locked in the pattern of trench warfare whose grim grip was to hold them fast for four long years. By the end of December, on the Western Front alone, 95,654 British soldiers of all ranks had been killed.[9] And this was a mere tithe of the grisly harvest other years held in store.

CHAPTER NINE

The Dardanelles

AS THE RIGIDITY AND HORROR OF THE TRENCH-BOUND 'SLOGGING MATCH' established itself on the Western Front, the War Council cast about for some alternative and more hopeful strategy than 'sending our armies to chew barbed wire in Flanders',[1] and a division of opinion developed between those – the 'Westerners' – who believed the war could be more swiftly won by a diversionary thrust across the North Sea and into the Baltic, and the 'Easterners', who believed in a breakthrough from the Eastern Mediterranean.

One of the Easterners' stratagems, the forcing of the Dardanelles Straits (between the Turkish mainland and the Gallipoli Peninsula), with the capture of Constantinople as its objective, soon gained promoters. Although Churchill had earlier on been a Westerner, the alternative strategy soon captured and held his attention and fired his imagination. And the particular relevance of this plan to the needs of the moment was vividly underlined when on 5 November 1914 Turkey entered the war as Germany's ally.

The original Dardanelles project involved the commitment of a considerable military force to take, and to hold, the Gallipoli Peninsula, as well as the naval forces required to force the Straits, pushing through into the Sea of Marmara, where Constantinople, which held the key to many doors, would lie within the expedition's grasp.

The search for an alternative strategy had continued all through the winter months of 1914, and at least four possibilities were seriously considered. Lord Kitchener having told the War Council that no troops could be made available for a Mediterranean operation from the western theatre of war, Churchill, as 1914 ended, was pressing for the Borkum plan (a scheme to invade Schleswig-Holstein from the island of Borkum off the Dutch coast). Then at the turn of the year came a desperate appeal to the Allies from the Russian Grand Duke Nicholas for some major demonstration to be made against the Turks, who were pressing the Russians hard in the Caucasus. The way to help them swiftly and effectively was by attacking the Turks through the Dardanelles.

Meanwhile, a vitally important change had taken place in the considerations governing the Dardanelles plan. Admiral Carden (the Commander in the Eastern Mediterranean) had given it as his considered professional opinion that the Dardanelles could be forced by ships alone, and on 12 January 1915 his detailed scheme was received at the Admiralty.

The strategic situation had thus been transformed. When Churchill eloquently presented this scheme to his weary colleagues the following day, Sir Maurice Hankey (Secretary to the War Council, later Lord Hankey) recalled how 'The whole atmosphere changed. Fatigue was forgotten. The War Council turned eagerly from the dreary vista of a "slogging match" on the Western Front to brighter prospects, as they seemed, in the Mediterranean.'[2] The Council resolved that the Admiralty should forthwith prepare for a naval expedition in February to bombard and take the Gallipoli Peninsula, with Constantinople as its objective.

One of Churchill's first actions on going to the Admiralty had been to persuade Admiral of the Fleet Lord Fisher, then aged seventy, to abandon his retirement to act (at first) as his unofficial adviser on naval matters. Lord Fisher had always been a controversial character; but his single-minded devotion to the Navy, and his undoubted brilliance, had placed him in a special position of power and influence. By 1910, when he retired after six years as First Sea Lord, he had become a legend – admired by many, and detested by not a few; for, to use his own words, he was 'ruthless, relentless and remorseless' in regard to anyone who opposed or gainsaid him.[3]

Churchill had first met Admiral Fisher in 1907, and over the succeeding years had come to know and esteem this strange, volcanic character. For three years after he was appointed to the Admiralty, the youthful First Lord and the veteran Admiral worked in the closest co-operation and warm friendship, and Churchill had shown unusual patience and tact in handling his eccentric but invaluable colleague. When at the end of October 1914 the First Sea Lord, Prince Louis of Battenberg, was forced to resign as the result of a shameful campaign of vilification based on his German family origins, Churchill pressed for Lord Fisher to be his successor: the appointment was duly made. It was to prove to be of ill omen to Churchill himself.

From the time the Dardanelles project was first mooted, Clementine knew about it, and became well acquainted with the arguments which surrounded the concept, and the rival schemes which were propounded. Winston then, as later, had complete confidence in her discretion, and to her he confided his hopes, his fears and his frustrations. She knew how deeply he believed in the strategic plan for the forcing of the Dardanelles, a belief he never abandoned – neither in the overturn of his most sanguine hopes and careful plans, nor afterwards, in the shambles of his own fortunes.

Many errors of judgement, both by the commanders on the spot and by the War Council in London, contributed to the failure of Churchill's grand strategy. Since an immediate success was expected after the first naval assault, there was no prior agreement on the number of men and ships that might ultimately be committed in the Eastern Mediterranean. And when there was no easy victory, the field was open to those whose opposition to the plan Churchill had so eloquently persuaded away – most particularly Lord Fisher, who from the beginning had blown hot then cold, and whose changing views had swung during the weeks of discussion and preparation from violent opposition to wild enthusiasm, and back again to grumbling disapproval.

These dramatic tergiversations in the old Admiral's views caused Clementine increasing uneasiness. Winston had told her that although Lord Fisher's original preference had been for the Borkum plan, he had been won over, and had thrown himself enthusiastically into the preparations for the Dardanelles operation. Nevertheless, Clementine was anxious. She knew how changeable and chancy 'Fisher weather' could be, and she realized that the elder man's real feelings and instincts had been overruled by Winston's passionate enthusiasm and by his capacity to put a case with irresistible force.

It had meanwhile been decided that a military force should be sent to the Dardanelles, and on 25 February the King reviewed the Royal Naval Division (who had received their baptism of fire at the siege of Antwerp) before their embarkation for Gallipoli. The review was held on the downs above Blandford Camp in Dorset. Clementine went with Winston; Margot Asquith and Violet were also there, for Violet's third brother Arthur ('Oc') was among those on parade. It was a day of brilliant sunshine sparkling with frost; Clementine and Violet were on horseback, and before the King's arrival they cantered along the lines of the battalions drawn up on the sweeping downland. It was a stirring sight, and although hopes were high, it must have had a poignant touch, for among those 'standing like rocks before their men'[4] were faces dear and familiar to Clementine and Violet, and to many of those who had gathered to watch the review. They were a fair pattern of the generation of gifted young men, so many of whom the war was to sweep away.

From the outset the Dardanelles force was ill-equipped; it was discovered at the last moment that the Hood Battalion was without sufficient medical personnel or drugs and, Asquith wrote to Venetia, 'Clemmie showed a good deal of resource, with the result that they will pick up some necessary "details" at Malta.'[5]

Two weeks later there was another farewell, when Winston and Clementine went to Charing Cross to bid Godspeed to Major-General Sir Ian Hamilton, who was leaving with his staff to take command at the seat of operations. Among his officers was Jack Churchill (now a major). The little band was full of hope and confidence in the enterprise. They did not guess – happily, could not know – the disasters that lay ahead.

During these anxious, fateful months of the spring of 1915 even the nursery world at Admiralty House, cocooned and sheltered in its upstairs existence, was aware of some threatening cloud which seemed to brood over the grown-ups' life. Hanging round the dining-room table, or crawling on the drawing-room floor, the children heard snippets of conversation. Diana remembered clearly sensing the atmosphere of anxiety, and every night she ended her prayers with an earnest plea: 'God bless the Dardanelles' – whatever they might be.

On 19 February the bombardment and reduction of the outer defences of the Dardanelles began, and although operations had to be suspended owing to bad weather, they were resumed a few days later, and were progressing well. But as time went on, Churchill became concerned that Admiral Carden did not seem to be pressing the attack with sufficient vigour. The Admiral was in fact ill, and in the middle of March he had to resign his command; his place was taken by his second-in-command, Admiral J. M. de Robeck.

On 18 March, British and French ships entered the Straits and bombarded the Turkish forts. It is now known that the ammunition in these forts was exhausted in this onslaught, and that further supplies were not at hand. Unfortunately, two British battleships (expendable, inasmuch as they were both out of date and due to be scrapped) and a French ship were sunk by mines. But instead of driving on, Admiral de Robeck, after consultation with Sir Ian Hamilton, called off the naval attack. It was never to be renewed.

The commanders at the scene of these events decided to withdraw and to prepare an amphibious landing with reinforcements on the Gallipoli Peninsula. It was not until 25 April that the landings were made by British and Anzac (Australian and New Zealand) troops. In the five weeks' interval between the main attack on the Straits and the amphibious assault on the Peninsula, the enemy had not been idle, and the landings were made in the face of blistering fire from the Turks. At the cost of heavy loss of life, the invaders gained a precarious foothold along the southern tip of the Peninsula. The Turks held the higher ground, and no further advance was possible; our forces dug themselves lines of trenches – it was a touch of bitter irony that the same hideous pattern as on the Western Front was repeating itself.

Churchill was consternated and did his utmost to get orders sent for the resumption of the attack, but there was general reluctance both at the Admiralty and at Downing Street to overrule the 'man on the spot'. Lord Fisher obstructed all Churchill's efforts to send naval reinforcements, being determined that the British Navy should not be endangered in the Dardanelles. Churchill was to write later: 'The "No" principle had become established in men's minds, and nothing could ever eradicate it.'[6]

Events at home now started to move towards crisis point. Gallipoli was not the only source of disturbing news; in France the Second Battle of Ypres had begun on 22 April. During the course of it the Germans used poison gas for the first time, adding a new and terrifying element to life at the Front. Sir John French complained of a shortage of shells; the shell 'scandal' was thereupon taken up by Lord Northcliffe, who fanned the controversy in his newspapers. Lord Kitchener, the hero of Omdurman and of the Boer War, hitherto beyond criticism in the eyes of the British public, was the principal target of his attacks, but blame and discredit fell upon the whole Liberal administration. There was a political truce in force, but many Conservative backbenchers were restive and critical; and above all, their hostility was directed upon that renegade Tory – Winston Churchill. Every naval disaster was laid at his door. They had blamed him for his part in the siege of Antwerp, and now, as the news made clear bit by bit the miscarrying of the Dardanelles plan, the volume of criticism directed personally against him increased every hour.

It was at this moment of rising political tension at home and bad news from abroad that Lord Fisher abruptly resigned as First Sea Lord.

In these last few months Churchill's relationship with Fisher had been particularly difficult, indeed explosive. Soon after the go-ahead had been given to the Dardanelles operation, rumblings of dissent and dissatisfaction began to emanate from the First Sea Lord; but he never voiced his objections in open Council. The disagreements between the two men were not confined to the Dardanelles, and on no fewer than seven occasions Lord Fisher had resigned (or threatened to resign) on other issues; but he had always, in the end, been persuaded to stay.

Clementine's doubts and anxieties about the old Admiral had meanwhile increased. She noticed that whenever Winston was away from the Admiralty for a few days, the First Sea Lord seemed profoundly uneasy at being 'in charge of the shop'. Despite his frequently expressed contempt for politicians, he showed every sign of nervousness when the moment-to-moment responsibility was his. Furthermore, a curious incident took place, which left a strong and unfavourable impression upon her.

Sometimes, when Winston was going to be away, he would say to Clementine, 'Just look after "the old boy" for me.' And so she would invite Lord Fisher to have luncheon with her from time to time during Winston's absences.

Early in May 1915, Winston went to Paris to take part in the delicate negotiations to bring Italy into the war. While he was away, Clementine invited Lord Fisher to luncheon, which passed very agreeably, and he took his leave of her in cheerful mood. A short while later she herself left the sitting-room and, to her astonishment, found the old Admiral lurking in the passage. She asked him what he wanted, whereupon, in a brusque and somewhat incoherent manner, he told her that, while she no doubt was under the impression that Winston was conferring with Sir John French, he was in fact frolicking with a mistress in Paris! Clementine was much taken aback, but treated this interesting piece of information with the scorn it deserved. 'Be quiet, you silly old man,' she said, 'and get out.' He went.

Clementine, of course, told Winston about this extraordinary episode, and of her deep-seated doubts of Lord Fisher's mental steadiness and loyalty – but he just brushed them aside. He was fully confident of his firm relationship with the Admiral and oblivious to anything except that agreement between them had always been reached.

But on 14 May the final outburst occurred. At a meeting of the War Council to discuss further reinforcements for Gallipoli, Lord Fisher's long-standing, but hitherto suppressed, opposition to the campaign exploded. Interrupting the Council's proceedings, he announced that he 'had been against the Dardanelles operations from the beginning, and that the Prime Minister and Lord Kitchener knew this fact well'.[7] This statement was received in silence. Outraged, Churchill wrote in a letter to the Prime Minister on the same day: 'The First Sea Lord has agreed in writing to every executive telegram on which the operations have been conducted.'[8]

Still, so accustomed was Churchill to Lord Fisher's erratic behaviour that when he saw him that same evening at the Admiralty, the two men had a long talk in which they came to an agreement on all outstanding points, and parted amicably. Early the next morning, 15 May, however, Churchill received a letter from the Admiral containing his resignation. Fisher's final and fatal decision was probably triggered by Churchill's tactless codicil to a late-night instruction: 'First Lord to see after action.'[9] Churchill immediately went to the Prime Minister, who had received a similar missive. Asquith was furious, and wrote out an order in the name of the King to Lord Fisher to return at once to his post. Winston wrote him a moving personal appeal – but it was all to no avail: Fisher steadfastly refused to work any longer with Churchill; the end of their road together had been reached. It was a stunning blow, but Churchill did not waste time lamenting, and speedily reconstituted his Board of Admiralty, the other Sea Lords remaining firm and steady. On the Sunday afternoon of this most harassing weekend, Winston and Clementine drove down to The Wharf, bearing his proposals for the new Board of Admiralty, which seemed to have Asquith's approval. Violet Bonham Carter wrote that 'Clemmie was naturally very upset and as ever very brave. We had quite a happy dinner and they both drove away in better spirits.'[10]

But the crisis was not over – it was only just beginning. Moreover, though none of them knew it at the time, Mr Asquith himself was reeling from a shattering personal shock. On 11 May, Venetia Stanley had written to tell him that she intended to marry Edwin Montagu*. 'This is too terrible; no hell could be so bad,' he had written to her a few days later.[11] At a moment, therefore, of deep personal distress, he was, as Prime Minister, confronted with a grave and complex crisis.

Early on the morning of Monday, 17 May, Bonar Law, the Leader of the Opposition, called on Lloyd George at No. 11 Downing Street to seek confirmation that Fisher had in fact resigned. He made it clear that a combination of the 'shell scandal' with this latest development would make it impossible for him to restrain his followers; that if Fisher resigned, Churchill could not stay as First Lord of the Admiralty; and that the Opposition was prepared to make the issue the subject of a full-scale parliamentary challenge.

Lloyd George begged Bonar Law to wait, and went through the

* The second son of the 1st Lord Swaythling. At this time Financial Secretary to the Treasury. The marriage took place in July 1915.

communicating door to see Asquith in No. 10. During his talk with the Prime Minister, Lloyd George argued forcefully in favour of a Coalition Government, and the Prime Minister acquiesced in his view. Lloyd George returned to No. 11, and brought Bonar Law back to join the discussion. Within a quarter of an hour the idea of a coalition grew from an idea to a possibility, and from a possibility to a probability. The price to be paid was the removal of Churchill from the Admiralty and Haldane from the Government (this last demand arising from a despicable smear campaign based on the Lord Chancellor's lifelong study of German thought and letters).

Churchill had always strongly supported the idea of a Coalition Government, but he hoped that such a large question could wait until he had dealt with the immediate crisis and had announced his new Board of Admiralty. But when he saw the Prime Minister in the afternoon he was consternated to learn that in the course of the wide reconstruction involved in the formation of a Coalition Government, he would have to leave the Admiralty. In the midst of this conversation with Asquith, which became at moments vehement, a message arrived from the Admiralty asking Churchill to return there at once – the German Grand Fleet was putting to sea.

This news put all considerations of the political crisis, and of his own fate which was so closely involved with it, to the back of Churchill's mind. Returning at once to the Admiralty and gathering around him the senior naval staff, he prepared to supervise what promised to be a major encounter between the two great naval powers. During the night, as he watched the development of this dramatic situation, Churchill received a letter from the Prime Minister (similar to those sent to all ministers) announcing to him Asquith's intention to form a National administration. Replying to his chief, Churchill wrote that he would gladly serve in the new Government, but would accept only a military department, and, if that were not convenient, would prefer to be found a post in the field.

During the morning of Tuesday, 18 May, it became clear from signals that the German fleet was turning for home; the crisis at sea was resolved.

For a week more there followed a period of behind-the-scenes activity and tension, while the mandarins of the political parties conferred, attended by a buzz of intrigue, public speculation and gossip. Churchill made no secret of the fact that he earnestly hoped to stay at the Admiralty, to see through much of what he had initiated, and above all to have a powerful say in the matter of the Dardanelles campaign, for which he bore so much responsibility. But, as the week wore on, it became clear that too many voices in public, in the press, and in private, in high governmental and administrative places, were against him. On Friday, 21 May, in answer to an emotional, deep-felt written appeal from Churchill to be allowed to stay at his post, the Prime Minister replied kindly, but briefly, that he 'must take it as settled that you are not to remain at the Admiralty'.[12] Churchill bowed with stoicism to this, for him, bitter decision. The next day Mr Asquith saw him briefly, writing afterwards to Venetia Stanley that it was 'a most painful interview to me: but he was good and in his best mood'.[13] Churchill drew some consolation from the fact that his successor was to be Balfour, one of his few friends among the Tories, and one who, as a member of the War Council, had always supported the plan for the Dardanelles.

Of his own feelings at this bitter hour, Churchill would write: 'The more serious physical wounds are often surprisingly endurable at the moment they are received. There is an interval of uncertain length before sensation is renewed. The shock numbs but does not paralyse: the wound bleeds but does not smart. So it is with the great reverses and losses of life.'[14]

Many years later, looking back over her life with Winston, Clementine was to say that, of all the events they had lived through together, none had been so agonizing as the drama of the Dardanelles. During these days of 1915, and for long weeks and months thereafter, she shared with Winston every anxiety, every brief hope, every twist and turn of the devious course of the crisis. Half a century later, Clementine told Martin Gilbert, Winston's biographer: 'I thought he would die of grief.'[15]

Clementine was always at her best when things were going badly, and we may be sure she rose to the level of these dire events. But as she watched them, as a helpless spectator, bitterness and anger welled up in her heart. Winston's sense of his betrayal was slower in developing: utterly loyal himself, and always courageous in the defence of his own friends or colleagues, it took him some time to grasp that the Prime Minister, who had supported and approved the Dardanelles plan all along, had sacrificed him to Tory prejudice and hatred with scarcely a struggle. Asquith's abandonment of Haldane was even more chilling, for the two men had been close and lifelong friends: but he let him go without a word.[16]

Unable to contain her feelings, Clementine wrote a remarkable letter to the Prime Minister on 20 May 1915. Her love for Winston, and a burning sense that he was being shabbily treated, made her both brave and eloquent:

> My dear Mr Asquith,
>
> For nearly four years Winston has worked to master every detail of naval science. There is no man in this country who possesses equal knowledge capacity & vigour. If he goes, the injury to Admiralty business will not be reparable for many months – if indeed it is ever made good during the war.
>
> Why do you part with Winston? unless indeed you have lost confidence in his work and ability?
>
> But I know that cannot be the reason. Is not the reason expediency – 'to restore public confidence'. I suggest to you that public confidence will be restored in Germany by Winston's downfall.
>
> There is no general desire here for a change, but it certainly is being fostered by the press who have apparently made up their minds. I trust they are not making up yours for you.
>
> All you have to do is stand by Winston and the Board of Admiralty, and [Admiral of the Fleet] Sir Arthur Wilson.
>
> If you throw Winston overboard you will be committing an act of

> weakness and your Coalition Government will not be as formidable a War machine as the present Government.
>
> Winston may in your eyes & in those with whom he has to work have faults but he has the supreme quality which I venture to say very few of your present or future Cabinet possess – the power, the imagination, the deadliness to fight Germany.
>
> If you send him to another place he will no longer be fighting – If you waste this valuable War material you will be doing an injury to this Country.
>
> Yours sincerely,
> Clementine S. Churchill[17]

The question remains: did Winston either inspire the letter or see it before it was sent? At No. 10 views differed: Maurice Bonham Carter (Asquith's Private Secretary), writing on 22 May to Violet (who had departed for Egypt on the 20th) about Clementine's letter, felt Winston 'could not have perpetrated such a bêtise, & I now learn that he did not inspire it, though he allowed it to go (perhaps without learning its contents)'.[18]

Margot Asquith was enraged. In her diary for 20 May she expressed herself with typical immoderation:

> When Henry was up he showed me the enclosed amazing letter from Clemmie Churchill.
>
> It shows the soul of a servant – that touch of blackmail & insolence & the revelation of black ingratitude & want of affection justifying everything I have thought of this shallow couple. I expressed myself with vigour to Henry.
>
> H: after all its the letter of a wife
>
> M: – a fish-wife you mean!! but of course this is Winston not Clemmie – you don't mean to say you think W. didn't see this letter?
>
> H: Yes I rather agree with you.

Margot reflected that 'Clemmie will be terribly sorry for this later on.' Lloyd George appeared and was shown the letter: 'He was amazed but saw Winston in it – I didn't think his soul was as much stirred by indignation to the degree it might have been.'[19]

Asquith never answered the letter. Clementine was later told that on several occasions he read her outburst aloud at the luncheon table, with amused relish. To Venetia Stanley he commented that 'she wrote me the letter of a maniac.'[20]

There is a poignant description of Clementine at this time in a letter to Venetia Stanley from Edwin Montagu (to whom she was now engaged). Montagu had called at Admiralty House on 26 May. 'She was so sweet but so miserable and crying all the time. I was very inarticulate, but how I feel for her and him.'[21] Later he wrote Clementine a moving and prophetic letter:

My dear Mrs Winston,

My heart bled to see you so unhappy and I came back from your house to write a line in the hope of atoning for my lack of capacity to express myself verbally.

It is a hard time and it is true that Winston has suffered a blow to prestige, reputation and happiness which counts above all. All that is not worth arguing.

But it is also indisputably true that Winston is far too great to be more than pulled up for a period. His courage is enormous, his genius understood even by his enemies and I am as confident that he will rise again as I am that the sun will rise tomorrow . . .

Be as miserable as you must about the present; have no misgivings as to the future; I have none, Winston I am sure has none and I know that in your heart and amid your gloom you have undaunted confidence in the man you love.[22]

In the Coalition Government announced on 26 May Winston was made Chancellor of the Duchy of Lancaster – a sinecure with no executive power (other than to appoint magistrates for the county of Lancashire), but in this case coupled with a place in the Cabinet and a seat on the War Council; this proviso alone made the offer of such an otherwise derisory post acceptable to Winston, for he felt he would still be able to influence the course of the Dardanelles operation.

That morning, as he left the Admiralty in gloom and despair, he could not guess that twenty-four years later, when Britain once more was in mortal peril, the message would be flashed to the Navy: 'Winston is back.'

Despite hard feelings, social amenities were resumed and the Churchills were bidden to dine with the Asquiths on 8 June. Margot's diary for the 9th describes an extraordinary scene.

Hearing that Clemmie Churchill was behaving like a lunatic & crying daily over Winston being turned out of the Admiralty I asked her if she wd. come to tea before dining with us yesterday. She answered pleasantly on the telephone & arrived looking cool & handsome in a muslin dress – She has lost her looks since her last baby [Sarah, born October 1914] and she is fatter wh. has taken away from her refinements & given prominence to her Xpressionless protruding eyes but she is quand même very good looking . . . My heart bled for her & I put my arms round her with a little squeeze 'Darling I've been thinking & feeling a lot for you.' She pushed me resolutely back & put up her veil – I saw I was in for it and scanned a very hard insolent young woman with little or no sense of humour . . . frivolous, bad-tempered <u>ungrateful</u> & common au fond . . .

After a half page of abuse of 'these sort of women', Margot resumed her account of the interview, which became increasingly heated, and in which the new Coalition, Lord Northcliffe and his newspapers, Lloyd George and the Prime Minister were all tossed around. Eventually Clementine said, ' "I think I had better go!!" '

> M: Sit down Clemmie & calm yourself. I'm very very sad & loathe coalition – don't think only of yourself & Winston. Think of poor Henry & all he has gone through.
> C: The P.M. !! That's good!! Why he has thrown his dearest friend & his most remarkable colleague, Haldane & Winston to the wolves!
> She got up held on to the tea table & harangued me in fish-wife style on Henry's defects till I got up & sat at my writing table quite calmly while she screamed on . . . till I stopped her & said 'Go Clemmie – leave the room – you are off your head – I had hoped to have a very different kind of talk' –
> C: I haven't come here to be forgiven over my letter I can tell you!
> M: Your letter was terrible Clemmie, if I had written it to so dear & old a friend as Henry I wd. never forgive myself to my dying day, & your letter of apology* was not at all nice. You are a hard little thing & very very foolish as you will do Winston harm in his career.

Finally, after a few more exchanges, Clementine

> walked to the door – Goodbye Margot
> M: looking at her quite calmly – Goodbye darling.[23]

Almost incredibly, only a few hours later they met for dinner! In a letter to Sylvia Henley†, Asquith wrote:

> 10 Downing St
> Wed. midnight 9 June 1915
> . . . Clemmie came here before dinner, & I gather from Margot's account they had a very bad & almost Billingsgate half-hour. She & Winston came to dinner (with a lot of others . . .) and before she left I inveigled her here into the Cabinet room, and spoke very faithfully [*sic*] to her. In the end, we parted on good & even affectionate terms: and I trust I have dispelled her rather hysterical mutiny against the Coalition and all its works.[24]

* * *

A mundane consideration, but nevertheless a very real one, was that in going to the Duchy of Lancaster Winston saw his ministerial salary cut by more than half. Balfour thoughtfully assured the Churchills that they need be in no haste to move out of Admiralty House. But Clementine could not wait

* No letter of apology has been found in the Asquith Papers.
† After Venetia's engagement, Asquith turned immediately to Sylvia Henley, her elder sister, as a confidante, and corresponded with her frequently until 1919; so much so that by March 1916 Lady Cynthia Asquith would refer to her in her diary as HHA's 'leading lady' (Cynthia Asquith, *Diaries 1915–1918*, p. 146, entry for 25 March 1916).

to be gone from a scene which was now so painful to her. Their own house in Eccleston Square was still let to Sir Edward Grey, but Ivor Guest (now Lord Wimborne) offered the homeless pair his house in Arlington Street, to which they moved in the middle of June. Here Winston and Clementine paused for a short space while they made more permanent plans.

These were hard, bleak days for both of them. The sting of public obloquy was added to their feelings of humiliation and sense of personal betrayal. Clementine had faced social ostracism and public criticism of Winston from the earliest days of their marriage – indeed, she had gloried in them. But this was different. There was the haunting knowledge that lives had been lost, and that many more must yet be sacrificed before the end of the story. This was an anguishing burden for Winston to sustain. So much could not be told now; so many voices that later would speak out in defence of the Dardanelles plan, and of Churchill's role in it, were at this fateful time perforce silent. Only the critics were unbridled, and the continual lash of criticism, much of it envenomed and ill-informed, had to be borne in silence.

As the weeks went by, their feelings of bitterness grew towards those who, in their eyes, had betrayed Winston and, above all, the 'plan', at the moment of greatest trial. Nevertheless, Winston was still a member of the Government, although his colleagues now included men who he knew disliked and mistrusted him, and the chief he still served with loyalty and friendship had, in the event, proved supine in his defence.

Whatever passed between Winston and Clementine in these desolate days and nights, to the world they displayed a proud, courageous and dignified aspect. But Winston's state of mind is painfully revealed in a letter to Sir Archibald Sinclair*, a young Scottish baronet and a regular officer in the Life Guards, who had recently become a friend – and would remain one for the rest of his life. Archie was out in France, and had evidently written to Winston, who replied on 9 June 1915:

> It was vy nice to get your letter. One must try to bear misfortune with a smile . . . But the hour is bitter: & idleness – torture. Here I am in a fat sinecure vy well received & treated by the new Cabinet who have adopted my policy & taken all the steps that I was pressing for & more . . . I thought it my duty to stand by the National government at the outset, & especially until the grave decisions about the Dardanelles had been taken. But now all that is settled and I cannot endure sitting here to wait for a turn of the political wind. I do not want office, but only war direction: that perhaps never again. Everything else – not that. At least so I feel in my evil moments. Those who take to the sword . . .[25]

It was a sore trial enduring the curiosity or even the sympathy of strangers, but the outside world had to be faced, and Clementine summoned

* Later 1st Viscount Thurso; Leader of the Liberal Party 1935–45; Secretary of State for Air, 1940–5.

up all her pride and fortitude to meet this ordeal. On 30 June she honoured an engagement to open a new wing at her old school, at Berkhamsted. Her former Headmistress, Miss Harris, still reigned, and received her with kindness. In her speech to the girls, Clementine spoke of the many women whom the war had bereaved, but some of her words had a poignantly personal ring: 'At this time I think one of the most wonderful things is the stoic fortitude with which women bear the most agonising sorrows . . . I think people are able to put aside their private griefs because the thoughts and energies of everyone are concentrated on one idea – how to win the war: how to make victory sure and safe in the end.'

After due thought and consultation within the family, the 'Winstons' and the 'Jacks' decided to pool their resources. Both families were under financial strain, and so it seemed a good arrangement all round. Jack and Goonie were living at No. 41 Cromwell Road, in South Kensington, but it was really too big for them, particularly now that Jack was serving abroad. It was quite a task amalgamating the two households, and especially combining the nurseries under the command of one nanny, who had the charge of the five cousins: three boys (John George, aged six, Randolph, four, and Peregrine, two) and two girls (Diana, now six, and the baby, Sarah). Winston and Clementine moved with their brood into Cromwell Road at the end of June.

* * *

One of the first actions of the new Coalition was to endorse the vigorous prosecution of the Dardanelles campaign. As a member of the War Council (now reconstituted and known as the Dardanelles Committee), Winston did all in his power to sustain and further the plan, in which he still wholeheartedly believed. But with no great department to administer, he suddenly found time on his hands; and it hung sad and heavy. Years later he would describe his feelings: 'The change from the intense executive activities of each day's work at the Admiralty to the narrowly measured duties of a counsellor left me gasping. Like a sea-beast fished up from the depths, or a diver too suddenly hoisted, my veins threatened to burst from the fall in pressure.'[26] It was during these months of gloom and despair that Winston first discovered the charm of painting.

Winston and Clementine had rented for the summer, as a weekend and holiday refuge, a small house, Hoe Farm, near Godalming. One day, at the end of May, wandering, brooding, in the garden, Winston came upon Goonie sketching with watercolours; watching her, he became fascinated, demanded a brush, and tried his hand. Goonie, delighted with this development, lent him John George's paintbox. After experimenting briefly with watercolours, however, Winston became convinced that 'La peinture à l'huile est bien difficile, mais c'est beaucoup plus beau que la peinture à l'eau.'[27] Clementine was so thrilled that Winston had found a distraction from his worries that she rushed off to Godalming and bought every variety of oil paints available. Alas, she was not aware that turpentine is essential to painting with oils, and the ensuing results were most unsatisfactory!

But help was at hand: John Lavery, the celebrated painter, and his beautiful wife, Hazel, lived round the corner from the Churchills in Cromwell Place, and they had made friends. Clementine, in a quandary with the oil paints, telephoned John Lavery, who was so overjoyed to hear of Winston's new ploy that he leapt into a hired car and drove immediately to Hoe Farm, bringing not only turpentine, but all his knowledge and skill. Hazel Lavery, too, was a gifted painter, and between them they gave Winston his first lessons in painting, an occupation of which he was never to tire, and a companion who kept him company almost to the end of the road.

It was during this summer that Sir William Orpen painted a remarkable full-length portrait of Winston. At the time Clementine did not care for the picture – it showed all too painfully the ravages wrought upon Winston by these days of crisis and personal torment. She minded its almost 'beaten and finished' look, but with the passage of time she came to think it was the truest portrait ever painted of him. That it was a truthful portrayal is borne out by Ellis Ashmead-Bartlett, the war correspondent*, who, returning home from Gallipoli in the early summer of 1915, noted in his diary: 'I am much surprised at the change in Winston Churchill. He looks years older, his face is pale, he seems very depressed and to feel keenly his retirement from the Admiralty.'[28]

* * *

During these summer months social relations with Downing Street were somewhat fragile. Winston naturally saw the Prime Minister often in the course of business, but Margot and Clementine were frequently at odds. Asquith, who hated any unpleasantness, was quite wary of Clementine's sharp tongue, and although he had made fun of the letter she had written him at the height of the crisis, he knew she had to be reckoned with; no doubt he was quite glad if Margot could stand between them. Later in August he was to report to Sylvia Henley: 'I sat next Clementine Churchill who was quite amiable, and showed neither teeth nor claws.'[29]

Among the many vexations and frustrations which were Churchill's lot at this time, none embittered him more than the occasion when in July 1915 he was prevented, on the very eve of his departure, from visiting Gallipoli. It was Kitchener who had suggested that he should make an official visit, and Asquith and Balfour both gave their approval. When, however, other Tory colleagues got wind of the proposal, they opposed it tooth and nail. Asquith bowed before the storm, and Churchill's proposed journey was cancelled. It was probably with this expedition in view that Winston wrote a letter which was found among his papers, dated 17 July 1915, and marked

*Son of the Sir Ellis Ashmead-Bartlett who had been one of Lady Blanche's lovers.

'To be sent to Mrs Churchill in the event of my death'. In it he outlined the financial provisions he had been able to make for her and the children – they would have been dismally slender. The letter shows what deep faith he had in her judgement and resolution, for he appointed her his sole literary executor, with the request that she should get hold of all his papers, especially those dealing with his work at the Admiralty.

> There is no hurry; but some day I shld like the truth to be known. Randolph will carry on the lamp. Do not grieve for me too much. I am a spirit confident of my rights. Death is only an incident, & not the most important wh happens to us in this state of being. On the whole, especially since I met you my darling one I have been happy, & you have taught me how noble a woman's heart can be. If there is anywhere else I shall be on the look out for you. Meanwhile look forward, feel free, rejoice in life, cherish the children, guard my memory. God bless you.
>
> Good-bye
> W

With Winston's departure from the Admiralty, Clementine was relieved of the task of running a large official house, and of the obligation to entertain extensively. As soon as she had reorganized her family and household, she looked around for some form of war work. As a result of the 'shell scandal', which had contributed to the downfall of the Liberal Government, a Ministry of Munitions had been set up, with Lloyd George at its head. New factories were started, and many hitherto occupied with peacetime production were now turned over to making munitions; and a vast civilian army of men and women was daily growing in numbers to produce arms. From the outbreak of the war a variety of organizations had devoted themselves to sustaining the fighting soldier and sailor, but now the needs of the civilian war workers were also pressing. The arms factories worked day and night, and for the most part the canteen arrangements organized by the managements were totally inadequate. In June 1915, therefore, the Young Men's Christian Association (YMCA) formed the Munition Workers' Auxiliary Committee, for the express purpose of providing canteens for the workers. Clementine was invited to join this committee, and undertook the task of organizing canteens in the north and north-eastern metropolitan area of London. During the ensuing months she became responsible for the opening, staffing and running of nine canteens, each one feeding anything up to five hundred workers at a time.

Like many other women who undertook quite large administrative jobs at this period, Clementine had no previous experience or training; none the less, she plunged into the work, undaunted and full of enthusiasm. Her new task was a challenge; and, more than that, for her at this moment it was an anodyne against the bitter thoughts and gnawing anxieties which beset her.

Driving herself, she toured her area, enlisting helpers and persuading factory managers to give the necessary co-operation and provide suitable accommodation. Once on its feet, each canteen had to be visited often, to encourage the staff (90 per cent of whom were unpaid volunteers), to deal with problems and complaints, and to ensure high standards of food and service.

Such a novelty was it for women to be employed on a large scale, side by side with men in industry, that one of the first difficulties Clementine had to deal with was a general complaint from the men that they objected to the women workers smoking in the canteens and recreation rooms! A special committee meeting had to be called to consider this point. After due deliberation, a ruling was given that, where and when smoking was allowed, women should be on a par with their male colleagues. This dictum was rather reluctantly accepted by the men, and thus another barrier in the sex war was demolished.

As the summer of 1915 wore on, the ill-starred campaign in Gallipoli ran into further difficulties. At home, the new Coalition was proving to be a worse war machine than the Government it had superseded. Within the Cabinet there were animosities and tensions which vitiated its power of decision, and the enlarged Dardanelles Committee was too big for effective action. Early in August British forces launched an attack at Suvla Bay: it was a disastrous failure, whose repercussions extended beyond the blood-soaked beaches to Greece and the Balkan states; and in October 1915 Bulgaria threw in its lot with Germany and Austria. As the weeks went by and events moved from bad to worse, the idea of evacuation and abandonment was – inevitably – mooted. In November the Dardanelles Committee was reduced in numbers; the smaller body – the War Committee or Council – did not include Churchill.

For many months the thought of resignation had been in Churchill's mind. The honourable alternative of taking up his commission (he was a major in the Oxfordshire Yeomanry) and joining his regiment in France was always open to him. As the weeks had gone by, he had found himself in the invidious position of bearing responsibility without the power to act according to his deeply held convictions. Increasingly frustrated, he had sent a letter of resignation to the Prime Minister at the end of October; but Asquith had persuaded him to stay on. Now, however, as he found that (although continuing in the Cabinet) he was to be excluded from the newly constructed Dardanelles Committee, the one condition which had made tolerable to him his acceptance of the Chancellorship of the Duchy of Lancaster was removed.

On 11 November 1915 Churchill wrote to the Prime Minister, once more, and finally, tendering his resignation. He told him:

> I could not accept a position of general responsibility for war policy without any effective share in its guidance and control . . . Nor do I feel in times like these able to remain in well-paid inactivity. I therefore ask you to submit my resignation to the King. I am an officer, and I place myself unreservedly at the disposal of the military authorities, observing that my regiment is in France.

> I have a clear conscience, which enables me to bear my responsibility for past events with composure.
>
> With much respect and unaltered personal friendship, I bid you good-bye.

Four days later, on 15 November, Churchill made a statement to the House of Commons, as he was by custom entitled to do as a resigning minister. There was much he could not say – either for reasons of security, or through loyalty to friends and former colleagues.

He then went home to Cromwell Road to prepare for his departure for France.

CHAPTER TEN

The Place of Honour

THE DECISION TAKEN AND THE DEED DONE, WINSTON MADE HASTE TO BE gone. Sir Max Aitken (later 1st Baron Beaverbrook), who visited the house on the eve of his departure, found that

> The whole household was upside down while the soldier-statesman was buckling on his sword. Downstairs, Mr 'Eddie' Marsh, his faithful secretary, was in tears . . . Upstairs, Lady Randolph was in a state of despair at the idea of her brilliant son being relegated to the trenches. Mrs Churchill seemed to be the only person who remained calm, collected and efficient.[1]

Even had she felt inclined to do so, Clementine had little time to brood or repine, for there were only a few days to gather together all the equipment and uniform Winston would need in France.

On Tuesday, 16 November there was a farewell luncheon at No. 41 Cromwell Road. Violet Asquith and her stepmother Margot were present, and Eddie Marsh; otherwise it was the family – Winston, Clementine, Nellie Hozier and Goonie. In the circumstances it may seem surprising that any Asquiths should have been present. But Violet was a close friend and political confidante, and despite the fact that Winston and Margot Asquith never really got on together, one feels she could not bear to miss this dramatic moment – and perhaps just to keep a watchful eye? Violet described what must have been a somewhat dismal occasion: 'Clemmie was admirably calm and brave, poor Eddie blinking back his tears, the rest of us trying to "play up" and hide our leaden hearts. Winston alone was at his gayest and his best.'[2]

On Thursday, 18 November, 'Major Churchill' left for France. On arriving at Boulogne he found the Commander-in-Chief, Sir John French, had sent a car to meet him. After reporting briefly to his own regiment (the Queen's Own Oxfordshire Hussars) at Bléquin, Winston went to GHQ at St Omer, where Sir John welcomed him with the utmost cordiality. Winston stayed a night or two at GHQ while his immediate future as a soldier was decided upon.

Sir John offered Winston the alternatives of remaining at GHQ as an ADC or the command of a brigade. Winston unhesitatingly chose the latter, but with the proviso that he should gain some experience of trench warfare as a regimental officer before taking up his command: and so it was arranged. Winston wrote to Clementine: 'I am staying tonight at GHQ in a fine chateau, with hot baths, beds, champagne & all the conveniences . . . I am sure I am going to be entirely happy out here & at peace. I must try to win my way as a good & sincere soldier. But do not suppose I shall run any foolish risks or do anything wh is not obviously required.'[3]

The following day it was arranged that Winston should at once join a battalion of the Grenadier Guards in the line, and he wrote to Clementine at midnight on Friday, 19 November:

> My dearest soul – (this is what the gt d of Marlborough used to write from the low countries to his cat) All is vy well arranged . . . I cannot tell the rota in wh we shall go into the trenches. But I do hope you will realise what a vy harmless thing this is. To my surprise I learn they only have about 15 killed & wounded each day out of 8000 men exposed! It will make me vy sulky if I think you are allowing yourself to be made anxious by any risk like that.

He himself had no backward-looking thoughts:

> I am vy happy here. I did not know what release from care meant. It is a blessed peace. How I ever cd have wasted so many months in impotent misery, wh might have been spent in war, I cannot tell.

Clementine's first letter already showed how acutely she felt their separation: 'Altho' it's only a few miles you seem to me as far away as the stars, lost among a million khaki figures.' She ended on a note of almost pathetic yearning: 'Write to me Winston. I want a letter from you badly.'[4]

She did not have long to wait, for by Sunday she had received his first letter telling her of his intention to go into the line before accepting a brigade, and she lost no time in answering it:

> 41 Cromwell Road
>
> My Darling,
>
> I was very much relieved when your letter reached me yesterday morning, but the news in it makes me terribly anxious. I feel very proud of you my Dear.
>
> I long for you to have a Brigade & yet not too soon for fear of partly dimming the 'blaze of glory' in which you have left the country.
>
> Wherever I go I find people awestruck at your sacrifice . . .
>
> I have had a charming letter from Sir John French, telling me that he was going to look after you . . .
>
> Quite apart from the danger, I do wish you had not gone at once into the

'line' – I fear so much that you may get pneumonia or an internal chill unless you get gradually hardened.

Your loving lonely
Clemmie
[sketch of a cat][5]

On Sunday, 21 November Winston joined the 2nd Battalion of the Grenadier Guards, stationed at that time near Laventie and about to go into the line at Neuve Chapelle. His sense of history and romance was gratified by the coincidence that John Churchill, 1st Duke of Marlborough, had once served in, and commanded, the same unit. The officers were suspicious and even resentful of his presence, and he was received with reserve and coldness, the commanding officer, Lieutenant-Colonel Jeffreys, remarking: 'I think I

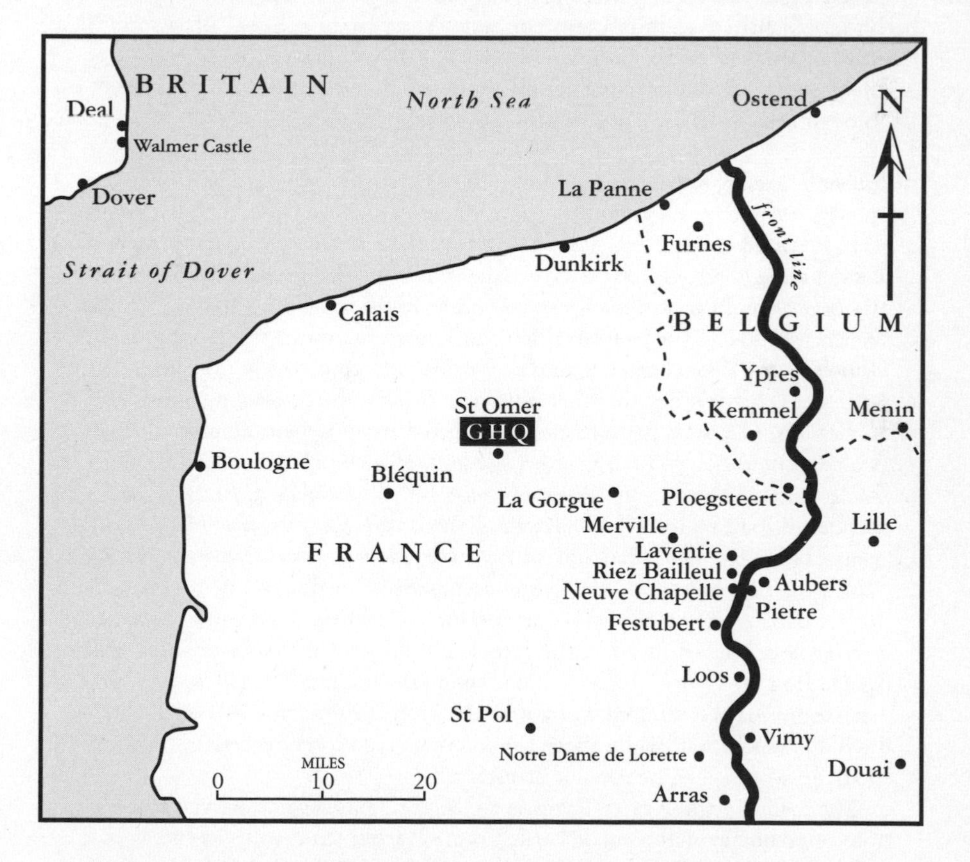

THE OSTEND-TO-ARRAS SECTOR OF THE WESTERN FRONT, NOVEMBER AND DECEMBER 1915

ought to tell you we were not at all consulted in the matter of your coming to join us.'[6] Winston knew that only time and his own bearing would break down their hostility.*

His equipment was evidently very deficient, for he sent Clementine a long list of urgent requirements, namely:

1. A warm brown leather waistcoat.
2. A pair of trench wading boots . . .
3. A periscope (most important) . . .
4. A sheepskin sleeping bag . . .
5. 2 pairs of khaki trousers . . .
6. 1 pair of brown buttoned boots –
7. Three small face towels

'Voila tout!' he added cheerfully at the end of this list. But one comfort he had with him was much appreciated: 'Your little pillow,' he wrote, 'is a boon & a pet.'

Clementine hastened to gather all these much-needed items, and wrote on 25 November (7.30 a.m.) to report progress:

> I went yesterday & sent you all the things you ask for in your letter from the trenches with the exception (Alas) of the trench wading boots. London seems to be emptied of these, but I am going to make a fresh try this morning & if I fail I shall send you pro tem, a pair of rubber waders which they say is the next best thing. I wake up in the night & think of you shivering in the trenches; it makes me so miserable (You know how warm the Kat has to be before she can sleep) I fear I should never sleep in a trench even in a sheep skin bag. Randolph wishes to come with me to choose the wading boots. He wishes to send you as personal presents from himself a photograph of himself & a spade. He says you must have a spade 'to dig out a little sideways, so if a bomb comes in the trench Papa won't be killed.' Don't you think it rather wonderful for a child of 4 to have thought of this. On Monday I dined with Venetia [Stanley, now Montagu] where everyone was thrilled at your having joined the Grenadiers . . . Nellie met a young soldier from a line regiment (a former partner) who said that it was common knowledge that you had refused a brigade & wished to go in the trenches – He said Everyone thought this splendid as the General Expectation among the rank & file of officers was that you would join your regiment for a week or a fortnight; that you would then be put on some staff while the regiment was given some interesting work & that you would then be given a Brigade . . .
>
> My darling I think of you constantly & I do hope that when you think of me, it is not a picture of a harsh arguing scold, but your loving & sad Clemmie. I love you very much more even than I thought I did – for seven

* For Winston Churchill's State Funeral on 30 January 1965, the Bearer Party would be found from the 2nd Battalion Grenadier Guards.

> years you have filled my whole life & now I feel more than half my life has vanished across the channel. I have cut out of the Daily Mirror a delightful snap-shot of you in uniform taken just as you left the house. There was a thick fog & the figure is misty & dim & so I feel you receding into the fog & mud of Flanders & not coming back for so long.

Apart from her anxieties for Winston on account of obvious dangers from bullets and shells, Clementine was fearful lest he should neglect some elementary health precaution, such as inoculation, and so fall a victim to typhoid or enteric. She begged him to be inoculated: '[Y]ou ought to be done twice running (the interval between the 2 inoculations must not be more than 10 days) as soon as possible . . . Please be done as soon as you come out of the line. Every officer in France & Flanders has been inoculated twice.'[7]

As soon as Winston finished his first forty-eight hours' spell in the trenches on 23 November he wrote to Clementine from the 'billets in support' where he and his comrades were resting. He described the trenches:

> It is a wild scene . . . Filth and rubbish everywhere, graves built into the defences . . . feet and clothing breaking through the soil, water & muck on all sides; & about this scene in the dazzling moonlight troops of enormous rats creep & glide, to the unceasing accompaniment of rifle & machine guns & the venomous whining & whirring of the bullets wh pass overhead. Amid these surroundings, aided by wet & cold, & every minor discomfort, I have found happiness & content such as I have not known for many months.

But on his return to the trenches a few days later Winston had a near escape from death. He had received a peremptory telegram summoning him to meet the Corps Commander at a certain point about three miles away on the main road late in the afternoon. Taking with him his soldier-servant to carry his coat, he arrived at the rendezvous after walking across wet fields under spasmodic fire, and waited vainly for the General. Finally a staff officer turned up, who informed him there had been a mistake, that it was too late now, and that the General had returned to his Headquarters. Winston, highly indignant at this cavalier treatment, returned the long wet way he had come. By this time darkness was falling over the dismal countryside; it was pouring with rain, and it took him over two hours to reach his unit. When he finally arrived back in his own lines, he discovered that shortly after he had left in the morning, the dugout in which he would have been sitting had been struck by a shell. The whole structure had been smashed, and the mess orderly who was in it had had his head blown off. 'When I saw the ruin,' Winston wrote to Clementine, 'I was not so angry with the General after all . . . Now see from this how vain it is to worry about things . . . One must yield oneself simply & naturally to the mood of the game, and trust in God wh is another way of saying the same thing.'[8] Winston had a thought for Clementine's peace of mind: he did not send her his letter describing this incident until he was safely returned to GHQ.

Resting with his unit after his second spell in the firing line, Winston wrote to Clementine on 27 November: 'We came out of the line last night without mishap, & marched in under brilliant moonlight while the men sang "Tipperary" and the "Farmer's Boy" and the guns boomed applause. It is like getting to a jolly good tavern after a long day's hunting, wet & cold & hungry, but not without having had sport.' His peace of mind and feeling of release from care still prevailed despite the dangers and physical discomforts of his daily life; and he had a feeling of detachment from the hurly-burly of politics and the machinations of government. Lloyd George, McKenna and Asquith seemed very far away and looked to him 'like the mandarins of some remote province of China'.[9]

But despite this feeling of distance he now had from the political scene, Winston none the less urged Clementine to keep in touch with politicians and the press: 'Show complete confidence in our fortunes. Hold your head vy high. You always do. Above all don't be worried about me. If my destiny has not already been accomplished, I shall be guarded surely.'

In practical ways Winston was not remiss, sending through his man of business £100 for any urgent bills and later £300 as 'a reserve in case of accidents'.[10] He urged Clementine and Goonie not to deny themselves any reasonable comfort: 'Keep a good table . . . entertain with discrimination, have a little amusement from time to time . . . With £140 a month there should be sufficient.'[11] That winter, too, Lady Randolph let her own house, and came to join the throng at No. 41 Cromwell Road. Clementine wrote appreciatively, 'Your Mother is being very generous & is contributing £40 a month to the upkeep of this establishment until she comes to live with us.'[12] Lady Randolph was writing literary articles to help 'keep the home fires burning'; she sent Winston gossipy letters full of pluck. She had many political contacts, and gave and attended luncheons and dinners, helping to keep open the links for her absent son.

Clementine's work with the YMCA was getting into its stride, and she spent long days touring her own north-easterly sector of London, where the number of canteens over which she presided eventually rose to nine. When Winston resigned and the two Churchill households amalgamated, as a further measure of economy Winston and Clementine presently decided to dispense with their car; this made her work even more tiring, as her canteens were widely scattered. Writing to Winston, Clementine said: 'Without a motor it's harder to get to Enfield than to France – ¼ of the day has been spent in tube, tram, train, the remainder grappling with committees.'[13]

Clementine was at this time much concerned about Nellie, who in August 1915 had become engaged to Bertram Romilly, a lieutenant-colonel in the Scots Guards. Romantically good-looking, charming and a gallant soldier (he had won the DSO), he had suffered a severe head wound, and his health was fragile. He and Nellie had met after he had been invalided home. Clementine feared for her sister's happiness, married as she would be to a semi-invalid. Moreover, as she wrote to Winston:

> I don't believe she loves him at all but is simply marrying him out of pity . . . She vacillated (for the last week) between breaking off entirely, postponement, & immediate marriage with every hour of the day, but now she has hardened into a sort of mule-like obstinacy & says with a drawn wretched face that she loves him, is divinely happy & will marry him on the 4th. She is now furious with me for my former support of her postponing intentions, & says that if I say one word against her marriage on Dec. 4th she will leave the house & never come near me when she is married. Goonie thinks the marriage ought not to take place, but we can do no more.[14]

Winston was devoted to Nellie – 'The Nellinita', as he called her – so he shared Clementine's anxiety, but agreed that nothing could be done to stop the marriage.

Another wedding was causing a stir in London society: Violet Asquith was to marry Maurice Bonham Carter, her father's Private Secretary. In a long letter, in which she tried to gather up every bit of news for Winston's diversion, Clementine commented somewhat acidly: 'rumour whispers darkly but insistently that we have sustained a very damaging defeat & have lost many men. But for comfort & confidence we all look on the P.M. & McKenna both looking happy, sleek & complacent & for distraction on Violet's forthcoming marriage which is turning the town topsy-turvy with excitement.'[15]

On 1 December she wrote an account of this great occasion:

> Yesterday was Violet's marriage – Great throng of crowds everywhere – Randolph officiated as one of the pages & looked quite beautiful in a little Russian velvet suit with fur. His looks made quite a sensation & at Downing Street afterwards he was surrounded & kissed & admired by dozens of lovely women . . .
>
> The great feature at yesterday's wedding was the re-appearance of K [Lord Kitchener, Secretary of State for War] who, while everyone still imagined him to be receiving the plaudits of the Italians stalked into the Church & afterwards signed the register. I am told that in his absence the W.O. had begun to be swept & garnished & that his sudden return caused some dismay.

But full as her letters were of news from all aspects of the home front, Clementine's deep and continual anxiety and concern for Winston's health and safety leaps up again and again. From Cromwell Road on Sunday, 28 November, she wrote:

> My Darling,
>
> I miss you terribly – I ache to see you. When do you think you will get a little leave? Shall I come & spend it with you in Paris or will you come home?
>
> I don't like to make any request which might worry or vex you, but it makes me very anxious to feel that you are staying longer in the trenches than your duty requires. All the other officers & the men have been hardened to

the wet & cold by their training but you have gone at one swoop from an atmosphere of hot rooms, sedentary work & Turkish baths to a life of the most cruel hardships & exposure. And besides thinking of me & the babies think of your duty to yourself & your reputation. If you were killed & you had over exposed yourself the world might think that you had sought death out of grief for your share in the Dardanelles. It is your duty to the country to try to live (consistent with your honour as a soldier) . . .

Tell me some more about your life – Have you a nice servant – What do you have to eat & do you eat it in a trench or in a farm or where? I shd mind the rats more even than the bullets – Can you kill them or wld that be wasting good ammunition? . . .

The Daily Mail rings me up nearly every day & asks if I have had any news from 'Major Churchill'. Major Churchill has a strange sound, but I am prouder of this title than of any other . . . when I think of you my Dearest Darling, I forget all disappointment, bitterness, or ambition & long to have you safe & warm & alive in my arms. Since you have re-become a soldier I look upon civilians of high or low degree with pity & indulgence. The wives of men over military age may be lucky but I am sorry for them being married to feeble & incompetent old men.

I think you will get this letter on your birthday & it brings you all my love & many passionate kisses.

I find my morning breakfast lonely without you so Sarah fills your place & does her best to look almost exactly like you.

I'm keeping the flag flying till you return by getting up Early & having breakfast down-stairs . . . Goodbye my Darling. I love your letters. I read them again & again.

On 28 November Lord Cavan (the Commander of the Guards' Division) had invited Winston to luncheon with him; he suggested that until he was offered a command of his own, Winston should remain at Brigade HQ. Winston refused. 'I said I wouldn't miss a day of it. Nor did I,' he wrote to Clementine two evenings later:

I also scorned the modest comfort of Battalion HQ & lived in the wet & the mud with the men in the firing line.

My physique is such that I support these conditions without the slightest ill effect. Of course I have seen vy little, but I have seen enough to be quite at my ease about all the ordinary things.[16]

On 30 November – Winston's forty-first birthday – his company came out of the trenches for an eight-day break; his unit had been subjected to a three-hour spell of shelling, but despite splinters and debris falling very close, only two men in the company had been hurt. They were all glad to be relieved and had celebrated his birthday behind the lines with a convivial dinner party.

Winston had received Clementine's 'dear letter' of the twenty-eighth. 'I reciprocate intensely the feelings of love & devotion you show to me,' he wrote back to her: 'My greatest good fortune in a life of brilliant experience has been to find you, & to lead my life with you. I don't feel far away from you out here at all. I feel vy near in my heart; and also I feel that the nearer I get to honour, the nearer I am to you.'[17]

* * *

With his company at present relieved from duty in the line, Winston now took leave of the Grenadiers. He described to Clementine, with pardonable satisfaction, the warmth and friendliness of the farewells which surrounded his departure, coupled with pressing invitations 'to return whenever I liked & stay as long as I liked'.[18] These were in strong contrast to the icy nature of the welcome he had received on his arrival among them two weeks before. Winston then returned to GHQ at St Omer to see Sir John French, for the next step for him as a soldier had now to be determined.

When Winston had visited the Commander-in-Chief just over two weeks before, on his arrival in France, Sir John had offered him the command of a brigade; Winston also had Asquith's private assurance (given him when he left the Admiralty) that the Prime Minister would support such an appointment. But when he arrived at GHQ he found that Sir John had been recalled to London for consultations, and it was by now fairly general knowledge that his days as Commander-in-Chief were numbered. Discontent with his direction of strategy had been growing since the failure of the British offensive at Loos in the early autumn.

While waiting for Sir John's return, Winston lunched again with Lord Cavan, and had a long talk with him about his affairs: Cavan's advice was that he should take a battalion before accepting command of a brigade. This advice accorded with Winston's own thoughts and wishes at this moment. Writing to Clementine about the meeting, he added: 'He [Lord Cavan] spoke of my having high command as if it were the natural thing, but urged the importance of going up step by step.'[19]

One piece of news must have reassured Clementine: while Winston was at GHQ he took the opportunity of being inoculated, and was 'consequently teetotal and housebound'.[20] Winston ended his letter of 4 December on a buoyant note: 'They all say I look 5 years younger: & certainly I have never been in better health & spirits. Christmas in Paris – I think for you & me. I shall probably be in the collar then; but 2 or 3 days shd not be an impossibility.'

During these days, Winston was able to visit different parts of the front line, and he had long conversations with friends and officers in command, among whom he found a general impression of the 'utter inability to take a decision on the part of the Government . . . The able soldiers there are miserable at the Government's drifting. Some urge me to return and try to break them up. I reply no – I will not go back unless I am wounded; or unless I have effective control.'[21] Nevertheless, the effect of his visit to GHQ was to unsettle Winston's calm detachment from politics. As he watched from afar the march of events, his special knowledge and experience enabled him to sense the drift of affairs, and his feelings grew daily more contemptuous and bitter against those with the power to act, whose hands should have been gripping and directing the situation. He poured out all his thoughts and feelings to Clementine, reminding her 'These letters are for you alone', but adding that she might read or copy out for his mother and others anything 'not purely for us two'.[22]

Through all the months Winston was to be in France, a period for the greater part of which he was in or near the front line, he and Clementine wrote to each other almost every day. Their letters form a uniquely interesting and deeply moving correspondence. For the most part the letters were concerned with matters of deep import to themselves personally, and with details of public figures and events which engrossed their interest and attention. Their biographers – and indeed posterity – must ever be thankful that both Winston and Clementine were able to take advantage of a special privilege: nearly all their letters were sent through GHQ, and were not subject to censorship; had they been, much fascinating information, and many frankly expressed opinions, would perforce never have been committed to paper.

Winston's letters nearly always included an urgent demand for food or other comforts, much like a schoolboy's letters home.

> Will you send now regularly once a week a small box of food to supplement the rations. Sardines, chocolate, potted meats, and other things wh may strike your fancy . . . Send me also a new Onoto [fountain] pen. I have stupidly lost mine. Send me also lots of love and many kisses.[23]

And two days later:

> Will you now send me 2 bottles of my old brandy & a bottle of peach brandy. This consignment might be repeated at intervals of ten days.[24]

And after another two days:

> I want 2 more pairs of thick Jaeger draws [*sic*], vests & socks (soft). 2 more pairs of brown leather gloves (warm) 1 more pair of field boots (like those I had from Fortnum & Mason) only from the fourth hole from the bottom instead of holes there shd be good strong tags for lacing quicker. One size larger than the last. Also one more pair of Fortnum & M's ankle boots only with tags right up from the bottom hole (the same size these as before).
>
> With these continual wettings and no means of drying one must have plenty of spares. I am so sorry to be so extravagant.[25]

And a week later:

Also send me a big bath towel. I now have to wipe myself all over with things that resemble pocket handkerchiefs.[26]

In conditions of real hardship small comforts and luxuries acquire a vast significance; moreover, food parcels were shared out among the mess, and Winston was anxious to make a suitable and generous contribution. Thus it was a minor but vexing inconvenience that the food parcels and various items of clothing and equipment he had asked Clementine to procure and send to him did not appear. Naturally she was much distressed when she learnt that none of the parcels she had hastened to send off had arrived by the end of the first week of December, because she had mistakenly sent them to the Grenadiers instead of to GHQ. She earned a mild scolding for this: 'How naughty of you not to send everything to G.H.Q. as I said. Now I do not know whether the boots have reached the Guards or not – or what has happened to the food. However, it will all come right – I daresay.'27

Winston retrieved the first food parcel on 10 December from Grenadier Headquarters, and two days later he announced with satisfaction that 'the most divine & glorious sleeping bag has arrived, & I spent last night in it in one long 11 hours purr. Also food boxes are now following steadily; & I get daily evidences of the Cat's untiring zeal on my behalf.'[28] So all was now well on the 'Parcels Front'!

Apart from this brief period, when the non-delivery of parcels arose from a misunderstanding as to where they should be sent, the postal services available to Winston and Clementine appear to have been nothing short of miraculous. Small parcels took only about three days to arrive, and letters were often received within forty-eight hours, although obviously these timings varied according to the conditions of warfare on the spot at any one time; a special privilege available to them was that their letters were often carried by King's Messenger from the Admiralty.

On the night of 3 December Sir John French returned to his own GHQ from his consultations with the Prime Minister. Winston and two or three others in his confidence dined with him that night and learnt the latest developments concerning the command of the armies. Asquith evidently wanted Sir John to go, but without any inconvenient commotion; French was reluctant to relinquish his command, but determined to behave with dignity. The result was a sort of agreed stalemate which could only have a damaging effect at a time when decision and action were required daily if not hourly. Winston commented to Clementine: 'For three weeks no one has thought of the enemy.'[29] Nevertheless, his understanding and sympathy for his friend in these last days of uncertainty were profound: 'I am so sorry for him. No man can sustain two different kinds of separate worries – a tremendous army in the face of the enemy: a gnawing intrigue at his back.'[30]

Winston's own future was also discussed that evening, and in this same letter to Clementine, dated the day after the dinner, he wrote: 'I proposed to French that I shd take a battalion; but he rejected it, & said "no a brigade at once" & that he wd settle it quickly in case any accident shd happen to him.

I have acquiesced.'[31] There is no mention in this letter of Sir John having discussed the question of Winston's military future in specific terms with the Prime Minister; but he told Winston that Asquith had spoken 'with emotion' about him. Winston remarked to Clementine that 'Asquith's sentiments are always governed by his interests. They are vy hearty & warm within limits wh cost nothing.'[32] This somewhat harsh judgement was to be amply borne out by future events.

Clementine, meanwhile, had received Winston's letter describing his lucky escape from almost certain death, owing to the General's caprice. She wrote to him on 4 December (as usual, early in the morning, at 6.30 a.m.).

> It is horrible to sit here in warmth & luxury while danger & suffering are so close to you. That dreadful walk across the fields there & back among falling shells was on Nov: 24th & now it is 10 days later & Heaven knows what narrow escapes you may have had since.

But she had strong views on the subject of Winston's promotion:

> My Darling, altho' I ache for you to have a brigade so that you may be in less danger I admire you so very much for taking a battalion first. I am sure it is the right & wise thing. General Bridges came to see me last night; (what a tremendous fellow he is to look at). He told me he had seen you & that you were then in high spirits. He said 'I suppose Winston is going to get a brigade.' I said I thought you would rather take a battalion first. His face lighted up & he said 'I am so glad'. It is nectar to me to feel & see generous admiration & appreciation of you, which for so long have been denied unjustly . . .
>
> I am so glad that you & Lord Cavan are friends – He seems to be a fine soldier & one of the few unblighted generals in the general mildew.
>
> What you tell me of Sir John F. grieves me but not really for your sake, for I prefer you to win your way than to be thought a favourite of the C. in C. I feel confidence in your star my Dearest & I know all soldiers who meet you will love you . . .

There was plenty of news from the home front as well:

> Margot & Elizabeth [Asquith, her daughter] came here yesterday to tea & were very friendly. The 'Block' [nickname for Mr Asquith] I have seen only at Downing Street after Violet's wedding; he ran across me & Lord Haldane(!) together in the hall, muttered a few civil words & shuffled off sniffing nervously. Goonie & I have been bidden for luncheon there on Sunday which is most unusual as they are generally away; I don't expect the old boy will be there . . .
>
> To-day is Nellie's marriage, but the whole house is still asleep. For the last few days the rooms have looked very odd, full of Nellie's presents & cardboard boxes & new clothes & tissue paper.

Since Clementine had confided her anxieties about this match to Winston, Nellie had become much calmer and more settled in her mind; and she and Bertram Romilly were married at the Guards' Chapel on 4 December. It was a marriage which was to last the test of time and divers tribulations. Two days after the wedding, Clementine described it all to Winston in one of her early-morning letters:

> My Darling,
>
> Nellie was married on Saturday – I feel much happier now about her, as from several people I have heard good reports of Bertram . . .
>
> I wish you could have been here – Nellie looked really lovely; her long train was carried by Diana, Johnny & Randolph. After the wedding we all went to Aunt Mary's house in Portland Place & many people came there to see the last of her as Miss Nellie Hozier! It is 7 years & nearly 3 months since you & I drove away from there to Blenheim.
>
> Yesterday Goonie & I lunched at Downing Street. No 'Block' (he had gone down to Munstead after attending a conference at Calais on Saturday). But Violet & Bongey [Maurice Bonham Carter] appeared from their honeymoon on their way to the Italian Riviera where they are to spend a month. They both looked rather dreary & blue-stockingey.
>
> I long for more news from G.H.Q. where you have now been for 6 days. I suppose the conference at Calais was partly about the possible change there. I cannot get used to not knowing what is really going on.[33]

Being 'out of the know' was a deprivation Clementine felt very much. At the end of November she had written to Winston: 'For the first time for seven years besides being parted from you I am cut off from the stream of private news & have to rely upon the newspapers & rumour, so that I am in a state of suspended animation.'[34] Quite apart from any source of private news, a general shortage of information for public consumption was a curious feature of the First World War, and in the absence of official news, every kind of rumour and speculation ran rife.

When Clementine received the letter in which Winston told her that he had allowed Sir John French to persuade him to take a brigade, she was greatly disturbed. She saw clearly the adverse comment and hostility such an appointment might arouse. Her love for Winston was ardent, but not blind, and she used all her powers of persuasion and argument to dissuade him from accepting such promotion at this moment.

> I hope so much my Darling that you may still decide to take a battalion first, much as I long for you to be not so much in the trenches. I am absolutely certain that whoever is C. in C., you will rise to high commands. I'm sure everyone feels that anything else would be wasting a valuable instrument. But everyone who really loves you & has your interest at heart wants you to go step by step whereas I notice the Downing Street tone is 'of course Winston will have a brigade in a fortnight'. Thus do they hope to ease their conscience

> from the wrong they have done you, and then hope to hear no more of you . . . Sir John loves you & wants himself to have the joy of doing something for you, but I believe in Lord Cavan's advice – you & he should make a very strong combination & if he gets a corps I feel sure you wld soon get a division under him. Do get a battalion now & a brigade later.[35]

After this eloquent plea for second (and wiser) thoughts, Clementine turned to her second preoccupation on Winston's behalf – his comforts.

> I am so distressed about the food, & to-day I have despatched a big box to G.H.Q . . . I have also sent boots (3 pairs) vests, pants, socks, gloves, sleeping cap, onoto pen.
>
> If you go back to the trenches shall I send you a trench stove – I saw a lovely one in a shop quite portable & takes to bits – You are meant to have it in your dug-out. It burns charcoal.

Clementine took a more detached view than Winston of the impending change of command:

> I feel so grieved for Sir J's anxieties, but I am sure a change is inevitable & tho' I am sure he is a good soldier if I were Prime Minister I would make the change; but I hope swiftly & with decision not wavering towards it in the way he does about everything. Do not my Dear be shocked & angry with me for saying this. I know he is your friend & I too feel much warmth, affection & gratitude towards him, but with the 2 disasters impending in the East, I would like a fresh un-tortured mind in the west.
>
> Goodbye my Darling – I feel sad to-day & as if we could not win the war. But perhaps this is becos' I long for you very much & I feel the reaction after Nellie's wedding & rain is falling in buckets. But in your star I have confidence.
>
> Your loving
> Clemmie
> [sketch of cat][36]

She added an endearing P.S. 'Do Majors or Colonels command battalions?'

During the next few days Winston remained at GHQ, passing the time in visiting units of the French army, where he was received with warmth. On one of these occasions he was presented with 'a fine steel helmet . . . wh I am going to wear, as it looks so nice & will perhaps protect my valuable cranium'.[37] During part of this time French was away in Paris, and so Winston's own personal affairs hung fire. His busy mind, however, played on the role he relished, and now with reason entertained – that of a brigadier in active command of a fighting brigade. When Clementine's letter of 6 December arrived, with its earnest plea to be guided by Lord Cavan's 'step by step' advice, Winston brushed her objections aside, writing: 'You must not suppose that anything that I can do will enable me to "win my way"*. Rising rapidly through the grades cd only be accomplished by favour, i.e., by a view being taken of my personal qualities apart from military services or local experience.'[38]

With so much uncertainty still prevailing about his personal future, Winston wrote that, failing a definite decision, he would return to his Grenadiers the next day for another cycle in the trenches. That same evening, he succeeded in speaking to Clementine on the telephone; this conversation, and a few others, were arranged through GHQ and the Admiralty, where chivalrous feelings still burned for the erstwhile First Lord. Clementine described how when she went to the Admiralty to speak to Winston on the telephone,

> Everything there [was] very quiet & solemn except old Page [a Messenger] . . . who frisked about like a faithful old dog & plied me with tea & arm chairs near the fire. Masterton [James Masterton-Smith, the First Lord's Private Secretary] very nice & amiable – I think he manages the whole show. Colonel Hankey came out of 'your' room where he had been speaking to Mr Balfour & was then wending his way to Downing Street. He said to me 'In these moments of anxiety & difficulty I miss your Husband's courage & power to take a decision.'[39]

She told Winston:

> It was wonderful yesterday to hear your voice on the telephone, But very tantalising, as there is so much I want to say to you which cannot be shouted into an unsympathetic receiver!
>
> It must be thrilling to see all the famous positions along the French line. I feel that you have had an interesting & happy week which consoles me for your absence. Here everyone is downcast & gloomy & there are rumours of a very bad reverse to our troops at Salonica . . . Mrs Greville, dining at [Sir Ernest] Cassell's [*sic*] last night sat next to the P.M. who she said was very depressed about Salonica; perhaps however his spirits were merely dashed by his neighbour being of mature years! My dear if only you can escape death or injury, how well out of all this you are . . . Goodbye my Darling. Think of me often – During this last week I feel as if I had missed your thoughts – or is it fancy?
>
> Clemmie
> [sketch of cat][40]

Clementine's 'antennae' were not out of tune: during this time Winston's train of thought and hers were not in harmony. She was in an agony of mind: if he were given a brigade he would be safer – and yet she felt deeply it would harm him in the eyes of his enemies and in the estimation of his true friends. But despite her counsels, the die was cast. On 10 December

* CSC's words in her letter to WSC of 4 December 1915.

Winston wrote that he was to be given the command of the 56th Brigade. He had lunched the previous day with Lord Cavan and had talked the whole question over with him. He assured Clementine that Lord Cavan seemed quite content with the arrangement. Winston was of course aware that his appointment would arouse 'criticism & carping'. But he pointed out to Clementine that even had he taken a battalion for a short period, he would still have been assailed for using this as a stepping stone to higher things. 'I am satisfied this is the right thing to do in the circumstances, & for the rest my attention will concentrate upon the Germans.'[41] In this same letter Winston asked Clementine to order him a new khaki tunic, with the insignia of a brigadier-general; he enjoined secrecy, suggesting she could pretend it was for Bertram Romilly.

Whatever their differences in point of view, the bond of tenderness between them was not weakened. He ended this letter:

> You have been much in my thoughts this week my darling. I rejoiced to hear your voice over the telephone . . . I cd not say much & even feared you might think I was abrupt. One cannot really talk down it. But nevertheless we will try it again.
>
> With tenderest love,
> Your devoted
> W

Winston had sent Clementine a letter and papers of a highly confidential nature from Lord Curzon, who was now Lord Privy Seal, instructing her, after she had herself read them, to keep them under lock and key. Curzon had now emerged as the strongest advocate in the Cabinet of a renewed effort in Gallipoli, and Clementine read the papers with passionate interest; on 12 December she told Winston that they had thrown her 'into a state of agitation and mild hope that at last for the first time since you left the Admiralty a courageous & bold decision was going to be taken . . . How clear & able Lord C's 2 papers are – I wish he could be Scty of State for War instead of that cowardly & base old K.'

Clementine also was much concerned at this time because Sarah, now just over a year old, was suffering from acute bouts of neuralgia:

> She becomes rigid & screams with agony for about 20 minutes & when the pain ceases she is quite done & exhausted & falls asleep. She has lost her lovely pink cheeks and is losing weight. Parkinson has been called in & has prescribed (as the last resort) bromide, but we have not yet opened the bottle. It is on the mantle piece [*sic*] & I'm putting it off as long as I can. Today she is a little better.

It is hardly surprising that this letter ended on a somewhat melancholic note:

Meanwhile I toil away at my working-men's restaurants. After Christmas I am opening 2 to seat 400 men each . . . It is snowing hard and makes me fear that you are wet & cold. Write & tell me what everyone thinks about Sir John's retirement. I fear you will perhaps not go to GHQ so often now? Do you know Haig & is he a friend?

Your own Clemmie
[sketch of cat][42]

Meanwhile, Winston returned to the Grenadiers, and wrote to tell Clementine that he would stay with them rather than wait about at GHQ. Lord Cavan had arranged for Winston to be instructed in the supply system, in preparation for his new appointment: 'I am to follow the course of a biscuit from the base to the trenches etc.'[43]

Once Clementine knew that the decision was taken, she did not continue her protestations, or indulge in warnings of the ill results which might flow from this step, which she still nevertheless deeply deplored. She knew how much Winston yearned for a command in keeping with his capacity, and, accepting this outcome of events, she tried to share in his satisfaction. This was to be her attitude on the many occasions in the years to come when she differed strongly from Winston on various issues. Earnest, even passionate in the advocacy of her point of view, once Winston had taken a decision she would swallow her objections and loyally support him through thick and thin, revealing only to her most intimate friends or closest relations that her instincts and advice had all been to the contrary.

Now, subduing her true feelings, Clementine wrote bravely on 15 December:

My Darling,

I am thrilled to hear that you are to have a Brigade, but I should rejoice more if I thought that in that position you would be in less danger than you are now. I suppose the danger from rifle fire will be less & that from shells greater.

The greater part of her letter, however, was finished later that day, when she had some interesting news to impart:

Later. Goonie has just returned from lunching at Downing Street where she sat between the P.M. & K. She reports that K looked very thin; he told her that he had seen your paper on Trench Warfare which he thought very good & he is having it circulated to the Staffs. The P.M. sniffed & asked after you & asked if you were happy to which Goonie replied acting according to consultation (for we had discussed it beforehand) that you said in all your letters that you were very happy. The old Boy looked rather uncomfortable & then passed on to me 'Why don't I ever see Clemmie, why doesn't she come here?' Goonie enquired if I had Ever been invited to which he replied I ought to propose myself. I think he feels thoroughly sheepish & uneasy, or perhaps he just pretends to be as he thinks it good taste! . . .

I am so relieved that at last the things I ordered are beginning to arrive. Now Christmas congestion is beginning, so if there is anything very special you want, let me know quickly & I will send it out by F.E. I spoke to him this morning on the telephone & he said he would come and see me before starting. The government attitude is that everything is now going splendidly at Salonika. Rumours of the Dardanelles Evacuation are of course all over London (the usual W.O. leakage I suppose) but the rumours differ. The latest Society canard is that the Government will fall this week & that for the duration of the war the Speaker will be made P.M.! This is the very newest, last week it was Lord Derby –

Christmas is coming very soon – shall we meet then?

Goodbye my Darling Winston. I think of you constantly especially at night. I hope you are not very cold.

Your
Clemmie
[sketch of cat]

Meanwhile, events which were to have direct and dramatic repercussions for Winston were moving fast. On 15 December French returned to London to surrender his command, and his replacement by Sir Douglas Haig was announced the following day. During his farewell interview with the Prime Minister, Sir John told him that he had given Churchill a brigade, and 'Asquith said he was delighted.'[44] But a few hours later the Prime Minister wrote a note to French, saying that 'with regard to our conversation about our friend – the appointment might cause some criticism' and should not therefore be made, adding: 'Perhaps you might give him a battalion.' It is probable that, during the interval between Asquith's conversation with French and the writing of this note, he had been told of the question tabled for the House of Commons on the following day by Sir Charles Hunter, a Conservative MP, asking the Under-Secretary of State for War 'if Major Winston Churchill has been promised the command of an Infantry brigade; if this officer has ever commanded a battalion of Infantry; and for how many weeks he has served at the front as an Infantry Officer?'[45] Rather than face a parliamentary challenge in defence of Churchill's promotion, Asquith bent before the prospect of a storm and vetoed the brigade. Naturally French was astonished and dismayed: astonished, no doubt, by the lack of loyalty and firmness shown by the Prime Minister, and dismayed because he had told many people that Winston was to have a brigade. His hour of influence and power was gone, however, and he could but acquiesce in a decision which caused him profound distress and must inflict mortification on his friend, to whom he hastened to communicate the distasteful news.

While these events were taking place in London, Winston had come out of the line with the Grenadiers and had gone to GHQ, from where he wrote a brief note to Clementine early on 15 December to tell her that he was safe and sound, and he remained at GHQ to bid farewell to 'my poor friend who returns to pack up tomorrow'.[46]

Although he knew nothing as yet of Asquith's volte-face, Winston realized that the changeover in command might have an effect on his personal affairs. 'Believe me,' he wrote to Clementine, 'I am superior to anything that can happen to me out here. My conviction that the greatest of my work is still to be done is strong within me.'[47] And, significantly, he suggested for the first time that it might be his duty to return early in the following year to take his place in Parliament, 'to procure the dismissal of Asquith and Kitchener', whose lack of direction and ineptitude provoked his rising contempt.

His long letter was closed up, but not sent, when 'stop press' news of an urgent character arrived. Another sheet of paper marked 'later' was slipped into the envelope. It read:

> My darling,
>
> I reopen my letter to say that French has telephoned from London that the P.M. has written to him that I am not to have a Brigade but a Battalion. I hope however to secure one that is now going into the line. You will cancel the order for the tunic!
>
> Do not allow the P.M. to discuss my affairs with you. Be vy cool & detached and avoid any sign of acquiescence in anything he may say.
>
> Your devoted
>
> W

On this crucial Wednesday, 15 December, Clementine knew nothing further than that Winston was to be given command of the 56th Brigade. But during the next two days various events were further to disturb her peace of mind. On Thursday, 16 December she met Lord Esher* (who had just returned from France, where he had seen Winston) in the Berkeley Grill at luncheon; he had expressed himself of views which greatly perturbed her, the more so as they echoed her own inmost feelings.

That same afternoon Sir Charles Hunter asked his 'loaded' question in the House of Commons. It brought into the open, once more, the hostility and malice of Winston's enemies. In answer to the question, the Under-Secretary of State for War replied: 'I have no knowledge myself, and have not been able to obtain any, of a promise of command of an Infantry brigade having been made to my right hon and gallant Friend.' Clementine would of course have read the report of the question and the exchanges it produced. It can only have confirmed her fears that Winston's promotion would arouse fierce criticism and opposition.

The following morning she wrote to Winston and confined herself to generalities, giving only a partial account of her conversation with Lord Esher:

* At that time a permanent member of the Committee of Imperial Defence. During the First World War he was a confidential liaison officer with the French.

I met Lord Esher at a restaurant yesterday fresh from France. He said you looked like a boy, all the lines of care gone from your face. He seemed to think there was going to be a row over your Brigade. I pointed out to him that if you had a Battalion first & then a Brigade there would have been 2 rows instead of one.

Jack Tennant denied knowledge of it in the House yesterday. When will it be announced? I feel very anxious & long for a sight of you.

Do you know Sir Douglas Haig? Did he agree to your appointment or was it finally settled before he supervened? He looks a superior man, but his expression is cold & prejudiced, & I fear he is narrow.

She ended her letter:

Randolph asks every day about you & also asks every day when the war will be over. I fear he's a peace crank. I feel like that too![48]

But later that day Clementine was in touch with Eddie Marsh (now one of the Prime Minister's Private Secretaries), and what he told her prompted her to write a second letter to Winston, in which she not only gave him a full account of her meeting with Lord Esher, but no longer hid from him her state of intense anxiety.

41 Cromwell Road
17th December [1915]
[Letter 2]

My Darling,

Since writing to you I have been wondering whether I ought not to have told you exactly what Lord Esher said. I did not do so becos' I thought your appointment was absolutely fixed; but now I hear from Eddie that you say that the recall of Sir John may affect your private fortunes. So I will tell you (for what it is worth) what Lord E. said. Of course you will know better than I can whether he is in touch with feeling in the Army or not & whether to attach importance & weight to his opinion. He said 'Of course you know Winston is taking a Brigade & as a personal friend of his I am very sorry about it; as I think he is making a great mistake. Of course it's not his fault, Sir John forced it upon him. All W's friends are very distressed about it as they hoped he would take a batallion [*sic*] first.' He said how tremendously popular & respected you had become in the short time you had been there & repeated to me the story you told me of the Colonel of the Grenadiers receiving you so disagreeably & then being entirely won over. This interview took place in the crowded grill-room of the Berkeley. I preserved a calm & composed demeanour, but I was astonished & hurt at his blurting all this out to me. He repeated again & again that the thing was a mistake; I tried at last to head him off by asking him personal questions about you, how you were looking, if you were well. He then launched forth again, saying that you had been in the greatest danger, in more than was necessary etc & that French had determined to give you this Brigade as he was convinced you wld otherwise be killed.

After this I crawled home quite stunned & heart-broken.

My Darling Love – I live from day to day in suspense and anguish. At night when I lie down I say to myself Thank God he is still alive. The 4 weeks of your absence seem to me like 4 years – If only My Dear you had no military ambitions. If only you would stay with the Oxfordshire Hussars in their billets –

I can just bear it – feeling that you are really happy. I have ceased to have ambitions for you – Just come back to me alive that's all.

Your loving
Clemmie

But the events and cares of that long day were not yet over. During the course of the evening F. E. Smith (now Solicitor-General) telephoned to say he was going over to France and would see Winston the next day; he offered to act as messenger or postman. Later still she received the letter from Winston written from GHQ on 15 December, with its last-minute enclosure telling her he was not, after all, to be given the promised brigade.

At one o'clock on the morning of Friday, 18 December, Clementine wrote her third letter to Winston in twenty-four hours. It is jumbled and disjointed. One senses her stunned dismay, as she sat wearily writing in those dark hours:

My Darling,

Late in the evening I had a telephone message from F. E. saying he was going to see you tomorrow. I would like to have sent you some delicacies but the shops were closed so all this letter brings is my tender love. My Dear – your letter has just come telling me that your hopes of a Brigade have vanished. I do trust that Haig will give you one later. If he does it may be all for the best – but if not it is cruel that the change at G.H.Q. came before all was fixed. You will receive later by King's messenger 2 letters written earlier in the afternoon [of the 17th]. Do not pay any attention to No. 2 written at a moment of sadness, uncertainty & agitation. Your letter with its firm & confident tone has restored me. I telephoned to F. E. where he was dining & he very good naturedly came round & saw me. He is your true & faithful friend & I felt so glad that you will now see each other . . .

I must say I am astounded at the P.M. not backing you for a brigade but I cannot help hoping that he has asked Haig to give you one later on after you have commanded a battalion for a little while.

My own Darling I feel such absolute confidence in your future – it is your present which causes me agony – I feel as if I had a tight band of pain round my heart.

It fills me with great pride to think that you have won the love & respect of those splendid Grenadiers & their austere Colonel.

In happier times you must let me see them all. Perhaps if any of them come home on leave they would come & see me.

Your loving
Clemmie
[sketch of cat]

When Sir John French returned to GHQ to pack up and take his leave he gave Winston a full account of his conversation with the Prime Minister and the events that flowed therefrom, and he showed him Asquith's note. Its tone, and particularly the last phrase – 'Perhaps you might give him a battalion' – rankled deeply. In giving Clementine all the details of the sorry affair, Winston commented: 'The almost contemptuous indifference of this note was a revelation to me.'[49] He went on to recall the sequence of events: when he left the Admiralty, Asquith had offered Winston a brigade in the event of his going out to France; in the following September Winston told the Prime Minister of offers of command made to him by French, and the Prime Minister had assured him of his support in any advancement thought fitting by the Commander-in-Chief. Winston called to mind the long tale of his connection, work and friendship with Asquith. 'Altogether I am inclined to think his conduct reaches the limit of meanness & ungenerousness . . . Personally I feel that every link is severed: & while I do not wish to decide in a hurry – my feeling is that all relationship shd cease.'[50]

Winston then discussed the question of whether in these changed circumstances he should accept command of a battalion; his mood was one of uncertainty and depression; he expressed himself as reluctant to accept the responsibilities of such an appointment unless it were under 'a C-in-C who believes in me'. He realized clearly that he would be an easy target for criticism. He was as yet uncertain of what his relationship with the new Commander-in-Chief would be, although he had known him in days gone by when Haig was a major and he, Winston, a young MP. In this gloomy hour, one bright vista gleamed ahead: he could always return to his Grenadiers – 'This at any rate is the place of honour.'[51]

Winston ended his letter with an urgent request: he had written two letters to Clementine on the previous day, and he now asked her to burn them – 'I was depressed & my thought was not organised. It is now quite clear & good again & I see plainly the steps to take.' Clementine faithfully did his bidding: the letters do not survive.

But these distressing events had a more cheerful sequel. Later that day he was able to tell Clementine of his first interview with Sir Douglas Haig, who had received him with the 'utmost kindness of manner & consideration'.[52] In the course of their conversation the new Commander-in-Chief had assured Winston of his sympathetic support, and Winston felt reassured that even under the new regime he would get a fair chance. 'In these circumstances I consented to take a battalion.' Altogether the meeting was a great success, and Winston endearingly commented: 'So I am back on my perch again with my feathers stroked down.'[53]

He had received Clementine's letters written on 17 December, and his ire was roused by her account of Lord Esher's outburst: 'Esher talks foolishly. It wd not have been a great mistake for me to take a Brigade. There was something to be said either way.'[54] As to her relationship with Asquith, Clementine received explicit instructions: she was to make no change, but was to show an unwillingness to discuss him or his affairs. 'Don't you tell me he was quite right – or let him persuade you,' he added.

Two days later, Clementine, who must have been eating her heart out for news, wrote:

20th December [1915]

My Darling,

Your two letters of the 17th [destroyed at his request] made me so sad & if your happier one of the 18th, (describing your interview with Haig) had not come at the same time I would have been absolutely heart-broken.

I will write to you fully tomorrow, but can now only send a scrawl to send you a thousand kisses & to assure you of my undying love for you. You must not beat your poor Kat so hard; it is very cruel & not the right treatment for mousers.

I am absolutely worn out to-night.

I gave F.E. a letter for you which will I hope soon arrive.

Your loving
Clemmie

Large posters just out:-
TROOPS WITHDRAWN FROM DARDANELLES
OFFICIAL

Pending his posting to a battalion, Winston moved two hundred yards down the street to a house occupied by Sir Max Aitken – 'a sort of Canadian war office',[55] as Winston described it – from where Sir Max operated as a Canadian eye-witness at the Front. Here the two men's lifelong friendship was cemented.

Winston awaited the outcome of events with surprising patience. 'It is odd to pass these days of absolute idleness,' he wrote. '. . . It does not fret me. In war one takes everything as it comes.'[56] He watched from afar the irresolution and incompetence of the men in power, and turned himself with relief to the 'tremendous little tasks wh my new work will give me'. He looked forward to this new experience and to building a good relationship with the men he would soon command; he hoped to come to them 'like a breeze . . . I shall give them my vy best.'[57]

He was at GHQ when French took his departure; it moved him very much. After seeing many generals and other commanders, Sir John had sent for Churchill, saying 'it is fitting my last quarter of an hour here shd be spent with you.' After that he had departed 'with a guard of honour, saluting officers, cheering soldiers & townsfolk – stepping swiftly from the stage of history into the dull humdrum of ordinary life'.[58]

In their letters a possible reunion in Paris had been canvassed, but in the event – and better still – Winston came home for Christmas. He arrived on Christmas Eve, and paid a visit to the Prime Minister: it cannot have been a comfortable interview.

But great were the rejoicings at No. 41 Cromwell Road, where the two families were packed in. The two elder children must have sensed the strain of the last weeks: but now it was Christmastime, and Papa was home to make it all quite perfect. As for their parents – neither the bitter disappointments and dramatic events of the last year, nor the anxieties and perils which loomed ahead, could dim the joy and thankfulness that they were together again.

Winston's leave was painfully short: he had arrived home on Friday, and he left for France again on the following Monday. Now, as always, his wife and family had to share him with events and people, for he did not omit to look up some of his political friends. During those three brief days, however, we may be quite sure Winston and Clementine found time to open their hearts to one another. They had many things to discuss: past griefs and mortifications; present political preoccupations; and the prospect before them of long, grey months of separation.

Then Winston went back, like so many more, to danger and squalor and, in his case, frustration too. Clementine saw him off; they cut it rather fine – 'I could not tell you how much I wanted you at the station,' she wrote in a short note on 28 December – 'I was so out of breath with running for the train.'

CHAPTER ELEVEN

Waiting in Silence

WHEN CHURCHILL WENT BACK TO FRANCE AT THE END OF DECEMBER 1915 HE was politically down-and-out; many people thought he was finished for ever – and some hoped he was.

He turned to soldiering, as we have seen, with relief; but there too the way to further advancement was temporarily barred. He had the worst of both worlds: he was at once too junior in a strictly military sense, and too formidable in another. But Churchill found real solace and satisfaction in the zealous performance of his duties as an officer in the field; and he devoted to his regiment the energy and thought he had once applied to larger matters. He enjoyed the company of his fellow officers, and soon won their respect and friendship, sharing with them to the full the daily hazards and discomforts of life at the Front.

At this low ebb in his fortunes Winston turned more and more to Clementine as the one person to whom he could express himself freely, and on whom he could unload the bitterness – and even hatred – he felt for those who had seen him sink unaided. She was also his link with the political world, whose hold upon him continued to be strong; again and again in his letters he was to urge her to keep in touch with events and people, and to go about in society for that purpose. Clementine rose valiantly to the occasion, but her days were already long and full. Apart from her family and household, she was committed to her canteen work, which at this time was increasingly demanding and arduous. It required a supreme effort on her part to find the time, and to summon up the extra energy for social life. But she was zealous in following Winston's instructions to move in political society, sending him long accounts of her conversations and impressions.

In all her letters during these early months of 1916 one is struck not only by her faith (at times prophetic) in Winston's star, but also by her wisdom. No one had shared more keenly his disappointments and mortifications; we know she could feel things passionately, and express herself even violently; but she seems to have set herself out to deflect and moderate Winston's wrath and bitterness against Asquith. Partly she feared Winston might blunt his talents, and sour his normally warm and generous nature, by too much brooding over past wrongs. But perhaps more than that, she did not want him to fall too much under the spell of Asquith's Liberal rival Lloyd George, about whom she had some rough and penetrating things to say.

Winston's brief Christmas leave had left Clementine worn out with emotion and excitement; so she went to stay with her Stanley cousins in Cheshire for a few days to recover herself. Before she left for Alderley, Lloyd George had lunched with her on 29 December, and had told her about the crisis then rending the Cabinet over the introduction of a limited form of compulsory military service, an issue over which three ministers were threatening to resign. Clementine had duly regaled Winston with all the intricate details in a long letter to him later that same day. From Alderley she wrote again on 30 December, having further pondered upon the crisis, and some of the personalities involved in it.

> I came here last night & intend to stay till Monday. I hope the change will cure the melancholia which was dispelled by your return, but which has now settled on me again! . . . I suppose that if compulsion is carried without a single resignation (which seems likely) it will be a feather in the P.M's cap & a vindication of his slow statecraft. I am very much afraid this is going to be a 'personal triumph' for him.
>
> I think my Darling you will have to be very patient – Do not burn any boats – The P.M. has not treated you worse than Ll. G has done, in fact not so badly for he is not as much in your debt as the other man, (i.e. Marconi).* On the other hand are the Dardanelles. I feel sure that if the choice were equal you would prefer to work with the P.M. than with Ll. G. – It's true that when association ceases with the P.M. he cools & congeals visibly, but all the time you were at the Admiralty he was loyal & steadfast while the other would barter you away at any time in any place – I assure you he is the direct descendant of Judas Iscariott [*sic*]. At this moment altho I hate the P.M, if he held out his hand I could take it, (tho' I would give it a nasty twist) but before taking Ll. G's I would have to safeguard myself with charms, touchwoods, exorcisms & by crossing myself –
>
> I always can get on with him & yesterday I had a good talk, but you can't hold his eyes, they shift away –
>
> You know I'm not good at pretending but I am going to put my pride in my pocket & reconnoitre Downing Street.
>
> Even this one night in the country has done me good & I feel able to sit up & take an interest in life. Yesterday I was bitterly dejected & could not quench an endless flow of tears.

'You are a vy sapient cat to write as you do,'[1] Winston commented when he received this letter with Clementine's dissection of Asquith and Lloyd

*Churchill had done much to help Lloyd George during the Marconi Scandal of 1912–13, when the revelation that Lloyd George owned shares in the American Marconi Company and had concealed this information from the House of Commons had nearly forced him to resign.

George. But he felt deeply that the former had shown himself to be a 'weak and disloyal chief', and that their days of working together were over, whereas Lloyd George, although Winston conceded (somewhat reluctantly) that he 'is no doubt all you say', had always been against the Dardanelles, and not, like Asquith, a 'co-adventurer' who had then ruined the policy by his slothfulness and procrastination.

At the dawn of this New Year, Winston's greetings were tinged with subdued optimism: 'All my tenderest wishes for a Happy New Year. I think it will be better for us than the last – wh after all was not so bad. At any rate our fortunes have more room to expand & less to decline than in Jany last.' His forecast for world affairs was less sanguine: 'I do not see how any end will be reached in 1916: and the probability is that 1917 will dawn like this new year in world wide bloodshed & devastation.'[2]

Clementine's New Year letter was rather depressed in tone: everywhere at home people were now feeling the weight of the war; bereavement, separation, shortages and toil had changed the almost ecstatic atmosphere of endeavour and sacrifice of the first few months of the conflict to a grim and dogged spirit of 'Carry on'. The family gathered at Alderley was no exception to the prevailing mood:

> The household here . . . is very subdued in spite of seven grand-children. Three daughters of the house, Margaret Goodenough & Blanche Serocold & Sylvia Henley look patient but war worn. Margaret has seen her Admiral 4 times since the beginning of the war (he is still cruising around in the Southampton). Blanche is going to have her 3rd child & her husband recovered from his serious wound has his brigade in the line somewhere – Anthony Henley is moderately safe on a Divisional Staff – Venetia the prosperous and the happy,* arrives this evening to enliven us & to lift us out of the Doldrums – We expect to hear from her 'the latest' concerning the crisis – she entertained the P.M. on New Year's Eve to Beer & Skittles.[3]

Clementine was much depressed by a temporary absence of news from Winston, and two days later she wrote this pathetic note:

Alderley Park,
Chelford, Cheshire
3rd January [1916]

My Darling,

Still no letter & you have been gone a week to-day.

I wonder where you are.

I have no news of the great world, the little intimate things are starved – So my pen runs dry –

I return home to-day hoping rather hopelessly to find something from you.

Adieu
Clemmie

I send you some timid & tentative kisses.

* Venetia Stanley had married Edwin Montagu, MP, in July 1915.

In fact Winston had not been neglectful; he had written every day (save one) long and newsful letters of his daily doings and reflections, and when Clementine reached home, two of these letters had arrived, 'so I feel revived, but not yet nourished,' she wrote in a letter marked '5 a.m.' on 5 January. But later that day two more letters arrived to assuage her hunger.

Winston's immediate future as a soldier had now been settled: he was appointed lieutenant-colonel in command of the 6th Battalion the Royal Scots Fusiliers in the IXth Division, under General Furse. The battalion was at present in the village of Moolenacker (not far from Armentières), licking its wounds after being severely mauled in battle. Winston made much of joining a Scots regiment, and asked her to send him a copy of Burns's poetry: 'I will soothe and cheer their spirits by quotations from it . . . You know I am a vy gt admirer of that race. A wife, a constituency, & now a regiment attest the sincerity of my choice!'[4] He also asked her to send him a new tunic with the insignia of a lieutenant-colonel and 'a Glengarry cap'.[5] Although always much addicted to dashing hats, Winston in the event very rarely wore his 'bonnet' – he (rightly) opined it did not 'become him', and he reverted to his French *poilu*'s helmet. This was greatly to Clementine's relief, since steel was safer than tartan.

Soon Clementine had some 'Downing Street' news to impart. On 7 January she had tea with Margot Asquith: 'She was in a very good mood, there being I am sure great relief in that quarter that the division on compulsion went so well . . . I met the P.M in the hall. – He looked shy when he first saw me, but thought better of it & I had a little talk to him – He looks as if he had been thro' a good deal. He asked after you with compunction in his voice.'[6]

The following Sunday she lunched with the Asquiths. It was the first time since Winston's resignation that she had spoken at any length with the Prime Minister. 'He talked a great deal about you & asked a great many questions. I was perfectly natural (except perhaps that I was a little too buoyant) & he tried to be natural too, but it was an effort. I think it is a good thing to keep up civil relations & it is always interesting to follow the Block's train of thought.'[7]

When Winston received her letter he was not satisfied with her account: 'You do not tell me in yr letter what the P.M. said. You only say he said a lot. But I shd like a verbatim report of the Kat's conversation with the old ruffian.'[8] So Clementine duly sat down to write an amplified version of her original account:

> If you received the impression from my letter that the P.M. 'said a lot' about politics and the situation I gave you a wrong impression – He talked a great deal about trivialities & femininities which you know he adores and he asked a good many questions about you & about the detail of your life out there – He wanted the answers to be reassuring, & my good manners as a guest forbade me making him uncomfortable which of course I could easily have done. He seemed grateful to me for sparing him – ! He is a sensualist & if I had

> depicted you in a tragic & sinister light it would have ruined his meal & I shd probably not be bidden again.[9]

Winston's almost daily letters told her of his first experiences with his new battalion. He found the work absorbing, and a challenge; practically every regular officer had been killed in recent actions, and the regiment was, in consequence, officered by very young men 'vy brave & willing & intelligent: but of course all quite new to soldiering'.[10] He had been allowed to bring with him as his second-in-command his friend Sir Archibald Sinclair: Winston had formed a high regard for him both as an experienced soldier and as a friend in whom he could confide. Their friendship forged in these hard times was to be an enduring one. But despite the absorbing preoccupations of his new command, Winston had moments of deep depression, and these he confided only to Clementine: 'I do not ever show anything but a smiling face to the military world: a proper complete detachment & contentment. But so it is a relief to write one's heart out to you. Bear with me.'[11]

Her letters to him were a source not only of news, but also of comfort and strength; and, like her, he became depressed and agitated if the flow of her letters was for any reason interrupted. 'You cannot write to me too often or too long – my dearest & sweetest. The beauty & strength of your character & the sagacity of yr judgment are more realised by me every day. I ought to have followed yr counsels in my days of prosperity. Only sometimes they are too negative. I shd have made nothing if I had not made mistakes.'[12]

On 11 January Clementine wrote in answer to several letters which had arrived simultaneously. Winston had dwelt on the conscription crisis, which had raised the possibility that the Government might fall:

> I knew my Darling that you would be feeling the political excitement of the last 10 days. Everybody here seemed rather thrilled, but I could not get worked up over it as it seemed impossible to me that the anti-compulsionists would face the racket. I'm afraid I can't agree that the P.M's position is only temporarily strengthened – He will always in the end tip down on the side of strong measures after delaying them & devitalising them so as to try & keep everybody together.
>
> His method of defeating the enemy is not by well-planned lightning strokes but by presenting to him a large stolid gelatinous mass which he the enemy is supposed to pommel in vain.
>
> I am afraid the war will drag slowly to its end with him & K still at the helm. No one trusts Ll-G. & Bonar is light metal. These are only my own reflections as I have seen no-one of consequence for some days – Yesterday I was very sad, the end of Gallipoli & a fine battleship gone down.

She also had practical and sagacious views on Winston's intention to press the Prime Minister for the publication of the papers relating to the Dardanelles:

If you ask the P.M to publish the Dardanelles papers let me know what happens. If he refuses or delays I beg you not to do anything without telling me first & giving me time to give you my valuable(!) opinion on it. It is an unequal match between the P.M. & an officer in the field in war-time – If he dissents I fear you will have to wait. If you insisted on publication agnst his wish you would have agnst you all the forces of cohesion & stability including every member of the Cabinet.

On the other hand when the papers are eventually published his refusal to do so earlier will have a very bad effect for him.

But of course the P.M. may consent – Are you quite certain however that this is the best time for publication, when you are away & not able to speak in the debate which is bound to take place. I am very anxious that you should not blunt this precious weapon prematurely.[13]

But her letters were not all views and political advice. The following day she wrote again; perhaps in the night she had brooded over the sad and depressed passages in his letters, and she sensed his feelings of frustration, bitterness and isolation.

41 Cromwell Road
Jan. 12th, 1916

My own Darling,

I long so to be able to comfort you. Later on when you are in danger in the trenches you will be equable & contented, while I who am now comparatively at ease will be in mortal anxiety. Try not to brood too much; I would be so unhappy if your naturally open and unsuspicious nature became embittered. Patience is the only grace you need. If you are not killed, as sure as day follows night you will come into your own again. I know you don't fear death, it is I who dread that. But I am almost glad to be suffering now, becos' I am sure no single soul will be allowed to live thro' this time without sorrow, so perhaps what we are enduring now will be counted & we shall be spared the greatest pain of all.

I remember quite well when we were at the Admiralty during those wonderful opening weeks of the war we were both so happy, you with the success of the Naval preparations & with the excitement of swiftly moving events and I with pride at the glamour surrounding you & the Navy – I remember feeling guilty & ashamed that the terrible casualties of those first battles did not sadden me more. I wondered how long we should continue to tread on air –

When it is all over, we shall be proud that you were a soldier & not a politician for the greater part of the war. Soldiers and soldiers' wives seem to me now the only real people.

I am glad you tell me all you feel my Darling, I want to know it all. I too shew a detached & smiling face to the world.

I have Goonie for a safety-valve & you must make use of Archie who is safe & loyal – Do not tell curious acquaintances your opinion of the P.M's character or policy . . .

We hope now that Cape Helles has been evacuated that Jack may come home on leave, but we have had no news.

You will see that Edwin Montagu is restored to the Duchy of Lancaster but keeps his work as financial secretary to the Treasury which saves him from the slur of well-paid inactivity –

I will try and see Lloyd-George again –

I wish I had some interesting news for you.

Goodbye my Darling. I send you many kisses –

Your letters are very precious.

Your loving
Clemmie
[sketch of cat]

Do you want any more cigars yet??

'Your letter of 12th is splendid,' Winston replied. 'I run in and out of moods; but I do not doubt the wisdom & necessity of my coming out here: nor do I repent at all my decision.'[14] He was insistent that Clementine should maintain contact with his political confidants, and commanded her in a long letter on 16 January: 'Tell them I am taking time to consider, method & occasion [for him to return to England], but that in principle I have decided. Also I think you might have a talk with Cawley – & even with the Fiend himself [Fisher].'

Although the declaration of Winston's decision presently to return home and re-enter the political arena must have struck a chill note for her, Clementine carried out his instructions – baulking only at Fisher, whom she flatly refused to contact. It must have been a source of reassurance to her that Archie Sinclair – Winston's closest confidant – agreed with her: 'Archie is a strong advocate of my staying here,' he wrote in the same letter. 'It is odd how similar are the standpoints from wh you and he both view my tiresome affairs.'

The reserve billets in which Winston's battalion found itself for the time being were a series of 'squalid little French farms rising from a sea of sopping fields & muddy lanes'.[15] He told Clementine, however, that he was quite comfortable in his little farmhouse. 'The guns boom away in the distance, & at night the sky to the Northward blinks & flickers with the wicked lights of war.'[16] All his thoughts and activity were concentrated on the detail of his new command: he attended a machine gun school, conducted drilling periods, practised bomb-throwing and, by night, accompanied by the faithful Archie, patrolled the regimental area. About four days before the battalion was due to move forward to start taking over its position in the line, Winston organized a combined sports day and concert for the men, which was a tremendous success and did much to raise their spirits. 'Poor fellows –', he wrote, 'nothing like this has ever been done for them before. They do not get much to brighten their lives – short though they may be.'[17]

A red-letter day was when the hamper Clementine had despatched him for Christmas suddenly arrived on 13 January. 'I never saw such dainties & such profusion,' Winston wrote appreciatively the same day. 'We shall eat them sparingly keeping the best for the trenches.'

Winston's letters gave lively and detailed accounts of his military activities: he was immersed in plans for his battalion's organization and welfare. Yet there is scarcely one in which his feeling of injustice, or his sense of Asquith's betrayal, does not emerge; and through them also runs the burning thread of constant anger at the utter squandering of men, opportunity and materials consequent on the failure to carry through the strategy of the Dardanelles operation. His feeling of frustration was sharp: 'I have no means of expression. I am impotent to give what there is to be given – of truth & value & urgency. I must wait in silence the sombre movement of events. Still it is better to be gagged than give unheeded counsel.'[18]

Although he was out of sight, Winston was determined that he should not be out of mind, and he never ceased to adjure Clementine to keep in touch with his friends, and even 'pseudo-friends'.[19] 'Don't neglect these matters,' he wrote, 'I have no one but you to act for me. I shd like you to make the seeing of my friends a regular business, like your canteens wh are going so well. It is fatal to let the threads drop . . . There is nothing to ask of them – only represent me in their circle.'[20]

Clementine, like many others, was feeling the burden and plod of life on the home front. Her work for the YMCA was ever on the increase, and she was just now greatly occupied with a new canteen for a big munitions factory which was to be opened on Hackney Marshes, where a thousand men would have to be fed at one sitting. In her own family circle she had worries also: her mother had hurt her leg and was laid up, while the formidable old Lady Airlie seemed at point of death; she in fact recovered, but Clementine at the time wrote to Winston: 'She is having a long struggle, the light flickers & burns quite low and then burns up brightly again. It is a strange contrast to the swift deaths of strong young men.'[21] In another letter she reported that everyone seemed rather war-worn, and apologized that her account was not more amusing.[22] A tempting suggestion arose in her mind: 'If I came to Dieppe could you get 2 days' leave? I do long to see you. I'm very very lonely.'[23]

Many women would have found a real consolation in such days of trial and separation in the company of their children. Clementine loved her children, and was devoted to their welfare, but it is marked how small a part they play in her letters. Her whole life's centre was out there in France, and all her thoughts and actions revolved around Winston, his welfare and his interests. 'I hope you love me very much Darling,' she wrote at the end of one letter – 'I long for you often – I wake up in the night & think of you in your squalid billet & of all the women in Europe who are lying awake praying for safety for their men.'[24]

The 6th Battalion the Royal Scots Fusiliers was due to go into the trenches on 27 January, and they would move into 'support' positions a few days before. Earlier Winston had visited the part of the line to be occupied by his battalion, near Ploegsteert (which had instantly been rechristened 'Plugstreet' by the Tommies), a village a mile or two over the Belgian border. He had been impressed by the condition of the trenches and defences, and had reported favourably upon them to Clementine. As Battalion Commander he would live alternately in two places, the Hospice (Support HQ) and Laurence Farm (Advance Battalion HQ): 'in fact I shall only move back about 3/4s of a mile from the front line when we are in support and supposed to be "resting". Therefore for the next 2 or 3 months we shall all dwell continuously in close range of the enemy's artillery.'[25] He hastened to reassure Clementine by saying that the houses in that neighbourhood had not been much knocked about, and that losses had been small. But the knowledge of the exact date that Winston was to move into the front line threw Clementine into a panic: 'Oh my dearest, I can't get used to the idea,' she wrote on 21 January.

A few days later he tried to calm her fears:

> Don't worry I beseech you. Women & children will be living under exactly the same conditions within a quarter of a mile of my advanced HQ: & 2 nuns are still residing in the support HQ.
>
> It is splendid having you at home to think about me & love me & share my inmost fancies. What shd I find to hold on to without you. All my gt political estate seems to have vanished away – all my friends are mute – all my own moyens are in abeyance. But there is the Kat with her kittens, supplied I trust adequately with cream & occasional mice. That is all my world in England.[26]

Clementine was conscientious about Winston's instructions for keeping in touch with people. She wrote: 'I will try & see your friends, but everyone "in office" seems to be unbelievably smug – Were we like that when you were in power? There is an atmospheric non conductive barrier between those whose men are in danger & those whose men are in powerful security at home.'[27]

She gave him some penetrating appraisals of political figures. Lloyd George was temporarily weakened by his unpopular views on labour questions and compulsory service, and she commented:

> I think it will take Ll-George (even allowing for his marvellous recuperative powers) a long time to recover – If tomorrow the P.M disappeared Bonar Law would be the successor. He has made a great impression in the House during these last weeks by his skilful handling of delicate topics & this impression will spread to the country. Myself I think Bonar is not a big man, but he is a very skilful one & does not miss his markets. I think Ll-George will remain 'perdu' for a bit & then gradually slide away from his 'compulsion' attitude towards the working men. Montagu after an absence of 6 months from the Cabinet finds very little change except a greater disinclination to action, the only Warrior is Curzon.
>
> God bless you my Darling and keep you safe
>
> Your loving
> Clemmie[28]

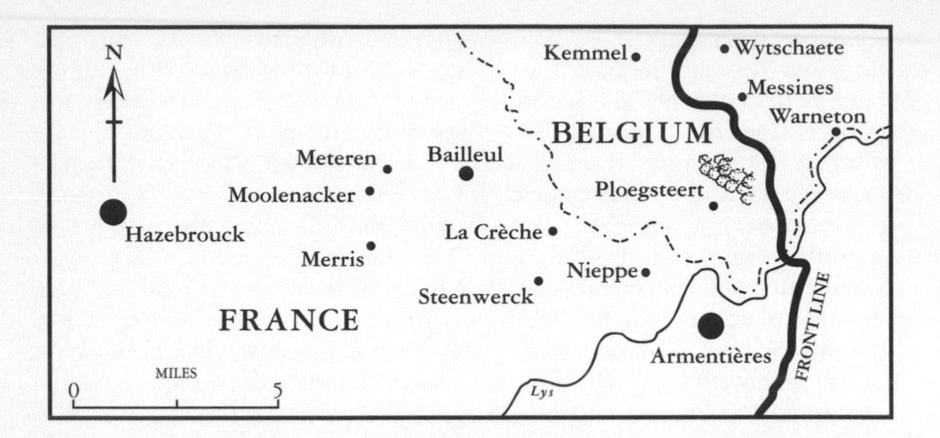

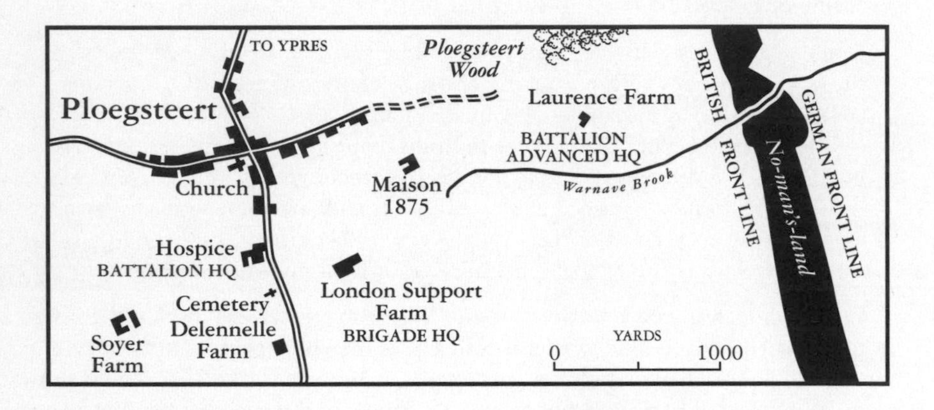

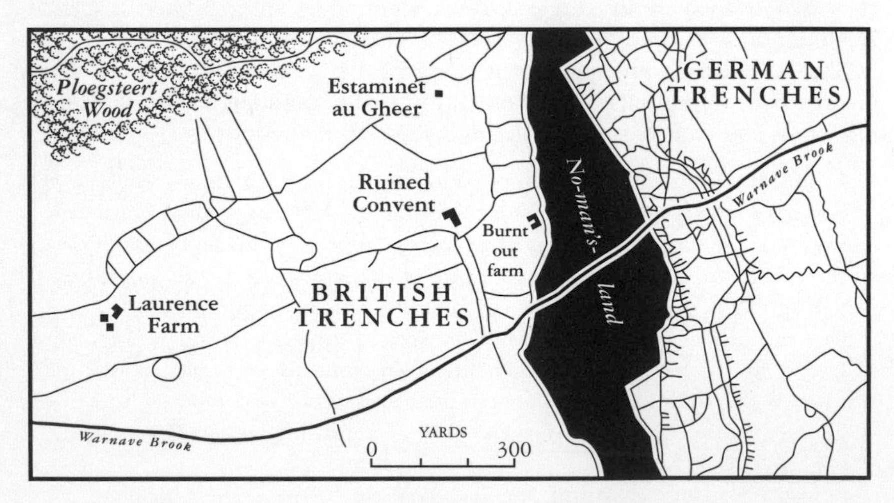

THE PLOEGSTEERT SECTOR OF THE WESTERN FRONT, 1916

Clementine had been for a long time wary of Lloyd George, and her suspicions were not mitigated by his charm: 'I get on so well with him & I know he likes me, but he is a sneak – I would never like you to be intimately connected becos tho' he seems to recover again & again from his muddles & mistakes I am not sure his partner would; he would instead be saddled with the whole lot while Ll-G skipped off laughing.'[29] Earlier on she had written of him: 'He is a barometer, but not a really useful one as he is always measuring his own temperature not yours!'[30]

But Winston continued to differ with her on this subject, writing: 'I am sorry for what you tell me about L.G. he has been vy faithless & is now friendless.' In justice he reminded her: 'Still he has been more on the true trail than anyone else in this war.'[31]

It had obviously occurred to Winston that he had been somewhat peremptory in his tone when urging Clementine not to lose touch with his friends and the political world generally, and he wrote almost apologetically: 'My beloved, I wrote a miauling letter yesterday, and I expect the Kat will be flustered by my directions to keep in touch with so many people. Do only just what comes easily & naturally to you my darling. On the other hand don't simply vanish out of the political circle & plunge into bed & canteens.'[32]

Clementine had responded vigorously to his 'miauling' letter: 'Now don't scold your Kat too much for being a hermit. Here in two days I have hob-nobbed with Montagu, Birrell, Lloyd George & a South African potentate [Sir Abe Bailey]! Tomorrow night I am dining with Cassel. Please send me home the Distinguished Conduct Medal at once & much praise.'[33] But she found most of the politicians she met cynical and remote. On 29 January she wrote a long letter:

> My darling,
>
> I dined with F.E. [now Attorney-General] & Lady F.E. on Thursday night . . . F.E. was most pleasant & mellow, but like Lord Robert Cecil has become an absolute mandarin & is enamoured of the Government & all its machinery – I suppose that a dignified position & heaps of money have inevitably a sedative effect. After dinner we dropped in for 5 minutes at the Empire & so to bed – Yesterday I put on 3 layers of armour & went to luncheon with the McKennas [he was now Chancellor of the Exchequer]. There I met Sir Ernest Cassel & my sense of humour was tickled by the contrast in their attitude towards the war, the red-hot patriotism of the German & the tepid counter-jumping calculation of the Englishman – McKenna's candour is astonishing – He wld reduce the size of the present army if he had the power. His plan is to pay our allies to do all the fighting while we do all the manufacturing here. He reeled out strings of figures none of which I was in the position to question; he really is a most noxious creature. Sir Ernest & I walked away from the house much depressed. I expressed myself rather forcibly which rather shocked him; (altho' he agreed) he is much impressed by the purple of authority & office. I am afraid the war will go on forever at this rate.

Winston was much delighted by these accounts of Clementine's social activities and showed his appreciation: 'You have indeed been active seeing all those people . . . Persevere, the D.C.M. is yours.'[34]

Enclosed in nearly all Clementine's letters were any newspaper cuttings she thought would interest him: she never sheltered him from the slings and arrows of public comment, any more than Winston spared her details of the dangers through which he passed. Commenting on one batch of cuttings, she wrote: 'I keep sending you the press-cuttings – It is quite out of the common to come across a sentence that is not ignorant & prejudiced.' She then added a remark which reveals her good and clear understanding in these matters: 'But all this does not really influence the public against you becos' the public is very fond of your personality.'[35] Through all the ups and downs of his long life this dictum remained largely true.

Her faith in Winston's future was unwavering, and she sought to reassure him:

> Do not fear, your political estate has not vanished, it is all waiting for you when the right moment comes which (Alas for the country) may not be till after the war – If only you come safely thro' . . .
>
> To-day is your first day in the trenches & how I pray you may be protected I hope you wear a steel helmet always & not the Glengarry.
>
> It seems so hard that I cannot come & see you – It would be so easy & I cld live with the poor French women in a ruined cottage & hoe turnips.[36]

The nuns, who were so gallantly sticking it out and keeping up the little chapel in the Hospice, received Winston and his colleagues 'most graciously' when they marched in. That evening, 26 January, Winston wrote:

> I am extremely well lodged here – with a fine bedroom looking out across fields to the German lines 3,000 yards away . . . On the right & left the guns are booming, & behind us a British field piece barks like a spaniel at frequent intervals. But the women & children still inhabit the little town & laugh at the shells wh occasionally buff into the old church.

On the eve of their going into the trenches Churchill delivered a short homily to his young officers, counselling them, among other sensible precautions, to have a spare pair of boots, and reiterating his theme that 'War is a game that is played with a smile' – to which maxim he added the rider: 'If you can't smile grin.'[37]

The next day the battalion moved up into the trenches for a tour of forty-eight hours (future spells of duty were to be of six days' duration). The two days turned out to be relatively calm and uneventful, but Winston wrote: 'You know a Colonel's day in the line is almost the greatest personal demand on a man's qualities – vy like being a Captain of a vy big ship in submarine infested waters.'[38] The battalion was relieved on 29 January, and he reported that it had all gone off with clockwork precision – the changeover being completed without casualties. He had been delighted with the bearing of his officers and men. When they got safely back some of them had attended mass in the chapel, where the old priest, despite a shattered church and a house frequently under shell-fire, was still at his post. But when Clementine wrote on 30 January, she did not know that Winston and his comrades were back in their support billets, and her letter breathes anxiety:

> My darling
> Today the newspapers contain very alarming news of a German attack – I wish I knew for certain that it was nowhere in your neighbourhood – No letters from you lately, the last is dated Jan: 24th. – The communiqué is brief but we have had them brief before and a very long casualty list to follow . . . I long for the return of Lloyd-George & F.E. for personal news of you – I pray this attack of the enemy may not lead to a foolish counter offensive on our part. The air is thick with rumours that our push is to be made quite soon.

The mood at home was subdued, and people were weighed down by the long and weary course of the war. Many people's lives had changed, and Clementine gave an account of a former gilded socialite, Irene Lawley, 'rather pretty & fluffy & used to be rather silly', who under the harrow of war had become a person of consequence and endurance. '[N]ow,' wrote Clementine,

> she gets up at crack of dawn & nurses all day at a hospital & in her off time she drives herself in a little open car – She took me home last night thro' the pitch black streets driving most skilfully . . . She wears white fur clothes & the car is pale blue & altogether she looked rather attractive but very over-worked and unhappy & I then recollected that Lord Vernon whom she loved and Charles Lister who loved her are both dead –
> The atmosphere here is very bleak & gloomy & chills my heart in the spare moments of my work – The Government people are unbelievably smug – I am seeing them occasionally to please you my Darling but I cannot take any interest in these soul-less cold blooded tortoises & I have ceased to feel any curiosity about them.

One feels in this long letter that Clementine had reached a still centre of depression and sadness. In it she opened her heart about her religious feelings – a subject she rarely spoke or wrote about:

> I fear this trivial letter will be no good to you my Darling; I am rather in the rut to-day & can see only miles & miles of uphill road – Are your men and officers religious? Tell me if being near danger makes you think of Christ. Being unhappy brings Him to my thoughts but only, I fear, becos' I want to be comforted not becos I want Him for Himself . . .
> Goodbye my Dearest love I send you tender kisses & fervent wishes for your safety. Your loving
>
> Clemmie

Enclosed with this sad letter, copied out in her own hand, was this poem by Christina Rossetti:

UP-HILL

Does the road wind up-hill all the way?
Yes, to the very end.
Will the day's journey take the whole long day?
From morn to night, my friend.

But is there for the night a resting-place?
A roof for when the slow, dark hours begin.
May not the darkness hide it from my face?
You cannot miss that inn.

Shall I meet other wayfarers at night?
Those who have gone before.
Then must I knock, or call when just in sight?
They will not keep you standing at [that] door.

Shall I find comfort, travel-sore and weak?
Of labour you shall find the sum.
Will there be beds for me and all who seek?
Yea, beds for all who come.

For the most part Clementine kept her anguish and anxiety to herself. But fortunately she had a loyal confidante in her sister-in-law, Goonie; and on many evenings during these dark months she would slip round the corner to No. 5 Cromwell Place, where the Laverys lived. They were staunch and understanding friends, and it soothed Clementine to sit quietly in John Lavery's studio while he painted. He did a portrait of her at this time, which seems to distil sadness and stillness. It hangs now, in its accustomed place, in Winston's study at Chartwell.

The news that Winston was once more out of the line, and in a situation of comparative safety, revived her spirits. Moreover, she was now deeply engrossed with the arrangements for the official opening by Lloyd George (Minister of Munitions) of the new canteen for which she was responsible. She had some anxieties about it: 'I am feeling rather nervous about Lloyd George addressing my Munition Workers; the skilled men are growling about his visit; they are angry about the "diluted labour"*. It will be very disagreeable if there is an uproar especially if it makes my nice canteen unpopular!'[39] She was also somewhat concerned 'if the Meeting gets widely reported, that people will think you & he are working together. You may have to work with him but never trust him – If he does not do you in he will at any rate "let you down".'[40]

Clementine (whose claws were evidently fully extended at this moment)

* The employment of women and unskilled men in the munitions factory, vehemently opposed by the trades unions.

also sent Winston a barbed account of a meeting she had with A. J. Balfour (First Lord of the Admiralty).

> I saw Mr Balfour yesterday playing tennis, looking more like a shabby maiden aunt's tabby cat than ever. He said 'remember me to Winston & tell him that Jacky Fisher is on the rampage.' I would have asked him to be more explicit but he took himself off – to rest before dinner I should think! I wish I had had your letter telling me of the German aeroplanes visits behind our lines (it arrived this morning). He was discussing the Foker [*sic*] machine & said it never ventured off its own ground but only attacked us when we raided. He is like the rest of them smug, purblind, indifferent, ignorant, casual![41]

During the last days of January Winston was cheered by seeing F.E. and Lloyd George, who visited GHQ during this time and made contact with him. He enjoyed meeting them, but these encounters had not aroused in him any feelings of envy or discontent: 'I must say I felt vy strong & self-reliant meeting these two men today: & did not envy their situation or regret at all my decision to quit them.'[42]

On 1 February his battalion again moved into the trenches, this time for a period of six days. He described to Clementine some of the 'disagreeables' they had exchanged on the first day, and broke off suddenly to ask her if such detailed accounts of these incidents caused her too much anxiety – 'But I think you like to know the dimensions of the dangers & what they are like.'[43]

On 4 February the opening of the canteen at Ponders End, the arrangements for which had been such a preoccupation for Clementine, took place; it was a resounding success. 'The great meeting is over & the whole thing went off brilliantly,'[44] Clementine began a long and lively letter to Winston that evening. The proceedings had started with a reception in the canteen:

> There was my Head Cook, Mr. Quinlan resplendent in white coat & hat, the paid staff in brown holland overalls & my voluntary workers (about 150) looking like blue and white angels. There was a great crowd of people from the neighbourhood & some from London and all the Y.M.C.A. swells . . . your mother & Violet Bonham Carter & Lady Horner & Lady Henry Grosvenor who runs Woolwich etc as I do north London.

Violet's presence had intrigued Clementine, because she had invited herself:

> she rang me up & insisted on coming. I think she came to see (on behalf of Downing St) what was up between you & Ll-G. She was very pleasant & agreeable.

Clementine escorted Mr and Mrs Lloyd George round the whole establishment, introducing them to numerous people, and they all then proceeded for the meeting to the new workshops, where the men were gathered:

> 2000 of them packed like sardines standing very silently. They did not cheer when they saw Ll-G but looked at him with interest & curiosity, (but don't say I said so), they gave me a beautiful cheer.

In fact, the Manager had told Clementine that the previous day 300 men had threatened to file out of the meeting in silent protest against Lloyd George's labour policies, and had promised not to do so only when the Manager had pointed out that such an act would be a great discourtesy to Clementine, whose guest Lloyd George was. 'Of course Ll-G does not know this,' she wrote, 'but I must say it made me feel very superior & protective towards him!!'

The Manager paid a warm tribute to Clementine and her band of workers, and then Lloyd George addressed the meeting. 'Ll-G made a quite undistinguished speech,' she wrote to Winston, 'and the shabby little tike altho' he said he had just returned from the Front never mentioned your name.' However, 'He was very well received but not enthusiastically.' For her own part,

> I was simply deluged with presents a bouquet, a cheque from the Directors of the firm for 100 guineas, (not for myself! but to spend as I like on canteens) sleeve-links for you from the men and a little brooch in the shape of a shell from the men enclosed in a really lovely gold box studded with turquoises, pearls & diamonds for me. I nearly fainted with emotion & my speech was wrecked as I had not expected these gifts – (except the cheque). But I just read out your message & the men were delighted – Don't tell anyone about all this as it sounds vain, but I want you to know about my small success. I really have worked hard but now I shall have to redouble my efforts to deserve all this. I feel I must give the men fat chickens every day to eat!

The events of the day had strengthened her contempt for Lloyd George:

> On the way home he said quite casually 'I'm so surprised, Curzon wants the "Air", I thought perhaps Winston might have done it – Do you think he would have liked it?' I said 'Winston would do it better than anyone else.' He did not reply – I don't hate him, but I feel contempt & almost pity for him. This ungenerous cautious streak in his nature will in his old age which is fast approaching leave him lonely & friendless – Ishmael! I do not think you will ever need him, he will need you when he is on the down gradient – & of course you will help him and he knows it. As we drove home silently in the dark he was very white shabby & tired & I felt young strong & vital & felt you out there young & strong & vital & I thought & I know he felt me thinking 'If only Winston is not killed you will need us both' – Meanwhile you won't strike a blow or speak a word for him becos' you know it won't make

any difference to his attitude later – You can always rely on him in any 'Marconi' affair.[45]

The triumphant success of the canteen opening had completely dispelled her depression and gloom of a short while before.

My Darling Dear One, to-day I feel happy & hopeful about you & your future. I know (D.V.) that you will come back rejuvenated & strengthened from the War & dominate all these decrepit exhausted politicians. Don't close your mind to the P.M. entirely. He is lazy but (or perhaps therefore) healthy & anyhow he is not a skunk tho' a wily old tortoise. I must meet him this week and tweak his ear – I feel full of beans as if you were Commander-in-Chief and not merely Lt. Colonel . . .

God bless you & I send you a thousand kisses.

Clemmie
[sketch of cat][46]

But her day of joy and triumph might well have ended in heartbreak, for on that same day Winston, Archie and some others were just finishing luncheon at Laurence Farm (Battalion Advanced HQ) when a shell burst close by. They were just discussing the prudence of repairing to a dugout when

there was a tremendous crash, dust & splinters came flying through the room, plates were smashed, chairs broken. Everyone was covered with debris and the Adjutant (he is only 18) hit on the finger. A shell had struck the roof and burst in the next room – mine & Archie's. We did not take long in reaching our shelter – wh is a good one! My bedroom presented a woe begone appearance, the nose of the shell passing clean through it smashed the floor and cut a hole in the rear wall. Luckily vy few of my things were damaged . . . The wonderful good luck is that the shell (a 4.2) did not – & cd not have – burst properly. Otherwise we shd have had the wall thrown in on us – & some wd surely have been hurt.

I have made them put up another still stronger dugout – quite close, on wh they are now hard at work. I slept peacefully in my tiny war-scarred room last night, after a prolonged tour of the trenches.[47]

Three days later, Winston was summoned to Brigade Headquarters to take over command in the temporary absence of the Brigadier, and he had a long talk with Lord Curzon (Lord Privy Seal), who was paying a visit to the Front. The possibility had been mooted in some quarters (indeed, Lloyd George had mentioned it to Clementine) that Winston might be called home and made responsible for air affairs. Clementine wrote about this to Winston on 7 February:

Before this letter reaches you, you will have seen Garvin's appeal in the Observer for you & Lord Fisher to return. You as Air Minister, Fisher as First

> Sea Lord. If only you had been given the 'air' last May something substantial might by now have been accomplished.
>
> Oh my Darling I long so for it to happen, & I feel that it would except for the competition for the post inside the Cabinet – There are, Alas, 12 Ministers with minor officers who probably all think themselves competent. I have been invited to stay at Walmer* next Sunday. How wonderful it would be if by then you had returned.

In the same letter she gives him some news of the children:

> You will be surprised when you see Diana [six and a half]. I have cut off her hair & she now looks like Peter Pan – A great improvement! Sarah [sixteen months] is on the verge of voluble speech & is only waiting for some teeth.

A few days later she took Diana and Randolph [four and three-quarters] to a children's party. Clementine reported, all aglow with maternal satisfaction, that 'They both looked quite beautiful, quite different from the other children. I felt very proud to have produced 2 such delicious beings.'48

On 12 February (at 5 a.m.) she wrote:

> Yesterday Goonie & I lunched with Lord Curzon at his handsome Mansion which (to indicate war economy even among the highest in the land) was swaddled with dust sheets – He was very affable & talked a great deal about you & also & especially a great deal about himself & his experiences in Flanders! He told me you were well & in high spirits & said he thought you would soon be a Brigadier – But he did not as I should have liked express indignation and surprise that you should be commanding a battallion [*sic*], but seemed to think it very natural, suitable & proper – What short memories these people have – ! He discussed the possibility of your being made a Brigadier from the point of view that Brigadiers have £1000 a year whereas Colonels receive but £500 & expressed the opinion that as I must be feeling rather poor the promotion would no doubt be a very welcome one to me! Altogether he was most superior & patronizing tho' genuinely friendly & Goonie & I laughed at him a good deal as we drove home in his motor which he kindly lent us for the afternoon – He gave me 3 bottles of brandy for you which I will despatch by a tame messenger.

Winston himself was under no illusions, however, as to either office at home (remarking that Asquith regarded him now more as a critic than as a colleague) or promotion in the field, about which last prospect he wrote:

> Neither do I expect any speedy promotion here. The kind of work I am doing is being done equally well by many others . . . Haig will no doubt eventually

* Walmer Castle, on the coast of Kent, is the official residence of the Lord Warden of the Cinque Ports. Lord Beauchamp, the then holder of this ancient and honorific office, had lent the castle to the Asquiths as a weekend retreat.

– if I survive – give me a Brigade. But he will be chiefly concerned at the impression such an appointment wd produce in the army: & he will certainly run no risks on my account.[49]

Clementine spent an agreeable weekend with the Asquiths at Walmer Castle, and on Sunday, 13 February she wrote to Winston:

My Darling,

The journey here yesterday was terribly long & tedious; we took four hours & the train stopped at each station.

But I am glad to have come as I think it is useful & also I am having a pleasant Sunday. A perfect day, glorious sunshine, the sea blue & without a ripple & no wind—

I sat in the garden most of the morning & found growing out of doors a pink rose, a white violet & a sprig of cherry pie!

Sitting on the bastion they had been able to hear the rumble of heavy guns. Clementine had played golf with 'the Prime who was very pleasant & mellow . . . at one moment [I] thought I was going to give the old boy a good beating (which I shd have relished) but Alas! I fell off towards the end & he won by a short length.'

Winston upbraided Clementine mildly when he replied to this letter (as on a previous occasion) for giving him so few details of her conversations with the Prime Minister – to which Clementine answered that her exchanges with Mr Asquith had been of the most frivolous and trivial nature: 'You know what the P.M. is – He loathes talking about the War or work of any sort – He asked anxiously if you were happy.'[50]

In this same letter Clementine gave Winston some sage advice about his relationship with the Prime Minister. It shows how well she understood the workings of the minds of the powerful:

I think if you could do it, you might write to him occasionally private interesting friendly letters. With him it is so much out of sight out of mind – I am sure that he feels affection for you & that he would like you to be in the Government again, if it could be done without a row. This sounds very cowardly, but few Prime Ministers would do more. Anyhow nothing is gained by letting him see that you consider he has behaved badly; he only waddles off as quickly as possible & avoids you in future.

On 16 February she told Winston that 'Three letters from you arrived in a covey.' One of these had been written in a sombre mood. Clementine wrote:

Do not I pray of you let this mood deepen & permanently tinge your heart & mind. There are only two things that can prevent you from being again the heart of action in this country – I mean your death or a serious wound. But I

> will not believe that either of these terrible things are going to happen. I am sure you will return to power after the war with increased prestige.

But while she looked forward with ardent hope to the brilliant future she was confident lay before him, Clementine made a heartfelt plea for a real place in his life for her and the children:

> [They] are becoming so grown up & intelligent – They will be very sweet companions & I look forward with longing to the time when you come back to me & we will have a little country basket & in the intervals of your work we will all curl up together in it & be so happy. Only you must not become too famous or you won't have time for these pastoral joys! You will have to promise me that in future however full of work & ideas you are you will keep out of every day an hour & every week a day & every year 6 weeks for the small things of Life. Things like painting . . . playing grizzly-bear, sitting on the grass with me & generally Leisure with a big L.

Although in their life together love never failed, 'Leisure' was always to be at a premium.

On 13 February Winston had gone back into the line for another tour of duty of six days, and every day he penned a long letter home, usually writing in the evening before setting out on his nightly patrol of the trenches with Archie Sinclair. 'I sit in a battered wicker chair within this shot scarred dwelling by the glowing coals of a brazier in the light of an acetyline lamp.'[51]

After two relatively quiet days the Headquarters was again the victim of a direct hit, when both the 'dining-room' and Winston's and Archie's bedroom were battered, and the signal office next door completely shattered; but, rather than move their Headquarters, they were busily piling sandbags inside all the walls on the upper floor. 'We have been hunted by shells during these last two days in "rest billets",' Winston wrote a little later; three times in a fortnight his bedroom (three different ones!) had been pierced by shells, and the church steeple which had withstood sixteen months of fighting had come down under shell-fire. 'One lives calmly on the brink of the abyss,' he wrote, but he now appreciated how people wearied under the strain, the excitement dying away leaving only a feeling of 'dull resentment'.[52]

The dream of a few days together in Dieppe had faded, but Winston now knew he would be able to take a week's leave at the beginning of March. He told Clementine that he would cross to Dover in a destroyer, and suggested she should meet him there at the Lord Warden Hotel. He gave her precise instructions for the plans for the short and precious time he would be at home: 'You must parcel out the days as well as possible. I will have one dinner at my mother's, at least 3 at home, 2 plays alone with you & one man's dinner out somewhere.' He added that he hoped she would try to 'work in all my friends' at lunches, and that he wanted to have at least one day's painting in John Lavery's studio. Even on paper it did not look as if seven days could hold all he eagerly hoped to cram into them; but he assured her he would be 'vy good & keep all my engagements punctually'. He stated a distinct preference for inviting friends to dine with them at home rather than going out; he asked her to have a servant ready to look after him, but, finally: 'I put it all in your hands my dearest soul.'[53]

During these last days he had been in the line, Winston had complained of lack of communication from home, for he hungered for every crumb of news from Clementine. But there had only been a short hold-up in the delivery of her letters. She had continued to be a faithful correspondent, writing him lively accounts of all her doings.

By far the greater part of Clementine's time during these months had been dedicated to her canteens. Energetic and conscientious, she found (perhaps to her surprise) that she was also highly efficient. She told Winston with pardonable pride:

> I must tell you that from living with you & watching you for 7 years I have assimilated (in a small way) your methods & habits of work. In this Canteen work I find people who do it with me are surprised that I do things quickly & that I expect them to drive along too, & altho' I get very tired I find the others get tired first. When I am in full swing I begin work at 9 a.m. & finish at 7.30 p.m. It's no use scolding me becos' it's all your fault – You have taught me to work outside office hours.[54]

The Ponders End canteen was to be enlarged, with another wing added for the men; in addition she was also to be responsible for feeding an extra 500 girls. Soon her canteens would be feeding 1,800 people.

While he was proud of her devotion and ability, Winston, like many (if not most) men then, tended to regard any time not spent in wifely or maternal duties as mere occupational therapy to pass away surplus time. While Winston and his affairs always commanded the cream of her heart and energy, Clementine now was experiencing for the first time the challenge and satisfaction of a job, and her thoughts were as much engrossed by its daily demands and problems as by the political events of the hour. Her days must have been very long. Letter after letter is marked as being written around five or six o'clock in the morning.

Her own demanding work gripped her, not only because she had a strong desire to serve, but also because it gave her an insight into the new phenomenon of the employment of women working alongside men in heavy industry. She had visited Woolwich Arsenal with Lady Henry Grosvenor, from whom she learnt a great deal. She gave a vivid account of her visit to Winston:

> Very soon 90,000 workers will be employed in the Arsenal & if the war goes on long enough I should think you could bury the Germans with the hail of Munitions – But Alas! the same thing is going on in Germany . . . I was

much interested in the girls at one big dining room – They earn between 18/- & 25/- weekly. They are nearly all quite young very fresh and pretty & rather hoydenish – Some of them were snowballing with boys outside the canteen. The fore-women have their meals in a separate room. They are of course older & they belong to the professional & suffragetty classes & some of them look as if they had seen 'life' – They were nearly all smoking & playing Bridge & nap after their meals! The quiet ones were reading rather advanced books, translations from the Russian Authors & such-like.

I understand the prejudice of the skilled male worker against diluted female labour. The women are full of beans & become terribly skilful very quickly.

They want good money too, but of course nothing like the men – A woman earning 35/- a week feels like a millionaire. They don't mind paying properly for their food now. They used, a few months ago to grudge 6d for a good dinner.[55]

Clementine was overjoyed when she got Winston's letter with the first inkling that he might soon have leave. 'My Darling,' she wrote on 19 February (at 5 a.m.),

Your last letter brings the joyful news that very soon you will be home on leave. This makes me very happy & I am now beginning to count the days. To-day you come out of the trenches so for a few days I feel more at ease – This last week has been a very anxious one, as although you have not been in the thick of the onslaught the Germans seem to have plenty of every sort of ammunition to spare for other parts of the line. You seem to have had a tremendous dose. Please do not become rash & prowl too much in the moonlight – I am sure it is a great risk.

The burden of anxiety was heavy during those days when Clementine knew that Winston was in the line, and earlier she described the sense of release from tension when she knew one such period of acute danger was safely over: 'Till I got your letter yesterday I had not realized you were now in rest-billets – The relief I feel is quite extraordinary – One doesn't realize how high the tension is until it is suddenly lowered.'[56] Now she had to endure one more anxious stretch of days while he was again in the trenches, and then, God willing, she would have him home again.

Meanwhile she was going about quite a bit, and promised she would report to Winston 'how the London world is wagging'.[57] In one week she had been out to dinner or entertained at home each evening. Her own guests had included Lord Rothermere and the F. E. Smiths. But her most ambitious social undertaking was a dinner for the Asquiths, when she had to

work like a beaver to get together the 8 indispensible [*sic*] bridge players which are necessary for their comfort & happiness . . . the party is: – P.M.,

> Margot, Ivor Wimborne, Lady Mainwaring, Bogey Harris, Goonie, Mr. Cazalet, myself, Sir E. Cassel. Nine people becos' Lady Mainwaring refuses to play bridge as she always bunny-hugs & fox-trots after her meals! . . . We invited her as we thought the P.M would like something quite new![58]

The evening turned out a great success:

> the old sybarite [Asquith] thoroughly enjoyed himself – Sunny [Marlborough] turned up at the last minute & made himself very agreeable & as Ivor was there too, the Churchill family presented a solid & prosperous appearance. The food was good & afterwards there were 2 tables of bridge – I sat between the P.M. & Sir Ernest Cassel – The P.M won a little money & went home in high good humour. Ivor remained behind & philosophised & moralised.[59]

Less than a week now remained until Winston's eagerly awaited leave. There was a last-minute scare that the threat of a new German offensive might lead to all leave being stopped – but all was well in the end, and on 2 March Clementine went to Dover to meet him.

One can conceive a more romantic setting than the Lord Warden Hotel on Dover Pier. But it would be churlish indeed to criticize the scenery, when, despite the malice of the enemy and the dislocation of war, 'journeys end in lovers' meetings.'

CHAPTER TWELVE

Day Must Dawn

WINSTON'S LEAVE IN MARCH 1916 LASTED TEN DAYS, AND AT THE END OF IT both he and Clementine were physically and mentally exhausted and in turmoil. Many of the plans she had made for him – and so skilfully dovetailed so that no precious minute would be wasted – had to be abandoned or changed, for almost as soon as he arrived home Winston became embroiled in political machinations, as a result of which he decided to take part in the debate on the Naval Estimates. Several friends and confidants he saw during his first few days at home took the view that the time was now ripe for him to intervene actively once more in politics. Among those whose views he sought were F. E. Smith, J. L. Garvin (Editor of *The Observer*) and Sir Max Aitken.

Clementine had never taken to Max Aitken, the dynamic Canadian who was shortly to acquire the *Daily Express* and would become the 1st Baron Beaverbrook*, and she viewed with uneasiness his steadily growing friendship with Winston. Her lifelong mistrust of him may well date from this period when Aitken was plying him with political advice with which she profoundly disagreed.

She was also dismayed that Winston was once more in close touch with Lord Fisher, who had been the immediate cause of his downfall at the Admiralty. The terrifying old man was once more on the rampage over the affairs of his beloved Navy, and urging Winston in his usual immoderate terms to attack the Government head-on about their mismanagement and ineptitude, assuring him that he alone could lead a united Opposition, powerful enough even to topple the Government.

Clementine's heart must have sunk when she saw Winston so deeply involved with people she neither liked nor trusted, and whose advice ran contrary to her deepest instincts. One day, during a luncheon at Cromwell Road at which Lord Fisher was present, she could contain herself no longer, and suddenly burst out: 'Keep your hands off my husband. You have all but ruined him once. Leave him alone now!'[1]

* Sir William Maxwell Aitken (1879–1964), politician and newspaper magnate, created 1st Baron Beaverbrook, 1917. He took over the *Daily Express* in 1919, and founded the *Sunday Express* in 1921.

These persuasive counsellors lent their weight to Winston's already awakened desire, and he determined to speak in the debate. The night before, the Asquiths dined with the Churchills, and Winston was perfectly open as to his intention of speaking the following day. After dinner the Prime Minister, in a private conversation, sought to dissuade him from a step he felt sure would be unwise, and do Winston nothing but harm. But it was to no avail.

On 7 March Churchill launched a broadside attack in the House of Commons on Balfour's ineffectual tenure at the Admiralty. He deplored the inactive role played by the Navy and called for a new shipbuilding programme. He made a powerful case and was listened to with serious attention, and his speech might have done much to repair his political fortunes had he not flung all away by his conclusion, which was nothing less than an ardent appeal for the recall of Lord Fisher as First Sea Lord to supply the driving force and mental energy which were so patently lacking. This appeal was greeted with derision and hostility; Churchill was stunned both by the violence of the reaction he provoked and by the lack of any general support.[2]

While he was out in France Winston had ever been pondering the question as to where his duty lay. Now renewed active participation in politics, and discussion with his friends and colleagues, brought the issue to a point where an immediate decision seemed inevitable. But there were two views on it, and, as on the 'brigade' question, Clementine again found herself in opposition to Winston's line of thought. She felt there were many reasons why his return to home politics should be postponed until a more opportune moment, reasons she was to beg him to consider now and in the following weeks. However, his view prevailed, and when she bade him farewell on the pier at Dover on 13 March, she had in her charge a letter from her husband to the Prime Minister, asking to be relieved of his military duties so that he could return home to take part in public affairs.

Clementine spent that night at the Lord Warden Hotel, from where she wrote to the Prime Minister, enclosing Winston's letter; she wrote also to one or two other colleagues who had been involved in the anxious discussions of the last few days, telling them what was afoot. 'I then crept into bed & thought about you and prayed for happier days & calm waters.'[3]

She was up early the next morning to go and see Sir Edward Carson, the Unionist leader and Attorney-General in the Coalition Government, who was ill at his home at Birchington in Kent, bearing a letter in which Winston sought this brilliant man's advice in his personal dilemma. She had a long and difficult time locating his house, but finally arrived, and was received by Sir Edward, who was still confined to his bed. They had a long talk, and that same evening she wrote Winston a full account of it. Carson took a grave view of Winston's course of action, and said of his part in the debate: 'Winston has probably done the country & the government a service by his intervention last Tuesday, but he has not done himself any good.'[4] Clementine felt strongly the wisdom of Sir Edward's view:

> I wish so much that we had planned things better & that you had seen him before sending the P.M. your letter – I am sure that tremendous decisions require time to resolve & a quiet atmosphere. The inferno of last week was not a favourable atmosphere. I hope I have not given the impression that Carson does not think you would be an asset here – He does but would like the asset in an undamaged condition. I feel sure that nothing will be lost by waiting.[5]

Clementine did not leave Sir Edward's house until half-past one, after which she had a long and weary drive home through blinding rain, made worse by the tedium and frustration of a puncture. She arrived back at Cromwell Road in the evening, worn out, to find a letter from Asquith's Private Secretary telling her that Winston had telegraphed the Prime Minister, asking him to disregard the contents of the letter forwarded by Clementine from Dover. Although Clementine was taken aback by this turn of events, she was also deeply relieved: 'I am anxiously waiting to hear further, but in the meanwhile if it means you are taking more time for consideration, I am sure my Darling that you are wise,' she wrote in a note she dashed off as soon as she had read the letter from Downing Street. 'My Darling I wish now I were near you. I feel that perhaps now you are in France you may revise your judgement. I long for you to come back, but I want you to come back welcomed & acclaimed by all, as you ought to be and as I know you will be very soon. Whatever you finally decide I loyally agree to.'[6]

But she was suddenly gripped by fear for the reason which might have prompted Winston to change his mind, and added:

> A dreadful anxiety has just seized me. I have recollected that you said that the only thing that would make you pause would be the knowledge that we are going to attack. Please God it is not the reason. Surely we are not going to make a counter-offensive? Let me know quickly the truth.
>
> My Dearest own Winston thro these tumultuous days we have been together I have never been able to tell you or shew you how deep & true my love is for you & how I know that ultimately what you decide will be right & good.
>
> God bless you
> Clemmie

Clementine spent the better part of the following day seeing various people who had knowledge of Winston's decision, and in putting them in the now somewhat altered picture, as best she could. After the visit of Sir Henry Dalziel (the owner of the *Pall Mall Gazette*) she wrote: 'He stayed some time and I think he is personally very friendly – But when I am re-incarnated and go into "Public Life" I have determined with great courtesy to hold all newspaper men at arm's length, so that at a distance they may gaze at me with curiosity interest & respect – If they came closer they might observe the flaws in my armour!'[7] Although writing half in jest, Clementine maintained this very policy when, in later years, she was eagerly sought after for interviews, which she rarely gave. Her relations with the press were invariably courteous, but distant.

Very early (5.30 a.m.) on 16 March, Clementine wrote to Winston again:

> My Darling
>
> Every time the post came yesterday I ran down to the hall for the letter explaining your telegram to the P.M. Every time I was disappointed – Late in the afternoon I retired to bed as my cold had gone down to my chest – I fell into a deep sleep & woke up only a few minutes ago – I immediately realized that while I had been sleeping, the best post of all (10 p.m) for letters from the front, had been & gone – I crept downstairs thro' the sleeping house & there on the hall table was the much wished-for blue envelope . . . I expect that my letters about the Carson interview will have seemed equally long in coming to you!

The anxiously awaited letter had been written on 13 March, and in it Winston explained that he felt he needed a little more time for thought 'in this vy different atmosphere' before committing himself to an irrevocable step. He had therefore telegraphed the Prime Minister to delay any action on the letter he had written at the end of his leave. There was no doubt that Winston found the choice before him a hard one to make and he wrote,

> My dearest soul –
>
> You have seen me vy weak & foolish & mentally infirm this week. Dual obligations, both honourable both weighty have rent me . . . I cannot tell you how much I love & honour you and how sweet & steadfast you have been through all my hesitations & perplexity.[8]

Clementine was deeply relieved that Winston had changed his mind, and in that same early-morning letter of 16 March she wrote an eloquent and moving explanation of her own thoughts and feelings in relation to the anguishing dilemma which confronted him:

> I think there are some solid qualities which English men & women value very highly – Virtues such as steadfastness & stability – After your speech in the House of C [November 1915] in which you placed yourself 'unreservedly at the disposal of the Military Authorities' it seems to me that more than your own conviction is needed that it is your duty to return to Parliament – I am convinced that sooner or later the demand will be made & that once made it will become insistent – Your speech has certainly animated & vivified [*sic*] the Admiralty but it has done you personally harm – I mean if you had been silent or put it differently the demand for your return would perhaps come sooner – But come it will – It must.
>
> I pray therefore my Darling Love that you may decide to bide the time – We are living on such a gigantic scale that I am sure everything ought to be simplified – our actions too, so that without explanation or justification they

& their motives can be understood and grasped by all – You have assumed the yoke of your own free will like many other men, tho' none of them are in your situation. The others, having assumed the yoke, cannot disengage themselves. You, owing to your exceptional circumstances have received the written promise of the head of the Government that in your speaking the word you shall be free. But that word must be spoken by others, if when free you are to be effective as an instrument to help the movement of events. Please forgive me my Darling if I express myself clumsily.

On 22 March, Archie Sinclair arrived on leave at Cromwell Road. Winston had confided him to Clementine's and Goonie's care, writing beforehand: 'I want him to stay at Cromwell & you & Goonie to cherish & nourish him. He is all alone in the world, & vy precious as a friend to me.'[9] Clementine found that he looked pale and careworn, and both the sisters-in-law did their utmost to look after him, and to give him a good time. Archie brought with him on his arrival only a short note from Winston, written in a somewhat morose vein: the battalion was back in the trenches; the weather had worsened; and he was losing his boon companion.

Knowing that Archie had earlier taken the view that he should remain in France, Winston wrote to him while he was in London: 'I can almost hear you and Clemmie arriving by the most noble of arguments at the conclusion that I must inevitably stay here till the Day of Judgement.'[10] But during these last days Winston's intentions about his future had been hardening, and his letter of 21 March to Clementine ended with this firm admonition: 'Be careful not to use arguments or take up an attitude in conflict with my general intention, & do nothing to discourage friends who wish for my return. On the contrary labour as opportunity serves to create favourable circumstances.'[11]

Various friends and colleagues of Winston's called on Clementine just now, with letters or messages for him, and she loyally transmitted his views to them. Among the callers was C. P. Scott, then Editor of the *Manchester Guardian*, of whom she wrote to Winston: 'He is a valuable champion becos' of his great integrity – & incorruptibility – but he is not a man of the world.'[12] She doubted whether he was a good judge of effective political action,[13] and she also wondered whether J. L. Garvin was a sound counsellor.[14] Clementine was meticulous in reporting their arguments, but adjured Winston to 'Keep a level mind my Darling, & a stout heart.'[15]

Although Winston's friends at home held differing views as to whether the time was ripe for his return, Clementine was convinced that he should wait. In the early hours (beginning 4 a.m.) on 24 March she wrote this long letter, once more advancing her own passionately held point of view and showing her keen awareness of his present political isolation:

My Dearest Love you know that you can rely upon my steadfastness & loyalty, but the anxiety & grief at the step you are about to take sinks deeper into my heart day by day. It seems to me such an awful risk to take – to come back just now so lonely & unprotected with no following in the House & no backing in the Country. Please do not be angry with me for writing plainly—

I think I know you very well & it seems to me that you are actuated by 2 motives – 1) You want to be in the place where your powers for helping to win the War will have fullest scope 2) You have a devouring thirst for 'War Direction' – Now I do beg you to reflect on this 2nd point – The war is (D.V.) 3/4 over, the corner is nearly turned – (Perhaps you will disagree) The end is a long way off but still I think we are going to win in spite of slowness & hesitation – In your present weakened condition shall you recover prestige & the necessary power in time to be of real use? I think perhaps you may if you wait a little longer – But it is a great risk – It is indeed a gamble – If you do not succeed you may gradually decline in the public opinion (tho' a speech would always attract attention). If that happened then your return from the battlefield in the middle of the War might be a serious handicap to you in the future. The Government is nerveless & helpless but it represents all there is practically in public life – If you come back & attack them they are bound to defend themselves & try to down you – And just now you are very defenceless. The Government may not wage war very vigorously but when on the defensive they are very strong.

Do not be anxious about my attitude – I do not tell my thoughts to any but you. When you were here last week I did not feel that there were any great or good elements of strength surrounding you. Fisher is a powerful but malevolent engine; you think him unimpaired by age but when the break-up comes, as soon it must it will come with a damaging explosion to all near him – Garvin and Scott, good time men & personal friends, but often wrong-headed, Dalziell [*sic*] curious & interested but corrupt & time-serving – in the Cabinet that Judas Lloyd-George never staunch in times of trial, always ready to injure secretly those with whom he is publicly associated . . .

Could you have the courage & the self denial if you returned to see no one, to help & not merely to criticize to refrain absolutely from personal attack. (You are vulnerable & your enemies would hope that you would indulge in this), to refrain also from all recriminations & attacks upon the past but only to apply your mind to the future – Not only the immediate future, but the great future beyond the war of which everyone is now (perhaps it is only the hopefulness of spring-time) perceiving the dawn.

I reflect upon you my Dear for hours together & lately I have had the time as I have had a little attack of bronchitis of which I have had to take care.

I was going away to the country for 3 days to try and shake it off when Archie was announced & so I put it off. I am not much fun for him as I can't go out.

The atmosphere here is wicked & stifling, out where you are it is clean & clear – I fear very much that you will be very sad & unhappy here.

You must forgive me for this letter. If I did not tell you my thoughts I could not write at all.

The War is a terrible searcher of character. One must try to plod & persevere and absolutely stamp self out. If at the end one is found grimly holding onto one's simple daily round one can't have failed utterly.

This long letter was followed only a few hours later – at 8.30 a.m. – by another one, which enclosed a letter from Sir Edward Carson, which she had read. 'You see how anxious he is that you should not blunt or break yourself as an instrument by a premature return,'[16] she wrote, again showing her concern for Winston's public image:

> When you do return the reason should be apparent to the man in the street, tho he need not necessarily agree with it.
>
> It would be damaging for instance if it got about that you had returned becos' you were dissatisfied with your prospects of promotion & irked by the smallness of your duties in your present position.

But whatever the counsels Winston received, for or against the wisdom of his returning, he was now firmly fixed in his own mind that his duty and his destiny lay in once more taking part in political life at home. The only point which remained to be decided was at what moment he should take this step. There had recently been a change of command at brigade level, which shut the door on any hopes he might have nurtured in that direction; but he did not want to leave his battalion while it was in contact with the enemy, and he also watched for an opportune moment as far as home affairs were concerned.

Winston's own analysis of the situation was that the broad facts concerning his resignation from the Army and return home could confidently be submitted to the judgement of the public. He set them out for Clementine when he wrote to her on 22 March: he had resigned his ministerial office and salary rather than hold a sinecure; he had served in France for nearly five months, 'almost always in the front line, certainly without discredit'; he had a recognized position in British politics, acquired over the years, 'enabling me to command the attention . . . of my fellow countrymen in a manner not exceeded by 3 or 4 living men', and in the present critical period of the national fortunes he felt he could not exclude himself from his responsibilities. 'Surely,' he wrote, 'these facts may stand by themselves in answer to sneers & cavillings.'[17]

Clementine was not convinced by his argument. 'The facts you mention in support of your immediate return are weighty & well expressed,' she wrote in reply, 'but it would be better if they were stated by others than yourself.'[18]

Meanwhile, the bronchitis she had already mentioned had worsened, and as soon as Archie's leave was over she intended to go to the sea for a few days to recover. This was a low moment for Clementine. Many women would have found unbearable the claims that politics made on Winston's whole existence. From the outset she had realized, and soon accepted, that they would always come first; but every now and then the desire that any woman might feel, to own a larger share of her husband's life, broke through. Now, lonely, tired, ill and beset with anxiety, she made an almost anguished plea: 'My Darling these grave public anxieties are very wearing –

When next I see you I hope there will be a little time for us both alone – We are still young, but Time flies stealing love away & leaving only friendship which is very peaceful but not stimulating or warming.'[19]

Nor was Winston oblivious to the extent to which his public life jostled their private relationship, for in a letter which must have crossed with hers, he wrote:

> I reproach myself so much with having got so involved in politics when I was home that all the comfort & joy of our meeting was spoiled. I had the share of events wh took an oddly exciting turn. And then the difficulty of reconciling two different kinds of lives & obligations or choosing definitely between them!

But he promised (one is sure with all sincerity) that

> next time I come home it will be with a set purpose & a clear course, & with no wild & anxious hurry of fleeting moments & uncertain plans. I am going to live calmly.[20]

And Clementine's heartfelt outburst drew from Winston a declaration of the depth of his feelings for her, and a rare (very rare) admission that he too knew a longing for tranquillity:

> Oh my darling do not write of 'friendship' to me – I love you more each month that passes and feel the need of you & all your beauty. My precious charming Clemmie – I too feel sometimes the longing for rest & peace. So much effort, so many years of ceaseless fighting & worry, so much excitement & now this rough fierce life here under the hammer of Thor, makes my older mind turn – for the first time I think to other things than action.

He went on to paint in his mind's eye the picture of an idyllic holiday: they would go to Spain or Italy

> & just paint & wander about together in bright warm sunlight far from the clash of arms or the bray of Parliaments . . .
>
> Sometimes also I think I wd not mind stopping living vy much – I am so devoured by egoism that I wd like to have another soul in another world & meet you in another setting, & pay you all the love & honour of the gt romances.[21]

Clementine received this letter in time for her birthday on 1 April. 'Your delicious loving letter arrived last night & warmed & comforted me. To-day is my birth-day & it is like the 1st of May in a poetry-book – so blue & sunshiny . . . I am 31, but if the war were nearly won & you were safe & had peace of heart I should not feel more than 20.'[22]

In order to recover from her bronchitis, Clementine spent four agreeable days staying with Sir Ernest Cassel at his house, Branksome Dene, in Bournemouth. She sat in the sun and took long walks along the beach, often in the company of Mrs Keppel, who was also recovering from an illness. Clementine had always liked her, but on this occasion she (rather touchingly) complained to Winston, 'she tries me by long political discussions about you & by arguing against your coming home – She is very fond of you personally – '[23]

On her return to Cromwell Road, Clementine found a number of letters that Winston asked her to forward, including one for Lord Northcliffe (founder of the *Daily Mail* and *Daily Mirror*, and proprietor of *The Times* since 1908), which he expressly told her to read first. Her cautious attitude to the press was again illustrated by her reply:

> I have dispatched all letters you sent me to forward with the exception of one that reached me this morning for Lord Northcliffe (which you said I might hold up for a couple of days) & which I earnestly beg you not to ask me to post, but to destroy. If it goes it will form part of your biography in after times, and after the way Lord N has flouted you I cannot bear that you should write to him in that vein. Besides I do not think it is as well expressed as some of your letters.
>
> I am sure it is no use writing private letters to great journalists – Even if they do, in consequence decide to run you, they feel patronizing & protective about it & the support then lacks in genuine ardour.[24]

Clementine's arguments must have played their part in Winston's decision to hold back indefinitely his letter to Northcliffe.[25]

She now knew that Winston was definitely determined, beyond any persuasion, upon returning home, and was only waiting to choose the moment to do so. Believing his best course was to remain for the time being at the Front, she found herself in an agonizing dilemma, and wrote on 6 April:

> My Darling own Dear Winston I am so torn and lacerated over you. If I say 'stay where you are' a wicked bullet may find you which you might but for me escape . . .
>
> If I were sure that you would come thro' unscathed I would say:- 'wait wait have patience, don't pluck the fruit before it is ripe – Everything will come to you if you don't snatch at it' – To be great one's actions must be able to be understood by simple people. Your motive for going to the Front was easy to understand – Your motive for coming back requires explanation.
>
> That is why your Fisher speech was not a success, people could not understand it. It required another speech to make it clear. I do long to see you so terribly – Your last visit was no help to me personally – I must see you soon . . .
>
> Darling Don't be vexed for me for writing so crudely – if to help you or make you great or happy I could give up my life it would be easy for me to

do it. I love you very much. By doing nothing you risk your life, by taking the action you contemplate you risk a life-long rankling regret which you might never admit even to yourself, & on which you would brood & spend much time in arguing to yourself that it was the right thing to do – And you would rehearse all the past events over & over again & gradually live in the past instead of in the present and in the great future . . .

If you will only listen a tiny bit to me I know (barring all tragic accidents) that you will prevail & that some day perhaps soon, perhaps not for 5 years you will have a great & commanding position in this country, you will be held in the people's hearts and in their respect. I have no originality or brilliancy but I feel within me the power to help you now if you will let me. Just becos' I am ordinary & love you I know what is right for you & good for you in the end.

Clementine's often expressed hope that Winston should wait to return until he was summoned by a call from a large body of public opinion was somewhat over-optimistic in the political circumstances of that moment; but her faith in his ultimate destiny – for which she ever strove to keep him untarnished – was not to be confounded. The fulfilment of her hopes expressed in these dark days took more than the five years she hazarded as a guess – but the day would come when she would see Winston indeed called by the voice of the people to a 'great and commanding position' and 'held in the people's hearts and in their respect'.

* * *

Winston, meanwhile, was still closely involved with his Army life and regular tours of duty in the front line. Clementine had been concerned lest his growing preoccupation with affairs and events at home should distract his interest and attention from his duties as a soldier, and she had written: 'I hope my Darling that you are giving your battalion as much care & thought now as you did before you had resolved to come home – I hope they don't feel that your interest in them has waned – I want them to miss you terribly.'[26] Winston reassured her on this point: 'I am doing my duty here vy thoroughly, and am touched often by small evidences of the partisanship of these young officers . . . The men put up my photograph in the trenches, & I am sure they will make an effort if I asked them and some big test came upon us.'[27]

Although Clementine knew that Winston had taken the decision she feared might so adversely affect his future, she none the less deployed yet once again both logic and emotion, in begging him to wait. In a letter on 12 April she argued that if the Government fell – as seemed possible in view of the renewed agitation for universal conscription – he would be included in any reconstructed administration, but that if Asquith survived, Winston would be at the mercy of his political enemies. 'The present Government may not be strong enough to beat the Germans,' she wrote, 'but I think they are powerful enough to do you in & I pray God you do not give the heartless brutes the chance – '[28]

To disagree with Winston on this issue caused Clementine untold anguish, and in her earnestness and distress her sense of reality seems for once to have deserted her:

> My Darling Love – For once only I pray be patient. It will come if you wait. Don't tear off the unripe fruit which is maturing tho' slowly or check its growth by the frost of a premature return –
>
> I could not bear you to lose your military halo. I have had cause during the 8 years we have lived together to be proud & glad for you so often, but it is this I cherish most of all. And it is this phase which when all is known will strike the imagination of the people: The man who prepared & mobilized the Fleet who really won the war for England in the trenches as a simple Colonel. It would be a great romance.
>
> You say you want to be where you can help the war most – If you come home & your return is not generally accepted as correct soldier-like conduct you will not be really able to help the war. You are helping it now by example. You are always an interesting figure, be a great one my Darling – You have the opportunity.[29]

By entrusting her with the letters to his various confidants and political allies, Winston had laid a distasteful task upon her; Clementine ended her long letter with a cry of protest:

> Oh Winston I do not like all these letters I have to forward – I prefer Charlotte Corday* – Shall I do it for you?
>
> Do not be alienated from me by what I write – If I hide what I feel from you the constraint wld be unbearable.
>
> Your loving
> Clemmie

These were indeed dark days for Clementine. She knew she was fighting a losing battle; and yet she dreaded the possible consequences if in the end Winston were to change his mind. On 14 April she wrote a heart-rending letter:

> My Darling
>
> It is a long while since I have had real news of you – Your last letter contained only instructions & enclosures. Keep me going with something more. These last weeks since you returned to France have been sad & cruel & I have not been well which always makes anxiety harder to bear. Sometimes when I have been out all day canteening I dread coming home to find a

* Charlotte Corday (1768–93), right-wing republican during the French Revolution, who stabbed the Jacobin leader Jean Paul Marat to death in his bath, July 1793. She was guillotined four days later.

telegram with terrible news. And now if the telephone rings it may be the War Office to say that you have been killed. And yet in spite of this I keep on writing urging you not to leave the scene of these awful dangers.

I am very unhappy.

But despite her anxiety and unhappiness, Clementine continued to go out and to keep herself in touch with the political world:

I saw Rufus Isaacs [Lord Chief Justice] the other day at a party after dinner at Downing Street – He was very friendly & I had a long talk to him. He fears coming back unless sent for would be injurious to your reputation. Nearly all the Ministers were there. Grey terribly aged & worn looking, Kitchener thinner & sad, A.J.B. [Balfour] wan & white, but still purring away.[30]

Another night she had dined with the Asquiths:

The P.M. seems quite unconcerned. The Pall Mall [Gazette] last night had an inspired article saying he had got a united Cabinet on recruiting measures. Can it be true? He is like morphia.[31]

There was one more letter from Winston in this remarkable series – written after he had received Clementine's further bid to make him reconsider his decision. But he was not prepared to argue the case any more – his mind was made up, and further discussion would be fruitless, perhaps painful. He dismissed the tortured subject in one sentence: 'Well there is no use going over the old ground again, nor in darkening this page with my reflections.'[32]

Winston came home on leave in the middle of April, and on 28 April he returned to France to go into the line with his battalion for the last time. He had always hoped he would find the opportunity finally to return home for some military reason, and in the event this hope was to be fulfilled. Because of the heavy losses in the Scots regiments, it was decided to amalgamate several of the battalions; in this reorganization Churchill had to give way to a senior colonel, and thus his connection with the Royal Scots Fusiliers came naturally to an end. Early in May 1916 he returned home to England, where his true destiny awaited him.

* * *

Whatever Clementine's feelings had been about the wisdom of Winston returning to politics in April 1916, they were eclipsed by the joy and relief of having him safe at home once more. In the House of Commons Winston took his seat on the back benches, where, allied with others both in and out of Parliament, he was part of a now vociferous, and increasingly strong, patriotic opposition, bent on achieving a more vigorous and efficient prosecution of the war. The Prime Minister was not only facing formidable opponents outside his Government: within the closest council of ministers Lloyd George was bent on achieving power, and took every advantage offered him by the increasing weakness of the administration to bring this about.

Churchill was determined to clear his name from the ignorant and unjust accusations which had been continuously levelled at him ever since the debacle of the Dardanelles. He sought no other defence for his conduct than the revelation of the truth. Although Asquith would not agree to the publication of the official papers, he appointed a Commission to investigate the Dardanelles operations. This Commission started its sessions in August 1916, and for nearly a year Churchill found himself involved in a continual and harassing defence of his own responsibilities.

This year of 1916 carried its share of portentous events on all war fronts. In May the indecisive Battle of Jutland saw the only engagement between the two Grand Fleets; and on 1 July began the Battle of the Somme, which dragged on until November. The appalling casualty lists increased the volume of protest against the inefficient direction of the war, and in December Lloyd George successfully manoeuvred Asquith from power and succeeded him as Prime Minister.

Throbbing with energy and feeling himself equal to great tasks, Churchill confidently expected office under the new regime, and he felt once more the pangs of bitter disappointment and frustration when his hopes were confounded. But he had under-estimated, as Clementine had not, the tenacious hostility of some of his Conservative opponents, who had stipulated on joining the Cabinet that Churchill should not be given office. Moreover, although the Dardanelles Commission had been sitting for some months, it had so far published no report, and so in the eyes of many the main responsibility for that disastrous episode still lay heavily upon his shoulders.

However, temporary distraction at least from brooding and bitter thoughts was near at hand: both Churchill families had been bidden to Blenheim for Christmas with their united nurseries. The chill of the quarrel between Sunny and Clementine had long since been forgotten, and it must have been a marvellous party, for in addition to the Duke's own family he had invited the F. E. Smiths with their three children. And so, if for Winston and Clementine the winds blew chill outside, and the leaden clouds seemed to hold no promise for the future, the interior scene was warm and glowing.

On the home front, 1917 was to prove the hardest year of the war. By now there was scarcely a household that had not suffered bereavement, and inexorably the war was penetrating everyday life. But for Clementine the anguishing tension and anxiety of the previous year were mercifully eased. She had Winston at home once more, and in politics his fortunes began to revive. During the spring months of 1917, Churchill was constantly active in Parliament, and by his knowledge of affairs and fearless, forceful advocacy he was a formidable element to be reckoned with. In March 1917 the Interim Report of the Dardanelles Commission was published and part, at least, of the truth was told. Churchill's reputation was cleared of the damaging charges which had up to then been voiced without official contradiction; furthermore, it was clear to what extent others had been involved in decisions the responsibility for which, until now, he had been made to bear almost alone. Lloyd George was now strong enough to defy Conservative objections, and on 17 July 1917 Churchill was made Minister of Munitions.

As was still the custom, he had, on being appointed a minister, to seek re-election in his constituency. Clementine campaigned energetically for him at Dundee, addressing meetings on his behalf when his ministerial duties kept him in London. It was a stormy battle, and at one meeting she had the unpleasant experience of being shouted down. But barracking did not seem to have bothered her: 'Rain was falling heavily, but under the shelter of an umbrella she stood wrestling with the hecklers, one of whom she invited to "come closer so that she might hear what he had to say." At least two accepted her invitation, and while the audience, or a portion of those present, engaged in a wrangle on their own, Mrs Churchill listened to their stories.'[33]

Winston was safely returned with a majority of over 5,000. He took up his new appointment with enthusiasm. It marked a new start, and once more his volcanic energy could be harnessed to a challenging task.

* * *

For some time after Winston's return from the Front, he and Clementine had continued to share No. 41 Cromwell Road with Jack and Goonie, but presently they moved back into No. 33 Eccleston Square, which had been let. Now, Clementine's dream of 'a little country basket' was revived.[34] Both of them felt the need to retreat from the hurly-burly of London, and Winston was so enthralled with painting that he sought every opportunity of being in the countryside.

They presently found a charming house, Lullenden, near East Grinstead. It was a low building, built in grey stone, some of it dating back to the fifteenth century; there was a large barn nearby, and a small farm. Downstairs, running the full length of the house, was a big, high room with a large open fireplace, which made a delightful 'studio-like' nursery, from where a staircase led to a gallery, off which were the children's bedrooms; Clementine set about converting the barn for the children's use as well. There was a good deal to be done to the place, including the laying on of piped water.

Although at first the children remained in London, going down to Lullenden for weekends and holidays, it was later decided, on account of the air raids, that it was better for them to be permanently based in the country. Not that the raids had alarmed the children; Randolph recounted that they were 'tremendously exciting, since we children would be woken up in the middle of the night, wrapped up in blankets and carried down to the basement where there would be a lot of grown-ups having supper and drinking champagne. We liked Zeppelins very much indeed and thought it a great treat to mix with grown-ups in the middle of the night.'[35] It must have been a lively nursery party, for the cousins, Johnny (nine) and Peregrine (four), also moved down to the freedom and safety of the country. Despite the lack of entertainment provided by the air raids, the children were all very happy at Lullenden. The elder ones went to school in the nearby village of Dormansland.

Churchill's work as Minister of Munitions took him a great deal to France, for it was of prime importance for him to be in close touch with the commanders in the field, who knew at first hand the needs of the armies. During the summer of 1918 Lord Haig put at his disposal Château Verchocq, near St Omer, to facilitate his operations. At home the pressure of his work was so great that for a time Winston actually moved into rooms in the Ministry of Munitions in Northumberland Avenue. Any weekends he was not in France he spent at Lullenden, which was conveniently near the airfields of Penshurst and Godstone.

In the spring of 1918 Clementine was again with child: it possibly may not have been the moment of her choice for, domestically, these coming months were ones of considerable upheaval. It was, of course, a relief to get the children out of London; but, delighted though she was by their acquisition of Lullenden, the move involved her in much to-ing and fro-ing, as well as the organization required to settle her nursery party into their new life. The lease on No. 33 Eccleston Square came to an end this spring, which would leave them without a London base, and her involvement with her canteen work was as heavy as ever. Like many others in this fifth year of the war, Clementine was profoundly tired. German Zeppelin (and later aeroplane) raids on London since May 1915, although not on the scale of the Blitz in the Second World War, had been an added strain, especially when night raids had started in September 1917 – and the early months of the new year had seen a flurry of them.

It was in this context, and at a moment of fatigue and depression, that a rather extraordinary conversation took place between Clementine and Jean Hamilton, wife of General Sir Ian Hamilton.

The friendship between Hamilton and Churchill went back to North-West Frontier days, when Winston was a cavalry subaltern and war correspondent, and Ian Hamilton a colonel with a brilliant military career ahead of him. They were to meet again soon in South Africa, where Churchill reported on Hamilton's famous march through the Orange Free State and the Transvaal to Pretoria, the Boer capital, subsequently writing a book, *Ian Hamilton's March* (1900). Fifteen years later, the Dardanelles and Gallipoli brought the two men into close contact again, with General Sir Ian Hamilton commanding the Mediterranean Expeditionary Force. In mid-October 1915 Hamilton was relieved of his command – effectively, he was made a scapegoat. The aftermath of the disastrous campaign, and the subsequent setting up of the Dardanelles Commission in 1916, meant that the

Churchills and the Hamiltons saw a good deal of each other, sharing a burden of the flood of public criticism for these disasters.

The two wives were never such close friends as their husbands; Jean rather disapproved of Winston's manners and character – though when he took up painting, this made a bond, as Jean also was an amateur painter. But she didn't really like Clementine, writing in her diary on 18 June 1918, when the Churchills (having just relinquished their Eccleston Square house) were staying with them in London: 'I can't like Clemmie Churchill, she is like glancing cold water to me, very shallow.' However, two nights later (20 June) she found that 'Clemmie was nicer and sweeter tonight.' And before the end of the visit the two women had a long private talk. Clementine must have known of Jean's heartbreak that she had never had a child (she was now in her late fifties) and that she was seriously thinking of adopting a sixteen-month-old boy, 'Harry', who had been abandoned outside the Paddington crèche, of which Lady Hamilton was President. On Friday, 21 June, Jean wrote:

> Clemmie . . . urged me on no account to adopt Harry as she thinks [here there are several words heavily crossed out and now illegible] . . . and she asked if I'd like to have her baby; of course I said I would, and asked her when she expected it. She said: 'In November', and I offered to have her here for it, as she had been telling me how expensive a Nursing Home would be – £25 for room alone, and she said she could not possibly afford it. She said if she had twins I would have one! I don't think I'd much like a son of Winston's, I'd smack him well when he was young, if I got him, hoping to insert some manners into him, poor Winston is utterly destitute of them, though in the main has, I believe, a good heart.[36]

Quite clearly Clementine was in an exhausted and wrought-up state, worrying about their now homeless situation in London, and how this would be resolved, and where her baby would be born. In the event Jean Hamilton did adopt Harry; and of course, come the autumn good plans had been made for the arrival of Clementine's baby: kindly Aunt Cornelia Wimborne lent the Churchills her house, No. 3 Tenterden Street, near Hanover Square.

Meanwhile, happier weeks and months ensued for Clementine: with the children now permanently in the country, she made Lullenden her main base, and during this summer gave herself over to country life, finding pleasure and interest in it. After making the house habitable and comfortable, she addressed herself to the problems of the small farm, much assisted by the labours of two German prisoners-of-war. She also became very garden-minded; writing to Winston in July 1918, she asked him 'to take me to Crawley in the motor (about 10 miles) to see Cheal's Nursery Gardens. I want to buy some little rock plants to put in the chinks everywhere. This cannot be done before the autumn but the point is to see them all in bloom before they fade for the effect.'[37]

During Winston's absences, the family depended for their transport on a pony and trap – a means of travel not without its own hazards: in August Clementine described to Winston how their horse panicked at a steam-roller, the trap being badly damaged. The horse escaped injury, and Clementine sustained only a badly bruised knee, but she wrote: '"The Chumbolly" (we have not got a new name for this one) objected & said that if he was disturbed again like this from his broodings he should come out & see what the matter was. He has however dozed off again & like Mahomet's coffin is suspended between 2 Worlds.'[38]

But Clementine was not completely rusticated: despite her pregnancy she still continued her work, and made long, tiring expeditions to London to grapple with her canteens. Her ability and devotion to this work were officially recognized when, in the New Year's Honours, she was made a Commander of the Order of the British Empire (CBE).

During the summer of 1918 the Germans made their last desperate bid on the Marne, but the Allied counter-offensive soon gained momentum and everywhere now the blood-red flood tide of victory rolled forward.

During the later part of the summer Clementine made some country visits but, as ever, Winston and his work took prime place in her thoughts, and she followed in minutest detail the progress of the war. From Mells in Somerset, where she was staying with old family friends, Sir John and Lady Horner, she wrote on 13 August: 'The War news continues good, but it seems to me that "the Victory" is now complete & that for the present nothing more is to be expected? I do hope we shall be careful not to waste our men in pushing now that the spurt is finished.' For these thoughts she was presently praised by Winston as showing herself to be 'a vy wise & sagacious military pussy cat'.[39]

Clementine relished hearing and reading all the news from him: 'How much better you describe things than the most brilliant Newspaper Correspondent. But I forget! You were one once – but that was before I knew you,' she wrote on 15 August, continuing:

> This is a delicious place to rest and dream & I feel my new little baby likes it – Full of comfort, beautiful things, sweet smelling flowers, peaches ripening on old walls, gentle flittings & hummings & pretty grandchildren. But under all this the sadness & melancholy of it all – Both the sons dead, one lying in the little churchyard next to the House carried away at sixteen by Scarlet Fever, the other sleeping in France* as does the Husband of the best loved daughter of the House Katherine Asquith†. Both their swords are hanging in the beautiful little Gothic Church beside long inscriptions commemorating long dead Horners who died in their beds.[40]

Winston's letters to her at this time were (not unnaturally in view of his task as Minister of Munitions) full of plans and details of weapons and

*Edward Horner, killed in action, 1917.
† Katherine Horner had married Raymond, Asquith's eldest son, killed in the Battle of the Somme, 1916.

engines of destruction. Clementine understood the necessity of this, but with the feeling that at last the war was drawing to its close, and the bright personal preoccupation of a new baby, her mind dwelt on the future, and on the role she longed for Winston to play in the reconstruction of peacetime life. '[D]o come home and look after what is to be done with the Munition Workers when the fighting really does stop,' she wrote to him. 'Even if the fighting is not over yet, your share of it must be & I would like you to be praised as a reconstructive genius as well as for a Mustard Gas fiend, a Tank juggernaut & a flying Terror – Besides the credit for all these Bogey parts will be given to subordinates.' Her Utopian ardour was thoroughly aroused, for she continued:

> Can't the men Munition Workers build lovely garden cities & pull down slums in places like Bethnal Green, Newcastle, Glasgow, Leeds etc & can't the women munition workers make all the lovely furniture for them Baby's cradles cupboards etc . . .
>
> Do come home & arrange all this.[41]

The 12th of September marked the tenth anniversary of Winston's and Clementine's wedding. He was in France, and wrote:

> Ten years ago my dearest one we were sliding down to Blenheim in our special train. Do you remember. It is a long stage on life's road. Do you think we have been less happy or more happy than the average married couple? I reproach myself vy much for not having been more to you. But at any rate in these ten years the sun has never yet gone down on our wrath. Never once have we closed our eyes in slumber with an unappeased difference. My dearest sweet I hope & pray that future years may bring you serene & smiling days, & full & fruitful occupation. I think that you will find real scope in the new world opening out to women, & find interests wh will enrich yr life. And always at yr side in true & tender friendship as long as he breathes will be your ever devoted, if only partially satisfactory, W.[42]

It was during this absence that Clementine was found for the first time seriously wanting as a correspondent. For a whole week Winston received no word from her, and on 15 September he wrote her a reproachful letter: 'Up to this moment I have not had a single letter. Really it is unkind. Mails have reached me with gt regularity & swiftness by aeroplane or messengers. They have comprised all manner of communications, but never one line from the Cat.' His letter continued with much interesting news, and was signed: 'Your ever devoted though vilely neglected Pig.'

However, we get a hint from her letter of 17 September (when his of the 15th had not yet arrived) that Clementine, normally so good and reasonable about the demands of his work, was feeling that Winston could give higher priority to hearth and home:

> My darling Winston,
> You have been away such a long time & I have not written to you once, & you have sent me two lovely letters.
> It seems ages since that evening when you disappeared like a swallow into the twilight over the sea . . .
> Do come home soon – You have been away for nearly a month with your 2 visits – you bad Vagrant. Next time I shall trip over to America & back. I shall just have time to be found dozing on the hearth at Lullenden when you return from your Château or Paris –
> Love from
> Clemmie

And then, writing from Lullenden the next day, she ended a letter about their immediate plans with a postscript:

> I was dozing peacefully & innocently in my basket with my tail over my eyes when your scolding about no letters arrived. I will write you some good ones again presently when you go on your next journey – This time my pen wouldn't work [small drawing of cat]

But the dialogue soon resumed.

* * *

Victory came at last, and on 11 November – Armistice Day – Clementine joined Winston at the Ministry of Munitions, to be with him at this historic moment. When they left to call on the Prime Minister at No. 10 on this day of triumph and rejoicing, their car was surrounded by a wildly cheering crowd.

Four days after the Armistice had been signed, Clementine gave birth to their fourth child: another red-headed girl, whom they named Marigold Frances.

CHAPTER THIRTEEN

Bleak Morning

THE GENERAL ELECTION WHICH WAS HELD IMMEDIATELY FOLLOWING THE END of the war was characterized by hysterical patriotism and extravagant demands for vengeance against Germany. Churchill did not share the feelings which engendered such slogans as 'Hang the Kaiser' and 'Squeeze them till the pips squeak'. His watchword then, as later, was 'In Victory – Magnanimity: in Peace – Goodwill'. At the Armistice, when he learnt that Germany was at the point of near-starvation, his immediate reaction was that ships 'crammed with provisions' should be sent immediately to Hamburg.[1] Even before the fighting ended Churchill had spoken of the importance of not seeking the utter ruin of Germany,[2] and during the election campaign he stressed that war reparations should not be pressed to the point where the German working classes would be reduced to 'a condition of sweated labour and servitude',[3] which could have only malign repercussions in Britain.

Lloyd George and Bonar Law had issued a joint manifesto, and the result of the election was a sweeping victory for the Coalition, in which the Conservative Party was the dominant partner. But it saw the final and fatal schism in the Liberal Party: Asquith and many of his followers lost their seats, and those who survived sat on the Opposition benches.

Clementine could take no part in this electoral battle, for Marigold was barely a month old on polling day, but Winston was re-elected at Dundee with a handsome majority.

That Christmas the Churchill family gathered once more at Blenheim, and Randolph, who was seven, had a vivid recollection of it: 'Christmas 1918 coincided with the end of the War and with the delayed celebrations of the twenty-first birthday of the son and heir, Lord Blandford . . . There was a paper-chase on horseback and a whole ox was roasted. The day concluded with a gigantic bonfire on top of which was placed an effigy of the Kaiser.'[4]

In the newly formed administration, Churchill was made Secretary of State for War, an office he was to hold for two years. The Air Ministry, which had first come into being in 1917, was now combined with the War Office under his aegis. This arrangement aroused criticism in several quarters, and Clementine herself thought the duality of the ministries unwise. After she had sat next to an eminent airman at dinner one night, she wrote to Winston:

> Darling really don't you think it would be better to give up the Air & continue concentrating as you are doing on the War Office. It would be a sign of real strength to do so, & people would admire it very much. It is weak to hang on to 2 offices. You really are only doing the one. Or again if you swallow the 2 you will have violent indigestion! It would be a tour de force to do the 2, like keeping a lot of balls in the air at the same time. After all, you want to be a statesman not a juggler.[5]

However, the two offices continued to be combined until Churchill left the War Office in 1921 to become Colonial Secretary. During his term as Air Minister he did much to forward the building up of this branch of defence, whose vital role and powerful future he saw with penetrating clarity.

When Winston had abandoned flying in 1914 at Clementine's anguished and repeated requests shortly before the birth of Sarah, he had already completed many of the flying hours required to obtain a pilot's licence. Since his appointment to the Ministry of Munitions he had frequently flown to and from France, and now, in the summer of 1919, he began once more to take flying lessons.

Flying was still a precarious business, and on 18 July Winston narrowly escaped death while piloting a dual-control aeroplane over Croydon airfield. The plane was 70–80 feet above ground and travelling at about 60 m.p.h. when the controls failed and the plane plunged towards the ground, striking it with terrific force. In the few seconds before impact Colonel Scott, his instructor, had the presence of mind to switch off the engine, thus preventing an explosion. Miraculously neither of the men was killed, although Colonel Scott suffered severe injuries (from which, happily, he recovered). Winston himself was badly shaken, his forehead scratched, and his legs covered with bruises. Nevertheless, two hours after this misadventure he presided at a dinner in honour of General Pershing (Commander-in-Chief of the American Expeditionary Force in Europe), and made a speech. But despite his debonair demeanour, this accident shook him severely, and, to the intense relief of Clementine and his friends, he abandoned further attempts to qualify as a pilot.

Churchill's work for the Army and Air Force, together with the demands of the Peace Conference which had opened early in 1919 at Versailles, continued to make his presence in France often necessary. Clementine therefore had much to cope with on her own, including the running of their small farm at Lullenden. In March of that year she had engaged a new gardener/bailiff, whose very energy and determination caused her anxiety. She wrote to Winston:

> If we sell Lullenden we can pass this tyrant on with the place! . . .
>
> What I feel about him is that either he will save Lullenden for us or we shall be positively forced to sell . . .
>
> I feel very anxious about our private affairs. It's clear we are far too much extended – [6]

In this same month a most tragic event took place in their domestic life, during the virulent influenza epidemic which swept across the world during the winter after the war, claiming 150,000 lives in England alone. Winston was away in France when their Scottish nanny, Isabelle, who had been with them for several years, was struck by this horrible illness. Its course was swift and violent; in her delirium the poor girl seized Marigold from her cot, taking the child into her own bed. Clementine took Marigold down to her own room, but because so many people were ill her efforts to get a doctor were unavailing. She spent a fearful night going up and down stairs between the dying Isabelle and the infant Marigold, who was frightened by these inexplicable events.

Isabelle died in the early hours. Clementine herself sickened with the influenza, and had a high temperature, but mercifully her own attack was not so serious. This experience affected her deeply, for she had been very fond of Isabelle; and for days there was acute anxiety lest Marigold should have caught the fell disease. However, all was well. Winston on his return to England was made to stay with Sunny Marlborough until all fear of infection was past.

In August 1919 Clementine accompanied Winston when he visited the British Army on the Rhine, where he inspected many British units in the occupied zone. In Cologne there was a great parade and march-past, and Winston addressed the troops. Clementine was unfavourably impressed by the arrogant manner of some of the senior Army wives towards the German population.[7]

September saw them once more apart, but the letters they exchanged on their eleventh wedding anniversary remain as a moving witness to their feelings for each other. Winston wrote from France:

> 11 Sept. 1919
>
> My darling one
>
> Only these few lines to mark the eleventh time we have seen the 12th Sept. together. How I rejoice to think of my gt good fortune on that day! There came to me the greatest happiness & the greatest honour of my life. My dear it is a rock of comfort to have yr love & companionship at my side. Every year we have formed more bonds of deep affection. I can never express my gratitude to you for all you have done for me & for all you have been to me.
>
> Yr ever loving & devoted
>
> W

Clementine, who had evidently not remembered the actual day in time, replied:

Sept 14. 1919
Lullenden,
East Grinstead

My Darling Winston

At Wynyard [in Co. Durham, home of Lord and Lady Londonderry] early in the morning of Sept 12th I woke up & remembered suddenly the importance of the day, & I had not written to you! And then your dear letter appeared upon my breakfast tray. It made me very happy and illumined my day.

I love to feel that I am a comfort in your rather tumultuous life. My Darling, you have been the great event in mine. You took me from the straitened little by-path I was treading and took me with you into the life & colour & jostle of the high-way. But how sad it is that Time slips along so fast. Eleven years more & we shall be quite middle-aged. But I have been happier every year since we started.

Despite the 'new broom' tactics of the bailiff at Lullenden, the property continued to be a heavy financial burden. Then a happy solution presented itself: the Hamiltons took a great liking to the place, rented it in the spring, and later that year bought it. Although both Winston and Clementine were relieved to have this burden off their shoulders, it was a wrench, as the whole family had loved it. However, the Hamiltons were most hospitable, and invited the Churchill family on visits to their former home.

For a few months after the sale of Lullenden they had neither a town nor a country house; so while Clementine house-hunted, they all stayed with the Freddie Guests at Templeton, at Roehampton, the families sharing the expenses. Soon, however, the Churchills found a house to their liking: No. 2 Sussex Square, which was admirably suited to the needs of their now considerable family. It was a high, handsome house (now demolished), just north of the Bayswater Road and within easy reach of Hyde Park. They also acquired the mews buildings at the back of the house, which provided a splendid studio for Winston.

Not all Winston's expeditions to France were for working purposes only. Through his letters to Clementine we have delightful accounts of a spring holiday he took in France towards the end of March 1920 with General Lord Rawlinson, with whom he shared a taste for hunting and painting, Lord Rawlinson being a watercolourist. The Duke of Westminster* put his charming house at Mimizan, south of Bordeaux, at their disposal, and there they spent some happy, carefree days *en garçon*. Clementine enjoyed the vivid descriptions he sent her of his hunting, particularly as she knew Mimizan, and had hunted there herself. She meanwhile was busy with the children's spring holidays, during which she supervised the kitting out of Randolph for his preparatory school: 'He looks such a thin shrimp in trousers & an Eton collar!'[8] she wrote to Winston.

* Hugh Grosvenor, DSO, 2nd Duke of Westminster (1879–1953), known from infancy as 'Bendor', after one of his grandfather's racehorses – Bend Or. A lifelong friend of Winston's since Boer War days. A great character and a superb host, he has been described as 'the last grand seigneur'.

Now nine, Randolph started at Sandroyd's, a preparatory school near Cobham in Surrey, that summer term. It appears that both Randolph and Diana were extremely naughty children, and in the early years nannies and nursery-maids came and went with inconvenient frequency. Perhaps the two older children were just full of super-abundant animal spirits: Randolph recalled that 'Diana was more docile than I was' but that he 'could never brook authority or discipline'.[9] During the greater part of the war, Winston and Clementine's nursery had been amalgamated with Jack and Goonie's, and Goonie's nanny had commanded (with varying degrees of success) the combined force of children. After the war, there was a nanny for the two little ones, and Diana and Randolph had a series of governesses, the most remembered being Miss Kinsey, who was amiable and tenacious, and who even after the children had started school used sometimes to come back to the family during the holidays.

After Randolph had gone to Sandroyd's, Clementine decided that school would be more satisfactory for both Diana and Sarah than being taught at home (she had not forgotten how happy she had been at Berkhamsted High School), and in the autumn of 1920, soon after they had all moved into their new home in Sussex Square, the two elder girls went as day pupils to Notting Hill High School, where they both stayed for a number of years.

* * *

In the opening days of 1921 Blanche, Countess of Airlie, Clementine's majestic and austere grandmother, died at the age of ninety. Clementine had been too much over-awed by her as a child to love her, and when she was grown up she had not come to like her much as a person. But the resilient, magnificent old lady was a landmark in Clementine's life, and her death brought back thoughts of long-ago summer holidays at Airlie Castle, when she and Kitty and the twins were children.

Lady Airlie's death was the first of a series which were to darken the year that lay ahead. Mercifully oblivious of what the coming months had in store, Winston and Clementine were able to enjoy a sunshine holiday in Nice during January. They stayed in a hotel with Sir Ernest Cassel and his granddaughter, Edwina Ashley, who the following year was to marry Lord Louis Mountbatten RN (later Admiral of the Fleet Earl Mountbatten of Burma).

This was the first real holiday Winston and Clementine had had together since the beginning of the war, and Clementine was both physically and mentally in need of it. After the years of tension and anxiety, arduous war work, the birth of Marigold, and, last but not least, the effort of moving into Sussex Square, this new year found her reserves of strength low, and her spirits depressed. Winston was much concerned about her, and when he returned to London, leaving her behind to continue her holiday, adjured her again and again to take great care of herself, to avoid over-fatigue and to follow the doctor's instructions. 'I do hope you will soon see sunshine & preen yr poor feathers in it,' he wrote on 1 February; and in another letter,

'I do hope you are having fun & tennis & above all recharging yr accumulators.'[10]

Clementine's letters must have reassured him, for they bubbled over with animation and enjoyment. Perhaps the most important feature of this much-needed holiday, and the most effective therapeutic remedy, lay in the fact that she was simply enjoying herself. All her life her overpowering sense of duty, and the conscientious performance of whatever that entailed, again and again dominated her existence to the exclusion of fun and pleasure. Now, in these few sunlit weeks, she gave herself over to doing what she liked, and in being what she was – a beautiful woman of nearly thirty-six at the height of her vitality and powers.

As her health revived, which it quickly did, she played tennis with increasing zest and ability. Winston was somewhat concerned lest she should overstrain herself, and thus retard her complete recovery; after talking to her doctor he wrote on 9 February:

> Please my darling think of nothing else but this, subordinate everything in yr life to regathering yr nervous energy, and recharging yr batteries. Don't throw away yr gains as you make them.
>
> Don't play in any tournament. If you play tennis – play for pleasure – not to excel.

However, as the weeks went on she felt so much stronger that she disregarded Winston's prudent counsels and took part in several high-class amateur mixed doubles tournaments, writing him long and excited accounts.

After her visit to Sir Ernest Cassel, Clementine went on to stay with Lady Essex (Adèle) at her villa, Lou Mas, at St Jean Cap Ferrat. Imbued with a carefree holiday spirit, she even went to the Gaming Rooms several times, observing the characters who frequented them with a mordant wit. Although no real gambler herself, she dabbled at the green baize tables lightheartedly (but ever cautiously), writing to Winston with great glee that she had learnt a new 'system' for roulette, which she proceeded to describe in minute technical detail, remarking, somewhat disarmingly, that 'It did not work at all well but I think it is better than playing without a system as one certainly loses one's money more slowly –'[11]

During her progress along the coast Clementine also met several well-known personalities, and she duly described these meetings to Winston. Lady Essex had invited J. L. Garvin to luncheon: 'He has a great belief in you,' Clementine wrote; later in the garden 'he released about ⅛ of an Observer Article on me.'[12] A less pleasant encounter was when she saw Ellis Ashmead-Bartlett, the war correspondent*, who had taken a hostile view of Winston's role in the Dardanelles campaign:

* See note to p. 147.

> Ashmead-Bartlett is at Beaulieu. I saw him playing tennis & could not remember whether his conduct merited 'the cut direct'. While I was revolving this he came up & nodded so I bowed coldly & as there was a rabbit wire fence between us I did not have to shake hands. He has a mean odious face & is the sort of creature who would haunt the Riviera.[13]

As always when they were apart, Winston was a good correspondent, and kept her up to date with life at home. Randolph was back at Sandroyd's, and Winston had twice visited him at school, reporting to Clementine that he had 'found him very well and very sprightly' and that the Headmaster seemed to be quite pleased with him, but 'described him as very combative'.[14] The two elder girls, who had been afflicted during most of the winter by a series of coughs, colds and other minor ailments, had been sent off by Clementine for a recuperative visit to Broadstairs in the charge of the nice cosy maid Annie, and this was so successful as to their health that Winston authorized a third week for them there.

Marigold had not gone with the seaside party but stayed at home keeping house with her father, who would have hated a solitary house denuded of all 'kittens'. She was now in her third year, and full of fascination and vitality. She loved to run, full pelt, round and round the dining-room table while the grown-ups were at luncheon, so that Clementine, fearing she might slip and crack her head against the sharp corners of the table, had soft pads affixed to them. Marigold had a sweet, true little voice, and learnt the popular song, 'I'm Forever Blowing Bubbles': she was always singing it, so that it became her signature tune. During this winter, and the spring that followed, she was particularly prone to catching coughs and suffering from sore throats. Twice during Clementine's absence in France Marigold was thus afflicted, and on one occasion the doctor had to be called.

It is touching to see how much pleasure, care and interest Winston took in his children, despite his manifold preoccupations, and how conscientiously he took his role of parent-in-charge. Of the elder girls he wrote appreciatively: 'The children are vy sweet. Diana is shaping into a beautiful being. Sarah full of life & human qualities.'[15] Warmth, vitality, and the free (and often noisy) expression of their mutual affection always constituted the prevailing atmosphere of Winston's and Clementine's family life. All the children had their pet-names: Randolph was 'The Rabbit' (a somewhat mystifying pseudonym for such a 'combative' character!); Diana was most aptly and charmingly called 'The Gold-cream Kitten'; Sarah, with her cloud of hectic red hair, was 'Bumble-bee'; and the darling Marigold, 'Duckadilly'.

In January and February of 1921, two events of major significance in Winston's life and career took place. On 26 January his first cousin once removed, Lord Herbert Vane-Tempest, was killed in a railway accident in Wales; as he was childless, an estate in Ireland known as Garron Tower, and a considerable sum of money, passed to Winston through the devious processes of entail and inheritance.

And on 13 February his appointment as Colonial Secretary was announced. This was a notable and upward step in Churchill's career, for the office of Colonial Secretary was particularly important at this period. Winston stayed for the first weekend in February with the Prime Minister at Chequers, which had just been presented to the nation by Lord and Lady Lee of Fareham as a country home for British Prime Ministers in perpetuity. From there he wrote to Clementine: 'Here I am. You wd like to see this place – Perhaps you will some day! It is just the kind of house you admire – a pannelled [*sic*] museum full of history, full of treasures – but insufficiently warmed – Anyhow a wonderful possession.'[16]

It is worth remarking in passing that when Clementine did see Chequers later – nearly twenty years on – it was still 'insufficiently warmed', and her taste for gloomy Elizabethan houses stuffed with 'treasures' had, in the intervening years, waned.

In the midst of her sunshine holiday Clementine thought constantly of Winston and of all the events which so closely touched their life. From Lou Mas on 7 February she wrote him a long letter:

> I can imagine all the various things which are filling your mind, i.e. your new post, your inheritance, the painting, Chequers, the P.M., the re-arrangement of the Government . . . the Book [*The World Crisis*], the poor old worn out War Office – make this little peninsula seem a pin point in the sea while you are soaring in an aeroplane above the great Corniche of life . . . I wish I were with you spinning round & round instead of sitting lazily here in the sun.

The unexpected windfall brought about by the death of his kinsman was of immense and immediate consequence to Winston's financial affairs. He was now assured of an income of around £4,000 a year. This meant that for the first time in his life he had an income independent of any parliamentary or ministerial salary, or from what he could earn by writing. Although Clementine was the least 'money-grubbing' of women, she of course rejoiced with all her heart at this unsought, unawaited shower of golden security. Their financial difficulties had been from the earliest days of their marriage a source of wearing anxiety to her, and she had struggled constantly between maintaining elegant and fastidious standards and the (to her) embarrassing horror of unpaid bills. Overnight, this situation had been changed. She wrote to Winston, on 8 February: 'I can't describe the blessed feeling of relief that we need never never be worried about money again. (Except thro' our own fault of course!) It is like floating in a bath of cream.' And then, a fortnight later, when Winston told her that the legacy was in fact appreciably smaller than he had been led to suppose:

> I am so sorry my Dearest that you have had a deception as to the amount of the inheritance – It is certainly very disappointing after you had worked it out so carefully & thought about how to lay it out to the best advantage; but don't let a 'crumpled rose leaf' like this spoil the really glorious fact which

rushed on us so suddenly a little time ago – that haunting care has vanished for ever from our lives –

It's so delicious to be easy – I hope I shall never take it for granted, but always feel like a cork bobbing on a sunny sea.[17]

One of Clementine's constant worries in the past had been that, in order to make ends more likely to meet, Winston might accept unworthy or unsuitable contracts for 'pot-boiling' newspaper and magazine articles; and even if the offers were above reproach, the labour involved in fulfilling such contracts was an extra load on top of his ministerial and parliamentary work. Now, in the light of recent events, Clementine tried to dissuade Winston from writing an article for the *Strand Magazine* about his painting. She developed her arguments in a letter to him on 10 February:

if <u>you</u> write the Article what are you going to write about
1) Art in general? I expect the professionals would be vexed & say you do not yet know enough about Art –
2) Your own pictures in particular? The danger there seems to me that either it may be thought naif or conceited –

I am as anxious as you are to snooker that £1000 & as proud as you can be that you have had the offer; but just now I do not think it would be wise to do anything which will cause you to be discussed trivially, as it were.

If there is to be an argument let it be as to whether you are going to be a good 'Imperial Minister' or not . . .

I think one of the reasons that your friends rejoice in your new fortune is that it frees you from the necessity of earning money from writing at inconvenient moments.

Winston promised to consider Clementine's views, but added: 'An article by Mr Balfour on golf or philosophy or by Mr Bonar Law on chess would be considered entirely proper. I think I can make it very light and amusing without in any way offending the professional painters.'[18] Eventually two articles, illustrated by reproductions of his paintings, were published in the *Strand Magazine*, in December 1921 and January 1922, under the title 'Painting as a Pastime'*. Although in this case, as in so many others, Winston's view prevailed, it is a striking example of Clementine's vigilance and sensitivity to anything that might harm his reputation or dim his lustre.

With characteristic warmth and generosity, Winston did not confine the benefits of his good fortune solely to himself and Clementine. Nellie Romilly was at this time in very difficult financial circumstances, with a severely war-wounded husband and two small children, and to help provide for them all she was proposing to open a hat shop. At the end of January, Winston wrote to tell Clementine that Nellie had found suitable premises

*These articles would eventually become the enchanting small volume, *Painting as a Pastime*, first published in 1948.

for her shop, and that he had lent her £500; and a little while later, when Nellie was still beset by worries, Winston assured her that he would help her through. He also looked into Lady Blanche's financial arrangements, and Clementine wrote on 13 February:

> I am so glad you are looking into Mother's affairs for her.
>
> I do hope you will be able to improve matters for her, & anyhow I would like her to have £100 a year out of the increased allowance you are so kindly making me.

Towards the end of February 1921 a new and enjoyable project took shape. Winston had to go to Cairo, where there was to be a Middle Eastern Conference to try to arrive at a working settlement for the affairs of that turbulent part of the world. The British delegation were travelling out by sea, and Winston proposed to Clementine that she should come with him, getting on board at Marseilles, where their ship, the *Sphinx*, was due to call. Clementine had by this time moved to the Hôtel Bristol at Beaulieu, and she wrote from there on 21 February: 'I am living in blissful contemplation of our smooth and care-free future; (I mean from a money point of view) & I am laying up stores of health by staying a lot in bed. The last two days I have been thinking chiefly of the great pleasure and excitement of going with you to Egypt & Palestine. I am thrilled by the idea & so so longing to see you.' Winston wrote to her with fuller details of their journey and programme: they would probably stay ten days in Cairo, and then go to Jerusalem; she would be able to sight-see, and probably play some tennis, 'so do not forget your racquet'.[19]

In order to see the two elder girls before leaving for his journey, Winston sent for them to return from Broadstairs two days early. 'They were in the pink,' he reported to Clementine on 27 February; 'The Duckadilly received them with joy & is quite free from cold . . . We are going down to see Randolph today. I gave the children the choice of Randolph or the Zoo. They screamed for Randolph in most loyal & gallant fashion. So we arranged for the Zoo too.' Of their forthcoming travels together he wrote: 'I am looking forward eagerly to seeing you again. We shall have a beautiful cabin together, if only it is not rough – then I shall hide in any old dog hole far from yr sight . . . If it is fine, it will be lovely & I shall write & paint & we will talk over all our affairs.' Among their travelling companions were to be Archie Sinclair (who had followed Winston into political life and was now his Private Secretary), and the famous romantic figure of Colonel T. E. Lawrence – 'Lawrence of Arabia' – with whom during these next weeks Winston and Clementine formed a firm friendship.

This journey to the Middle East was, in personal terms, a great success; and, on a wider canvas, at the Cairo Conference Churchill was instrumental in bringing about a political settlement in Iraq and Transjordan. In intervals between sessions Winston found time to paint, and he and Clementine paid a visit to the Pyramids and the Sphinx, escorted by Lawrence of Arabia and

Gertrude Bell, the famous traveller and Arabist. In Jerusalem, Churchill reasserted the British Government's policy of allowing a Jewish National Home to be established in Palestine.

* * *

A few days after their homecoming a tragic event took place. On 14 April Bill Hozier was found shot dead in a hotel bedroom in Paris: he had killed himself. He was only thirty-four – debonair, good-looking and gallant. After the war he had retired from the Navy and gone into business, but, like his mother and Nellie his twin, he had a weakness for gambling, which inevitably had led him into financial difficulties. Some little time before Bill's death, Winston, who was extremely fond of him, had extracted from him a solemn promise that he would never gamble at cards again. Now Winston was filled with remorse as well as grief, for he feared that Bill's promise to him might have contributed to the reason for his suicide. But no evidence was found that Bill had in fact gambled again; nothing was known of any losses or heavy debts, and indeed he had just paid 10,000 francs into his bank account. His death was ever to remain a tragic mystery.

Clementine and Nellie travelled at once to Dieppe, where Lady Blanche had once more established herself several years earlier. On Sunday, 17 April Clementine wrote to Winston:

> Your beautiful letter has come –
>
> My poor Mamma is so brave and dignified, but I do not think that she will recover from the shock & the grief. She sits in her chair shrunk and small. When we saw her she did not yet know that Bill had killed himself, but I saw by the look of agony & fear that she half guessed.

There was some doubt as to whether Bill could be buried in consecrated ground, or the body allowed to rest before the funeral in the church. The British Vice-Consul rallied round and went to see the clergyman, and all was decently and compassionately arranged.

The funeral was to take place on Monday, and the body would lie from that morning in the church. Clementine had the service delayed until four o'clock, in the hope that Winston might be able to come. 'Oh Winston my Dear do come tomorrow,' she wrote, '& dignify by your presence Bill's poor Suicide's Funeral.' She had sent Winston's letter to the clergyman for him to see, 'as I wanted him to realize what we all felt for Bill & that he was not a mere scapegrace disowned by his family'. Despite the very short notice, Winston managed to arrive in time for the funeral. In his will Bill had bequeathed him his elegant gold-topped malacca cane: Winston used it for the rest of his life.

* * *

At the end of June Winston's mother died at her home in Westbourne Street, Bayswater; she was sixty-seven. Her marriage to George Cornwallis-West had (perhaps predictably) ended in divorce in 1913, and he had promptly married the famous actress, Mrs Patrick Campbell. Four years later, Lady Randolph (then sixty-four) was married for the third time, and again to a man over twenty years younger than herself – Montagu Porch, a member of the Northern Nigerian Civil Service whom she had met several years before. Not unnaturally, Lady Randolph's family had tried to dissuade her from this marriage; but Montagu Porch's devotion was real, and Jennie Churchill was very lonely, and in June 1918 they had married. Soon afterwards, Montagu Porch resigned from the Nigerian Civil Service to live in London, and despite the many forebodings they were happy together.

Now, in the summer of 1921, Lady Randolph had tripped in a perilously high-heeled pair of shoes and fallen down the staircase, while staying in the country at Mells with her old friend Lady Horner. The local doctor diagnosed a broken ankle, and Lady Randolph was taken back to her house in London. But then gangrene set in, and her leg had to be amputated.

Sussex Square is just round the corner from Westbourne Street, and in these weeks of painful illness Clementine would often go and sit with her mother-in-law. Although Clementine and Lady Randolph had never been very close – indeed, both she and Goonie had been exasperated many times by 'Belle-Maman's' extravagances – understanding and affection had grown over the years. The sisters-in-law had hated to see her humiliated by George Cornwallis-West's vagaries, and had been warmly sympathetic. Now, as Clementine witnessed the indomitable courage with which this vital, worldly woman faced the ordeal and pain of amputation, she felt a new admiration for her.

Montagu Porch was away on business in Africa when his wife had her accident; Winston kept him in close touch, and on 23 June had cabled him to say that she was out of danger. Then, early on the morning of 29 June Lady Randolph had a sudden, violent haemorrhage. Winston was sent for, and went running through the streets to his mother. But when he arrived, she was already dead.

Moving letters of condolence poured in – Jennie had become a legend in her lifetime. Winston was deeply grieved, for he had loved his mother dearly. He always spoke of her with uncritical admiration and devotion. Of all the tributes to her, perhaps his own words in his answer to Lord Curzon's letter strike the truest note: 'I do not feel a sense of tragedy, but only of loss. Her life was a full one. The wine of life was in her veins. Sorrows and storms were conquered by her nature & on the whole it was a life of sunshine.'[20]

* * *

Plans for the summer holidays of 1921 promised to be agreeable and fun for all the family. All four children went into lodgings, in a house called

Overblow at Broadstairs, for the first half of August, in the charge of Mlle Rose, a young French nursery governess. It was planned that about the middle of the month the elder children would go up to Scotland to join their parents, who would then be staying with the Duke and Duchess of Westminster at their beautiful and remote estate in Sutherland. In the meantime, Clementine went on her own to stay with the Westminsters at Eaton, their home near Chester, to take part in a tennis tournament. She intended to join the train on which the children would be travelling from the south, so that they would all journey on to Scotland together: Winston was to join them a few days later.

Clementine received happy news from the children, but in retrospect there are references in their naïve letters which sound a warning note. On 2 August, Randolph wrote to his mother:

> Dear Mummie,
>
> I hope that you are quite well. My legs are suffering from the sun, it has made them go all red. The other day Falkner lent us his big shrimping net, so we went out and caught lots of shrimps for tea. On Sunday we went out in a little rowing boat, which was great fun! Marigold has been rather ill, but is ever so much better today. We are all quite well here except for our legs.
>
> With much love
> Randolph
> XXXXXXXXXXXXXXXXXX

Sarah also wrote to her mother: 'Baba is very sweet and quite well now. We are enjoying ourselves very much here. We bathe every day.'[21]

So from the start of their seaside holiday Marigold had not been very well; and since in the last year she had been subject to 'throats' and coughs, perhaps not enough significance was attached at first to what was in fact the onset of a mortal illness. The local doctor was very good, but – alas for the 'Duckadilly' – the age of antibiotics had not yet dawned, and her painful sore throat progressed into a fatal septicaemia. The child became really ill about 14 August, but it was a day or two before Mlle Rose, stimulated into action by the kind and watchful landlady of Overblow, sent for her mother. Clementine left Eaton at once and rushed to Broadstairs; the three elder children travelled to Scotland according to plan, with Bessie, Clementine's maid. By the time her mother arrived, Marigold was gravely ill. Winston came down from London; a specialist was sent for, but he could do nothing. Clementine wrote to the children in Scotland to tell them of the gravity of Marigold's illness. Diana replied:

> 20.8.21
>
> My darling Mummy,
>
> We received your letter last night and I was very sorry to hear poor little Marigold is so ill. We have had a very unlucky year . . .
>
> Yesterday we went to an Island called Handa, and we took our tea with us.

> We all go riding every day. I do wish you could be here; but it is impossible.
>
> Much love and kisses from
>
> Diana

On the evening of 22 August, Clementine was sitting by Marigold's bedside. The child was sinking. Suddenly she said to her mother: 'Sing me "Bubbles".' Clementine, summoning all the control she knew, began the haunting, wistful little song that Marigold loved so much; she had not struggled very far, when the child put out her hand and whispered, 'Not tonight . . . finish it tomorrow.' The next day Marigold died. She was two years and nine months old; both her parents were with her. Clementine in her agony gave a succession of wild shrieks, like an animal in mortal pain.[22]

After the funeral in London, Winston and Clementine boarded the night train in a stupor of grief and went up to join the children at Lochmore. They stayed there for nearly a fortnight, at the end of which Winston went on to Dunrobin to stay with the Duke and Duchess of Sutherland, and Clementine took the children home, to make the necessary preparations for their return to school.

Margot Asquith wrote in her memoirs that 'No true woman ever gets over the loss of a child.'[23] It was certainly so with Clementine: but she did not indulge her grief; rather, she battened it down, and got on with life with all its persistent, trivial demands.

At Dunrobin, a very large party of 'swells and notabilities' was gathered. Winston found some beautiful scenes to paint, but painful thoughts persisted: 'Many tender thoughts my darling one of you & yr sweet kittens. Alas I keep on feeling the hurt of the Duckadilly. I expect you will have all made a pilgrimage yesterday.'[24]

They had; and Clementine wrote: 'I took the children on Sunday to Marigold's grave and as we knelt round it – would you believe that a little white butterfly . . . fluttered down & settled on the flowers which are now growing on it. We took some little bunches. The children were very silent all the way home.'[25] But she made heroic efforts to distract them, and made Randolph's return to Sandroyd's the occasion for an excursion and picnic. 'Yesterday we hired a car . . . & we all escorted Randolph in triumph back to school . . . We stopped on the road and had a splendid pic-nic & arrived torn & dishevelled at Sandroyds.'[26]

In late September, Sir Ernest Cassel died at his London house; Clementine wrote to Winston: 'I have been through so much lately that I thought I had little feeling left, but I wept for our dear old friend; he was a feature in our life and he cared deeply for you . . . I feel the poorer becos' he is gone. He was a true & loyal friend & a good man.'[27]

The end of this heavy year found Clementine on the edge of exhaustion, and she and Winston planned a Riviera holiday after Christmas at home. Winston went on ahead of her to the South of France, in company with Lloyd George. He had hardly left the house on Boxing Day when, one after the other, the children and servants started going down with influenza, of which there was currently another epidemic in London (but fortunately not of so violent a nature as the 'Spanish influenza' which had been so devastating two years before). Within twenty-four hours Clementine had four patients with roaring temperatures and one with pneumonia on her hands. Diana and Sarah, who at first seemed unaffected, were given asylum by their Great-Aunt Maude (Lady Maude Whyte, one of Lady Blanche's sisters); Diana developed the illness the following day and was returned to Sussex Square, which now, with two nurses, resembled a private nursing home. Indeed, Clementine remarked in a long letter she wrote to Winston on 27 December: 'I feel quite calm as now we are thoroughly organised, & could in fact with the same personnel deal with 2 more patients. Am thinking of advertising "2 empty beds for rich patients at 2, Sussex Square 50 guineas a week" – This would pay for the whole thing!'

But although she wrote in such a debonair fashion (Winston described her conduct and letter as 'Napoleonic'[28]), Clementine herself, after organizing the sick hordes, collapsed. On examination her blood pressure was found to be dramatically low; she was suffering from breathlessness, and in a state of utter nervous exhaustion, and the doctor ordered her to bed for a week. From there, she continued to give bulletins on all the patients to Winston.

Her plan to join him had to be postponed until everyone, herself included, was once more on their feet; nevertheless, Winston urged her to come out to France as soon as she was well enough 'to recuperate in this delicious sunshine, and let me mount guard in yr place over the kittens'.[29] He was having an agreeable holiday, painting and writing – and, as he confessed to Clementine, he 'played' (gambled) a little. 'It excites me so much to play – foolish moth,' he wrote, adding, 'But I have earned many times what I have lost by the work I have done here on my book.'[30]

But both Winston in the bright sunshine, and Clementine in her bed of sickness, had thoughts of the dark year through which they had just lived. Clementine wrote: 'the week that I was in bed I wandered in the miserable valley too tired to read much & all the sad events of last year culminating in Marigold passing and re-passing like a stage Army through my sad heart'.[31] And on New Year's Day, 1922, Winston had written: 'What changes in a year! What gaps! What a sense of fleeting shadows! But your sweet love and comradeship is a light that burns. The stronger as our brief years pass.'

CHAPTER FOURTEEN

The Twenties

THE YEAR 1921 HAD BEEN ONE OF GRIEF AND STRAIN FOR CLEMENTINE, BUT the New Year brought fresh hope. Towards the end of January she joined Winston in Cannes for a much-needed holiday, and she remained there for some weeks after he had to return to his parliamentary and ministerial duties in London. Venetia Montagu was with her for a time, which made agreeable company.

While she was in Cannes, Clementine's expectations were confirmed: she was once more with child. Nevertheless, it was such early days that she continued to play tennis. This caused Winston some anxiety, but she wrote reassuringly to tell him that she was

> really very well considering . . . I am playing a little tennis. There is a mild club Tournament going on here & I am playing in it . . . Don't think I am doing too much. On the contrary since the sun has come out & I have been playing I feel better & after playing I go to my Bunny & eschew the Casino & its heat & tobacco smoke, not to speak of its financial danger. The poor Cat has lost £10 at Chemin de Fer & is much annoyed –

She was evidently in good health and spirits; her letter ends on a gay note:

> Goodbye Darling – Kiss the two red haired kittens for me. I wonder if the new one will have red hair. Shall we have a bet about it 'Rouge ou Noir'?[1]

Clementine's tennis season ended in triumph, for she and her partner won the Mixed Doubles Handicap in the Cannes Lawn Tennis Tournament, and she returned much revived to London.

A source of recurring disappointment to Clementine was that, fond though Winston was of the children, she never could lure him to spend more than a few days with them all during the summer holidays. It was understandable: many fathers who are neither statesmen nor geniuses do not take kindly to rented holiday houses or lodgings at the seaside. Winston felt cooped up after a very few days, and now that painting was such a joy and occupation to him, he continually felt drawn to 'paintatious' (his own adjective) places, where the sun might be expected to shine brightly and continuously. But Clementine was forever trying to find somewhere which would be fun and suitable for the children, and where also Winston would be content to stay.

Hoping to achieve this very end, in the summer holidays of 1920, for instance, Clementine had taken one of the school houses at Rugby School, despite the fact that she greatly preferred to be by the sea. Rugby, however, was near Ashby St Ledgers, the Wimbornes' country house, and the plan was that Winston would stay with them all, and be diverted by polo with his Guest cousins. In the event, however, he made only spasmodic appearances: to do him justice, there was a political crisis on the brew, but the children were bitterly disappointed, and Clementine did not conceal her disgust. Her sole purpose had been to try to reconcile everyone's pleasures, and she found herself stranded 'inland', in a neighbourhood she did not particularly like, and in close proximity to the Guests, for whom she did not care.

However, the holiday plans for the summer of 1922 were arranged to everybody's satisfaction: they took a large house at Frinton-on-Sea, where Clementine installed herself with thc staff and children. She was expecting her child in September, and had no wish to go abroad. Winston was to go to France for the first part of the holidays, coming to join them at Frinton during the later part of August – and this time he did not defect. Clementine was perfectly content meantime and wrote to him happily on 8 August: 'I am so happy here – It's so comfortable & delicious. I hardly ever go outside the garden but just bask. The children scamper all over the place but I am not (just now) nimble enough to chase after them.'

The children had plenty of companions on the beach and at the tennis club, where Randolph and Diana competed as partners in the tournament; Clementine sent photographs of them playing: 'Randolph looks like a young champion but I'm sorry to say that playing together the children secured the Booby Prize, which I had the humiliation of presenting before a large crowd!'[2]

The holiday was such a success that Clementine pressed on Winston the idea of buying a house there: 'I do love this place & if we had a "permanent" little bug walk here we could use it for my tennis week-ends and children's recovery from illnesses at odd times . . . & be sure of always letting it well for the summer holidays if we fancied Deauville.'[3] Winston was glad she was so happy, but lukewarm about buying a house at Frinton.

Thoughts of the new baby occupied both their minds, but these were inextricably and painfully intertwined with memories of their beloved Marigold, the anniversary of whose illness and death fell in these days. 'I feel quite excited at the approach of a new kitten,' wrote Clementine on 8 August,

> Only 5 weeks now & a new being – perhaps a genius – anyhow very precious to us – will make its appearance & demand our attention. Darling, I hope it will be like you.

> Three days from now August the 11th our Marigold began to fade; She died on the 23rd.

Winston was particularly tender and understanding of her feelings in these days: 'I think a gt deal of the coming kitten & about you my sweet pet. I feel it will enrich yr life and brighten our home to have the nursery started again. I pray to God to watch over us all.'[4]

Again, a few days later, on 14 August, Winston wrote from Mimizan, where he was staying with Bendor Westminster*:

> My darling, I have thought vy often of you all & most of all of you. Yes I pass through again those sad scenes of last year when we lost our dear Duckadilly. Poor lamb – It is a gaping wound, whenever one touches it & removes the bandages & plasters of daily life. I do hope & pray all is well with you . . .
>
> Your own adventure is vy near now & I look forward so much to seeing you safe & well with a new darling kitten to cherish . . .

He joined Clementine and the children on 24 August, and they were all very happy together during the last lap of the holidays. Assisted by the children, their nanny and his detectives, Winston constructed a large and imposing sand fortress, which was the admiration of all on the beach, and which made a long and gallant stand against wind and tide. Presently Clementine returned to London to await the birth of her baby. Their fifth and last child, a girl, was born early on the morning of 15 September; they named her Mary.

It was in this summer that Winston first saw Chartwell, near Westerham in Kent. He was immediately captivated by it; but the story of its purchase and their life together there belongs to another chapter. It is enough to record here that Chartwell was bought in September 1922, and that it was to be Winston and Clementine's home for over forty years.

* * *

During the summer of 1921 a truce was signed with Sinn Fein, which at last marked the possible beginning of the end of the fruitless deadlock in Ireland. Violence and bloodshed had been rife on both sides since the previous spring. Clementine, who as ever watched public affairs with vigilance, had written to Winston in February 1921 about the agonizing situation:

> Do my darling use your influence now for some sort of moderation or at any rate justice in Ireland. Put yourself in the place of the Irish – If you were their leader you would not be cowed by severity & certainly not by reprisals which

* See note p. 220.

> fall like the rain from Heaven upon the Just & upon the Unjust . . .
> It always makes me unhappy & disappointed when I see you inclined to take for granted that the rough iron-fisted 'hunnish' way will prevail.[5]

Despite continuing strife, events moved towards a settlement, and Churchill took a prominent part in these negotiations, establishing close and helpful contacts with several of the Republican leaders; and he was one of the signatories of the Treaty which was signed in December 1921, by which Ireland was granted Dominion status, and partition was accepted, with separate parliaments in Dublin and Stormont (for the six Northern counties). There was immediate opposition to the Treaty from the Irish Republican Party, which was determined to attempt to secure the complete independence of a united Ireland, and a period of civil war ensued before the terms of settlement were imposed, by Irishmen themselves, upon their distracted country.

With the granting of Dominion status, Ireland became part of Churchill's responsibility as Colonial Secretary, and in consequence he himself stood in danger of his life. As he was known to be on the 'list' for assassination, he was for some time guarded by armed detectives. For many months he always slept with a revolver at hand, and an armchair in his room was reinforced with metal, so as to afford him protection should he have to 'fight it out'. These precautions, and the necessity for them, added a constant tension to their everyday existence; but Clementine never made much of it, either at the time or afterwards.

Throughout 1922, dissatisfaction with Lloyd George's leadership had been growing among the Conservatives, who formed the majority in the Coalition Government. Plot and counter-plot were the order of the day in high political circles, culminating in the now famous meeting at the Carlton Club on 17 October, when the Conservatives voted by a large majority to withdraw from the Coalition. The King sent for Bonar Law; a Conservative Government was formed; and a General Election took place forthwith.

In the midst of all this, on 16 October, Winston was taken gravely ill, and two days later he was operated on for appendicitis. In those days this was a serious operation, and although he weathered it well, he was not sufficiently recovered to take part in the early and crucial stages of the election campaign in Dundee, which opened on 30 October. Although Clementine was still feeding their new baby, Mary (only seven weeks old), she was determined to do what she could in the fight, and travelled to Dundee, arriving on 5 November with (as the local press somewhat spitefully put it) 'her unbaptized infant'. She stayed with supporters whose house was in a street ominously named Dudhope Crescent. Dundee was a two-member constituency, and there were six candidates: two National Liberals, one Labour (E. D. Morel), one Communist (W. Gallacher), one Liberal and a 'Prohibition' candidate, Edwin Scrymgeour, who had already contested the seat on five previous occasions.

Clementine flung herself into the campaign with energy, and in the course of the following week addressed at least six meetings, making spirited speeches on Winston's behalf. Although she was frequently shouted down, she was undaunted by the general uproar. General Louis Spears (a friend of Winston's since the Western Front days), who had come to help in the campaign, described how, when she appeared at one meeting wearing a string of pearls, women spat on her. He commented admiringly: 'Clemmie's bearing was magnificent – like an aristocrat going to the guillotine in a tumbril.'[6]

At one meeting, when she was consistently interrupted, the *Dundee Courier* reported: 'In the course of further efforts to destroy her speech Mrs Churchill appealed, "Do be kind and let me get on." She then treated the whole affair laughingly, and her smiles won the interrupters to comparative silence once again.'[7]

On 9 November, Clementine wrote Winston a long letter, telling him candidly,

> The situation here is an anxious one . . . If we win (which I pray & believe we will) we really must put in some time & work here & re-organise the whole organisation which was in chaos. Of course I feel the minute you arrive the atmosphere will change & the people will be roused. If you bring Sergeant Thompson etc tell him to conceal himself tactfully as it would not do if the populace thought you were afraid of them. The papers are so vile, they would misrepresent it & say you have brought detectives because you were afraid of the rowdy element. They are capable of anything.
>
> If you feel strong enough I think besides the Drill Hall Meeting which is pretty sure to be broken up, you should address one or two small open meetings. Every rowdy meeting rouses sympathy & brings votes & will especially as you have been so ill. Even in the rowdiest foulest place of all the people tho' abusive were really good-natured . . .
>
> I am longing to see you & so is Dundee – I shall be heartbroken if you don't get in. I find what the people like best is the settlement of the Irish Question. So I trot that out & also your share in giving the Boers self government. The idea against you seems to be that you are a 'War Monger' but I am exhibiting you as a Cherub Peace Maker with little fluffy wings round your chubby face. I think the line is not so much 'Smash the Socialists' as to try with your great abilities to help in finding a solution of the Capital & Labour problem & I tell them that now you are free from the cares & labours of office you will have time to think that out & work for it in the next Parliament.
>
> My darling, the misery here is appalling. Some of the people look absolutely starving. Morel's Election address just out [is] very moderate & in favour of only constitutional methods. So one cannot compare him with Gallacher.
>
> Your loving
> Clemmie

On 11 November Winston himself arrived in Dundee for the last few days of the campaign; he was barely convalescent, and delivered the greater part of his speeches sitting down. The meetings which he attended in these last stages of the campaign were particularly violent and noisy; the tide was flowing inexorably against him, and when the poll was announced on 15 November, Winston's massive majority had been swept away – Morel and Scrymgeour were overwhelmingly the victors. The national result gave the Conservatives a strong majority; in the Liberal Party there was a decisive verdict against Lloyd George but, more significant, the Labour Party polled over four million votes compared with five and a half million for the Conservatives.

Winston and Clementine returned wearily to London. Freed from all public responsibilities, he could now at least plan a long convalescence; they took a villa in Cannes, and the whole family moved out there for Christmas and the New Year.

Winston was once again at a turning-point in his career, and while he painted in the winter sunshine, he no doubt pondered it all deeply. He had been a wholehearted Coalitionist, and during the last months of the Lloyd George Government he had striven unavailingly to perpetuate the Coalition spirit in the formation of a Centre Party, which would fuse together liberal and radical opinion to form an effective movement against the rising threat of socialism. He now found himself a wanderer in a political 'No Man's Land' between die-hard Conservatism and an enfeebled and disunited Liberal Party, whose faith and purpose seemed to have vanished. As the New Year opened Churchill was, as he later wrote of himself: 'without an office, without a seat, without a party, and without an appendix'.[8]

The dismal condition of the Liberal Party after the First World War was a source of bewilderment and dismay to many of its ardent supporters. The chasm between the two wings of the party could not be bridged, and after the fall of Lloyd George and the end of the Coalition, the Liberals found their popularity waning, while the Labour Party grew in strength. Winston continued to hope that the forces of anti-socialism might be welded together effectively, and Clementine strongly approved of all his efforts to this end; but there was no way of preventing the new polarization of right and left.

Churchill stood for the last time as a 'Liberal Free Trader' at West Leicester in 1923: he was defeated, and Asquith's subsequent decision to maintain a socialist Government in power brought him to breaking point with the Liberals. Winston could not, however, actually bring himself at this moment to rejoin the Conservatives, and in March 1924 he stood as an 'Independent anti-Socialist' in a by-election in the Abbey Division of Westminster. Many prominent Conservatives supported him, and the campaign aroused much excitement. But it ended with Churchill being defeated by forty-three votes. In the following June, Stanley Baldwin publicly abandoned his party's pledge to introduce protective tariffs; this removed the last great obstacle to Churchill's reconciliation with the Conservative Party.

In September 1924 Churchill was adopted by the West Essex (Epping)

Conservative Association, and at the General Election following the fall of the Labour Government a month later, he won this seat with a handsome majority. He was to represent this constituency – later renamed the Wanstead and Woodford Division of Essex – for forty years. The return of the Conservatives to power saw Stanley Baldwin as Prime Minister, and the immediate return of Churchill to high office as Chancellor of the Exchequer from 1924 to 1929.

In all this logical sequence of thoughts and events, Clementine, dismayed by the plight of the Liberals, found herself broadly at one with Winston. But she did not want the Tory Party to win him back too easily. In February 1924, when his candidature for the Abbey by-election was being mooted, she was in the South of France, and Winston kept her closely informed about the situation at home. She was excited at the prospect of this new contest for him, assured as he would be of solid Conservative support. But in a letter on 24 February she sounded a note of warning: 'Do not however let the Tories get you too cheap – They have treated you so badly in the past & they ought to be made to feel it.' Later in the same letter she begged him not to stand 'unless you are reasonably sure of getting in – The movement inside the Tory Party to try & get you back is only just born & requires nursing & nourishing & educating to bring it to full strength. And there are of course counter influences as none of the Tory Leaders want you back as they see you would leap over their heads.' But in the by-election Clementine campaigned energetically for Winston, sending a personal letter to all the electors describing his twenty years of public service, and proudly proclaiming: 'With the single exception of Mr Lloyd George, no public man alive has been responsible for more important acts of social legislation than Mr Churchill.'

In the three General Elections which took place between 1922 and 1924 Clementine was ever in the thick of the fight. When Winston reunited himself with the Conservative Party, she understood his political evolution and followed his line of thought. Indeed, thereafter (as my mother told me) she herself did not vote Liberal again. But if with her reason Clementine may have 'crossed the floor of the House' with Winston, in her heart of hearts she remained to the end of her days a rather old-fashioned radical. Winston recognized and respected this, and there is a charming story of how, in the late spring of 1924, when he was still in a transitional stage between parties, he addressed a purely Conservative rally in Liverpool; Clementine accompanied him and sat on the platform. After the meeting Winston told his hosts: 'She's a Liberal, and always has been. It's all very strange for her. But to me, of course, it's just like coming home.'[9]

It is interesting that despite the fact that Clementine was intensely politically aware and well-informed, she did not choose, either now or later, to play a more active or prominent role in politics than do most dutiful wives of Members of Parliament – though she would have had ample opportunity to do so. She, in the main, confined her activity to Winston's own constituency, and rarely engaged in party politics at area or national level.

This may have sprung in part from her natural shyness – all her life she was a very 'private' person; but one feels that her underlying radicalism and a latent, almost subconscious, hostility to the Tories were also contributory reasons.

But within the bounds of the Epping Division she worked hard, and was frequently in the constituency, opening bazaars, visiting schools for speech days, attending Conservative women's meetings and taking a most active part in the work of the constituency Women's Advisory Committee. She nearly always accompanied Winston on his tours and visits, and her charm, her genuine interest in people and their lives, and her capacity for remembering names and faces (not one of Winston's conspicuous talents) endeared her to his supporters and workers and, over the years, built up a fund of affection and loyalty for them both.

Although the support Clementine gave Winston in his public life was of great value, the real sphere of her influence and usefulness to him politically was in private, for the most part unseen, although not unrealized by those who knew them both well. Her keen interest, her somewhat detached point of view, her highly developed critical faculty and her lofty integrity – all made her a valuable counsellor. Winston trusted her completely, and although he might at times disagree with her views, he always wanted to know what they were. He showed Clementine the drafts of his speeches, and although frequently her criticism caused him irritation, that he did not always take her advice was immaterial; the very fact that Clementine had the will, courage and capacity to battle with Winston in argument was an element throughout their life together of incalculable importance.

In one of Clementine's letters to Winston while he was at the Front in 1916 she had described how Sir Ernest Cassel was 'much impressed by the purple of authority & office'.[10] This was never the case with her: she was no respecter of persons – indeed, she had a streak of 'inverted' snobbishness – and she was generally much warier, and much less trusting of people, than was Winston. But these characteristics were of immense relevance in her role as the wife of a man with a brilliant public career, lived in the main among the great and the powerful.

An example of Clementine's 'wariness' occurred in 1925 when Winston, as Chancellor of the Exchequer, was about to be involved in a struggle over the Naval Estimates with Lord Beatty, Commander-in-Chief Grand Fleet of Jutland fame, and now the First Sea Lord. She wrote warningly from Lou Sueil, where she was staying with Jacques and Consuelo Balsan*:

> Now my lovely one – stand up to the Admiralty & don't be fascinated or flattered or cajoled by Beatty. I assure you the country doesn't care two pins about him. This may be very unfair to our only War Hero, but it's a fact. Consuelo tells me that he lunched here a little time ago & that he said 'I'm on my way home to fight my big battle with Churchill.' Consuelo said: 'Oh,

* See note on p. 78.

> I expect Winston will win all right –' to which Beatty retorted: 'I'm not so sure –' Of course I think it would be not good to score a sensational Winstonian triumph over your former love, but do not get sentimental & too soft hearted. Beatty is a tight little screw & he will bargain with you & cheat you as tho he were selling you a dud horse which is I fear what the Navy is.[11]

In the event Winston failed to win this tussle, for in 1927 Baldwin decided to accept Beatty's demands.

On one occasion, however, Clementine's judgement seems to have gone astray. While she was staying at the British Embassy in Rome in the spring of 1926, with the British Ambassador Sir Ronald Graham and his wife, Lady Sybil, she briefly met Mussolini, who was then the leader of the Italian Fascist government, and was to become absolute ruler of Italy in November of that year. Clementine at once became aware that she was in the presence of a considerable personality; she wrote to Winston:

> He is most impressive – quite simple & natural, very dignified, has a charming smile & the most beautiful golden brown piercing eyes which you see but can't look at. When he came in everyone (women too) got up as if he were a king – You couldn't help doing it. It seemed the natural thing to do . . . I had a few minutes talk with him – He sent you friendly messages & said he would like to meet you.[12]

After this tea party, Mussolini sent Clementine a signed photograph, and she reported: 'All the Embassy ladies are dying of jealousy.'[13] Winston commented rather dryly on her description of this meeting: 'What a picture you draw of Mussolini! I feel sure you are right in regarding him as a prodigy. But as old Birrell says "It is better to read about a world figure, than to live under his rule." '[14] The photograph of the Duce stood for a short time in the drawing-room at Chartwell; but it was not long before it was banished for ever.

* * *

When Winston became Chancellor of the Exchequer they moved into No. 11 Downing Street, selling the house in Sussex Square. 'Number Eleven' was then a charming house for a family, having the (discreetly) shared use of the 'Number Ten' garden. Clementine loved the house, and always remembered with pleasure the five years they lived there. The three elder children were passing through their teens; Diana made her debut and 'did' the London season from No. 11. Chartwell provided the ideal solution for weekends and holidays.

Churchill's tenure as Chancellor of the Exchequer was principally remembered for the Budget which returned Britain to the Gold Standard in April 1925. Although in retrospect Churchill was severely criticized for this policy, and various ills which subsequently afflicted Britain's economy were laid at its door, at the time the doctrine of the return to the Gold Standard found general acceptance among financial and economic pundits in all three parties.

In the following year a national crisis – the General Strike – brought Britain very near the precipitous edge of civil strife, when for nine days in May there was a nationwide stoppage in transport, printing, heavy industry, gas and electricity. Hundreds of thousands of mainly middle-class volunteers kept the essential services and supplies going. There was amazingly little violence: but the General Strike shook the country to the core.

Churchill's principal role in this upheaval was to keep the nation informed, and a small team headed by himself and J. C. C. Davidson, a junior minister, produced an emergency newspaper called *The British Gazette*; its circulation reached two and a half million. The tension and excitement of those days must have been intense, and once more Clementine found herself at the hub of the crisis. With a group of friends she helped to organize canteen arrangements for the teams of *British Gazette* workers at the printing office.

* * *

At the end of March 1925 Lady Blanche Hozier died; she was seventy-seven. Clementine was staying at Lou Sueil with the Balsans when she was sent for in haste, as her mother had been taken seriously ill. Both her daughters hurried to Dieppe to be with her, and kept vigil for the last days of her life. At Easter the previous year Clementine had spent two weeks with her mother, and had written a long letter to Winston, full of melancholy reflections:

> This is a sad old place & to me it is extraordinary that Mother should make it her home. To me it is haunted & decayed & melancholy. My sister Kitty died here of typhoid fever, & Bill is buried here in the cemetery at the top of the hill . . .
>
> It is extraordinary to reflect that if Mother had not on that first occasion come to Dieppe both Kitty & Bill might be alive today.
>
> I say Bill as well as Kitty, because it is here that Mother first began her regular gambling habits & it is here that Bill saw gambling from his childhood & used to come after he was grown up when on leave from his ship on week-end gambling expeditions.
>
> I went with Nellie to the Casino the other night & I was astounded at the reckless manner in which both Mother & Nellie gambled. Nellie very intelligently & dashingly, Mother in a superstitious and groping manner. It made me feel quite ill & ashamed to watch them & I went home to bed . . . Bill's grim & lonely end has not made the slightest difference – I don't feel I ought to criticize because gambling is not a temptation to me. It just seems to me a morbid mania . . .
>
> Mother who is getting very old & very feeble totters down every afternoon & plays – I believe only for an hour – But Bancos 2000 francs again & again without turning a hair.[15]

For twenty-five years Lady Blanche had lived the greater part of the time in Dieppe: she was one of the most colourful members of the British colony there, and many stories survive of her eccentric and sometimes malicious ways, though her charm and originality earned her many friends. Her gambling and drinking aroused comment; but although she was known to be badly off, she never borrowed money from her friends. She economized by letting her two charming houses in the rue des Fontaines and living herself for months on end in a modest hotel or pension. In this way she spent only about half of her income, leaving the rest for gambling. Indeed, when she died she was living in a hotel on the seafront, within easy walking distance of the Casino. Despite the many difficulties which beset her, Lady Blanche had an unquenchably debonair and gallant spirit – and throughout her life she was fascinating to men.

Nellie shared many of her mother's characteristics, but Clementine, although inheriting much of Lady Blanche's elegance and fastidiousness, was too unlike her in character and temperament for there to have been any real closeness between them. Also, in earlier years the knowledge that Kitty had been her mother's idolized favourite came between them, albeit unbidden, like a shadow. But with Clementine's marriage to Winston, her relationship with her mother had improved; Lady Blanche had taken to him from the start, and between her and her son-in-law there grew up a mutual regard and affection; she believed unwaveringly in his ability and destiny. Now, Winston wrote warmly and movingly of her to Clementine:

> Yr Mamma is a gt woman: & her life has been a noble life. When I think of all the courage & tenacity & self denial that she showed during the long hard years when she was fighting to bring up you & Nellie & Bill, I feel what a true mother & grand woman she proved herself, & I am the more glad & proud to think her blood flows in the veins of our children.
>
> My darling I grieve for you . . . the loss of a mother severs a chord in the heart and makes life seem lonely & its duration fleeting. I know the sense of amputation from my own experience three years ago . . .
>
> I greatly admired & liked yr mother. She was an ideal mother-in-law. Never shall I allow that relationship to be spoken of with mockery – for her sake. I am pleased to think that perhaps she wd also have given me a good character. At any rate I am sure our marriage & life together were one of the gt satisfactions of her life.[16]

* * *

In May 1929 there was a General Election, and Clementine as usual took an active part in the campaign, enthusiastically assisted by her children – Randolph, aged eighteen, speaking at a meeting, and Diana, Sarah and myself (now nearly seven) sporting our father's election colours in our hats. At this election the Conservatives failed to achieve an overall majority, and Baldwin resigned, preferring to go into opposition rather than form a coalition with the Liberals, who held the balance between the Labour and Conservative parties; Ramsay MacDonald formed his second Labour Government, depending upon the Liberals for survival. Winston therefore ceased to be Chancellor of the Exchequer early in June, and the family had yet again to move house; but now they had Chartwell as a country base to which they could retreat.

Clementine had been gravely ill with mastoid disease in 1928, and since then she had suffered recurring bouts of ill-health; now, only a month after the General Election, she had to undergo another operation – this time for tonsillitis. She had her operation in the first week of July, and was full of hope that she would be able to go with Winston a month later on a tour of Canada and the United States which he had planned to make now that he was freed from ministerial office. It would have been a cheerful party, for as well as Winston, Clementine and Randolph, Jack Churchill and his elder son Johnny were also to go on the journey.

To Clementine's bitter disappointment she was not sufficiently recovered to travel, and was left behind in a low and forlorn state. From the *Empress of Australia* on his first day out at sea, Winston wrote to her:

> All departures from home even on pleasure are sad. The vessel drifts away from the shore & an ever-widening gulf opens between one and the citadel of ones life & soul. But most of all I was distressed to think of you being lonely & unhappy & left behind. My dearest it wd have been madness for you to face this rocketting journey until you had regained full normal strength. This I trust & pray may be the result of 6 or 7 weeks of real quietness & calm.[17]

A few days later, one day out from Quebec, he wrote her a lively account of the voyage, but ended: 'My darling I have been rather sad at times thinking of you in low spirits at home. Do send me some messages. I love you so much & it grieves me to feel you are lonely.'[18]

Clementine was not quite deserted, however, having her three daughters at home – two of whom were of a highly companionable age. Diana (now twenty) spent much time with her mother this summer. They paid a visit to Dieppe, staying with Nellie, to whom Lady Blanche had bequeathed the St Antoines. Gradually Clementine improved in both health and spirits, and made plans to go to Italy later in October.

The news from the travellers in Canada and then America had been good, Randolph and Winston sharing the task of keeping Clementine and the family at home fed with accounts of their movements and doings. Interspersed among the lively accounts Winston gave of their travels were mentions of money matters.

The markets were booming, and even before he left England Winston had (on very sound advice) been indulging in speculation in the American stock market. These operations had reaped a most satisfactory harvest, and he reported all this to Clementine when he wrote to her from Santa Barbara, California, on 19 September. In addition, advance payment on his life of

Marlborough, royalties from his other books, and fees for various articles and lectures all combined to make the financial horizon brighter than it had been for some years, despite the loss of his ministerial salary. Winston set it all out for his careful 'Kat' on a single page, adding: 'So we have really recovered in a few weeks a small fortune.'

Since his departure, Clementine had been making arrangements for their London life that autumn and winter, and the current idea was that they should rent Venetia Montagu's house. In the same letter in which Winston set out their finances so clearly for Clementine, he wrote approvingly of the plans she had been making:

> I am glad you are taking Venetia's house for the [parliamentary] session. Do not hesitate to engage one or two extra servants. Now that we are in opposition we must gather colleagues & M.P.s together a little at lunch & dinner, also I have now a few business people who are of importance. We ought to be able to have lunches of 8–10 often, & dinners of the same size about twice a week. You should have a staff equal to this.[19]

The outlook seemed rosy.

Winston and his party spent three weeks in California, and then continued their journeyings through Arizona and via the Grand Canyon to Chicago. From there they went to New York, where they were the guests of Bernard Baruch, the American financier and speculator who became an elder statesman, and whose long friendship with Winston had started in 1918 when he was Minister of Munitions and Baruch a Commissioner on the American War Industries Board.

Clementine had fulfilled her holiday plan to go with Diana to Italy; they had a lovely time staying both in Rome and in Florence. Clementine was in good spirits, fully restored in her health and looking forward to the return of the other travellers. Early in October Randolph and Johnny left Winston to return to Oxford, and they brought her good news from New York. Winston's letters and cables continued to be couched in buoyant terms. He sailed for home on 30 October.

It had been a long separation, and Clementine went to the station to welcome Winston home on 5 November. But the joy of their meeting was marred: nothing had prepared her for the bad news he bore – and could not withhold from her for an instant. There and then, on the platform, Winston told her that he had lost a small fortune in the Great Crash of the American stock market, which had taken place during the last week of October.

It must have seemed a far, far cry from the elation and excitement they had both felt when Winston had received his windfall inheritance in 1921, and Clementine had written of the 'blessed feeling of relief that we need never be worried about money again'.[20] Through the succeeding years, the inheritance had been eroded both by Chartwell, which (as Clementine had early predicted) was costing much more than had been calculated, and also by their ample way of life.

But although this financial disaster had befallen him in the markets, Winston had, during his stay in the United States, signed contracts for articles in weekly magazines which would earn him £40,000. He could, as he had always done, keep them all by his pen and his prodigious industry; but the loss of such a capital sum was a body blow. They retreated to lick their wounds at Chartwell, where a regime of stringent economies was promulgated. As for their London life, for a year or two they either took furnished houses for a few months at a time, or, with more economy and convenience, stayed at the Goring Hotel near Victoria Station.

CHAPTER FIFTEEN

Chartwell

EVER SINCE THEY HAD SOLD LULLENDEN AT THE END OF 1919, WINSTON AND Clementine had cherished the hope of once more possessing a 'country basket', and his inheritance from his Vane-Tempest cousin in 1921 brought the dream within the realm of possibility. They both kept their eyes and ears open, and studied details of houses and even small estates coming on the market. Winston's ideas were braver and larger than Clementine's, envisaging a farm as well as a house and garden. But although she too longed for a country retreat, she was more cautious, and their only experience of farming, at Lullenden, had frightened her off another such venture. She had put her point of view to Winston in a letter in July 1921:

> Darling let us beware of risking our newly come fortune in operations which we do not understand & have not the time to learn & to practise when learnt. Politics are absolutely engrossing to you really, or should be, & now you have Painting for your Leisure & Polo for excitement & danger.
>
> I long for a country home but I would like it to be a rest and joy Bunny not a fresh preoccupation.
>
> I do think that if we really lived in the country it would be the greatest fun & also a life occupation to own & develop so varied a property. I should simply love it, but it would be engrossing & to make a success of it would need real hard work & concentration & just now I am for relaxation –
>
> I want to lie in the sun & blink & wake up now & then to eat a mouse caught by someone else & drink a little cream & doze off again.[1]

In July 1921 Winston saw a property near Westerham in Kent, about twenty-five miles from London, called Chartwell Manor. It had belonged to the Campbell Colquhoun family since the middle of the nineteenth century. The house itself was unprepossessing and dilapidated, but it stood on a hill commanding wide and beautiful views over the Weald of Kent. Before it to the south a valley fell dramatically away, and in the bottom was a lake, fed by a spring – the Chart Well. Winston was at once captivated by the beauty of the view, and quickly grasped the possibilities of the lie of the land, and of the private source of crystal water. On the far side of the valley the park-like fields sloped up to a wide belt of beeches which marked the boundary of the property and cast sheltering arms about the valley to the north and east.* This beautiful place, which combined a feeling of space and liberation with a sense of almost secret seclusion, cast at once and for ever its spell over Winston.

Thrilled with his discovery, he hastened to show it to Clementine, whose first reaction was enthusiastic. From nearby Fairlawne, where she was staying with the Cazalets for a tennis party, she wrote:

> I can think of nothing but that heavenly tree-crowned hill – it is like a view from an aeroplane being up there.
>
> I do hope we shall get it – If we do I feel we shall live there a great deal & be very very happy.[2]

But a closer and more thoughtful inspection of the house and property quickly moderated her first impression. The house appalled her: it was a grey Victorian mansion built round, and on the site of, a much older house; hemmed in by noisome laurels, it was damp, dreary and ugly. These surface defects, however, could have been coped with – but Clementine had detected a major, and unalterable, fault. Although to the south lies such a wide and pleasing prospect, the house had been built facing west, on the extreme edge of the property, and only eighty yards from the public road: from the other side of this road rose up a steep, wooded bank, which dominated the house, and effectively screened it from the afternoon sunshine. The bank was a mass of wild rhododendrons, which in the spring formed a cataract of mauve and purple. Unfortunately, these were colours which Clementine particularly disliked; moreover, neither the overshadowing trees nor the rhododendrons could ever be banished, as they were not part of the property, but were on common land. Another disadvantage which Clementine spotted was that, although some distance from the house, a public right of way ran right through the garden, slicing it inconveniently in two, and diminishing its privacy.

She saw at once that in order to correct (for to cure was impossible) the faults inherent in the siting and aspect of the house, it would be necessary to all intents and purposes to rebuild it. But Winston was blind to drawbacks, and deaf to reason. It was as effective to point out to him these radical faults, and to calculate the enormous cost and difficulty of correcting them, as it is to tell a man who is in love that the object of his affections has bandy legs or rabbit teeth.

However, Clementine's strong opposition to buying Chartwell had a delaying effect. Indeed, for some time nothing more was said about the

*Present-day (2002) visitors to Chartwell cannot appreciate the 'enclosing' effect of this belt of trees because it was largely destroyed by the great gale in October 1987. The National Trust has carried out mass replanting, but only time will restore the original effect.

project, and she assumed that, however unwillingly, Winston had accepted her arguments and abandoned the whole idea. But he had done no such thing: he had merely 'gone to ground' for the time being, and remained absolutely determined to acquire Chartwell. He was sure that the difficulties could be dismissed or surmounted, that Clementine would come to see this beautiful valley as the place of enchantment it was for him, and that together they would make there a home where they and their children would find happiness and contentment.

During the course of fifty-seven years of marriage, Winston and Clementine often had disagreements: on many occasions he disregarded her advice, or took a course which was contrary to her wishes – but the purchase of Chartwell was the only issue over which Clementine felt Winston had acted with less than candour towards her. The subject was hardly mentioned again between them during the ensuing year, which was to prove full of events which obliterated other thoughts and plans. It was the year of heavy tidings, culminating with the death of Marigold; and then, after those desolate days, came the hope and expectation of another child.

It was about the middle of September 1922, just after my birth, that Winston suddenly swept the three elder children off for a 'mystery' expedition to the country. On the way down in the car he told them that he was taking them to see a house he was thinking of buying, and he wanted to know what they thought of it. Sarah described their first impressions:

> Chartwell was wildly overgrown and untidy, and contained all the mystery of houses that had not been lived in for many years. We did a complete tour of the house and grounds, my father asking anxiously – it is still clear in my mind – 'Do you like it?' Did we like it? We were delirious. 'Oh, do buy it! Do buy it!' we exclaimed. 'Well, I'm not sure . . .' He kept us in anxious suspense.[3]

It was not until they had nearly reached home that Winston told them that he had already bought it. The children were ecstatic.

Perhaps he thought their unbounded enthusiasm would carry the day with Clementine; but when he confessed to her that he had – without further consulting her views – made an offer for the house, she was frankly devastated. After short negotiations the house and eighty acres were bought for £5,000.

Winston was not indifferent to Clementine's feelings in this matter – on the contrary, he longed for her approval over this major step in their life. But he never doubted that he could bring her to share his feelings for the place: everything, he assured her, would be done to transform the house according to her wishes; difficulties and defects would melt away like morning dew: his confidence and enthusiasm were boundless and touching.

But he made a long-term strategic error: for Clementine, the beauty of the valley, the eventual charm and comfort of the house, and the possibilities of the garden, never outweighed the worries and difficulties of running the house and property. Moreover, her gloomy prophecy of the cost of making it the delightful home it became was more than fulfilled. Clementine's fundamental and continuing disapproval of the whole Chartwell project formed a very important factor in her life from this time on: she accepted the *fait accompli*, but it never acquired for her the nature of a venture shared – rather, it was an extra duty, gallantly undertaken and doggedly carried through.

Had Clementine come to love Chartwell, and had it been the fulfilment of some of her dreams and wishes, she might have found, despite all the drawbacks, contentment and release from pressure there, and taken pleasure in all those domestic and garden pursuits which hold a compelling charm for so many Englishwomen. But although Chartwell was to be her principal home for the next forty years, she never became deeply attached to the place – although she devoted much time, thought and energy on both the house and the garden, and leaving her own 'stamp' upon them, still visible today.

But for Winston, Chartwell represented sheer pleasure, occupation and happiness – from the very first days, when he was slaughtering laurels and building dams, to later on when he was constructing walls and cottages. It became his workshop – his 'factory' – from whence poured his books, articles and speeches, which alone kept Chartwell and the whole fabric of our family life going. Here he found endless subjects to paint; here his friends and colleagues, and visitors from afar, foregathered. Here he planned his political forays, and here he retreated to relish his victories, or to lick his political wounds.

Chartwell never failed him, in good times or in bad. Not even in the last sad years when, silent and remote from us all, he would sit for hours in the golden sunshine of the summer days, gazing out over his enchanted valley and lakes; down over the lushly green and tufted Wealden landscape beyond, which melts at last into the faint blue-grey line of the South Downs.

* * *

Once Clementine had accepted the situation, she determined to make the very best of it, despite her deep-seated foreboding that they had bitten off more than they could chew. The first essential was to find a good, practical architect, combining withal an imaginative approach. Their choice fell upon Philip Tilden, who was well recommended to them: he had worked for Sir Philip Sassoon at Port Lympne in Kent, and also for Winston's aunt, Lady Leslie; and he was currently working on a house at Churt in Surrey for Lloyd George.

Mr Tilden's opinion of the house was much the same as Clementine's, his trained eye seeing the disadvantages of its natural aspect; but he was at one with them both in admiring the setting and the prospect. His greatest concern (which was amply justified) was the sinister presence of dry rot; but even his professional skill did not estimate how deep were its inroads upon the oldest part of the building.

It is sad to record that although they worked (or struggled) together for a number of years, the Churchills' intercourse with Tilden became increasingly strained and acrimonious. One cause of dissatisfaction was the recurrence of the dry rot which had so permeated the original fabric that its total eradication proved a lengthy and expensive affair. It would be unprofitable to recite the tale of reproach, sufficient perhaps merely to recall that the history of the relationship between architects and their patrons is a notably stormy one. A precedent certainly existed in the Churchill family in the explosive exchanges over a long period of years between Sarah, Duchess of Marlborough, and Sir John Vanbrugh over the building of Blenheim Palace.

The autumn and early winter of 1922 had been singularly exhausting for both Winston and Clementine; but after the General Election, liberated as he then was from both ministerial and parliamentary obligations, Winston determined to make the most of this unexpected lease of freedom, and swept the whole family off to the South of France. They let Sussex Square and took a delightful house, the Villa Rêve d'Or, perched high on the hillside above Cannes.

There for several months they enjoyed a sunlit and carefree existence. Winston wrote and painted to his heart's content, fully regaining his health after his operation for appendicitis; and Clementine, recovered from my birth, indulged to the full her enthusiasm for playing tennis – indeed, she and her partner won the local tournament. In due course the older children had to return to England for school, but they came out to France again for the Easter holidays. Winston made several journeys to London to attend to his personal affairs, and to visit Chartwell to discuss with the architect the plans which were now taking shape.

After one such visit at the end of January, he wrote a full report to Clementine. The chief challenge confronting the architect was to turn the whole 'sense' of the house through ninety degrees, so as to have the full benefit of the wonderful view and the southern aspect. It was proposed therefore to build out on the southern side of the original house a large four-storeyed wing. The western frontage of the house was to be extended on both sides of the core of the old house. It was further proposed to make full use of the good interior features of the oldest part of the former house, which, when later additions had been stripped away, revealed some fine beamed ceilings.

Winston went on to describe in detail how these plans were taking shape; he mentioned an improved pantry-kitchen arrangement, the plans for the linen and box-rooms, and other important domestic features. Tilden had conceived an idea for an additional reception room, which would be 'in the direct line of the drawing room and the boudoir. When the doors are opened all through, the length of these three rooms will be over 80 feet. Tilden is very keen on this. It undoubtedly makes a very fine sweep,' he wrote.[4]

'The Chartwell plans sound charming,' Clementine replied a day or two later. 'I hope in the new Tower arrangement the "sewing room" has not been eliminated? . . . Tell Mr Tilden the Sewing-room should be as large and nice as the Lingerie here as 2 or 3 maids will sit and sew there every day. The extra Reception room making a vista of 80 feet sounds very lovely and grand. We must have a State Festival there!'[5] (Can one detect a gentle note of irony in the last sentence?)

It must have been somewhat frustrating for Clementine to be out of direct control of the plans, and she was evidently pondering over them in a good deal of detail. 'I have been thinking about that extra sitting-room,' she wrote a few days later,

> It is <u>most</u> tempting to do this, but I fear the result will be to introduce into modern Chartwell the inconveniences of so many lovely old houses i.e. a wrong balance of living & sleeping accommodation. Without this extra room we already have ample living accommodation for ourselves & our potential 5 or 6 guests. We each of us have our private sitting-room, the children have Day Nursery & Schoolroom, there is an office for business, and for the entertainment of the few guests there are 2 very beautiful rooms, the Drawing-room & the Library . . .
>
> Also an extra sitting-room means more furniture, more fires, more flowers & more housemaiding. I do not want to have more than 2 housemaids at Chartwell.[6]

Clementine's eloquent plea won this particular argument, and the additional sitting-room was not embarked upon.

These details read somewhat curiously three-quarters of a century later, when the plans for Chartwell as discussed in these letters sound expansive, not to say grandiose. Then, even the cautious Clementine thought the possible five or six house guests 'a few', and, while counselling against any *folie de grandeur*, accepted as quite normal the proposed basic accommodation.

When they left the Villa Rêve d'Or and returned home in the early summer, Clementine went down to Chartwell to wrestle with the problems on the spot. In order to be able to supervise the work constantly, and to provide some country life for them all during the summer holidays, Winston took a house on the outskirts of Westerham called 'Hosey Rigge' (soon, of course, rechristened 'The Rosy Pig'). The older children spent most of their time at Chartwell, and were blissfully happy discovering the glorious possibilities for every kind of activity and game which the wild gardens and grounds provided; Winston made a wonderful tree-house for them in the great lime in the front drive. Although the house was far from ready for occupation, the stables were in action, and Winston's polo ponies had taken up residence, so the children could ride as well.

As the work on Chartwell progressed, the inevitably mounting expense of the alterations and the prospect of how much it would cost to run the house and garden preyed heavily on Clementine's mind. Winston sought to reassure her:

> My beloved, I do beg you not to worry about money, or to feel insecure. On the contrary the policy we are pursuing aims above all at stability (like Bonar Law!) Chartwell is to be our home. It will have cost us £20,000 and will be worth at least £15,000 apart from a fancy price. We must endeavour to live there for many years & hand it on to Randolph afterwards. We must make it in every way charming & as far as possible economically self contained. It will be cheaper than London . . .
>
> Eventually – though there is no hurry – we must sell Sussex [Square] & find a small flat for you and me . . . If we go into office we will live in Downing Street!

He was confident that the literary projects he had under way would see the work at Chartwell finished and leave them with six or seven months' living in hand, and he ended with an endearing plea:

> Add to this my darling yr courage & good will and I am certain that we can make ourselves a permanent resting place, so far as the money side of this uncertain & transitory world is concerned.
>
> But if you set yourself against Chartwell, or lose heart, or bite your bread & butter & yr pig then it only means further instability, recasting of plans & further expense & worry.[7]

Progress on the house seemed maddeningly slow during the winter, but after the usual delays and discouragements which always seem to attend house-building operations, the alterations and additions gradually neared completion. Still, it was not until the Easter holidays of 1924 that they were able to move in, and then only in 'camping' style. Winston and the children formed the advance guard who first took possession of Chartwell, for Clementine was staying with her mother in Dieppe.

Winston had persuaded her to spend a restful Easter with Lady Blanche, as she was thoroughly exhausted, not only by the constant prodding over a long period that had been required to get Chartwell to an even barely habitable state, but also by the efforts and excitements of the Abbey Division by-election which they had just fought. None the less, her absence at a moment which most women would regard as crucial, and requiring their personal participation and close supervision, seems highly significant. One cannot but feel that the real reason for her absence from the move into Chartwell lay deeper than fatigue, however genuine that was. Somewhere deep down, her basic mistrust of and disbelief in the 'Chartwell dream' must have played its part in allowing Winston to persuade her to have a holiday at this moment.

Winston himself slept at Chartwell for the first time on 17 April, and that night he wrote to Clementine:

My Darling,

This is the first letter I have ever written from this place, & it is right that it shd be to you. I am in bed in your bedroom (wh I have annexed temporarily) & wh is sparsely but comfortably furnished with the pick of yr two van loads.

They had had two glorious days moving in, and he reported that the children had 'worked like blacks'. He went on to describe the various activities which were in progress both in the house and garden, and, bursting into verse, declared:

> Only one thing lack these banks of green –
> The Pussy Cat who is their Queen . . .

He was thrilled by the spaciousness of the rooms: 'You cannot imagine the size of these rooms till you put furniture into them. This bedroom of yours is a magnificent aerial bower. Come as soon as you feel well enough to share it.'

Clementine's visit to Dieppe was not a happy one. She was utterly exhausted; she found Dieppe full of unhappy memories, and her mother was now ageing visibly. And part of her must have longed to be sharing with Winston and the children the first moments in their new home, into which she had poured so much thought and energy. On 21 April she wrote to Winston rather wistfully:

16 rue des Fontaines
Dieppe
April 21st [1924]

My Darling Winston

I was so much delighted to get your long letter describing all that you are doing at Chartwell I read it again & again & it made me long to be with you there.

If we are able in the future to live happily & peacefully there it will make up for all the effort you have poured out for it.

I have had sweet letters from all the children. They are blissfully happy. They will get to love it very much & it would be sad to have to part with it. I will do everything I can to help you to keep it . . .

I am not enjoying this, but I am resting – I find I am enormously & unbelievably tired & the strong air makes me drunk with sleep.

Despite Clementine's avowal of regret for her absence from Winston and the Chartwell scene at this moment, the hope she expresses for a long and happy future there, and her touching pledge of personal endeavour, one seems to detect almost a feeling of detachment from it all – as if she herself were a mere onlooker in this important new aspect of their life together.

Notwithstanding her deep-seated reservations about Chartwell, Clementine was never one to 'drag her feet', and there are few women who do not enjoy arranging and decorating a house. She relied on her own good taste and straightforward sense of practicality, rather than employing a professional decorator, which she would have regarded as a gross extravagance. Even when she could afford them, she always disliked contrived effects: the rooms on view at Chartwell today faithfully reflect her simple, unaffected taste, which relied on clear, clean colours, chintzes with bold flower designs, which well suited the large, rather gaunt rooms, and charming pieces of furniture either inherited or picked up 'for a song'. An exception to this was the long, low dining-room with its arcaded windows looking out over the valley and lakes, where the unstained oak tables and chairs were made to special order by Heal's; the great rush mats were also specially woven. The curtains here were a brilliant green glazed cotton.

The fashion then was for plain painted walls (the era of patterned wallpapers belonged to a later generation, and at that time one found them – ravishing dimity and flowery ones – in the servants' bedrooms). At Chartwell in the early twenties there were acres of palest cream paint (which went very well with the old beams), a pale blue for Clementine's sitting-room, and an ethereal cerulean blue for her barrel-domed bedroom, where the curtains and her four-poster bed covered in tomato-coloured moiré silk were sensational.

The scale of the garden from the first always somewhat daunted Clementine, and shortage of labour was a perennial problem. She had a destructive element in her approach to trees – no trees were better than trees she did not like; and the Chartwell valley was never planted or replanted to a consistent plan. But her taste in colours for shrubs and flowers was exquisite. Occasionally she became ambitious, launching out on a glade of *Lilium giganteum*, or a great planting of Himalayan blue poppies; but these did not long survive, and no trace remains. In the grey-walled rose garden, however, her taste and preferences are faithfully preserved: the plan was originally drawn up by the orderly minded Venetia Montagu (a very good gardener herself), and the pattern of the beds has hardly changed. A central feature were four standard wistarias; round the sides of the rose garden were borders which were a jumble of plants and shrubs in soft colours – cherry-pie, pentstemons, catmint, *Lilium regale*, ceanothus, fuchsias and paeonies. Up all the walls climbed roses and clematis; the pergola leading from the stone loggia to the rose garden was covered with vines.

Beyond the water garden was an azalea glade, which smelt heavenly, and in which there was not one distressing puce or orange. Above the goldfish pool was a mass planting of blue anchusas and white foxgloves which is still faithfully perpetuated today.

* * *

'Living off the land' and 'being self-sufficient' are phrases which have a compelling charm to many, and more especially to those with the least practical experience of country life. Winston was no exception, and no sooner were they installed than he launched out with enthusiasm upon farming activities. They were to prove (as they had done at Lullenden) on the whole unsuccessful and very costly. Neither Winston nor Clementine knew the first thing about farming or stock, and she regarded all this with the utmost apprehension. Cows, sheep, pigs and chickens were all tried in turn, and Winston's letters to her were full of accounts of the births and deaths of the livestock and the other events, mostly dramatic and disastrous, which attended the obscure lives of these animals and birds.

Apart from livestock which might be deemed useful, and hopefully profitable, there was an ever-changing population of pets. The black Australian swans were much prized by Winston, and he went to endless trouble (and expense) to protect them from the depredations of foxes, as he did the ruddy sheldrakes and the exquisitely 'enamelled' Carolina ducks. One of my bottle-fed lambs (misnamed Friendly) grew on into dreary and bad-tempered adulthood, and was banished only when it tactlessly butted Winston himself (to our secret delight). The marmalade cat was greatly beloved, and when he died in a week of heavy news during the war, Clementine would not allow Winston to be told until things looked brighter. The golden orfe lurked mysteriously in the water garden – a luscious prey for herons, and the subject of one of Winston's most beautiful pictures.

Quite soon the expense of the original alterations to Chartwell, and the cost of living there, fully justified Clementine's gloomiest prognostications. Rather less than a year after they had moved in, they were seriously discussing the advantages of letting it for a few months every year. Winston was offered eighty guineas a week to rent the house, and Clementine suggested that they could 'establish the children in a comfortable but economical hotel near Dinard, go there ourselves for part of the time & travel about painting for you, sight seeing for me, or we could go to Tours & do the "Chateaux" again – And we could go to Florence & Venice.'[8] It seems extraordinary now – the suggestion that Chartwell should be let, and that Winston and Clementine could go to a hotel (with four children, a nanny, a valet and a lady's maid), plus the extra hotel bills and fares if they both travelled – and that this could possibly be a saving. However, the idea did not materialize – perhaps for that very reason.

In 1926 a crisis point was reached in the economy of Chartwell, and two long memoranda survive, one from Clementine and one from Winston, proposing economies. Winston had been toying with the idea of building a dairy and 'going into milk'. Clementine viewed this idea with the utmost dismay, and in April she wrote seven pages of cogent argument which not only covered the undesirability of changing from beef to milk but reminded him of the troubles they had already experienced. '[Y]ou will remember that the chickens and the chicken houses got full of red mite and vermin; and you will also remember that one sow was covered with lice.'[9] Winston himself was alarmed by the general level of expenditure, and in September 1926 he addressed a memorandum to Clementine containing fourteen points covering every aspect of household and farm administration. Chartwell was to be let the following summer; the cattle and pigs and all but two of the ponies were to be sold; the family was to remain in one place for longer periods, and during the winter would visit Chartwell only for 'picnics with hampers'. 'Item 14' was headed 'BILLS', and covered everything domestic from the wines and cigars to be offered at meals and the number of courses, down to the amount of boot-polish used and the number of clean evening shirts Winston should wear each week. Visitors too were to be rationed: 'We must invite visitors very rarely, if at all, other than Jack and Goonie.'

Like all dogmatic manifestos which fly in the face of natural desires and instincts, this ruthless document was soon considerably modified. The family, it is true, spent the greater part of that winter in London, but over Christmas and the New Year eleven people signed the Visitors' Book at Chartwell. Some of the 'reforms' were carried out, however: no more was heard of dairy cattle; and the ponies and groom disappeared. The pigs diminished in number (only to increase again shortly through natural causes). But henceforward the farming activities at Chartwell were on a more modest scale.

There were to be other years of acute economic strain, notably after the Wall Street Crash in 1929, when Winston suffered such a great financial loss. To meet this crisis, Chartwell was run down to a very low ebb that winter; Winston's study was the only room left open in the house, so that he could work at weekends. Clementine stayed rather more in London, where they had taken Venetia Montagu's house. Winston would come down at weekends and join me and my governess in the small house he had recently built overlooking the kitchen garden. Originally intended for a married butler, this charming house (Wellstreet) became a useful retreat in this 'slump' period. Later it was let until the war to Horatia Seymour, Clementine's great friend.*

But Winston found it very difficult to economize for long, and lean times were followed by expansionist periods. From the day he bought Chartwell he engaged on operations of various kinds, a large number of which he aided or achieved himself. Beginning by damming a small stream in the valley to make two lakes connected by a waterfall, he progressed to make a swimming-pool. These ambitious works were the cause of endless vexations, with leaking dams and slipping banks; for Winston had a propensity for constructing lakes or pools on the sides of hills or slopes, thus ever contradicting the law of nature – that water flows downhill. Between 1929 and 1932 he built, largely with his own hands, the greater part of the red-brick wall which enclosed the vegetable garden, including in it a charming one-roomed cottage for Sarah and me. He also constructed a vast woodshed, the filling of which gave him extra employment and enjoyment in the form of cutting down and uprooting trees.

All these wonderful works required extra manpower at one stage or another. Since Clementine greatly deplored her gardeners being whisked off to assist in these labours, extra help had to be hired – at additional expense.

* Wellstreet now belongs to the National Trust and is the home of the Property Manager.

A letter from Clementine to Margery Street ('Streetie'), her secretary for many years, who went to Australia in the spring of 1933, gives a rather rueful account of expensive constructional works later that year: 'Mr. Pug [Winston] as usual is working hard not only at his Book but he is rapidly converting the whole valley at Chartwell into a Rock Garden. It's rather alarming & the "outdoor economies" have gone West as you can imagine. The fat boy now earns 35/- a week & Kurn is a permanence & there are 2 other men who have been there 6 weeks.'[10] Kurn, who had originally been engaged on a temporary basis, had indeed 'become a permanence' – he was still employed at Chartwell twenty-five years later!

* * *

Winston and Clementine's first recorded house party was in late June 1924, when Nellie and Goonie were the 'guinea-pigs'. Thereafter weekend parties of three to five people became the usual pattern, and since Chartwell was only an hour's drive from London, guests frequently came down for luncheon, returning during the late afternoon. The Churchills were on the best of terms with their immediate neighbours whose properties marched with Chartwell; and the family had agreeable contacts with the Astors* at Hever (where Winston often painted), with the Sidneys at Penshurst, and with the Cazalets at Fairlawne near Tonbridge.

Looking through the Chartwell Visitors' Book, one is immediately struck by the constant repetition of the same names. The 'stayers' were nearly all from the close family circle of the Jack Churchills and the Romillys, with their respective children, and Venetia Montagu and her sister, Sylvia Henley; or from that tight little band of faithful friends and colleagues (past or present) whose composition altered so little over the course of years.

But if it was, on the whole, a small, select group who came and went at Chartwell, it was also an intensely brilliant and diverse one. Centred on politics, it included such varied personalities as the Birkenheads, Brendan Bracken, Bob Boothby, the Duff Coopers, the Archie Sinclairs, the Cranbornes (later Salisburys), Max Beaverbrook, the Camroses (he was proprietor of the *Daily Telegraph*), the Bonham Carters and Lloyd George. Bernard Baruch, the American financier-statesman, never failed to pay a visit when over in Europe.

Among the regulars, letters and learning were represented by Eddie Marsh – of the bushy eyebrows and the slate-squeak voice – whose appearances as a perennial and faithful Private Secretary earlier in this account have masked his qualities as a delightful companion, and his brilliant achievements in classics and in *belles-lettres* (notably his translations of the Odes of Horace and the Fables of La Fontaine). He was the invaluable 'in-house' critic of Winston's writings. Although not a rich man, he built up an

* John Jacob Astor, later 1st Baron Astor of Hever. Younger brother of Waldorf, 2nd Viscount Astor (see note p. 97). He owned *The Times*.

astonishing collection of pictures, which plastered the walls of his tiny house in London. He was a Trustee of the Tate Gallery, and a Chairman of the Contemporary Art Society. On his death he bequeathed his pictures to the National Gallery, the Tate and other collections.

The classics, romance and adventure – all three – came in the person of the intense and strange Lawrence of Arabia. He would arrive on his (alas fatal) motorcycle as Aircraftman Shaw, and descend to dinner – to the amazed wonderment of my childish eyes – in the robes of a prince of Arabia.

Over the years Winston's growing and enduring passion for painting brought artists. The Laverys, who as wartime neighbours had started him off, came quite often in the earlier Chartwell days. And then, after twenty-five years, an old friend of Clementine's – Walter Sickert – made a sudden reappearance in her life. In June 1927 she had the misfortune to be knocked down by a bus in the Brompton Road; although she suffered only shock and severe bruising, the accident was reported in the newspapers, and a few days afterwards Mr Sickert called at No. 11 Downing Street anxious to know how she was faring. Clementine was delighted to see him again after so many years, and Winston and he took instantly to each other. The Sickerts thereafter came on several occasions to stay at Chartwell, and he took a lively and most helpful interest in Winston's painting.

Paul Maze, the ebullient and remarkable French painter, who had met Winston first on the Western Front, and who had the distinction of holding the DCM and the MM, as well as the *Croix de Guerre* and *Légion d'Honneur*, strongly supported Churchill's fight against appeasement and came a good deal to Chartwell in the mid-thirties; but he was not one of Clementine's favourites.

Dearly loved by us all was William Nicholson, a most charming companion and friend; dubbed the 'Cher Maître', he was a family favourite. Commissioned by some of their friends to paint a conversation piece for Winston's and Clementine's Silver Wedding in 1933, he paid many visits to Chartwell, and painted some enchanting studies of the swans, the lakes and the valley. Apart from his immense charm, Nicholson's influence on Winston's painting was thoroughly approved of by Clementine, chiefly for its softening and lightening effect on his palette, and Winston himself said he was 'the person who taught me most about painting'.[11]

Almost always present was a (usually young, and always exceptionally clever) literary researcher/assistant working with Winston on whatever major tome was on the stocks at the time. Of these, Bill Deakin, a Fellow and Lecturer at Wadham College, Oxford, came to Chartwell on and off from 1936 for three years; he was to reappear after the war to assist Winston with his war memoirs. Always admired and liked by both Winston and Clementine, Bill and his wife 'Pussy' became close friends of Clementine.

Among this Chartwell galaxy, too, from time to time 'shooting stars' from other spheres would appear: Charlie Chaplin; Albert Einstein; Tilly Losch; Ethel Barrymore.

By far the most regular visitor outside the family was 'The Prof'*. From 1925 until the outbreak of the Second World War, when his signature was transferred to the Chequers Book, he signed the Visitors' Book at Chartwell 112 times. He was a son of a German father and a half Russian-American mother, who lived and reared their family in Sidmouth, Devon; educated in England, Germany and France, he grew to be passionately English and patriotic. A brilliant scientist and curious personality, 'The Prof' was for nearly forty years Winston's close friend and his guide on all scientific matters, becoming during the Second World War his 'Personal Adviser' in an official scientific capacity. It was a strange friendship, for on the surface the men could not have been more different. 'The Prof' was a teetotaller, a non-smoker and a strict vegetarian; his nature was enclosed rather than outgoing; nevertheless, he was a constant visitor at Chartwell, welcomed as much by Clementine as by Winston. Fully self-occupied, he was an easy guest; he liked to play golf, and he also played very good lawn tennis. Clementine took endless pains with his food: he said he liked eggs, but we would watch with dismay as the yolks were removed from the specially prepared eggs, while he consumed only the whites! Winston, to whom the eating and drinking habits of 'The Prof' were an unfathomable mystery, would always press upon him a 'cubic centimetre' of brandy after meals; which 'The Prof' would imbibe, without much relish, to be obliging.

Whoever was staying at Chartwell, Winston's programme of work and play was independent and undisturbed. He usually worked in his room all the morning, leaving Clementine to entertain the company. Many of the male guests had work to do of their own, but there were plenty of diversions; there was an excellent golf course at Tandridge nearby, and in the summer tennis and swimming. During the afternoons Winston would paint or attend to whatever works of construction (or destruction) were to hand; in all these activities there was scope for appreciative spectators or volunteer unskilled labour, and guests and children alike would heave to.

In the winter afternoons and evenings Winston very much enjoyed playing cards. Bridge was never a Chartwell game, but mahjong, bezique, backgammon, and later Oklahoma† and gin rummy all had their vogue: of these, bezique and backgammon were the most enduring, and until the very last years he enjoyed playing these companionable games, both of which Clementine also played well and with pleasure. But when others were thinking of going to bed, Winston would retreat to his study and, summoning his secretary, might well dictate for two or three hours.

Mealtimes were the highlights of the day at Chartwell, from the point of view of both food and entertainment. The 'basic' house party, enlarged by other guests, usually formed a gathering it would be hard to beat for value. There was little warming up; the conversation plunged straight into some burning or vital question. But the talk was by no means confined to politics

* Frederick Alexander Lindemann, later 1st Baron Cherwell.

† An American card game, akin to gin rummy, much enjoyed post-war by the Churchill family and their guests.

– it ranged over history, art and literature; it toyed with philosophical themes; it visited the past and explored the future. 'The Prof' and his slide-rule were much in demand on all scientific problems. Sometimes the conversation was a ding-dong battle of wits and words between, say, Winston and Duff Cooper, with the rest of the company skirmishing on the sidelines and keeping the score. The verbal pyrotechnics waxed hot and fierce, usually dissolving in an instant into gales of laughter. General conversation usually dwindled, as nearly everyone wanted to share the main 'entertainment', whether it was a discussion or a dramatic and compelling monologue from Winston. Perhaps most enjoyable of all, particularly to us children (who from an early age partook of dining-room life), were the days when the Muse of Poetry and of Song held sway: when from the incredible store of our father's memory would flow verse, sacred and profane, heroic and frivolous, in glorious profusion. Many a time a luncheon party broke up only after the completion of a near word-perfect recitation of Macaulay's 'Horatius'. (Very popular with the children, as we could join in 'the brave days of old' bits.)

If 'high thinking' could be said to be a feature of life at Chartwell, it did not go hand-in-hand with 'plain living'; the hospitality was lavish and the food delicious. Clementine rarely had a highly trained cook – she could not afford one; more often than not she had a talented kitchen-maid, whom she would 'develop' herself. This was long before the period when cook-hostesses became a commonplace, but Clementine knew a great deal about delicious food and the theory of cooking, and she was highly successful in imparting her knowledge, never grudging time or trouble spent on planning and discussing the food with her cook. Naturally, what with the food and the talk, mealtimes tended to prolong themselves far into the afternoon or evening. It was unusual at a weekend to leave the dining-room after luncheon before half-past three or even four o'clock, and in the evenings the men stayed on endlessly after the women had gone to the drawing-room. Clementine found this habit increasingly trying as the years went by, and on one or two occasions marked her displeasure by taking the women off to bed, leaving a deserted drawing-room to greet the men when they eventually appeared, headed by Winston, full of guilty apologies.

Clementine rarely tried to compete against Winston in conversation during these long, verbose meals. She was, fortunately, a good listener, and for the most part enjoyed the arguments; and however riveting the topics, she was always alert and attentive to the food and the service. But she sometimes tired of the monologues and would firmly try, with varying degrees of success, to resurrect some form of general conversation. She could herself be a sparkling talker, with a sense of wit, and she laughed uproariously at times. In argument she often waxed passionate and partisan, and sometimes if the talk took a tone or direction of which she disapproved, she would after a time suddenly 'erupt', and could maul most savagely those of whose views or characters she either temporarily or habitually disapproved; the fact that they were her guests at the time afforded them no protection. Very occasionally, having delivered herself of her opinion, she simply rose and swept out, leaving her children puce with embarrassment, and any women guests flustered and uncertain as to whether they should 'sweep' too.

When Clementine was really roused, even Winston could not restrain her; and although these violent outbursts embarrassed him at the time, he would often recount the incident afterwards with a sort of rueful pride: 'Clemmie gave poor — a most fearful mauling.' Or, after one such occasion: 'She dropped on him like a jaguar out of a tree!' Her victims were never the timid, however tedious; but the brash and powerful, and those whose influence on Winston she deplored, were at sure risk – and Clementine's basic and undying radicalism also made high Tories and most very rich people potential targets for her contumely.

* * *

This wonderfully agreeable and enriching life, enjoyed equally by Winston, his family and his guests, was made possible only by his prodigious literary output. For life at Chartwell to be easy and comfortable, eight or nine indoor servants were needed: two in the kitchen; two in the pantry; two housemaids; a personal maid for Clementine (who also did a good deal of family sewing); a nursery-maid; and an 'odd-man' (boots, boilers, dustbins) – these made up the tally of personnel necessary to run this enlarged manor house in the twenties. In addition there was a nanny or governess and always two secretaries – one being available for late-night work whenever necessary.

Outside they employed a chauffeur, three gardeners, a groom for the polo ponies (until they were sold) and a working bailiff for what passed as a farm. Although most of the outdoor staff stayed for long periods, the turnover in the house was fairly constant – one reason certainly being the remoteness of the house from any bus stop. The process of procuring, and initiating, a stream of new servants was expensive, tiring and frustrating. To these problems, Winston, like so many men, was not particularly attentive or sympathetic. In September 1928 he wrote somewhat breezily to Clementine:

> do not worry about household matters. Let them crash if they will. All will be well. Servants exist to save one trouble, & shd never be allowed to disturb ones inner peace. There will always be food to eat, & sleep will come even if the beds are not made. Nothing is worse than worrying about trifles. The big things do not chafe as much & if these are rightly settled the rest will fall in its place.[12]

It is doubtful whether these Olympian sentiments had the cheering and calming effect intended.

The problem of running Chartwell efficiently and economically would have daunted even a woman with less exacting standards than Clementine. Over the years she was assisted in all these matters by two successive secretaries – first Margery Street, and then Grace Hamblin, a local girl who came to 'help out' in the office over a holiday period in 1932, and succeeded 'Streetie' when she returned to Australia the following year. More than anyone else, 'Hambone' (as we children early nicknamed her) helped Clementine to grapple with the ups and downs of Chartwell life. With the outbreak of war she came to London to be Clementine's personal Private Secretary, returning again to Chartwell after 1945. Depended upon and beloved not only by Clementine but by Winston and all of us children, she became and has remained a close personal friend. No account of Chartwell or of my parents' life there would be complete without a prominent 'Mention in Despatches' for Grace. After Winston's death, when Chartwell was handed over to the National Trust, she became its first Administrator.

From the first, Chartwell had spelt anxiety and difficulty for Clementine: their precarious financial state had a wearing effect on her nerves, and the perpetual backlog of bills, and the struggle she had to get even the local tradesmen paid, were the cause of gnawing worry and mortification. Winston's optimism, and the fertile flow of new and money-consuming projects which he was always considering, had an abrasive effect on her disposition, and were the cause of many arguments, which sometimes waxed hot and angry. On one occasion she became so enraged that she hurled a dish of spinach at Winston's head (she missed, and the dish hit the wall, leaving a tell-tale mark). She sometimes abandoned the unequal struggle of arguing with him, and resorted to writing calmly considered minutes or memoranda setting forth her point of view. It was a method she was to pursue increasingly, for she found she often succeeded in making her case in this way, when verbal argument utterly failed.

Despite all these cares and worries, which persisted throughout the fifteen between-the-war years that Winston and Clementine lived at Chartwell, the manner of the life they led belied the insecurity and fragility of their financial situation. They lived and entertained elegantly; they travelled; they brought up and educated their four surviving children handsomely. But had Winston's diligence, health or genius failed, the whole fabric of their life would have crashed, for they literally lived from book to book, and from one article to the next.

Of course, we children were aware from time to time that 'Papa and Mummie were economizing': we were lectured about the necessity for turning off lights, and the older ones were very much scolded for lengthy telephone conversations. But when we took refuge in Wellstreet Cottage, I thought it was great fun, and very cosy. And when on another occasion the house was partially shut up to save on heating and servants, and the big ground-floor rooms swathed in dust-sheets, the dining-room downstairs divided perfectly into a charming sitting-room-cum-dining-room, and it seemed an excellent 'winter' arrangement with only a few of us at home.

It was only as I grew older, and my mother began to confide some of her troubles and anxieties to me, that I came to understand that for her Chartwell was not – and never had been – the joy and refuge it was for my father. At first I found this difficult to grasp, and I almost resented her critical, unappreciative attitude to what was, for me, a Garden of Eden. It was only much later, when I read my mother's letters, that I really began to understand: then I realized how fragile was the raft which supported that seemingly so solid way of life.

But it is not the memory of the lean times that springs to my mind on looking back. Much more I recall the spaciousness of our life; the comfort and prettiness of the rooms; the masses of flowers (always clear, soft colours, and always arranged with great simplicity); the crackling log fires, and the delicious comfort of a warm, scented house on bleak winter days. And in the summer, all the joys of life in the garden, voices calling up from the tennis court, long happy hours by the swimming-pool, and tea in the loggia with strawberries and cream.

No times glow so vividly in my own memory as the wonderful Christmases. I remember, as if it were yesterday, the exquisite, tantalizing thrill of walking past a poky closet (known as the 'Genie's cupboard') which was strictly 'out of bounds' for weeks before, where my mother and Nana (my governess) rustled about with paper, and held whispered conferences. I used to watch with mounting excitement as the gardeners transformed the house: Clementine was never very 'tinsel' minded, and although modern Christmas decorations are exquisite in their glittering sophistication, I still recall how beautiful were the more sombre, mysterious effects produced by the different textures of the leaves, and the gradations of the lustrous greens of the ivy, laurel, yew and holly with which the house was hung. The centrepiece of the decorations was a large and beautiful della Robbia plaque hung aloft in the front hall, depicting the Christ-child who, swaddled and pensive, gazed down on us all.

Part of the potency of the Christmas magic lies in its unchangingness, and every family seems to evolve its own particular ritual, usually fiercely guarded from change by the children. It was certainly so with our family, and year after year the same party gathered at Chartwell: Jack and Goonie Churchill, with their three children, Johnny, Peregrine and Clarissa; Clementine's sister, Nellie, with her husband, Bertram Romilly, and their two sons, Giles and Esmond (curiously known as 'The Lambs'). One or other of the two lonely bachelors, 'The Prof' or Eddie Marsh, would be the only member of the party outside the family. The crowd of young cousins were close in age, and provided the necessary raw material for the jollity, special friendships and in-fighting inseparable from, and indeed, indispensable on, such occasions. Trailing far behind this closely knit group came myself: the 'baby', the 'Benjamin', alternately petted or excluded according to moods and circumstances.

When we were all assembled on Christmas Eve, the double doors between the drawing-room and the library were flung open to reveal the Christmas tree, glowing with light, and radiating warmth and a piny, waxy smell from a hundred real white wax candles. Electric lights can never distil for me the magic cast by the glimmering, sputtering beauty of the Christmas tree of my childhood days. Not surprisingly, one year the tree caught fire, and only Randolph's presence of mind, and speed in fetching an extinguisher, saved us from catastrophe.

Amusements and occupations were not hard to come by: in the year of the 'great snow', 1927, the bigger children built a wonderful igloo; there was skating on the lakes, and the lane below the house was so deep in snow that a tunnel had to be dug to allow the traffic to pass.

Amateur theatricals were a great ploy; the dining-room made a natural theatre, as there were dividing curtains. My mother greatly enjoyed these dramatic activities, and threw herself into them with enthusiasm. She wrote a lively account of it all to Streetie in January 1934:

> We all acted at Christmas – even me! You will hardly believe that. We (Sarah Mary & myself) acted a short play by Gertrude Jennings called 'Mother of Pearl' . . . I took the part of a dirty old tramp & I enjoyed it enormously . . . Mr. Gurnell the Chemist came & made us all up. Then there was a thriller called 'The Hand in the Dark' in which Mrs. Romilly, Sarah & Esmond performed. They wrote it themselves & it was really quite gruesome.
>
> We gave 3 performances 3 days running & you can imagine the disorder of the house. Nana prompted & produced & was very severe! Mr. Pug was too sweet. He said it was all lovely & that I was so tragic that I made him cry![13]

The Christmas of 1933 saw the last of the glorious family parties at Chartwell. Time was spinning on, and all the cousins were growing up and beginning to go their own ways. Of the five peacetime Christmases which remained, only one more, 1936, was to be at Chartwell. For all of the others the family for one reason or another was divided, either Winston or Clementine being on their travels. But the stay-at-homes received warm hospitality at Blenheim. 'Sunny' Marlborough died in 1934, and his son Bert and daughter-in-law Mary, with their younger family, continued the tradition of friendship and hospitality to Winston and Clementine and their children.

As the years went on, inevitably threads of anxiety, sorrow, disappointment and misunderstanding crept here and there into the tapestry of our family life. But when one contemplates the finished piece, it is not these muted and sombre tones, but the glowing, happy colours that predominate. There was so much laughter, activity and high spirits; and, above all, the golden skein of warmly expressed love and loyalty gleams throughout, strong and unfading.

61. Winston and Clementine electioneering on a rainy day in the thirties.

62. Charlie Chaplin visits Chartwell, September 1931. Left to right: Mary's pug; Tom Mitford; Freddie Birkenhead (F.E.'s son); Winston; Clementine; Diana; Randolph; Charlie Chaplin.

63. Sarah in the 1930s.

64. Diana married Duncan Sandys, 16 September 1935.

65. Vic Oliver, *c.*1936, whom Sarah married in December that year.

66. Diana with her mother at the first night of the revue *Follow the Sun* at the Adelphi Theatre, Manchester, in February 1936. Sarah was one of 'Mr Cochran's Young Ladies' and Vic Oliver the star of the show.

67. Major (later Sir) Desmond Morton. Director of the Industrial Intelligence Centre in the 1930s, he supplied WSC with vital information on Germany's undercover rearmament.

68. 'The Prof' (Professor F. A. Lindemann, later 1st Viscount Cherwell). He and WSC met in 1921 and quickly formed a friendship. He was a constant visitor to Chartwell, and instructed WSC on many technical and scientific matters.

69. M. Leon Blum, the French socialist statesman and former Prime Minister, visited Chartwell in May 1939.

70. Brendan Bracken. An adventurer: he met WSC in 1923 and increasingly became his political confidant, champion and crony. At first disliked by CSC, he won her trust and affection by his fidelity.

71. M.Y. *Rosaura*, 1934.

72. Lady Broughton (Vera) and Lord Moyne, on Komodo, inspecting a dragon lizard in a cage-trap.

73. Bathing party on board the *Rosaura*.

74. One of the captured Komodo 'dragons' at the London Zoo, *c.*1935.

75. Terence Philip with Clementine, in Madras, early January 1935.

76. All part of a parliamentary wife's life! CSC opens the bowling for the 'Lyons' Girls' cricket team *versus* Woodford Police Athletic Club, September 1936.

77. Aged fifty, Clementine took up skiing. Here she is in St Moritz, January 1937.

78. Clementine, to keep me company, dug out her old hunting clothes, and we had some good days with the Old Surrey and Burstow foxhounds, spring 1939.

79. Clementine, photographed by Cecil Beaton at No.10 Downing Street in September 1940, just before the beginning of the 'Blitz'.

80. Clementine, painted by Winston (after the war) from a photograph of her launching HMS *Indomitable* in March 1940.

81 and 82. The drawing room, and Clementine's bedroom, in the No.10 Annexe flat at Storey's Gate.

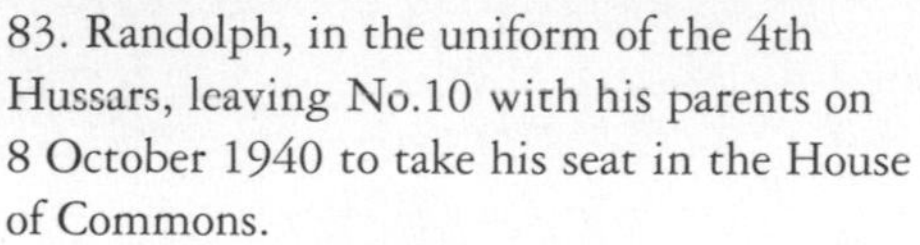

83. Randolph, in the uniform of the 4th Hussars, leaving No.10 with his parents on 8 October 1940 to take his seat in the House of Commons.

84. Winston and Clementine going downriver to the East End by launch after heavy bombing raids at the end of September 1940: the roads to the City were closed because of unexploded bombs.

85. Winston, with Clementine, visiting bomb-damaged Swansea, 11 April 1941.

86. Clementine, with Mary and Sarah, having witnessed the House of Commons' heart-warming welcome to Winston on 18 January 1944 on his return from North Africa after his serious illness.

87

88

87. Averell Harriman (FDR's special Lease-Lend envoy) and Lord Beaverbrook (at this time Minister of Supply) on board the *Duke of York* en route for Washington, December 1941.

89

90

88. Winston with Gil Winant, US Ambassador to London, 1941–46. Charming and unusual, he became a great friend of the Churchill family.

89. Sir Charles Wilson, MC, 1st Baron Moran. Winston's doctor from 1940 to the end of his life. His care and judgement, and his ability to summon specialized help from anywhere at any moment, were crucial to Winston's health.

90. Winston with his brother Major John Churchill (Jack), to whom he was very close. They are watching a mock battle in July 1941.

91. Nellie Romilly, Clementine's sister, *c.*1941. She often stayed at Chequers.

92. Clementine at Chequers, *c.*1941, with Jock Colville, one of Winston's Private Secretaries, who became and remained a close family friend.

93. Family group at Chequers, *c.*1943. At back, left to right: Pamela (Randolph's wife); Peregrine Spencer Churchill (Jack's son); Sarah; Jack Churchill; Duncan Sandys. In front: Clementine; Winston; Diana Sandys.

94. A line-up of babies at Fulmer Chase Maternity Home for the wives of officers of all three services, at Gerrards Cross, Buckinghamshire. Clementine was Chairman of the House Committee throughout the war.

95. Clementine addressing the crowd at the England *v.* Scotland International Soccer Match at Wembley, in aid of 'Mrs Churchill's Aid to Russia Fund', 17 January 1942.

96. Clementine sorting letters with contributions to her appeal for Aid to Russia.

97. In April–May 1945 Clementine travelled thousands of miles in the Soviet Union on behalf of her Aid to Russia Fund. Here she is in Leningrad, surrounded by children recovering from the effects of starvation during the blockade of the city.

98. Clementine in Moscow, at a reception given in her honour by the Anti-Fascist Committee of Soviet Women; with her is their Chairman, Mrs V. Grizodubova, a famous Russian 'ace'.

99. In Odessa, Clementine and her colleague, Miss Mabel Johnson, visited about 250 British prisoners-of-war who had been liberated by the Red Army and were in a camp awaiting repatriation.

100. 1st Quebec Conference, August 1943. Winston and Clementine, with myself (ADC to WSC), on the train between Halifax and Quebec. At back right: Detective Inspector W. H. Thompson, WSC's detective since before the war.

101. 2nd Quebec Conference, September 1944. Mrs Eleanor Roosevelt and Clementine make a joint broadcast.

102. Here is a charming picture of Winston and FDR greeting each other, September 1944.

103. Winston and Clementine went to Paris for the Armistice commemorations, 11 November 1944. Here Winston and General de Gaulle walk down the Champs Elysées from the Arc de Triomphe, cheered to the echo by thousands of Parisians.

104. Clementine, after receiving a Doctorate of Law from Glasgow University, with (left) Sir Hector Hetherington, the Principal of the University, and (right) Sir Kenneth Clark (later Lord Clark, OM), 19 June 1946.

105. Accompanied by two very proud daughters, Clementine after her investiture as a Dame Grand Cross of the Order of the British Empire, 9 July 1946.

106 Winston congratulating his first horse, Colonist II, after winning the Victor Wild Stakes, 13 May 1950. In the background (left), can be seen Clementine, and (right) Christopher Soames, who largely organized WSC's racing activities.

107. Sarah married Antony Beauchamp on Sea Island, Georgia, on 18 October 1949.

108. Anthony Eden and Clarissa Churchill, with her uncle and aunt, after their marriage on 14 August 1952, in the garden of No.10 Downing Street, where the wedding reception was held.

109. Jeremy Bernard, Christopher and Mary Soames's third child, was christened at St Mary's, Westerham, on 17 August 1952. Left to right: Clementine; Christopher with Emma Soames in front; Mary with Jeremy; Winston with Nicholas in front; 'Monty' – a godfather.

106

108

107

109

110. The 1951 General Election was won by the Conservative Party. Winston and Clementine leaving the count at Woodford Green, Essex, after Winston had been returned again as MP on 26 October. The following day he became Prime Minister once more – in his 77th year.

CHAPTER SIXTEEN

Herself and Others

FROM THE EARLIEST DAYS OF FAMILY LIFE A CONFLICT DEVELOPED IN Clementine's life between wifehood and motherhood. Many women today find themselves involved in a somewhat sterile competition between marriage, motherhood and a career, and politicians' wives particularly are nearly always unwilling participants in a sort of domestic 'tug-of-war', in which economic factors usually play a decisive role. Since Clementine and Winston nearly always had a London house, and always a nanny or a governess for the children, the cause of the competition between her two roles lay, not in economics, but partly in her own character, and largely in her husband's. The demands he himself made upon her, and the extraordinary circumstances of his career, created a sharp division between her two vocations. There was probably no single moment when Clementine made a conscious choice between her wifely and maternal interests, but a study of her life leaves one in no doubt as to what that declared choice would have been.

From the day she married him until his death fifty-seven years later, Winston dominated her whole life; and once this priority had been established, her children, personal pleasures, friends and outside interests competed for what was left. That she never lost her own individuality, despite the vocational quality of her devotion to her husband, and the demands his overpowering personality made on her, is a measure of Clementine's strength of character. Devoted and fiercely loyal, she never became a 'yes-woman' or lost her capacity for independent thought. But her dedication to Winston supplies the answer to the question why a woman whose social conscience had been so thoroughly aroused in the great radical reforming days, and whose organizing ability would be twice proved in wartime, did not in the intervening period of peace take up, to any marked degree, some form of community service, or develop any independent personal interest or occupation. She herself commented in a letter to Winston in 1920: 'The Canteens – I sometimes wonder now if it was all a dream. One thing is certain, I couldn't manage them again. I began to think I had real organizing ability, but it died with the war – if there ever was any!'[1] This self-deprecating remark was to be proved wrong many times over in another testing time.

Unlike many women in a similar situation, Clementine showed no desire to play a great part in charitable work. She was not ungenerous to good causes, but she fought shy of fund-raising committees, and her name was rarely to be found on the patronage or committee lists of charity balls or play and film premières. It was not 'her scene', and she had no desire for a platform to project either a cause or her own personality. The considerable amount of work she did in Winston's constituency seems to have absorbed most of the energy and time she had for public work.

However, in 1934 Clementine was made a 'Guardian' for Maternity and Child Welfare by the Kent County Council, and she attended meetings of the Public Health and Housing Committee for several years. She also became the County Council's representative on the Management Committee of the Alexandra Hospital at Swanley. A fellow committee member said that, although she was assiduous in all matters relating to her particular subject, he never felt Clementine really relished this sort of work. It never gave her the feeling that she was achieving any concrete results; moreover, she found the attitude of many of the other committee members hard and unsympathetic to the human problems involved. Since one of her chief characteristics was to react quickly to any situation and seek a direct solution, it is easy to understand that public committee work did not appeal very much to her.

Winston was to be Clementine's life-work. Her concentration on him and his career consumed the cream of her thought and energy. She said to me once: 'It took me all my time and strength just to keep up with him. I never had anything left over.' Asked how her husband managed to combine so many different interests and activities – politics, writing, building walls, painting – Clementine answered quite simply that Winston 'never did anything he didn't want to do, and left someone else to clear up the mess afterwards'.[2]

Clementine was a devoted and conscientious mother, but her priorities were never in doubt: Winston came first – always. It was not unusual in those days for upper-class children to lead much of their lives apart from their parents, but Winston's demanding nature, combined with the claims of his career, meant that even the school holidays had to take second or third place. When the children were babies, the competition between Clementine's nursery world and Winston's life was not so keen, but she always tried to share their seaside holidays, and suffered some unhappiness and much frustration that Winston could not be persuaded that lodgings at Westgate or Frinton-on-Sea were just as enjoyable as the South of France.

When the children had to have different arrangements, Clementine was always at pains to make suitable and enjoyable plans for them, but, to her own great sadness, time and again she could not share treats or expeditions with them. In later years, she realized what a 'missing factor' this had been in her relationship with her children. It would also have been a fortunate circumstance if, particularly in those early nursery years, there had been an enduring nanny or governess. But Clementine was not lucky enough to find such a treasure until much later on, and the three older children experienced many changes of rule in the nursery.

To her children, Clementine was a mixture of tenderness and severity. She loved her babies with a fierce physical relish, and she enjoyed them tumbling about in her room while she dressed or conducted the business of the household. As a small child I was vividly aware of her beauty and fastidiousness; of the velvety texture of her skin, the softness of her lingerie, and of how delicious she smelt. But as the children grew older, the same shyness and reserve which inhibited her friendships with her own contemporaries formed a barrier, tending to make the relationship between them somewhat unspontaneous and formal.

Clementine had no real understanding of the childish mind or outlook, and applied her own perfectionist standards not only to manners and morals, but to picnics or garden clothes. Consequently, although her children loved and revered her, they did not find in her a fun-maker, or a companion for their more untidy, knock-about activities. Both the circumstances of her life and her own nature combined to place her goddess-like upon a pedestal. But on any big issue, once the courage was screwed up to broach the subject, her children could be sure of her attention, astonishingly quick action and, if justified, the championing of their cause.

Winston was more relaxed and indulgent, greatly enjoying his children in the brief moments he could spend with them. His letters show a touching interest and concern for his nursery. His children, and later his grandchildren, were always conscious that he loved to have them around. 'Come to luncheon,' he would say to someone, extending an invitation to Chartwell, adding with obvious relish: 'You'll find us all bunged up with brats.' But the burden of bringing up and disciplining their brood was always left to Clementine.

When Chartwell was bought, school holidays together at last became a reality. Diana was then thirteen, Randolph eleven and Sarah nearly eight. The three elder children revelled in the opportunities and freedom their new home offered them, and they unreservedly shared their father's enthusiasm for the place. Moreover, as they were able to help with the outside constructions and improvements, it brought them many hours of relaxed companionship with him. Clementine's time and energies were chiefly concentrated on the house and its problems, and so she missed out on these enjoyable and companionable occupations. Earlier than many children, however, we were promoted to participation in dining-room meals, enjoying our parents' company and mingling freely with their guests.

From a very early age public events played their part in nursery life. Diana, aged five, had been conscious of the feeling of impending disaster which permeated the house in the spring of 1915, and the two older children had vague memories of walks in the park escorted by detectives to protect them from the suffragettes. The Irish troubles had brought armed detectives into the house: Randolph remembered how he and Diana, on returning one day (22 June 1922) from roller-skating in Holland Park, 'found the house surrounded by policemen. Indoors all sorts of tough-looking men were running up and down the stairs, looking in cupboards, attics and cellars. It was explained to us that earlier in the afternoon Sir Henry Wilson* had been assassinated by Irish gunmen on the steps of his house.'[3] It was feared by the authorities that Winston might be next on the list; hence these alarms and excursions.

From our earliest years we were all passionate partisans of our family's electoral fortunes. Diana's and Randolph's childhood letters, in particular, show a keen awareness of political affairs. In the General Election of November 1923 Diana, then fourteen, kept her parents (who were fighting West Leicester) well informed about the course of the campaign in their home constituency of Paddington, reporting that one of the candidates 'has not the slightest chance of getting in, & he is quite willing to pay the fine [lose his deposit] if he does not get a certain number of votes'.[4] In the same series of letters, she informed her father: 'I have got to write an essay for the "Chick Memorial Essay Competition". There were twelve subjects to choose from, & the one I have chosen is "The advantages and disadvantages of Tariff Reform as opposed to Free Trade". Perhaps when you come home & are not so busy, you will give me a few hints.'[5]

During the General Strike, when Randolph was at Eton, he was determined to keep himself abreast of events, writing to his mother:

> At the beginning of the Strike I asked the 'Sheep' [Mr Sheepshanks, his Housemaster] if I could install a wireless set, in order to hear the news bulletins. However he would not let me. So I have fitted up a secret one in the bottom of my armchair. It works extraordinarily well and I can hear London quite easily . . .
>
> I am so glad the 'British Gazette' was such a success. I found it impossible to secure a copy down here.[6]

As children, we soon became aware that our parents' main interest and time were consumed by immensely important tasks, beside which our own demands and concerns were trivial. We never expected either of them to attend our school plays, prize-givings or sports days. We knew they were both more urgently occupied, and any feelings of self-pity were overborne by a sense of gratification that their presence was so much required elsewhere. When our mother did manage to grace any of these (to children) important occasions, we were ecstatically grateful. Her elegance and beauty were a source of great pride to us, just as her high standards in all things were a cause of anxiety, lest we should fall short of her expectations.

Once the school years were over, Clementine was able to establish a closer relationship with her daughters, taking them, each in their turn, on various journeys and holidays. It was at these times that we saw our mother at her

* Field-Marshal Sir Henry Wilson (1864–1922), a prominent Ulsterman and recently Chief of the Imperial General Staff.

gayest and best, for she loved travelling and sightseeing. And such expeditions were always pleasurable in her company, for she never made visits to galleries or museums too long or tedious.

In the autumn of 1926, when Diana was seventeen, she spent some months with a French family in Paris. Before leaving her there, however, Diana and her mother had a week's holiday together in France, of which Diana wrote a happy account to her father. Four days at Dinard were followed by three 'delightful days together' in Paris, mainly spent sightseeing. 'One day we went and lunched at Voisin's, which I believe you sometimes frequent,'[7] wrote Diana. This was history repeating itself, for Lady Blanche had taken Clementine as a girl to this same restaurant. Diana was very happy during the winter and spring months she spent in Paris, and she wrote a series of charming and lively letters to her parents. Among all the items of news from home, one in particular seized her attention, and she was swiftly to turn it to her own advantage. On 15 November 1926 Diana wrote:

> Darling Mummy,
>
> I have just had a letter from Randolph, and he sends me a very interesting piece of information; namely that you have shingled your hair again!!!
>
> Well, Mummy, you really must let me cut mine off now, as it is quite absurd for me to have mine long and you to have yours short. Besides every one that I have met in Paris so far has advised & persuaded me to cut it off; as they nearly all say that my hair is much too overwhelming for my size. Also, I do really still suffer agonies with my hats, even with the new one.
>
> Everyone here has got shingled hair . . . and I am very démodé and old fashioned.
>
> Do write & say that I can have it shingled, Mummy Darling, and then Madame Bellaigue*, who is as anxious as anybody, will take me to a good coiffeur immediately. It would be better to have it off immediately, and then it will have a chance to settle down in time for Xmas.
>
> As for Papa, I am sure he does not really mind. I would write and ask him, only I am sure that your arguments will have more weight than mine.
>
> Do say yes, and I will try and be very good and tidy & work hard; – I will any how, but if I did not have so much hair to carry about with me it would be easier. Do not bother to write a letter in reply; because a post card will be quite large enough to write an immense YES.
>
> All my love – in anxious expectation –
> Your loving Diana

Such winning advocacy could not be resisted, and evidently Clementine must have sent the earnestly-sought-for permission, because four days later Diana wrote again to her: 'I have just had my hair shingled; and I don't regret it a bit. I thought I had better have it off at once, as if Papa comes

* With whose family Diana was staying.

over to Paris with you, he might have changed his mind. As it is, I am sure he will like it much better, when he sees it. Everyone here says it is a great improvement.'8 Winston, it seems, like so many fathers, had a sentimental attachment for the long and lovely burnished tresses of his 'gold-cream kitten'. Even Diana did not say goodbye to them without a pang, for in the same letter she wrote: 'I have just said Good-Night to my cut off tail, and when I look at it, I feel quite poetic. I have just composed a poem which begins; "Good-bye my golden glory!"'

* * *

With Randolph, Clementine never established a close relationship; perhaps it would never have been possible to do so, for as his personality developed, it produced features of character and outlook too dissimilar from his mother's nature and attitude to life. This lively boy manifestly needed a father's hand; but the main task of controlling him fell almost entirely upon Clementine, and so, right from the early days, Randolph and she were at loggerheads. Winston was ever conscious of how remote and cold a figure his own father (greatly though he had loved and admired him) had seemed to him, both as a child and as a young man. He was determined it should not be the same with him and his son, and throughout Randolph's boyhood and early manhood Winston spoilt and indulged him, lavishing affection and praise upon his attractive and precocious only son.

Winston had never been happy at any of the schools to which he had been sent, nor, with very few exceptions, had he formed a high opinion of any of the masters who taught him. His school reports had been far from satisfactory, earning him Lord Randolph's anger and frigid contempt. Winston was therefore apt to take lightly any disagreeable comments in his own son's reports, regaling him with tales of pedagogic folly and ineptitude, and generally encouraging him in saucy opinions about his pastors and masters.

Clementine greatly disapproved of this line of talk, for she was well aware that Randolph needed no encouragement in impertinent or rebellious behaviour. For his part, Randolph soon realized that his mother disapproved in general of the attitudes in which he was encouraged in the manly world to which his father's indulgence soon admitted him. Moreover, he must have felt that his mother did not share his father's uncritical admiration of his character and childish achievements. Thus early on a lasting barrier was formed between them.

Randolph greatly admired his brilliant godfather, Lord Birkenhead (F. E. Smith), and was brought up by his father 'on all the famous anecdotes illustrating his wit, brilliance and arrogance'.[9] Clementine, who ever deplored F.E.'s influence on Winston, feared even more for its heady effect on their immature son. During his formative years Randolph had many opportunities of seeing his godfather, both in his own parents' house and in Lord Birkenhead's home near Oxford, for the latter was always most kind to Randolph and undoubtedly had a strong influence upon him. But, as

Randolph himself later wrote with objective perception: 'Without F.E.'s learning or his majestic command of the language, I sought to emulate his style of polished repartee. It didn't work in my case. I did not have my godfather's shining abilities and could not aspire to his brilliant gift of repartee.'[10]

Winston himself had doubts, as time went on, about certain aspects of Randolph's character. Writing to Clementine, who was in Florence, in April 1928, he forwarded on to her Randolph's school report. Randolph was now in his seventeenth year and nearing the end of his Eton days; the report was evidently most satisfactory, but Winston remarked: 'There is no doubt he is developing fast, & in those directions wh will enable him to make his way in the world – by writing & speaking – in politics, at the bar, or in journalism. There are some vy strange & even formidable traits in his character . . . He is far more advanced than I was at his age, & quite out of the common – for good or ill.'[11]

Clementine's response was a mixture of immediate pleasure and future foreboding: 'Your delightful letter (with Randolph's report) arrived to-day & I read it with joy – I am so glad about Randolph. He is certainly going to be an interest, an anxiety & an excitement in our lives. I do hope he will always care for us –'[12]

After he left Eton, Randolph went to Christ Church, Oxford, but had been there only four terms when, on being offered a contract to give a series of lectures in the United States, he obtained a term's leave from the university and accepted this unexpected invitation. He disarmingly acknowledged that the idea of 'teaching, rather than learning, appealed to me strongly'.[13] Since he was at this time considerably encumbered with debts, and as the contract appeared highly advantageous financially, he eagerly grasped at this opportunity. Everyone, except Winston, thought the scheme a harebrained venture. Clementine feared it would, in fact, mean the end of his studies at Oxford, and that he would never get his degree: in this she was proved perfectly right, for he never returned to Christ Church. Randolph was of course aware once more of his mother's opposition to his wishes and intentions, which were, as on so many other occasions, encouraged by his father.

Strangely enough, however, this lecture tour, of which his mother so much disapproved, provided an interlude in which she and Randolph were for some time in each other's company, and upon which both in after years looked back with pleasure. Randolph left for the United States at the beginning of October 1930; he was nineteen years old, and combined with his unusual character and budding abilities stunning good looks.

Although there are several letters between Clementine and Randolph full of news, views and affection, by Christmas Clementine was feeling out of touch, and perhaps a little uneasy as to what Randolph was up to in the brave new world. Winston had just given her some money to buy herself a car; she had a brainwave, and wrote to Randolph:

> if you liked it I would spend the money on paying you & New York a flying visit?
>
> Shall I hop on a boat one day at the end of January or beginning of February & come to New York for a week . . . ?[14]

On receiving a swift and enthusiastic reply in the affirmative, Clementine booked her passage and set off in early February 1931 for New York in the Europa.

Soon after her arrival, she wrote a long letter to Winston, beginning with an account of the heart-warming welcome she had received from Randolph: 'The Europa docked early on Saturday & before 8 Randolph was on board looking splendid and beaming. His joy at seeing me was really sweet & I felt much moved.'[15]

They had not been together two hours before Randolph told his mother that he was in love with a young lady from Cleveland, Ohio – Miss Kay Halle* – and that he hoped to marry her. Clementine was somewhat taken aback, but soon had the opportunity of meeting Kay, and reported favourably to Winston. But in the course of several long, heart-to-heart talks with Randolph, she pointed out the possible disadvantages of marrying so young, and before he had in any way established himself in the world. These cogent maternal arguments were, needless to say, discounted by Randolph, who said he meant to marry the young lady if she would have him. In fact they never married, but remained lifelong friends. Meanwhile reports began appearing in the American press about Randolph's romance, and Clementine's arrival was directly linked with it. Several newspapers reported that she had arrived from England to put a stop to the whole thing.

From the moment she set foot in New York, Clementine was swept up in a whirl of plans by Winston's or Randolph's friends, and her intention of staying for only a few days was soon abandoned. Randolph persuaded her to go with him on part of his lecture tour. She gladly agreed, as she was unreservedly enjoying herself. But the key to her happiness lay in Randolph's warmth and kindness; she wrote to Winston: 'He [Randolph] is a darling. He has quite captivated me. He is a most sweet companion & he seems to enjoy my company . . . It is quite like a honeymoon.'[16]

For the best part of six weeks, Clementine travelled about with Randolph, visiting Palm Beach, Washington and Cleveland (where she met, and liked, Kay Halle's family). During this journey she kept a day-to-day diary, which tells how she and Randolph were welcomed and entertained wherever they went, as only the Americans know how. Far from being exhausted or bored by the succession of parties and entertainments, Clementine seems to have loved every moment of this happy and unexpected episode in her life. As time went on, this American interlude acquired a golden glow, and in after years, when the relationship between Clementine and Randolph had deteriorated, she would recall it with wistful nostalgia.

*Pronounced 'Halley'.

Randolph was determined to blaze his own trail, both as a journalist and as a politician. He was an aggressively loyal proponent of his father's views, and was constantly drawing public attention to Winston's scandalous exclusion from office. Randolph's political skirmishings, however, were nearly always the cause of parental dismay or embarrassment – sometimes both – and a constant source of noisy rows, for he would brook no moderating counsels, and forged ahead regardless.

In 1934 in particular Randolph set cats among the political pigeons by twice taking a prominent part in by-elections – the first at Wavertree, where he actually stood as an Independent Conservative candidate, thereby splitting the Tory vote and letting the socialist candidate in, and a few months later at Norwood, where he espoused the cause of the Independent Conservative candidate there. But this time the 'official' candidate, Duncan Sandys, won hands down, Randolph's protégé losing his deposit. The main topic of controversy in both campaigns was the Government's India policy, and although Winston was firmly and openly opposed to them over this, he was much embarrassed by Randolph's actions, which served only to divide Tory support. Clementine was far from home on a Far Eastern cruise at this time, and so she had the news from the electoral battlefronts only in batches of letters from home. But she understood immediately how awkward Randolph's interventions must be for Winston. From Auckland on 22 February 1935 she wrote: 'How anxious & agonising & thrilling Wavertree must have been. I wish I had been there to help – Darling Winston, I hope it has not queered the pitch at Epping. I do think the Rabbit ought to have consulted you before rushing into the fray. However he seems to have done very well –'[17]

Although Winston had disapproved of Randolph's candidature at Wavertree, he had spoken for him at his eve-of-poll meeting; but Randolph's renewed activity barely a month later at Norwood had been the cause of a violent row between him and his father, all of which the latter duly recounted to the absent Clementine. Winston did not lend Randolph his support in any way on this second occasion; nevertheless it was suspected in party circles that he had his father's approval. Altogether it was very tiresome, and Winston commented ruefully to Clementine: 'I shall of course have nothing to do with it, except to bear a good deal of the blame.'[18]

In the same letter Winston told Clementine how he had encountered Mr Baldwin about this time in the House of Commons, who remarked sympathetically that Mrs Baldwin had said: 'One's children are like a lot of live bombs. One never knows when they will go off, or in what direction.'[19]

* * *

During her childhood years Sarah caused her parents much anxiety on account of her health. Later on she developed an exceptionally strong constitution, and was able to stand up to all the rigorous demands of professional dancing and acting; but when she was five or six she alarmed everybody by producing tubercular glands in her neck, necessitating an operation. Too young and nervous to accept or understand soothing explanations, Sarah panicked when the chloroform mask was held over her face: becoming hysterical, she fought like a demon, and ran round and round the operating theatre until she was caught and subdued by sheer force. Neither she nor Clementine (who was the horrified witness of the scene) ever forgot this harrowing experience. The operation left a scar on Sarah's neck which she bore all her life, and it also made a deep impression on her inner being, making her dread all forms of physical restraint.

Sarah's seemingly frail health during these years caused Clementine much concern, haunted as she was by Marigold's death and the memory of her sister Kitty. When she was thirteen, Sarah left Notting Hill High School and went to North Foreland Lodge, a boarding school by the sea at Broadstairs in Kent, which Clementine selected very largely on the grounds of its bracing situation. After her time there, Sarah, like Diana, went to Paris for about a year, to a finishing school kept by the Mesdemoiselles Ozanne. After this she completed the programme for girls at that time by making her debut in the social world and 'doing the season'. Sarah was beginning to be strikingly lovely, with her auburn hair, green eyes and ivory complexion. But Clementine was candid about her appearance, writing to Margery Street, to whom she confided many of her thoughts and feelings about her children, for 'Streetie' had known them so well: 'Sometimes she looks absolutely lovely – but on the other hand she can look like a moping raven.'[20]

Sarah was very shy and, with her dreamy temperament, did not really enjoy the endless succession of dances. In June 1933, Clementine, again writing to Streetie, told her:

> Sarah is sweet & very good to me.
>
> We are half-way through the so-called 'Season' & she and I are really longing for it to be over.
>
> She is slowly overcoming her shyness & has plenty of partners but she makes few friends – She is nearly always ready to come home by two o'clock, so that I do not have very late nights.[21]

For Clementine, with her unsocial nature and early-to-bed habits, bringing out her daughters must have been a real test of maternal devotion; she also confided to Streetie her views on her fellow mothers: 'They are really rather a depressing back-biting tribe & I have to sit for hours with them on the Chaperons' Bench. I'm thinking of taking a cookery book to Balls. I could be hunting up tasty dishes for Margaret & Elizabeth to try, instead of listening to their gossip.'[22]

The end of Sarah's first season saw also the end of her compliance with a conventional programme. She begged and persuaded, until her parents reluctantly gave her permission to train to be a professional dancer and actress. No doubt they hoped her passionate longing to 'go on the stage' would melt away. Diana had for a time studied at the Royal Academy of Dramatic Art; but she had no real gift for acting, and after a while she gave up any dramatic aspirations. But Sarah had real talent, and her desire to act or dance professionally became an enduring passion, which dominated all others. Nevertheless, although her ambition to make the stage her career did not please her parents, once Clementine saw how much in earnest she was, she took trouble to see that Sarah went to a good school of dancing, and also insisted that she should have a rigorous medical examination to make sure that physically she was up to the demands of the training.

But Clementine's fundamental lack of sympathy with either of her daughters' histrionic ambitions was expressed when she wrote somewhat testily to Streetie in September 1934: 'It's very strange that both she [Sarah] and Diana should have this passionate wish to go on the stage without the slightest talent or even aptitude.'[23]

Between Sarah and her mother there grew up a loving and companionable understanding, which would survive some major emotional crises. But sadly, the relationship between Clementine and Diana, which was so sunny and open in the early days, became clouded by misunderstanding. The chilling seems to date roughly from 1932. In December of that year Diana, who was twenty-three, married John Bailey, the son of Winston's long-standing friend Sir Abe Bailey, a wealthy South African; the match, however, did not arouse much enthusiasm in Diana's family, although there was a brilliant wedding at St Margaret's, Westminster, where Winston and Clementine had themselves been married nearly twenty-five years before.

Any happiness between Diana and John Bailey was short-lived: they separated after barely a year, and were divorced early in 1935. Diana thereafter lived alone in London; but although she came very often at weekends to Chartwell, an element of strain was present between her mother and herself, and differences in their emotional make-up and misunderstandings marred their relationship from now on; both were thereby the losers.

Diana, however, was not to be solitary and sad in her personal life for long. During the summer of 1935 she met Duncan Sandys – a diplomat who had left the Foreign Office in order to go into politics, and the victor at Norwood earlier that year. While she was helping Randolph with his disastrous campaign at the by-election, Diana had observed Duncan across the political barricades. Towards the end of August Clementine wrote to tell Streetie of their engagement. Both Winston and she had obviously been taken by surprise: 'Mr Pug & I were rather staggered at first but now we rather like him (no money) & HOPE all is for the best . . . they are to be married quietly on Sept. 16th.'[24] A few months later she was more enthusiastic: 'I like my new son-in-law. He & Diana are very much in love. That is all a great relief.'[25]

In the autumn of that same year, 1935, Sarah started her stage career by becoming one of Mr C. B. Cochran's 'Young Ladies' and taking part, as a member of the chorus, in a revue called *Follow the Sun*. The show opened in Manchester in December, and Clementine travelled up to see the first night;

afterwards she wrote to Winston, who was in North Africa: 'You really would have been proud of Sarah. The dancing performed by the chorus was difficult & intricate & she was certainly in the first flight. She looked graceful & distinguished.'[26] To Streetie she avowed: 'I'm really very proud of her tho I would rather she were not on the stage.'[27] Neither Winston nor Clementine could ever really reconcile themselves to Sarah's stage career, and as to her being 'in the chorus' – they both belonged to the generation of which several 'ladies' of the chorus became duchesses but, on the other hand, ladies simply did not become chorus girls.

It was while she was in *Follow the Sun* that Sarah met the principal star of the show – the comedian, Vic Oliver. He was thirty-seven, seventeen years older than Sarah, and of Viennese Jewish extraction; he had already been married twice and was a man of great charm and talent. He and Sarah fell deeply in love, and it was not long before Winston and Clementine were confronted with the unpalatable fact that they wished to be married. An unhappy and unsettled year followed, during which time her parents used every means within their power to dissuade her from this marriage. But Sarah had not earned the nickname 'Mule' for nothing, and she clung tenaciously to her intention.

In September 1936, without prior warning, convinced she would never win her parents' approval (their consent was not necessary as she was over twenty-one), she bolted off to New York to join Vic Oliver, who was working there. Great was the dismay of her parents, for she had promised them to wait a year. The inevitable publicity attracted by her sudden flight added to the general pain and embarrassment at home. Public interest was further fanned by Randolph's following hot on Sarah's heels, at Winston's behest, to see what he could do to put a stop to a marriage they were sure would lead to much unhappiness for Sarah.

This was a bitter period. Both Winston and Clementine felt deeply wounded and ill-used; Sarah felt misunderstood and, not unnaturally, resented the various legal devices which were employed to delay or prevent her marriage. But through all the bitterness the bond of love between Sarah and her parents, though strained, never broke. Sarah and Vic Oliver were finally married on Christmas Eve 1936 in New York. Christmas that year at Chartwell was very quiet. But a cheerful note was struck by the presence of Diana and Duncan, who brought their three-month-old son, Julian – Winston and Clementine's first grandchild.

* * *

Meanwhile, a long way behind the others, I was also growing up. Apart from the preoccupations of her life with my father, my mother was naturally more taken up with the varied needs and demands of the three elder children. But I was blessed in having throughout my entire childhood and girlhood the same steady, loving presence and influence of one nanny-cum-governess. After Marigold's death, Mlle Rose, the young French nursery governess, had departed. No actual blame was ever laid at her door, but Clementine's confidence was utterly shaken (and she always felt she should have been sent for sooner). Since her life was such that she had to spend much of her time away from her children, she looked round for someone to be with them in whose judgement and reliability she could have complete trust. She did not have to look far: her own first cousin, Maryott Whyte (Lady Maude's daughter), then aged twenty-two, was a trained Norland nurse.

Called variously 'Cousin Moppet', 'Nana' and (by a later generation) 'Grandnana', Maryott Whyte came to rule over the schoolroom and, soon, the new nursery. She stayed with our family for over twenty years, giving a blessed sense of security to Clementine, and a stability and orderliness in the children's lives which had been greatly lacking until now through the constant change-around of nurses and governesses.

I was very much the 'Chartwell child', although before I started school I lived also at No. 11 Downing Street when my father was Chancellor of the Exchequer. But after 1929 I spent nearly all my time at Chartwell, going to local day schools. Eight years separated me from Sarah, and my life and upbringing were in some ways like that of an only child. But I can just remember Sarah having some connection with nursery and schoolroom life, and this link became a bond of lifelong affection and loyalty between us. Diana and Randolph were god-like beings, inhabiting the higher slopes of Olympus, and I never really caught up with them.

I had a lovely life, designed by my mother, but administered in all its details by the faithful (and sometimes fierce) Nana, to whom as a child I turned for everything. I lived chiefly among grown-ups, with whom I was on the whole more at ease than with children of my own age. The lack of such companionship was largely supplied by a procession of pets: dogs, cats, orphan lambs, chickens, bantams and goats. I also at one time or another possessed a monkey, budgerigars and two tame fox cubs. After my father gave up polo, there was no 'pony' life at Chartwell, but I rode very often at a nearby riding school.

My mother found a wonderful Frenchwoman, Mme Gabrielle l'Honoré, who from 1929 for nearly ten years came every summer during the school holidays to teach me French. She was minute, vivacious, talented and chic, and, as my mother wrote to my father: 'She looks like Madame de Pompadour, takes away all Diana's young men from her & would lure a deaf & dumb orang-outang to speak French.'[28]

Clementine was still full of physical energy, and, as we have seen, was a keen tennis player. She had enjoyed to the full the riding and hunting that had come her way earlier in her married life, and still enjoyed boar-hunting (as did Winston) when they stayed with Bendor Westminster in France; but with the stables closed at Chartwell, riding gave place to tennis, which became a great feature of our life there. Although Clementine gradually gave up participating in tournaments, she continued to play very well, right up to the war. She encouraged us all in the game, organizing private coaching at home, and spending much time playing unselfishly with us herself.

Diana's and Sarah's young men were much in favour if they were good tennis players; among the Chartwell regulars 'The Prof' was an excellent player, as was Desmond Morton* who lived nearby at Edenbridge, and our other tennis-playing neighbours included Victor and Peter Cazalet at Fairlawne.

In 1935, when she was fifty, Clementine suddenly became attracted by, and then positively addicted to, skiing. Several of her friends were winter-sports enthusiasts, and they fired her with the desire to learn to ski. Clementine also saw in this occupation an opportunity to 'make contact' with me: it was something we could do together, away from home.

I was now thirteen, and engrossed by my work at school (where I was a persevering plodder) and with my life at Chartwell, which was centred on animals and my riding, and dominated by Nana. My relationship with my mother was respectful and admiring rather than close. If she made a suggestion, my natural reaction was to say: 'I must ask Nana.' My mother not unnaturally felt the time had come to change the orbit of my loyalty, and to try to forge a closer relationship with me. It was of course a great thrill for me when she announced we were going to Austria in the Christmas holidays of 1935–6. Up to that time I had hardly ever been alone with her for any period of time, for Nana even took me away to Scotland or the seaside for my summer holidays. This was to be, therefore, a new experience for us both.

We went to Zürs in the Arlberg mountains in Austria, in a skiing party that included Goonie Churchill and Clarissa (fifteen), and Venetia Montagu and her daughter Judy (twelve). It was a great success, and Clementine threw herself into this strenuous and hazardous sport with zeal and perseverance. For three years running, this Christmas holiday treat was repeated: twice we went to Zürs, and once to Lenzerheide, in Switzerland. Clementine simply loved it, and, despite a late start, she became remarkably proficient. Although speed was never a characteristic of her skiing, she was the epitome of the maxim 'dogged does it'. On this first visit to Zürs, she wrote to Winston that: 'The difficulties of ski-ing seem to me insurmountable! Perhaps I shall feel differently in a week.'[29]

Ten days later, we all ventured on our first expedition (as opposed to classes in the skiing school), and my mother reported: 'We had a charming guide who carried our food our coats 1st aid apparatus & poor fellow, in the end almost had to carry me! . . . Mary fell down 19 times . . . Today I am going in a sleigh as I am really bored with tumbling down!'[30] This letter was signed: 'Your bruised & struggling but undaunted Clemmie'.

Each year she stayed on after I had to go back to England to return to school. She usually joined up with a friend, and persevered for a week or two more with this pleasurable if exhausting new-found sport.

It was chiefly during these lovely skiing holidays that I started to know

* Major Sir Desmond Morton (1891–1971). Head of the Industrial Intelligence Centre in the 1930s, he made available to WSC much information which the latter used to good effect when warning the Government of the danger ahead. In the early years of the Second World War he served as WSC's liaison with the Foreign Office, the Allied governments in London and the Secret Service. He was knighted in 1945.

my mother more as a person than a deity, and the foundation was laid for a long and loving relationship which was to be precious to us both. I began not only to understand her better, but also to enjoy her companionship. She was delightful to be with, and during the long *après ski* evenings, she would regale me with accounts of her childhood; most of all I recall the long reading-aloud sessions we had when, concentrating chiefly at that time on poetry, we would take it in turns to choose and read from anthologies or collected works. As time went on my mother's taste veered away from poetry to prose: her preference was for memoirs, histories and classical novels; long books of several volumes never daunted her.

The development of our new relationship was not always easy, and there were tempestuous passages between us from time to time, when I found my mother most difficult to understand, and immensely demanding. I dreaded her displeasure, and the emotional, electric storms that could brew. From her point of view, I must have been at times tiresome, graceless and priggish: I was in the 'terrible teens', or, as the French more elegantly put it – *l'âge ingrat*.

While my mother and I were away disporting ourselves in the snow, my father remained at Chartwell, totally engrossed in his writing; or he would seek warmth and sunshine on the Riviera, where, of course, he continued to work. Clementine longed for him to join her on her sunny, snowy Alps. 'I wish you liked the snow,' she wrote wistfully on 11 January 1937. 'If you saw it glistening under a royal sun & a limpid blue sky, I think you might.'

The highlight of Clementine's skiing career was when, with caution and dignity, she negotiated the Parsenn Run. Afterwards she wrote to Winston that it was

> Quite easy from a professional point of view, but the excitement & the beauty were almost too much for me & I did long for you to be with me.
>
> Do you know, I believe you would <u>love</u> this. I wish you could try. Only I'm afraid you would be too bold & come the most frightful purlers.[31]

From St Moritz about a week later, she telegraphed: 'Heavenly if you could come here. Lovely colours in the snow for painting. Love. Clemmie.'[32] But Winston was not to be lured, and the only snowscapes he ever painted were views of the garden at Chartwell, as seen from the windows of the house.

* * *

One of Clementine's great contributions to Winston was her judgement of people; both her cousin Sylvia Henley and Violet Bonham Carter agreed on this, the latter writing of Clementine: 'His cause was her cause, his enemies were her enemies, though (to her credit) his friends were not invariably her friends. Her appraisal of people was often more discriminating than his own.'[33] Like many women, Clementine formed swift, instinctive judgements, which she often could not justify in words. In the long run she was hardly ever wrong about somebody's character or capacity, and she rarely changed her mind. But time sometimes modified her views, and this was certainly the case with Brendan Bracken*. This extraordinary man, who shrouded his origins in mystery, appeared on the scene of Churchill's life in 1923, and determinedly attached himself to Winston's wagon. Robert Rhodes James has given a vivid picture of him: 'He was a big man, with flaming carrot-red hair in a tousled mop, with gig-lamp spectacles.'[34] He talked with a polyglot Irish–Australian–Cockney accent, and words would cascade from him in a torrent. Bracken's undeniable charm competed with a superficial toughness and brashness, which from the first aroused Clementine's antipathy and suspicion, and she took an instant dislike to him. She thought him an adventurer – which he was – and she mistrusted his influence on Winston. So little was she able to conceal her feelings of dislike and irritation that for a time Bracken rationed his visits, and arranged to meet Winston away from his home.

A further cause of Clementine's dislike of Brendan Bracken was, almost certainly, her knowledge of the rumour which persisted for many years that he was Winston's illegitimate son. Brendan did nothing to contradict this totally unsubstantiated theory, and Winston treated it with amused indifference. It was not, however, the sort of 'joke' that appealed to Clementine. The mystery surrounding Bracken's origins was only finally resolved by Andrew Boyle in his biography of 1974, when he established beyond all doubt that he was the son of a builder-stonemason and his wife, born in 1901 in Templemore, Tipperary.[35]

Barely two years after the beginning of their friendship, Churchill and Bracken quarrelled over the publication of a magazine article. The clash coincided in time with Churchill's appointment as Chancellor of the Exchequer, and for the next few years Bracken's place as political acolyte and confidant was largely taken by Bob Boothby, Winston's Parliamentary Private Secretary. But in 1929 the breach was healed: Brendan Bracken had by this time established himself both financially and socially, and in that year he had also become a Member of Parliament. From then on he moved into the close circle of Winston's friendship, fixing his political fortunes on Churchill's star, in good times and in bad.

Slowly Clementine's hostility was softened. In January 1931, we find her writing to Randolph:

> Mr Bracken has been here once or twice.
>
> I am giving up my vendetta against him & shall probably end by quite liking him. I'm not sure if this is broadmindedness or old age with its tolerance creeping in.[36]

It is much more likely that the 'sea-change' which took place in her feelings

* Brendan Bracken, MP (1901–58). Later Parliamentary Private Secretary to the Prime Minister (WSC), 1940–1; Minister of Information, 1941–5; 1st Lord of the Admiralty, 1945. Chairman of Union Corporation and *Financial Times*. Created 1st Viscount Bracken, 1952.

over the next years was due to Brendan's unusual qualities of heart and mind, and her appreciation of his passionate loyalty to Winston – although she was nearly always to doubt the wisdom of his political advice. But slowly he won her regard and affection, and later her abiding gratitude: for after the war, Brendan was to become a member of the Chartwell Literary Trust, and an adviser on whom Clementine learnt to lean.

Another of Winston's friends whom Clementine had viewed with disfavour from the first was Lord Birkenhead – F. E. Smith: and in his case the passing of time brought no alteration to her feelings. He had become Winston's closest friend, and Clementine's dislike of him did not affect the relationship. She did not 'take' to his personality, and moreover she deeply mistrusted his influence with Winston generally. Also, as we have seen, she disapproved strongly of the effect of F.E.'s arrogant brilliance on Randolph.

But Clementine had always liked and admired his wife, Margaret; and the Smith children – Freddie, Eleanor and Pamela – were contemporaries and friends of Diana, Randolph and Sarah. Moreover, Freddie and Pamela both married children of Lord and Lady Camrose – Sheila and Michael Berry. The Camroses were long-standing friends of the Churchills, and these links made easier and more natural a *modus vivendi* as far as Clementine and F.E. were concerned.

When F.E. died aged only fifty-eight in 1930 Winston was deeply grieved, and felt a real and abiding sense of loss. Clementine's warm and feeling letter to Margaret Birkenhead shows sensitivity and understanding.

Chartwell Manor,
Westerham, Kent
October the 1st 1930

Dearest Margaret,

Last night Winston wept for his friend. He said several times 'I feel so lonely'.

I think of the tension, the fatigue, the pain, the disappointed hopes of your long agonising vigil. But also of the pride you must feel.

Margaret Dear – I am sure Life still holds much joy for you. It must, for you have been – and are gallant and brave.

I am sure he must like you to think of that.

Your affectionate
Clementine S. Churchill

The third among Winston's close friends whom Clementine regarded with the utmost distrust and, at times, dislike was Max Beaverbrook. And, as with Brendan Bracken, she nearly always disagreed with the political advice he gave Winston.

Despite many differences of political view, and several periods of 'chilliness', the friendship between Winston and Max Beaverbrook was to prove enduring. Clementine was not unconscious of the Canadian's gnome-like charm but, apart from her fears for his political influence, the puritan streak in her was affronted by certain aspects of his life. From Grantully Castle in Perthshire, where she was staying with the Balsans in September 1926, she wrote to Winston concerning a luncheon party at Chartwell to which he had invited Lord Beaverbrook and his great friend, Mrs Norton: 'Please do not allow any very low conversation before the Children . . . & I hope the relationship between him & Mrs Norton will not be apparent to Randolph & Diana's inquisitive marmoset-like eyes and ears.'[37]

Max Beaverbrook must have been perfectly aware of Clementine's hostility towards him, and the situation would have appealed to his sense of irony: he may even have found a challenge and a piquancy in it. The amenities of surface friendship were always maintained between the two adversaries, and Max throughout the years was most attentive in many little ways – showering Clementine with fruit, flowers and sympathy when she lay ill. And during the Second World War he never failed to contribute generously to her charities. Clementine was not insensible to all this, and there is a series of letters from her expressing warm and genuine thanks for all these kindnesses.

There were also times of truce: nobody admired more than Clementine the prodigies Beaverbrook achieved at the Ministry of Aircraft Production in 1940–1; and in political storms she was the first to acknowledge in Max a true 'foul weather friend'. As the years rolled by, her dislike of him diminished. But as long as Winston continued in public life, so long did Clementine fight Max Beaverbrook's influence over him. Only in the last years did she lay aside her arms: they were all old, and anyway by then Max could neither help nor harm Winston.

It grieved Winston very much that three of his closest friends should meet with so little approval or appreciation from Clementine, and he would ever try to bring about a better measure of liking and understanding of them on her side. There is no doubt that Clementine was hypercritical, and possibly over-protective where Winston's career or reputation was concerned: for all their life together, she was to be the scabbard to his sword. But with hindsight one can see that these three men, Bracken, Birkenhead and Beaverbrook – 'The Three Terrible B's', as Clementine called them – all shared to some extent similar characteristics. In each of them there was a touch of the buccaneer, and a streak of brashness and vulgarity, which jarred with her, and made her fear for the influence they might have with Winston.

It must not be thought that because Clementine was less than enthusiastic about certain of Winston's close friends, and undeniably had a highly developed critical sense, she was not friendly and welcoming in general to guests young and old, and those who were part of our family circle. Peregrine (Jack and Goonie's younger son) was a year older than Sarah, and they were boon companions from early childhood. Peregrine loved his aunt, and found her sweet and affectionate, whereas his uncle and his cousin Randolph alarmed him. As adolescence advanced, Clementine thought Sarah and 'Pebbin' (Peregrine's nickname) should see less of each other – his mother explained to him that Aunt Clemmie thought they were so fond of each other that 'they might have married' – so his visits to Chartwell became less frequent. But the affection between Peregrine and Sarah would be lifelong.

Their other cousins, Tom and Diana Mitford*, were great friends of Diana and Randolph. Tom and Randolph were at Eton together (although Tom was the elder by two years); Randolph fell in love with Diana, who at sixteen was already ravishing with her large blue eyes and blonde hair. The Mitford cousins came often to stay in the late twenties and early thirties, and Diana loved her Chartwell visits; Cousin Winston dazzled her, and for the first time in her life she heard political talk which enthralled her; he was very fond of her and dubbed her 'Dina-mite'. Diana was also much impressed by the creature comforts, the prettiness of the rooms and the delicious food. She has fond memories of her Cousin Clementine: over fifty years later she would tell Randolph's son, Winston: 'I loved all the family. I loved your grandfather and grandmother. When people say Clementine was so cold, well, she was extremely kind to us and, to me particularly, wonderful.'[38]

Clementine's highly strung nature made her capable of rising to the demands of her exacting life, and the heroic streak in her enabled her to meet undaunted the challenges, triumphs and disasters of Winston's career. But she paid a high price in nervous wear and tear. Although she gave for the most part an impression of cool poise, Clementine was both shy and passionate; the face of serenity and calm she presented to the world at large over the years was an artefact of self-control, and when that control cracked – tempests could rage.

She was a perfectionist, and at times she sacrificed too much on the altar of that stern goddess. The timid, tearful child who had sobbed over the dirty marks on the dazzling white of her pinafore was truly the precursor of the immaculate woman who strove for perfection in all departments of her life, driving others hard – but herself hardest of all.

From their earliest days together, Clementine had been enthralled by politics, and was accustomed to debate, controversy and loud arguments. But occasionally she wearied of the general brouhaha. Johnnie Churchill (Jack and Goonie's elder son), an artist, spent much time from 1933 at Chartwell, while he was painting frescoes in the loggia depicting the triumphant campaigns of John, 1st Duke of Marlborough. He recalled on one occasion driving up to London with his Aunt Clemmie after a typical Chartwell weekend, when she burst out, 'I just can't stand it any longer.'[39] Whether it was at this time or a little earlier is not clear, but Peregrine was told by his mother, Goonie, that Clementine had been to her in a distressed state, and said she was contemplating leaving Winston; divorce was mentioned.[40] I would have been about ten or eleven at the time, and was blissfully unaware of these tensions; nor did my 'grown-up' confidante,

* Tom Mitford died of wounds in Burma in March 1945; he was thirty-six. Diana married Bryan Guinness in 1929; they were divorced in 1933. In 1936 she married Sir Oswald Mosley (d. 1980). As I write (2002) she is still alive and living in Paris.

Sarah, ever speak to me (then or later) about such a situation. In the early 1930s Clementine was in her mid-to-late forties, and was certainly in the menopausal age band. I can recall disagreements between my parents – raised voices – occasional histrionics (always most baffling and embarrassing to a child): but I also recall the swift and loving reconciliations that followed.

Sadly for Clementine, Chartwell was never the Garden of Eden it always was for Winston: she rarely found relaxation or peace of mind there; and she had no outlet for tension and anxiety in some form of creative occupation such as gardening, painting or needlework. Chartwell continued to be the financial burden she had so truly forecast, and exhausting to run to those standards of perfection she set herself. For most of her life Clementine suffered from anxiety: anxiety for Winston; anxiety for her children; almost continual anxiety over money matters. Again and again she felt over-burdened by the demands life – and Winston – made upon her, and the sheer weight of trivial troubles could lay her low. Yet real crises – personal or public – found her robust, resourceful and possessed of astonishing reserves of stamina.

My mother was not unconscious of these facets of her character, and she would sometimes recite, in wry exasperation, lines from an elegy written for a hard-pressed governess, but which she said would do very well as her own epitaph:

> Here lies a woman who always was tired,
> For she lived in a world where too much was required.

And there is this touching admission in one of her letters to Winston (obviously written after they had had 'words'), of her tendency to create mountains out of molehills:

> My dearest Love I hope you have forgiven me for being such a scold when you were here.
>
> When I arrived & saw you a flood of happiness spread through my being. It is a great fault in me that small things should have the power to harass & agonise me. I do not think Voronoff monkey glands* would do me any good! What I need is to be inoculated with vegetable marrow or cucumber juice![41]

Although Clementine took real pleasure in the company of many of the guests who came to Chartwell, and apparently had many friends, there were, in fact, very few people in whom she confided. Ever since she was a young woman she had shunned intimate friendships, and even with those truly close and trusted friends such as Goonie, Sylvia Henley and Horatia Seymour, she was loath to discuss her most besetting worries. Consequently

* A reference to Dr Serge Voronoff, a Russian-born physiologist, who specialized in the grafting of animal glands into the human body and the effect of gland secretions on senility, and whose book, *The Study of Old Age and My Method of Gland Grafting*, was published in 1927.

she denied herself the relief of 'unburdening', which is such an essential safety-valve for most people.

Winston and his life filled Clementine's whole existence, and when she was well and in good form, she neither desired nor needed other companions or distractions; but when she was low and fretted, it was often difficult to find a good solution. As her daughters grew up, she turned to each one and found company and companionship; but daughters soon have their own lives, careers, husbands and children. Clementine's diffidence in personal relationships made her hesitate to propose herself to people who would probably have been delighted to see her. All through her life Clementine missed chances with people, because she was too shy, and could not bring herself to be a little more natural and carefree in her approach. Thus, in a life full of people, she knew much loneliness.

Although Winston was endowed with a superb constitution himself, and was supplied with an immense store of stamina, he was always most understanding and solicitous in all physical illness. Swift to initiate action, and always seeking the best advice, he insisted that 'doctor's orders' should be strictly obeyed. Although Clementine carried and bore her babies with relative ease and lack of complication, Winston readily accepted that child-bearing entailed, particularly after the event, much care and rest; and his letters to Clementine were always full of tender solicitude and care for herself and the 'new kitten'.

Indeed, illness always evoked all his capacity for sympathy and understanding. Early in 1928 Clementine was gravely ill with mastoid; as was quite common in those days, the operation was performed in her own home (No. 11 Downing Street), and for some weeks thereafter she was in great pain and very low in spirit. Winston was alarmed and deeply concerned for her; he spent much time at her bedside, and used to read passages from the Psalms to her, and this time of pain and anxiety brought them very close. From Lou Sueil, where she went to convalesce, she wrote to him: 'My Darling One – I do not regret my illness for it has brought you so close to me . . . you are always deep down in my heart and now your tenderness has unlocked it.'[42] He replied: 'I loved to read what you wrote . . . I am always "there"; but I am afraid that vy often my business & my toys have made me a poor companion. Anyhow my darling I care for no one else in the world but you – & the kittens; & in spite of the anxiety of yr illness I was glad to feel that you relied on me & that I cd help & comfort you a little.'[43]

But Winston found it difficult to understand and cope with Clementine's periodic bouts of fatigue and nervous tension, the apparent causes of which seemed in themselves relatively trivial. Although 'Black Dog' had haunted him as a younger man, the beast had never undermined his power of action, and with his essential resilience, he was genuinely puzzled that the mere process of daily life should be able to lay his valiant and spirited 'Kat' so low. He had to force himself to accept – even when he could not understand – that when Clementine was affected by what a perceptive friend in later years would call 'High Metal Fatigue', she needed to distance herself for a space.

Nor was he oblivious to the fact that he himself (however unwittingly), through his innate self-centredness, and by the sheer pace and demands of their life, contributed to her troubles.

Looking back, one sees that it became quite usual for Clementine to take a week here or a fortnight there, during any year, at some spa or health establishment in England or abroad (of the kind nowadays known generically as 'health farms'), to 'do a cure', and to recharge her batteries. Although Winston hated her leaving him, he ruefully came to recognize that a 'let-up' away from home responsibilities – and away from him – was from time to time necessary for her poise and general health. His letters to her at these times not only express his solicitude and an element of self-reproach, but also reveal the extent to which he missed her, and how much he depended upon her. 'When do you think you will return my dear one,' he wrote when she was staying with the Balsans in the South of France in March 1925,

> Do not abridge yr holiday if it is doing you good – But of course I feel far safer from worry and depressions when you are with me & when I can confide in yr sweet soul. It has given me so much joy to see you becoming stronger & settling down in this new abode [Chartwell]. Health & nerves are the first requisites of happiness. I do think you have made great progress since the year began, in spite of all the work & burdens I have put on you.

And the letter ends with a moving avowal of all that Clementine meant to him: 'The most precious thing I have in life is yr love for me. I reproach myself for many shortcomings. You are a rock & I depend on you & rest on you.'[44] When, many years later, I, their youngest child, read this passage, not a word jarred. It was true when it was written, and it remained true for all the forty more years Winston and Clementine were to be together.

CHAPTER SEVENTEEN

Holiday Time

ONCE THEY HAD SETTLED AT CHARTWELL, WINSTON AND CLEMENTINE TENDED to stay away from home less. Their own pattern of weekend entertaining became established and, particularly in the summer, they enjoyed having their friends down to visit them.

Nevertheless they paid short visits together from time to time within a small circle of friends and relations. Among their ports of call were the Marlboroughs at Blenheim; Sir Philip Sassoon at Trent Park or Lympne; and Lord and Lady Desborough at Taplow. For several years they visited Lord and Lady Pembroke at Wilton House for Whitsun; in the Visitors' Book there, the weekend was labelled 'Winston-tide'. And many were the visits to the Duke of Westminster (Bendor) at one or other of his homes in England, Scotland or France.

But when the question of holidays or travel abroad arose, it revealed a wide difference in Winston and Clementine's tastes and temperaments. Winston was happiest of all at Chartwell, but the charm of landscapes lit by a brilliant and more constant sunshine was also strong. The South of France, with its wonderfully 'paintatious' scenery, both by the coast and in the hinterland, provided him with endless subjects, with all the amenities offered by hospitable friends in splendidly comfortable villas.

Miss Maxine Elliott, a former actress, and a considerable figure in late and 'neo-Edwardian' society, was consistently hospitable at the Château de l'Horizon at Golfe Juan; Mrs Reginald Fellowes – Daisy, the daughter of the Duc de Decazes, who had married a Churchill kinsman – also offered agreeable entertainment at Les Zoraïdes; and Lord Rothermere made them welcome at La Dragonnière at Cap Martin.

Winston was happy at all these places. The greater part of his days was filled with painting or writing, and he almost invariably took a secretary with him. He loved swimming in the translucent water, and in the evenings he briefly enjoyed the gilded company which always congregated along that sunshine coast; and of course the lure of the gaming tables was always, for him, potent.

Small talk and social chit-chat bored Winston profoundly – so usually he simply ignored it, resolutely pursuing his own themes. And although he greatly liked the company of beautiful, lively women, he tended to tire quickly of their conversation, and would always prefer to talk (or even shout!) across them, to some man, if one happened to be present; if he was trapped with no relief in sight, he would sink into a total and daunting silence.

Since Clementine did not write books or paint she found herself imprisoned in these luxurious cages with very little to do, and surrounded by people whom, for the most part, she found shallow, vulgar and boring. The expeditions to the Casino were for her a combination of boredom and anxiety: she did not enjoy gambling herself, and she was always consumed with anxiety lest Winston should lose a lot of money – which he quite often did. There is, however, this charming story: they were staying in the South of France, and she, tired and bored, had left the tables early and gone to bed, leaving him hard at it; when she awoke in the morning, she found every inch of her counterpane covered with banknotes! Winston had been in a lucky vein, and had bestrewn her bed with his winnings – as the birds covered the Babes in the Wood with leaves.

But there was one place that Clementine went to with real pleasure: Lou Sueil, the beautiful home, perched high above Eze, of Consuelo and Jacques Balsan.* The life and company at Lou Sueil were rather different from those in the other houses along that coast to which Winston and Clementine were so often invited. Although the Balsans also had a beautiful house at Dreux in the north of France (St Georges Motel), they led a settled and 'rooted' existence in both places. Consuelo had a lively social conscience, and wherever she lived she took an active interest in local community affairs. Both she and her husband were persons of culture and distinction, and their friends reflected their tastes and characters; with them Clementine found herself in an atmosphere in which she was at ease. Writing to Winston from Lou Sueil on 24 February 1924, she told him: 'Nothing could be kinder & more charming than Consuelo & Balsan. They love each other very much & it is a most peaceful & restful atmosphere.' Although Winston and Clementine sometimes stayed together at Lou Sueil, it became the custom for Clementine to pay an annual visit to the Balsans, usually in February, on her own. She also used to stay with them in the summertime in Scotland, where they had leased Grantully Castle, near Aberfeldy in Perthshire.

Clementine had greatly enjoyed the winter months of 1922–3 which they had all spent in Cannes as a family, in the Villa Rêve d'Or: then she had not been a guest, but able to lead her own life, playing a lot of tennis. But as time went on she opted out of the Riviera invitations (except those to Lou Sueil), and preferred to make plans of her own, or simply to stay at Chartwell.

Her favourite type of holiday was to stay in a modest but comfortable hotel in some beautiful and interesting place, with either a woman friend,

* See note p. 78.

or a party of congenial people, or (as they grew older) one of her children, to spend a week or two sightseeing and gallery-visiting in an unhurried fashion. Winston did not really enjoy sightseeing: he could enjoy a brisk expedition to some glorious panorama, or the scene of dramatic historical events (particularly if they afforded subjects to paint); but galleries and museums were not for him.

Winston could never really understand Clementine's antipathy to the Riviera, and he would always try to persuade her to go there with him, deploying all his skill to depict this, to him, delightful region in the most tempting guise. But she rarely relented; only occasionally making brief visits as *actes de présence*, lest his hospitable hosts and hostesses should be offended by her repeated absences.

But there were occasions when they achieved highly enjoyable and successful holidays together, such as when they visited Florence in 1925, and Venice two years later. Both these places afforded pleasures and occupations for each of them: Winston painting to his heart's content, while Clementine roamed the sights and galleries.

Two major expeditions on which they embarked together were marred, one by an accident and the other by a serious illness. The first was in the winter of 1931, when Winston had contracted to make a lecture tour (of forty lectures) in the United States. This he undertook largely to help recoup some of the money he had lost in the Wall Street Crash two years before. To make this exhausting enterprise less severe, Clementine and Diana went with him; they all sailed to New York in the *Europa* on 5 December, staying on their arrival at the Waldorf Astoria.

On the night of Sunday, 13 December, Winston was knocked down by a car while crossing Fifth Avenue to visit his friend Bernard Baruch on East 66th Street; he was taken at once to the Lennox Hill Hospital. When Clementine was told the news she was just going to bed; in her alarm she dressed so quickly that she arrived at the hospital in her stockinged feet.

Had Winston not been wearing a heavy fur-lined overcoat he might well have been killed, as the car was travelling at speed. As it was, he suffered severe shock and bruising, and developed pleurisy. He was released from hospital after about ten days, but had to spend a further fortnight in bed in the hotel, where they all passed a somewhat muted Christmas.

In the New Year Clementine and Diana took their invalid to convalesce in the sunshine of Nassau. Clementine wrote a long letter to Randolph after they had been there about ten days. Winston was terribly depressed by the slowness of his recovery, and suffering from severe pains in his arms and shoulders. He really needed three or four months of relaxation, but this was not possible, for as soon as he was at all fit, he was determined to finish the series of lectures, of which he had delivered only one. But he himself was seriously worried he might not be able to stay the course. Clementine wrote wisely to Randolph, 'I hope however they [the lectures] may actually help his recovery, especially if he starts off with a big success in New York.'[1] But, she told her son, Winston was very low:

> Last night he was very sad & said that he had now in the last 2 years had 3 very heavy blows. First the loss of all that money in the crash, then the loss of his political position in the Conservative Party* and now this terrible physical injury – He said he did not think he would ever recover completely from the three events.
>
> You can imagine how anxious I am & I often wish you were here to help & advise me.[2]

They remained for three weeks in the Bahamas, and then sailed for New York. Winston was much better, but still weak; nevertheless, he gave his first lecture since his accident on 28 January 1932, before two thousand people in Brooklyn. Thereafter, for the first three weeks in February, he travelled throughout the United States, lecturing every day in a different place: he had galvanized himself into action, and dominated his lassitude and weakness. Clementine stayed long enough to see that Winston was standing up to the tempo and strain of the tour, and then returned home. Winston arrived back at Plymouth on 17 March in the *Majestic*, and at Paddington Station in London was greeted by a small group of friends who presented him with a beautiful Daimler to celebrate his recovery.

In August Winston, Clementine and Sarah set out on what promised to be a fascinating journey. Accompanied by 'The Prof' and Colonel Pakenham-Walsh, the military historian who was helping Winston with his life of Marlborough, they toured the battlefields of Ramillies, Oudenarde and Malplaquet, and then followed the course of the long and fateful march of Marlborough's army, which had culminated in the Battle of Blenheim.

Winston then intended to go to Venice with his family, but during the first week in September, in Salzburg, he developed a persistent fever, which was diagnosed as paratyphoid. He was at once admitted to a clinic there, where he remained for over a fortnight, when, although still rather weak, he was allowed to return to Chartwell. Only four days after his homecoming, he suffered a severe internal haemorrhage, and had to be rushed by ambulance to a London nursing home. He took some time recovering from this illness.

* * *

In the late summer of 1934, Winston and Clementine went for a month's holiday with Lord Moyne (Walter Guinness), a former Financial Secretary to the Treasury, and a most intelligent, agreeable and unusual friend, aboard his yacht, MY *Rosaura*. Sailing from Marseilles, they visited Athens and Cyprus and the southern coast of Turkey. At Beirut they left the *Rosaura* and drove to the ruins of Palmyra, and from there to Damascus and on to Nazareth and Nablus, eventually meeting up again with their host in

* Churchill had resigned from Mr Baldwin's Business Committee (equivalent to a Shadow Cabinet) in January 1931 over the Conservative Party's Indian policy.

Jerusalem. Before rejoining the *Rosaura* at Alexandria, Winston and Clementine visited Jericho, Amman, Petra, Akaba and Cairo. It was a wonderful holiday, combining all the things they both enjoyed. But the active and enterprising element in Clementine's nature was not wholly satisfied with visits to places that had known civilization for thousands of years, and this cruise was to prove the prelude to another, much longer journey which would take her literally to the ends of the earth.

Lord Moyne had undertaken to try to capture alive for the London Zoo specimens of the species of strange, dragon-like, giant 'monitor' lizards which inhabit the island of Komodo, one of the Lesser Sunda Islands of Indonesia (then the Dutch East Indies). Both Winston and Clementine were invited to go with him in the *Rosaura* on this thrilling and interesting journey, which would last at least four months.

Winston saw how much Clementine longed to go, but he himself did not really enjoy sea travel for long periods of time – he felt imprisoned – and just now, moreover, he was engrossed by the successive volumes of the life of Marlborough, the first of which had been published the previous year. He was loath to let her go so far, for so long, without him; but perhaps he recognized that again and again in their life together her plans or preferences had been subordinated to his needs and wishes, and here was a chance for her which would be unlikely to occur again. And so, somewhat ruefully, he agreed that she should accept Walter Moyne's invitation.

This journey was to be one of the great events of Clementine's life. She recorded it all in long and vividly descriptive letters to Winston, the journalistic portions of which he had copied and circulated round the family (the tender passages he kept for himself alone). Winston, too, was a faithful correspondent: his long, minutely detailed 'Chartwell Bulletins', consecutively numbered (there were twelve in all), invariably dictated and typewritten, not only covered Chartwell news, but also gave brief accounts of parliamentary happenings, and, of course, news of the children. Each missive had its covering letter, written in his 'own paw', with any 'stop press' news, and relating his own private thoughts.

Clementine could not know it then, but looking back one realizes that this odyssey through tropical seas and coral-fringed islands came as a sunshine interlude before the darkening, anxious ending of the decade ushered in a period when, yet again, the utmost effort and commitment would be demanded of her.

She left London by train on 17 December 1934, travelling overland to join the *Rosaura* at Messina. As the train rumbled across Europe, she wrote on the eighteenth:

> My darling. As my train steamed away from Victoria and I saw you all collected on the platform, I thought how much I love you all, and above and more than them all you my sweet and darling Winston. You all looked so sweet and beautiful standing there, and I thought how fortunate I am to have such a family. Do not be vexed with your vagabond Cat – She has gone off

towards the jungle with her tail in the air, but she will return presently to her basket and curl down comfortably.

Happily installed in the *Rosaura* a few days later, she wrote: 'As I lie in bed I contemplate the photographs of my family erected on the chest of drawers against the opposite wall. I don't generally take photographs away with me, but this time is such a long absence I felt I could not do without them.'[3] In the same letter Clementine reported that she was 'beginning an enormous piece of needlework which Venetia [Montagu] made me bring. I have got 144 reels of silk with which to quilt it & I calculate that even if I sew all day & never catch a butterfly or a dragon I could not finish it before my return!' It was, in fact, to remain forever incomplete.

Until they reached Rangoon, the party on board the *Rosaura*, apart from Clementine, consisted of Lee and Posy Guinness (he was a cousin of Walter Moyne) and Terence Philip (a very personable and agreeable man in his forties). At Rangoon their host, Walter Moyne, and Lady Broughton (Vera, the wife of Sir Delves Broughton, from whom she was later divorced), were to join the yacht, having flown out from England; a flight of such length in those days was considered quite hazardous. But from Messina to Rangoon, a distance of some six thousand miles and twenty days' steaming, the Guinnesses, Terence Philip and Clementine were *à quatre*, and they combined most agreeably. On New Year's Day 1935, Clementine's thoughts turned homewards. 'Oh my Darling,' she wrote, 'I'm thinking so much of you & how you have enriched my life. I have loved you very much but I wish I had been a more amusing wife to you. How nice it would be if we were both young again.'

It was three weeks before Winston replied to this letter, for the distances were now immense and mails slow; but when she at last received his response, her heart must have been warmed by his grateful, touching recognition of all she meant to him:

In your letter from Madras you wrote some words vy dear to me, about my having enriched yr life. I cannot tell you what pleasure this gave me, because I always feel so overwhelmingly in yr debt, if there can be accounts in love. It was sweet of you to write this to me, & I hope & pray I shall be able to make you happy & secure during my remaining years, and cherish you my darling one as you deserve, & leave you in comfort when my race is run. What it has been to me to live all these years in yr heart & companionship no phrases can convey. Time passes swiftly, but is it not joyous to see how great and growing is the treasure we have gathered together, amid the storms & stresses of so many eventful & to millions tragic & terrible years?[4]

When Walter Moyne and Vera Broughton arrived safely as planned at Rangoon on 9 January, they brought more letters with them from home. On the Chartwell front the bulletin was, on the whole, satisfactory: the family was well, and Winston was continuing his wall-building activities. But the

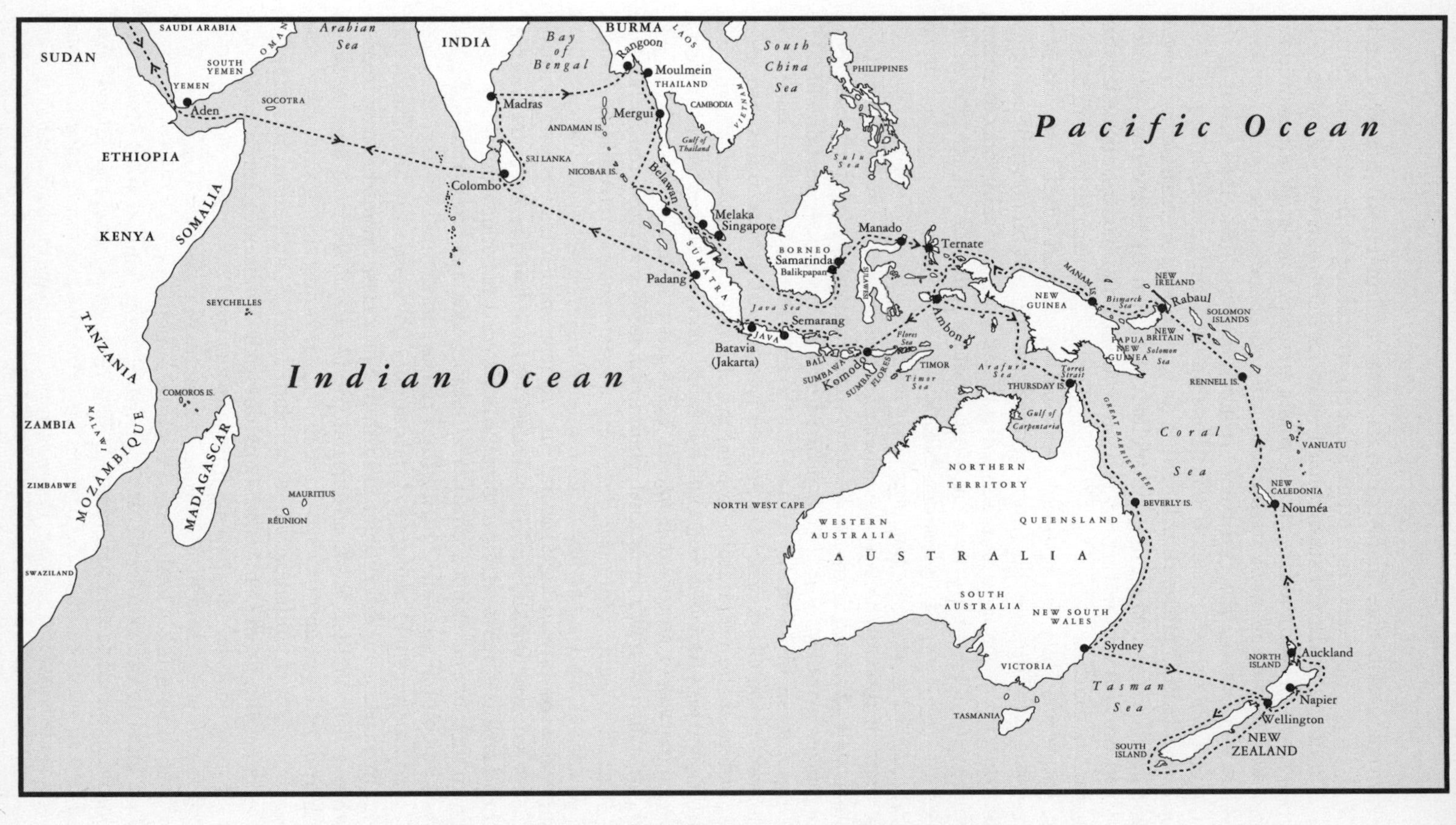

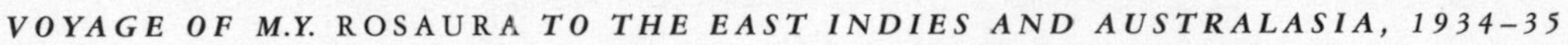

VOYAGE OF M.Y. ROSAURA *TO THE EAST INDIES AND AUSTRALASIA, 1934–35*

principal news in this New Year's Day letter was the arrival of a mechanical digger to perform two major landscaping jobs: the first was the creation of a ha-ha to the north of the swimming-pool, so that 'Your eye will plunge, as you desire, across a valley of unbroken green.'[5] The second task for the monster was to make the peninsula on the bottom lake into an island. Winston was evidently delighted with his new toy. The plan was that the digger should complete these two tasks in a week, at the cost of £25. 'In this one week he [the digger] will do more than forty men do,' Winston reported with enthusiasm. 'There is no difficulty about bringing him in as he is a caterpillar and can walk over the most sloppy fields without doing any harm.'

Unfortunately, these splendid plans went somewhat awry, and successive letters from Winston described the 'digger drama' in all its prolonged horror. On 31 January, he admitted that 'Making the island has proved a bigger and longer business than I expected. It will take a fortnight in all. But the results will be pleasing.' Nearly a month later, the digger was not only still at Chartwell, but hopelessly bogged. On 23 February Winston wrote: 'for the last three weeks it has done nothing as its works broke, and further, it got itself into a hole from which the greatest efforts have been necessary to extricate it.' And on 2 March he owned disconsolately, 'The digger has involved me in a chapter of accidents and I doubt if I shall get out of it under £150.' The weather had been appallingly wet, and the digger 'sunk deeper into the mud and finally wallowed himself into an awful pit. It became necessary to bring four hydraulic jacks . . . However after nearly a week the animal emerged from his hole and practically finished the job, though there is still a fortnight's tidying up for five men.'

The ill-starred digger had obviously acquired all the characteristics of a live animal with a perverse personality. Winston explained the creature's difficulties with great understanding: 'This animal is very strong with his hands but very feeble with his caterpillar legs, and as the fields are sopping, they had the greatest difficulty in taking him away. They will have to lay down sleepers all the way from the lake to the gate over which he will waddle on Monday. I shall be glad to see the last of him.'[6]

But as Clementine approached Singapore, she knew only of the arrival of the marvellous monster which was to perform such wonders in so short a time at so small a cost. 'What tremendous works you are doing at Chartwell,' she wrote,

> I'm delighted about the Ha-ha. Please do not throw back too much earth on the garden side or the slope might look too sudden . . .
>
> How lovely it will be when your beautiful wall reaches the end.[7]

Winston was able to report the departure of the 'animal' on 8 March, and one can almost hear his sigh of relief: 'The digger has gone, thank God.' And he was also able to reassure Clementine, who must certainly have had some twinges of anxiety about its depredations: 'We have had a successful week tidying up. The island will look delightful when all is green again and the lake refills. Next week we shall finish the haha, in a way which I am sure you will like.' This news, however, did not catch up with her until she reached Batavia on 1 April.

Another faithful correspondent was Cousin Moppet, who acted as 'fort-holder-in-chief' during any prolonged absence of Clementine. She kept the wanderer up to date with all the minutiae of domestic and family life. The school holidays were coming to an end, and I was spending the last lap at a nearby livery stable learning stable work; I had been to the circus, a pantomime, and several parties. Sarah had been in Wales hunting, but was soon to return to her theatrical dance training. Diana, who was living in London, was waiting rather bleakly and sadly for her divorce case from John Bailey to be heard. When that grey day came in mid-February, the staunch Moppet accompanied her to court.

We get a vivid glimpse of Winston in one of Cousin Moppet's letters: 'Winston has so many irons in the fire that the day is not nearly long enough – what with the new wall & the wonderful mechanical digger which does the work of 40 men – rebuilding Howes' [the chauffeur's] cottage, films & India, and when there is nothing else, Marlborough – so you see we are very busy.'[8]

Sarah's letters were very loving and gay:

[22 January 1935]

DARLING, DARLING Mummy,

While I imagine you, with blue sky above your head gaily chasing after butterflies and dragons, with Mr Terence Phillips [*sic*] – panting 'Hi, not so fast, I can't keep up' life goes on here at Chartwell much the same, except of course that we miss you very much, and in all the hundreds of little ways that you would like being missed.[9]

And a month later: 'Don't forget to come home sometime. Papa is miserable and frightfully naughty without you! Your children however are model in every way – and Mary is getting prettier and terribly horsey – and Cousin Mop is getting fat again.'[10]

At Singapore, where they arrived on 17 January, Clementine – despite the torrid heat – took the opportunity of going round the Dockyard. She gave Winston a remarkably detailed account of its construction, and the reduction of its scope and facilities due to the Government's cheese-paring in defence spending.[11] She had called at the principal bookshop to ask 'if there were a brisk demand for your books. They said that the compressed World Crisis had gone very well. The second volume of Marlborough was doing better than the first. They had sold 12 sets of Marlborough & had 5 more on order. I think that this is rather good when you realise how expensive it is.'[12]

After leaving Singapore, they visited Borneo briefly, and called in at the Celebes and the Moluccas, eventually reaching Amboina, from where they were to explore the south coast of New Guinea. On 30 January, Clementine wrote a thrilling account of an adventurous expedition they had made up the estuary of the Eilanden river, in an area where, for a thousand miles, the coast was practically unknown. 'This is the "genuine article"!' she wrote, 'unchartered [*sic*] seas, unexplored territory, stark naked savages.'

Leaving New Guinea, they steamed down the eastern coast of Australia along the Great Barrier Reef and reached Sydney early on 6 February; here Clementine had a happy reunion with Margery Street, the faithful and much-loved 'Streetie'. And at Auckland on 20 February Clementine was greeted by a further batch of family letters, full of accounts of Randolph's exploits in the Wavertree by-election*.

The yacht sailed on to Wellington, and around 21 February they were at Deep Water Cove in the Bay of Islands, where Clementine had her first (and last) experience of deep-sea fishing. She and Terence Philip were in one fishing boat, while Walter Moyne and Vera Broughton (old hands at the game) were in another. They spent hours trying to catch bait in a choppy sea before they even started 'really fishing'. After three hours of this, Clementine said she felt faint and wanted food, so they hove to in a little bay and had luncheon. 'But I felt so sick I could only drink claret and suck some very strong peppermints.'[13] In the afternoon, despite torrents of rain, they tried again, but with no success; they got back to the yacht drenched and exhausted.

After this stern experience Clementine and Terence Philip gave up fishing, and on one day when there was a gleam of sunshine they picnicked on an uninhabited island instead. Undeterred, however, by either the weather or the heaving seas, Walter Moyne and Vera Broughton persevered, catching a couple of swordfish and two or three sharks. Of her host Clementine wrote: 'Poor Walter caught 1 shark only. But he did not mind – He is the most unselfish man in the world and thinks the whole time of making pleasure & fun for his guests.'[14]

She had one rather frightening experience. The whole party were on a small uninhabited island hunting for lizards called tuataras, a few of which Walter Moyne hoped to capture for the Zoo. The island was in reality a sheer cliff, only one side of which was accessible by climbing up from rock to rock and tree root to tree root; where the hill flattened out at the top it was densely covered with jungle growth.

Writing about her adventure to Winston, Clementine described how she enjoyed the first part of the expedition:

> I love rock climbing; and you know I always start with zest any excursion or expedition – Where I fall short is that I soon get physically exhausted. Scrambling on hands and knees thro' the Bush soon tired me out & I thought I would go back & wait in the dinghy at the foot of the cliff . . . I was sure I could retrace my steps. But presently I realised that I was hopelessly lost.

* See p. 273.

She tried shouting, but there was no answer; she then tried to retrace her footsteps, but became exhausted struggling with the thick undergrowth. Presently there was a tropical rainstorm, and she was soaked despite the thick canopy of vegetation. Tired and disconsolate, she suddenly realized she was not alone:

> Suddenly I saw one of the lizards, quite close looking at me with his agate eyes. He was motionless. I sat down near him & we watched each other – Then I started shouting again & thought, Now the lizard will scuttle off but he did not move. They are stone deaf. But my voice did not carry far . . . Presently the lizard moved off and left me. Suddenly I heard Walter's voice far away . . . I called back but I felt he could not hear – Still I felt comforted – Presently I heard a loud crackling of branches & there was the second officer – I almost kissed him. He blew a whistle which was the signal Walter had arranged for whichever found me first. There were about 8 looking for me by now. Presently Walter appeared white with anxiety – I was really lost for only 1 hour, but it felt like much more in that dense enchanted wood – Of course there was no danger really I suppose but I thought of lying there & dying of hunger as far away from you as I can be on this earth.[15]

The *Rosaura* continued on her voyage, heading northwards now from New Caledonia to the New Hebrides, the Solomon Islands and New Britain. From the wastes of the Pacific Ocean, Clementine wrote on 7 March:

> Since Nouméa [the capital of New Caledonia] we have been crossing the trackless ocean calling at various islands. We are quite out of touch, the waters round some of these islands are uncharted, they are surrounded by cruel coral reefs and often no bottom can be found. Captain Laidlaw is a tower of strength for which I magnify and praise the Lord daily.

The sense of the distance that separated them was very real also to Winston, who earlier had written:

> it makes one gasp to look at the map & see what enormous distances you have covered since I saw the last of your dear waving hand at Victoria Station; and it depresses me to feel the <u>weight</u> of all that space pressing down upon us both. How glad I shall be when you turn homewards, & when the mails will be closing up together, instead of lagging & widening apart![16]

On 18 March the *Rosaura* arrived off the island of Komodo – the home of the huge monitor lizards whose capture was the object of the entire journey. The large trap was landed in sections, and a goat caught to act as bait. The dragons' lair was presently located. 'Halfway up a green velvet hill is a huge upended rock about 40 foot square – At the base are 2 cavernous holes & we could in the dust at the opening distinctly see the track of the monster's tail.

So he was at home!' wrote Clementine on 18 March, in her first instalment of her account of the great hunt.

On 26 March she again took up the tale:

> This has been an enchanting bewildering & exciting week – I meant to keep this letter in the form of a diary adding to it day by day but failed to do it – In this way I could have made you feel the excitement the suspense the heat the repeated disappointments the fearful smell of the decaying baits – (which of course had always to be approached and watched with the wind blowing towards one). Side by side with all this 'big game' 'Boys Own Annual' world is the enchantment of this island which is the most beautiful thing I have ever seen & is perhaps one of the loveliest wildest strangest spots in the world. It is deeply indented with bays and lagoons. It has innumerable paradise beaches – some of the finest sand (there is a pink one of powdered coral) some of wild rocks with coral gardens far lovelier than at Nassau, & accessible. That is if you are not afraid of being observed by a giant polyp or tickled by a sea snake 12 foot long.

The total result of the week's hunt was five small lizards, of which two were liberated; the longest was about six feet long and, wrote Clementine, 'they have not yet developed a very dragon like appearance. This comes with hoary age.' Apparently it was very difficult to trap a real monster, they being few, and very cunning, but the expedition's photographer had managed to photograph a twelve-foot lizard going off with half a pig in its jaws. 'He & Vera Broughton have lain in wait for hours in the long grass in the blazing sun near the traps, and near dead exposed & rotting animals to get good pictures. I spent one hot morning watching & decided upon the other life – But I did see one prehistoric peep of a dragon on a stony beach looking for crabs.'[17]

On his return, Lord Moyne presented two Komodo dragons to the London Zoo; they both lived until 1946, which is considered quite a good survival rate. Of the several tuataras that were captured, one survived the voyage home, and was duly given to the Zoo; but it lived only five or six months.[18]

For all this time, and on all these thrilling expeditions among strange peoples and enchanted islands, Clementine's constant companion was Terence Philip. She was in her fiftieth year, but still exceptionally beautiful, slender and graceful. Terence was about seven years younger; he had been born of British parents in Russia, his father being a wealthy merchant there. Much of his youth had been spent in Russia, and he spoke the language fluently. His family had returned to England when the Revolution broke out in 1917, and Terence made a career in the art world. At the period when he knew Clementine he was working for Knoedler's Art Gallery in Bond Street. He was much sought after by London hostesses as the perfect permanently available single man for any social occasion; he particularly enjoyed (and was enjoyed by) slightly older women. He was the ideal shipboard companion – suave, good-looking, charming and cultivated – and Clementine fell romantically in love with him.*

For a brief period in her life, which for the most part was rooted in obligations, convention and reality, she lived in a dream world of beauty and adventure. She tasted the heady elixir of admiration, and knew the pleasures of companionship in trivial doings and sayings. Looking back in after years on these five months of her life, she viewed the episode with detached amusement, in which one could detect only the faintest nostalgia. Acknowledging that Terence Philip had never really been in love with her, she added: 'But he made me like him.'[19]

After the voyage was over, their friendship continued for a little while. Terence Philip came several times to Chartwell in the following two years; but then he came no more. There was no quarrel – there was nothing to quarrel about – only agreeable things they had shared to be remembered. But their relationship was like a fragile tropical flower which cannot survive in greyer, colder climes. Once home, Clementine's real life, and the one and only great love in it, claimed her back entirely, as Terence's cosmopolitan, worldly life and ambivalent nature – so different from hers – claimed him. He presently went to New York, where he worked at Wildenstein's Gallery, until his early death during the Second World War.

Perhaps the very difference in their ways of life and outlooks had been as much part of the magic spell as were the blue seas, waving palms and tropical islands: Clementine had to awake from the unreality of a dream. The French nearly always have the right words for these matters of the heart, and she summed it all up in a saying which seems to breathe the Edwardian world of her youth: '*C'était une vraie connaissance de ville d'eau*.'[20]

* * *

There was one more great treat in store before the *Rosaura* and her company turned towards home: they spent two days in Bali. Clementine was entranced: '[It] is an enchanted island. Lovely temples embedded in green vegetation in every village. Lovely dancers. The inhabitants lead an Elysian life. They work for about 2 hours a day – The rest of the time they play musical instruments, dance, make offerings in the Temples to the gods, attend cock fights & make love! Perfect? Isn't it?'[21]

At Batavia, the capital of Java, where they arrived on 1 April, Clementine's fiftieth birthday, there were letters from home to greet them. Winston's enclosed a handsome cheque, which presently bought a beautiful pair of diamond clips. He wrote of her return: 'This has been a vy long separation for us; but now that "Your nose is turned homeward", we can look forward to the end of it . . . Now on yr return journey you will be meeting yr mails, instead of them lagging behind you.'[22] This letter was decorated

* A few years ago a woman who in her younger days had known Terence Philip, and his world, told me she had no doubt that he had homosexual leanings, but in those days, of course, discreetly so. (See also Pearson, *Citadel of the Heart*.)

with a charming pig, labelled 'On guard & waiting'. Clementine's letter in reply was signed 'Your loving Homeing Clemmie'.[23]

Winston's letters continued to bring her all the news from home. The digger had departed, the orchard had been grubbed up and re-turfed, and the chauffeur's cottage renovated. In the 'big house' new bookcases had been installed in the drawing-room, which had also been redecorated.

The news on the animal front was varied. A black swan was sitting on six eggs: a prudent bird, she had made her nest on the new island, and therefore escaped the fate of an unfortunate goose, which had been devoured by a fox. In the house the dog population had been exceedingly troublesome; my pug was the chief culprit. 'He commits at least three indiscretions a day,'[24] and this set a bad example to Trouble, Sarah's chocolate spaniel (the present of an admirer), and Harvey, Randolph's wire-haired terrier, who was also a guest at this time. In early April Winston could bear it no longer, and in an exasperated postscript announced that he had 'banished all the dogs from our part of the house . . . I really think you will have to buy a new strip of carpet outside my landing.'[25] Despite all these other preoccupations, Winston worked away at building his wall, and he promised it would be finished by her return. 'How paltry you must consider these domestic tales of peaceful England compared to your dragons and tuataras. But I think it is very important to have animals, flowers and plants in one's life while it lasts.'[26]

Familial tribulations were also weighing upon Winston. After the efforts and excitements of the election at Norwood, the children proceeded to cause some anxiety in matters of health: Randolph had a severe bout of jaundice; I had caught whooping-cough; and Sarah was looking tired and pale, which was not surprising considering, her father reported, she was 'dancing practically four hours a day as well as going to balls . . . I have therefore told her that she must not go to a ball on any night when she practises her dancing.'[27] Although Winston grappled with all these domestic situations gallantly, he nevertheless looked forward to Clementine being able to resume her control of such matters.

But his letters were not only concerned with 'domestic tales of peaceful England'. Each one contained an account of the political affairs which daily occupied his mind. During the whole of Clementine's absence, Winston had been embattled and battling with the India Bill. The debates tired him so much that he often could not face the drive to Chartwell late at night, staying instead at the flat which they had acquired a few years before in Morpeth Mansions, near Westminster Cathedral. On 1 January 1935 he had described how, in a debate in the House of Lords before Christmas: 'The poor old Tory Lords came up in hundreds from the country and voted as the Government Whips directed. They would certainly have much rather voted the other way but they know their place since the Parliament Act.' Now, three months later, they were still at it: 'I have had a very hard time with the India Bill,' he wrote on 5 April. 'They have taken it four days this week, and there are to be four days next week. I have to be there practically all the time and the only time I was away the Government got sixty clauses through in a twinkling.'

Quite apart from the India question, the political cauldron was brewing. There were at this time continuous rumours that Ramsay MacDonald would shortly resign, and that his successor, Baldwin, would wish to include Churchill in a reconstructed Government. He wrote to Clementine: 'I am not at all keen on this, and there is a very desperate General Election which lies ahead. One would simply have to take the disgrace of the Government for all the opportunities they have missed.'[28]

But emerging in importance over all other issues was the growing might of Nazi Germany. On 8 March, Winston told Clementine: 'The German situation is increasingly sombre.' A month later, he wrote: 'The political sensation of course is the statement by Hitler that his air force is already as strong as ours. This completely stultifies everything that Baldwin has said and incidentally vindicates all the assertions that I have made.'[29]

As the distance slowly diminished between them, Clementine's thoughts turned more and more towards Winston and home. From Suez, on 20 April, she wrote:

> Oh my darling Winston
>
> The Air Mail is just flitting & I send you this like John the Baptist to prepare the way before me, to tell you I love you & that I long to be folded in your arms . . .

Winston's last letter had told her how much he had missed her:

> I think a lot about you my darling Pussie . . . and rejoice that we have lived our lives together; and have still some years of expectation in this pleasant vale. I have been sometimes a little depressed about politics and would have liked to have been comforted by you. But I feel this has been a gt experience and adventure to you, & that it has introduced a new background to yr life, & a larger proportion; and so I have not grudged you yr long excursion; but now I do want you back.[30]

Clementine arrived home on 30 April 1935: and great were the family rejoicings. She had traversed in all thirty thousand miles of water.

The Silver Jubilee Celebrations of King George V and Queen Mary were very shortly to begin, and Winston and Clementine had been bidden to all the great occasions. Fortunately for her, she had lost a good deal of weight, and so she was able to step easily into model-sized dresses.

But *Wanderlust* had entered Clementine's blood, for in a letter to Margery Street later in the summer she wrote: 'It's very nice to be back but Oh Dear I want to start out again very badly! Mr Pug is very sweet but now he says "NO".'[31]

Like all travellers, Clementine returned with a cargo of strange bric-à-brac, some of which found a place at Chartwell, like the giant shells, which formed exotic-looking goblets for the smaller waterfalls (although in time they became a rather depressing shade of greeny-yellow). But the most charming of her mementos from the East was a Bali dove, an enchanting pinky-beige little bird with coral beak and feet, who lived in a beautiful wicker cage rather like a glorified lobster-pot. He was a great pet, and would 'crou-crou' and bow with exquisite oriental politeness to people he liked. He survived for two or three years, when, no doubt homesick for his enchanted island home, he died. Clementine had him buried under the sundial in the centre of the kitchen garden. Round the base are carved the following lines:

HERE LIES THE BALI DOVE
It does not do to wander
Too far from sober men,
But there's an island yonder,
I think of it again.*

* From a poem by W. P. Ker. The lines were suggested to Clementine by the writer and traveller Freya Stark.

CHAPTER EIGHTEEN

The Thirties

THE TEN YEARS THAT WINSTON CHURCHILL SPENT OUT OF OFFICE IN THE political wilderness have been regarded by many, in retrospect, as the continued working of that guiding hand of destiny which seems to have been with him throughout his life. Although during all that time Churchill himself was far from being resigned to his exclusion from the exercise of power, his gradual estrangement from the Conservative Party's leadership, due to his disagreement over their Indian policy, and then his campaign for rearmament and confrontation of the growing power of Germany, made the probability of his being offered office increasingly remote. This can now be seen as providential – for when the hour struck which Churchill had so long and so eloquently foretold, he was in an unrivalled position, untarnished by involvement in policies he might well not have been able either to change or to moderate, given the prevailing mood of the country in the thirties.

Even at the time Clementine understood this, for she wrote in one of her letters from Austria in January 1936: 'The political situation at home is depressing. I really would not like you to serve under Baldwin, unless he really gave you a great deal of power & you were able to inspire and vivify the Government. But as you say we are in it up to our necks & that cannot be altered now. All you could do would be to organise our armed forces.'[1]

When the Conservatives went into opposition in 1929, Churchill had been, at first, a member of Baldwin's Business Committee, but he had become increasingly disturbed by the Conservative Party's proposals for Indian self-government, which he regarded as rash and premature, and he had resigned from the Committee in protest in January 1931. This step excluded him from office when later that year a National Government of all three parties was formed with Ramsay MacDonald still as Prime Minister.

During the next few years the Indian question continued to occupy much of Churchill's thought and activity, and he was vigorous in opposing a policy with which he strongly disagreed. Despite Clementine's inherent radical tendencies, she agreed in the main with Winston's point of view on the matter of India's self-rule. Indeed, it was on this subject that she made one of her rare speeches in a controversial vein. At the Annual Conference held by the Central Women's Advisory Committee of the Conservative Party in May 1933, she spoke in support of an amendment to a motion from the platform congratulating the Government 'on the success of the policy of maintaining law and order in India, while continuing constitutional reform . . .'[2] The Duchess of Atholl, a Scottish MP and an outspoken supporter of Churchill's views, moved an amendment incorporating the following words: 'but would view with grave anxiety proposals to transfer at the present time responsibility at the Centre of Government in India [etc.]'. The amendment was debated, and Clementine made a short but effective speech in its support. Despite a powerful intervention by Lady Iveagh (one of the Party's 'big guns') urging the Conference not to accept the amendment, it was carried by a clear majority. Clementine returned home flushed and triumphant from her foray, to be much congratulated by Winston.

This same year of 1933 was full of fateful events: at home, unemployment figures reached the peak of three million; in January, Hitler became Chancellor of Germany; and in October, Germany withdrew from the League of Nations. Churchill had the previous year criticized the Government's inadequate defence programme, particularly with regard to air defence; now he attacked their proposals to accelerate disarmament, knowing that Germany was secretly rearming.

Soon the threat of the growing power of Nazi Germany, and Britain's unprepared state, came to dominate Churchill's thoughts, and he concentrated all his powers to the task of warning and awakening. His warnings fell largely on deaf ears. But there was a band, both in and out of Parliament, who fought this unequal fight with him. It was unequal not only because Churchill was at odds with both the hierarchy and most of the rank and file of his own party, but also because, in the main, people did not want to hear his message at a time when pacifism was growing as a mood, and appeasement as a policy, throughout the country. But this did not daunt him, and around him gathered politicians from different parties, civil servants and servicemen (as far as their position allowed) who supported his views. Chartwell was their meeting-place; and here also from time to time came harassed-looking Germans, Austrians, Poles and Czechs who, often at great personal risk, came to urge Churchill on in his efforts to awake the British people to the danger, which yearly loomed larger, before it was too late.

In all this, Clementine was heart and soul with Winston; and she sometimes waxed very fierce in argument with those who maintained the official party line, or took a 'pussy-foot' (pacifist) view of these matters. But on one major issue Clementine found herself at variance with Winston – the question of the abdication of King Edward VIII.

Early in December 1936 the British public was made aware of what the world's press at large had known and commented upon for many months past – namely, that King Edward VIII (who had succeeded his father in January of that same year) was deeply in love with an American, Wallis Simpson. The British press, by common agreement (unthinkable nowadays), had made no mention of this romantic situation, but it had been the subject of discussion and grave speculation in Court and Government circles after Mrs Simpson's divorce from her second husband, Ernest Simpson, in October 1936. When the news broke, people of all classes and all shades of opinion joined in the great argument as to whether a twice-divorced woman could be Queen Consort of England.

A large body of opinion, chiefly among the young and romantically minded, saw no reason why the King should not marry the woman he loved. There was a larger but less vocal section of the public which thought, quite simply, that the King should sacrifice his personal happiness to his royal duty. During the series of interviews between the King and Stanley Baldwin (who had succeeded Ramsay MacDonald in June 1935), the Prime Minister had to tell his Sovereign that Mrs Simpson would not be acceptable as Queen. The possibility of a morganatic marriage* was canvassed, but after Baldwin had consulted opinion both in Britain and in the Dominions (none of whose governments were prepared to accept Mrs Simpson on any terms), it became quite clear that either the King must renounce his plan to make her his wife, or he must abdicate his throne. He chose to do the latter.

During the course of these harassing and painful days, the King turned for advice to (among others) Churchill, who was a friend of long standing, and who sought by every means within his power to play for time. He felt this momentous decision was being resolved with wanton haste, and that, given more time and consultation, a happier issue might be devised. Churchill espoused the King's cause with loyalty and vigour. Always a monarchist, every instinct of personal loyalty and chivalry was aroused by the dilemma of his Sovereign, for whom he had always had a true regard. Moreover, Churchill tended to take the romantic view, and felt that some arrangement could have been devised whereby the King could marry Mrs Simpson, and the country be spared the convulsion of the crisis which would be caused by his abdication. Furthermore, Churchill was certain the King commanded wide sympathy among his subjects, who would rally to support him.

Clementine disagreed profoundly with Winston, both in his view of the problem and in his estimation of what public feeling really was. She thought Baldwin had correctly interpreted public opinion, and that in no circumstances would Mrs Simpson be acceptable to the British people or the Commonwealth, either as Queen Consort or even merely as the wife of the King. Clementine saw something else very clearly, too: she realized that Winston's championship of the King's cause would do him great harm, and that he would be accused of making political capital out of this crisis. In all this she was proved right: Churchill's intervention aroused a wave of hostility towards him, both in Parliament and in the country. When he rose in the House of Commons after the Prime Minister had made a statement

* By this device the King could have married Mrs Simpson, but she would have remained a private citizen, and any children of the marriage would not have been in the line of succession. Morganatic marriage is at the present time unknown in England, although occasionally practised among Continental royalty, and an Act of Parliament would have been necessary to make this solution to the impasse possible.

on 7 December, Churchill had barely begun his plea that 'no irrevocable step should be taken' before he was silenced by cries from all sides of 'No' and 'Sit down'; it was described by an observer as the 'most striking rebuff of modern parliamentary history'.[3] Four days later, King Edward VIII abdicated.

Although Clementine took the view that the King should have sacrificed his happiness to his duty, she was wholeheartedly at one with Winston in his anger against those (many of them in high places) who had turned against their Sovereign in his hour of anguished decision. And she was particularly scornful of those in the social world who had been only too willing to entertain King Edward and Mrs Simpson during the months that led up to the final crisis, but who turned cool when the day of reckoning came. In the February following the Abdication the Churchills dined with Henry ('Chips') Channon MP, the society host and diarist; among the other guests were Lord and Lady Granard (he had held several royal appointments and had been Master of the Horse). Channon wrote in his diary:

> Perhaps my guests were a little too Edward VIII for the Granards? It was, I admit, a thoroughly 'Cavalier' collection. Lord Granard tactlessly attacked the late King and Mrs Simpson, to his neighbour, Clemmie Churchill, who turned on him and asked crushingly, 'If you feel that way, why did you invite Mrs Simpson to your house and put her on your right?' A long and embarrassed pause followed.[4]

There is also an interesting and moving postscript to Winston's attitude towards King Edward and Mrs Simpson. On 12 May 1937, he and Clementine were in Westminster Abbey attending the Coronation of King George VI: at the moment when Queen Elizabeth was crowned as Consort, after making vows of the utmost solemnity, and receiving tokens of grace for her special task, Winston turned to Clementine and, his eyes full of tears, said: 'You were right; I see now the other one wouldn't have done.'

* * *

The period following the Abdication was probably the lowest point of Winston's political fortunes since the dark days immediately following the Dardanelles disaster. His warnings of the national peril ahead went largely unheeded, and now discredit was cast on him by the feeling that his support of the King sprang from ulterior motives, and was largely prompted by antipathy to Baldwin. To those who knew him well this was malicious nonsense: Churchill had reacted to the King's plight spontaneously and naturally, as one would expect of someone possessing his instincts of loyalty and chivalry. Nevertheless, this unjust appraisal blunted, at a most critical moment, his power to act and to persuade. Clementine had argued fiercely with Winston at the time and, privately, she now grieved at the price he had to pay for his intervention. In public, her attitude was robust and loyal, but realistic as always. Asked at this period if she thought Winston would ever be Prime Minister, she said: 'No, unless some great disaster were to sweep the country, and no one could wish for that.'[5]

Winston himself suffered at this time from feelings of almost fatalistic depression; the burden of Chartwell was telling, and he was definitely contemplating selling the property. He told Clementine that 'we can quite well carry on for a year or two more. But no good offer should be refused, having regard to the fact that our children are almost all flown, and my life is probably in its closing decade.'[6] Those last words make strange reading now: he had nearly three more decades to run, and his life's supreme task and achievement lay ahead.

But if Winston's star was burning dimly in his own country, he had a strong and admiring following in France; and this was demonstrated at the time of the State Visit of King George and Queen Elizabeth to Paris in July 1938. Although holding no official position, and despite being out of favour with his own Government, the French Government invited the Churchills to attend all the splendid functions which had been arranged to entertain the royal guests. Moreover, at all these elegant occasions, for which the French have such exceptional genius, Winston and Clementine were remarkably *bien placés*. This did not escape the notice of other English guests and interested observers from the British Embassy. Faded programmes still remain in Clementine's album, and one in particular catches the eye – the menu for the great luncheon at Versailles on 21 July. It was signed for her by her neighbours at the luncheon: one was the distinguished author Gabriel Hanotaux, the other Maréchal Philippe Pétain, both of whom made themselves exceedingly agreeable. This golden visit remained long in Winston's and Clementine's memories. In retrospect it would seem like the last gleaming of sumptuous hospitality and elegance before the gloom and squalor of war once more engulfed all Europe.

From 1935 onwards events full of portent followed fast one after another. The invasion and subsequent annexation of Abyssinia by Italy was followed the next year, in 1936, by the German occupation of the Rhineland. That same year also came the outbreak of the Spanish Civil War, in which both Germany and Italy intervened, and in which the bombing of civilian populations was seen for the first time in all its brutality and horror. And in the spring of 1938 the *Anschluss* saw the swallowing up of Austria. Each successive act of aggression by the dictators revealed the disunity and feebleness of the democracies, particularly Britain and France. Although in 1934 Britain embarked on measures of rearmament, the next year Hitler announced that the German air force was at least equal to that of Britain. In 1937 Stanley Baldwin retired as Prime Minister, and was succeeded by Neville Chamberlain.

In September 1938, after the now familiar propaganda campaign of lies, semi-truths and threats, Hitler demanded the incorporation in the Reich of those parts of Czechoslovakia where the population was predominantly German. Each act of aggression hitherto had found the democracies divided and futile, but now it was generally realized in both France and Britain that the moment had come when at last a stand must be made. Moreover, France was bound by treaty to support Czechoslovakia if it were attacked, and although Britain had no direct obligation to Czechoslovakia, we were pledged to support France. War seemed inevitable. In Britain, preparations began: the Fleet was mobilized; air-raid trenches were dug in the London parks; gas-masks were issued. The Prime Minister had been twice to parley with Hitler; now, on 29 September, he flew once more to Germany, this time to Munich, where he met both Hitler and Mussolini, with Daladier, the French Prime Minister. Chamberlain concluded with Hitler the Munich Agreement by which the greater part of Hitler's demands was conceded, and Czechoslovakia sacrificed. Britain and Germany entered into a solemn agreement never to go to war with one another again.

With tragic naïveté, Chamberlain genuinely believed he had come to terms with Hitler. When he returned to England, he was received with hysterical relief, joy and gratitude. Waving a scrap of paper, he said he had brought with him 'Peace with honour . . . Peace in our time'. (It was neither.) He appeared with the King and Queen on the balcony of Buckingham Palace; they were cheered vociferously by a huge, euphoric crowd. Throughout the country nearly everyone joined in the rejoicings, feeling only that the terrible moment of peril had passed, and not caring or choosing to count the cost.

But to Winston Churchill, and those who followed him, the Munich Agreement was an act of shame and betrayal; it represented the final, fatal grovelling to the insatiable appetite of the dictators. Looking back, it is difficult to describe the feelings of anger and bitterness felt by those who opposed the Agreement. Most people shared in the general sense of relief, or joined the chorus of praise and adulation which for a short period surrounded Chamberlain; but the little world of like-minded friends and colleagues which centred on Churchill throbbed with anger and emotion. It was almost impossible for those taking the opposing views in this fierce argument to meet without there being high words. Clementine felt this crisis with passionate indignation, and I well recall her setting upon one quite formidable lady – Eva, the wife of Admiral Sir Roger Keyes – who had expressed pro-Chamberlain views during a luncheon at Chartwell, and reducing her to tears. But whether people supported the Munich Agreement or not, only the most purblind thought it was anything more than time dearly bought.

* * *

The year of 1938 had not been a good one for Clementine's health, and in early July she had gone, alone, to a watering-place at Cauterets near Lourdes, in the French Pyrenees, where she took the cure for nearly three weeks before joining Winston in Paris for the very enjoyable State Visit. During her short stay in Paris she stubbed and broke her toe against the claw foot of an

Empire table; although the injury was not serious, this small accident was for many weeks the cause of considerable pain and hindrance. On top of this came the tensions and emotions of the Munich crisis, and by the autumn Clementine was again low and debilitated. However, at this moment a delightful invitation came her way: Lord Moyne once more invited her to go for a voyage in the *Rosaura*, this time to the Caribbean. The Government had set up a Royal Commission to enquire into the social conditions of the West Indies, and Lord Moyne had been appointed Chairman; he proposed to take his yacht, and as well as accommodating about half the members of the Commission in turn, he also invited his son, Murtogh Guinness, Vera Broughton and Clementine, who gratefully accepted this chance to restore her health and spirits.

Although she enjoyed this journey, and found it deeply interesting, it was not as carefree or idyllic as the long cruise she had made to the East Indies some three years before. In the first place, the whole purpose of this voyage was to carry out the exhaustive enquiries of the Royal Commission, and all purely pleasurable considerations were of secondary importance. Moreover, the enjoyment of the beauty of the islands and the pleasures of life on board were to some extent tempered, even for the guests, by the depressing effect induced by a study of the social conditions existing among the native populations, which had given rise to the demand for an official investigation.

Clementine felt keenly the contrast between the beauty of the island scenery and the squalor of the human conditions. From Barbados, she wrote to Winston:

> soon I think I will come home. These islands are beautiful in themselves but have been desecrated & fouled by man. These green hills covered with tropical bush & trees rise straight out of the sea, & fringing the coasts are hideous dilapidated crazy houses, unpainted for years with rusty, corrugated iron roofs – Trade stagnating, enough starchy food to keep the population alive but under nourished – Eighty per cent of the population is illegitimate, seventy per cent (in several islands) have syphilis and yaws. The homes of the labourers are small sheds full of holes stuffed with rags or patched with old tin – There is no sanitation of any sort, not even earth latrines; in some places the women have to walk 3 miles to get water – In many places the proportion of doctors to the population is one doctor to 30,000 persons. Labourers' wages are 1/- a day for men & 6d. for women – There is much unemployment & no system of Insurance – And this is a sample of the British Empire upon which the Sun never sets.[7]

Writing to Margery Street in a similar vein a month or two after her return home, she told her that 'the Commission disagreed bitterly among themselves, the Tories wanting to do nothing but introduce birth-control & even sterilization, & the Labour members stirring up trouble wherever we went (All the same, I felt much more sympathy with their point of view).'[8]

Clementine's thoughts had constantly turned towards home during her

Far Eastern journey, but on this voyage there is a note of almost poignant homesickness in her letters. Although she was half the world away under blue skies, events in Europe cast their long cold shadows even there. Bound up closely as she was with Winston's frustrating battle to awake people to reality, and caring passionately for the issues involved, Clementine found she longed to be with him in these anxious, fateful months.

While staying in Jamaica for a few days with the newly arrived Governor and his wife (Sir Arthur and Lady Richards) she had a rather strange and moving experience. The Governor had invited her to come with them on a tour of the island. 'I must tell you,' wrote Clementine,

> that in a tiny highland village where the Foundation Stone of a school was being laid the Chairman welcomed me as the wife of the future Prime Minister of England upon which the whole pitch black Audience burst into 'loud & prolonged cheers'. The Chairman told me that they all know about you & follow your doings in the Jamaican Press. I was thrilled & moved but rather embarrassed, as this rather took off from the Welcome to the New Governor who had never been to the spot before.[9]

Half the world away, Winston thought a lot about Clementine; but much though it interested him to hear about her doings, and the politics and problems of the Commission, what he really yearned to hear was news about her own health and well-being, and in his first letter on 19 December he wrote:

> I send you telegrams frequently: but in yr answers you do not tell me what I want to know – How are you? Are you better & more braced up? How is the voice? Have the rest & repose given you the means of recharging yr batteries. That is what I want to know. And even more – do you love me? I feel so deeply interwoven with you that I follow yr movements at every turn & in all circumstances. I wonder what you are doing now. Yr dear letter from the ship arrived after my chiding telegram had been sent off. Do cable every few days, just to let me know all is well & that you are happy when you think of me.

But Clementine's thoughts also turned to him, for two days after Winston wrote asking so lovingly about her health, she cabled him: 'My Darling. My thoughts are with you nearly all the time and though basking in lovely sunshine and blue seas I miss you and home terribly. Tender love. Clemmie.'

Clementine spent Christmas in the Leeward Islands, and once again this year, the family at home was somewhat depleted. Diana was happily ensconced in London with her second baby, Edwina, born on 22 December, and Sarah and Vic were away. So just three of us, Winston, Randolph and myself (now sixteen) went to Blenheim, where a large and convivial party of cousins and friends was gathered.

On Christmas Eve, although she had been away only a month, Clementine confessed to Winston: 'It feels much more. I miss you & Mary

& home terribly; & although it is a boon to miss the English Winter & to bask in this warmth; I really think I should come home – only that I hope this prolonged voyage in warm weather will really set me up in health – I do not yet feel very strong, but I am sure I shall.'[10]

On New Year's Day 1939, Clementine was in Antigua, where she visited Nelson's old dockyard:

> There is an historic & romantic old Naval Dockyard set in a beautiful little bay called English Harbour. Rodney, Hood & Nelson all were here. It is not deep enough for modern ships and the Navy gave it up 50 years ago. Nelson served here as a young Captain of 25 . . . It is a beautiful but sad and eerie spot with the deserted Dockyard Buildings falling to pieces. The old capstans are there rotting into the ground, & huge iron rings. There is a tiny dock for small ships and great stone pillars which used to support the roof of a gigantic boathouse, all blown away in a Hurricane. I was alone for a while . . . and I wondered if I should see the shades of departed heroes & hear the men in the Stores and Workshops.[11]

The dockyard made a lasting impression upon Clementine, and years later, in 1954, she became President of an appeal which raised enough money to restore this beautiful and historic place.

Although Winston sent frequent telegrams, almost for the first time in all their life together he was found wanting as a correspondent. It must have been the sheer volume and pressure of work, for he was in the throes of writing *A History of the English-Speaking Peoples* – a task which absorbed him completely. His first letter to Clementine, a month after she had left home, told her how intensively he had been working: 'I have been toiling double shifts . . . It is laborious: & I resent it & the pressure.'

But poor Clementine was feeling thoroughly neglected. Winston had taken to dictating most of his letters – even personal ones, for he found it burdensome writing in his own hand. Clementine had always resented this now established habit, and he nearly always wrote a long postscript or a covering note in his 'own paw' to mollify her. But now, on New Year's Day 1939, she begged him:

> Do you think you could dictate a few words every day to a secretary & she could send it off twice a week. Never mind about writing yourself – I used to mind about that, but I'm accustomed to type written letters now & I would rather have them than nothing. I feel quite cut off.
>
> Your loving but sad
> Clemmie

The series of telegrams Winston had sent had been a further cause of frustration to her, for she added in a rather cross postscript: 'Please don't telegraph – I hate telegrams just saying "all well rainy weather Love Winston".'

But even before he could have received this ruffled letter, Winston had resumed his habit of writing often, and his next letter (29 December) enlarged upon the happy news of the birth of Edwina: 'I cabled to you about Diana's baby daughter. It came most unexpectedly, was less than eight months old and weighed just over four and a half pounds. She, D, was perfectly well in the afternoon, sitting up and could see me. The baby is tiny but perfect, and by my latest news, thriving.' The long newsy letter contained further family, Chartwell and political news, and reported book progress: dictating until three in the morning, he had reached the Wars of the Roses and had 'just finished writing about Joan of Arc. I think she is the winner in the whole of French history. The leading women of those days were more remarkable and forceful than the men.'

The last paragraph, written in his own hand, told her of an event which profoundly moved her: 'My dearest Clemmie, You will be saddened to know that Sidney Peel has died. I do not know the cause. Many are dying that I knew when we were young. It is quite astonishing to reach the end of life & feel just as you did fifty years before. One must always hope for a sudden end, before faculties decay.'

In fact Clementine had already learnt this news, and from Barbados on 19 January 1939 she wrote:

> My darling. Four days ago I was sitting in the Public Library at Dominica (this was before getting your last letter)* reading up the back copies of the 'Times' and suddenly there was Sidney Peel looking at me from the middle of the Obituary page – A young photograph, as I used to know him – I closed my eyes; Time stood still, fell away, and I lived again those four years during which I saw him nearly every day – He was good to me and made my difficult rather arid life interesting – But I couldn't care for him & I was not kind or even very grateful – And then my Darling you came and in that moment I knew the difference –
>
> I am glad you wrote to me about it because at that moment I longed for you – I wanted to put my arms round you and cry and cry.

Scanning the newspapers, Clementine sensed that the whole political situation was deteriorating from day to day, and she felt deeply for Winston in his lonely battle:

> I have been reading Prescott's Conquest of Mexico – At the end of his life Cortez was treated with coldness & ingratitude by the Spanish Government. 'He found like Columbus that it was possible to deserve too greatly.' So you see you are in good company. And Oh Winston are we drifting into War? without the wit to avoid it or the will to prepare for it – God bless you my Darling.[12]

* WSC's letter of 29 December 1938.

It is clear from Clementine's letters that this holiday had not the quality of enchantment or the sense of escape from reality of her other voyage, when times were less harassing and anxious. Moreover, there were now tensions on board the yacht arising from political differences, and there was a real blow-up one evening while they were in Barbados. The company was listening to a broadcast from England, in which Churchill (and the anti-Government point of view in general) was attacked: Vera Broughton exclaimed in a loud voice, 'Hear, Hear!' – which angered Clementine, as it was evidently intended for her benefit. But what really wounded and enraged her was that Walter Moyne remained ostentatiously silent – thereby implying his tacit agreement with Lady Broughton. Clementine was already yearning for home, and her mind was particularly full of Winston and of his embattled position – and this was the last straw. She decided on the spur of the moment to return home. She went at once to her cabin and wrote a letter of explanation to her host; the next morning she went ashore and booked herself a berth on the *Cuba*, which was leaving Barbados the very next day bound for England. On her return aboard the *Rosaura* Vera Broughton came to her cabin, and although they had a somewhat heated exchange, she begged Clementine not to leave. But Clementine was not to be moved – she had made her plans, and politely but firmly she stuck to them. Accordingly, the next day she sailed for Britain, arriving home on 7 February. Despite her sudden departure under these strained circumstances, her long friendship with Walter Moyne remained unaltered.

* * *

There were seven months of uneasy peace left to run, and although one could not know the day or hour, all of us knew that time was short. Despite the feeling that the great storm was gathering on the near horizon, from a family point of view the spring and summer of 1939 were particularly happy months, and my mother, who was in much better health, felt in a mood to relish life at home. She spent a lot of time at Chartwell with me: I was basking in the glow of parental praise and approval at this time, having, to everyone's astonishment (including my own), passed my School Certificate with considerable success. This normal and modest achievement was rewarded by the present of a horse of my own, which caused me ecstatic delight. My mother dug out her own old hunting clothes, and later in the season we went out with the Old Surrey and Burstow hounds together on several occasions. A further treat was when she took me for a wonderful weekend to Paris at my summer half-term, where we packed into three lovely days an orgy of sightseeing and several plays.

In August my father made a tour of the Maginot Line at the invitation of the French Government, and afterwards he and my mother and I went to stay with Consuelo and Jacques Balsan at St Georges Motel, near Dreux. Appreciation of those halcyon summer days was heightened by our consciousness that the sands of peace were fast running out: there was swimming and tennis (so greatly enjoyed by Clementine) and *fraises des bois*; Winston painted several lovely pictures of the beautiful old rose-brick house and grounds, in company with Paul Maze, who was staying with his family at Le Moulin on the estate. We visited Chartres Cathedral and were drenched in the cool blueness of the windows: 'Look thy last on all things lovely every hour.'*

But agreeable though the company was, here again political tensions lurked only just below the surface. Paul Maze recounted in his diary on 21 August 1939 how he had

> Dined at the Château. Winston was fuming but with reason as the assemblée didn't see any danger ahead.
>
> As Charteris† was walking upstairs to go to his bedroom he shouted to me, 'don't listen to him. He is a warmonger.' He [Winston] was depressed as he left.[13]

On 23 August Russia and Germany signed a pact of mutual non-aggression. Hitler, his Eastern frontier now made safe, could turn on Poland; but Chamberlain had issued a solemn warning that in this eventuality Britain would stand by its guarantee to Poland. We had to go home: my father left at once for London by air. My mother and I followed the next day, and as we passed through Paris on that golden summer evening, the Gare du Nord teemed with soldiers – the French army was mobilizing.

* From 'Fare Well' by Walter de la Mare.
† The Hon. Sir Evan Charteris, KC (1864–1940).

CHAPTER NINETEEN

Year of Destiny

CLEMENTINE ARRIVED HOME WITH TEN DAYS OF PEACE TO RUN. WAR preparations were now afoot in earnest. Many people still believed that some miraculous, last-minute happening might yet stave off the terrible reckoning, but in the early hours of 1 September Germany fell upon Poland with all its might by land and air. The next day the mobilization of British troops was ordered, and, acknowledging that war now seemed certain, the Prime Minister sent for Churchill and invited him to join his small War Cabinet.

The final ultimatum to Germany expired at 11 a.m. on Sunday, 3 September 1939, and at 11.15 a.m. on a bright, breezy morning Chamberlain broadcast to the country and to the world that Britain was at war. Winston and Clementine listened to this dread yet, to them at least, long-expected news at Morpeth Mansions. The Prime Minister had hardly ceased speaking when the banshee wailing of the air-raid sirens, soon to become so familiar, was heard. Winston described how Clementine 'came into the room braced by the crisis and commented favourably upon German promptitude and precision'.[1] They went up onto the roof to see what was going on, and noted thirty or forty anti-aircraft barrage balloons slowly rising. 'We gave the Government a good mark for this evident sign of preparation, and as the quarter of an hour's notice which we had been led to expect we should receive was now running out we made our way to the shelter assigned to us, armed with a bottle of brandy and other appropriate medical comforts.'[2] However, this dramatically timed alarm proved to be a false one, and the 'All Clear' sounded about ten minutes later.

The House of Commons assembled briefly, and afterwards the Prime Minister again sent for Churchill, and offered him the office of First Lord of the Admiralty with a seat in the War Cabinet.

Winston reported to the Admiralty at six o'clock that evening. It was a moment fraught with emotions and memories, of which Winston wrote: 'So it was that I came again into the room I had quitted in pain and sorrow almost exactly a quarter of a century before.'[3] The Admiralty saluted his return as First Lord in the terse signal sent to the Fleet: 'Winston is back!'[4]

Admiralty House was once again to be the Churchills' home. When they had last lived there, for reasons of economy they had used only part of this beautiful, grand house; now the exigencies of another war made it sensible that they should once more inhabit only part of it. The Office of Works speedily made ready the top two floors of the house, and there Winston and Clementine and I were soon installed, in the nurseries and attics. Indeed, their drawing-room still had the chintz curtains with the gay design of red and blue seahorses, hung by Diana Cooper to divert their young son during Duff Cooper's tenure as First Lord a year or two earlier.

Not only had the flat in Morpeth Mansions to be disposed of, but a decision had soon to be taken about Chartwell. At the end of August the evacuation of women and children from London had begun, and a few days later Chartwell received its party of evacuees – two mothers with seven children between them, from the East End. Like nearly all those families who left their homes in the first official evacuation, these drifted back to London after about three weeks, preferring their own homes, husbands and familiar scenes, notwithstanding the risk of bombing, to the loneliness and intolerable boredom of country life. It was obviously impracticable to keep the big house open now, but the three-bedroomed cottage which Winston had been building at the bottom of the orchard during the last two years (with this situation partly in view) was nearly finished; they therefore decided to shut up the big house, and were soon able to use Orchard Cottage for fleeting weekends. Cousin Moppet was put in charge of the property, and she moved into the former chauffeur's cottage with Diana's two small children, Julian and Edwina, and their nanny, who had been evacuated from London.

When war was declared people braced themselves for sudden and terrible events to overtake the country. From afar we witnessed the grim annihilation of Poland; and then there ensued months of almost uncanny inactivity on land and in the air, causing this period to be known as the 'twilight war'. At sea, however, hostilities began at once: on the night of 3 September, a German U-boat torpedoed the outward-bound liner *Athenia*, which sank with heavy loss of life; in October the *Royal Oak* was torpedoed as she lay in the supposed safety of Scapa Flow. By the middle of November 60,000 tons of British shipping had been sunk by deadly magnetic mines. These dire events focused attention on the man who headed naval affairs. At the end of September, Clementine wrote to her sister Nellie:

> The war news is grim beyond words. One must fortify oneself by remembering that whereas the Germans are (we hope) at their peak, we have only just begun.
>
> Winston works night & day – He is well Thank God & gets tired only if he does not get his 8 hours sleep – He does not need it at a stretch but if he does not get that amount in the 24 then he gets weary.[5]

Winston's need for sleep to recoup his energies, and his capacity for going off quickly into a deep slumber, formed from the earliest weeks of the war a habit which he was to continue throughout the crisis years, namely, his insistence upon taking an hour's rest some time during the course of every afternoon or early evening. He undressed and went right to bed – and awoke a giant refreshed, and able to work after dinner long into the night. This regular siesta undoubtedly contributed to his ability to support the burdens of five long years of war: it enabled him, as he wrote himself, 'to press a day and a half's work into one'.6 But what was for him a life-saver placed many of those who worked closest to him under additional fatigue and strain. Nevertheless, most of those affected by this unusual habit accepted the inconvenience to themselves with understanding and loyalty, and Clementine and his close entourage protected his rest hours with determination.

With remarkable speed my mother seemed to get our new home running smoothly and comfortably. From the first days there was a constant stream of people for lunch or dinner; very few of them were social acquaintances or friends, nearly all political and service colleagues. For Winston, mealtimes were merely an agreeable extension of working hours – a golden opportunity to get to know a new colleague better, or to continue to pursue a line of thought.

But now over all conversations hung the shadow of secrecy: the servants had to be vetted by the security service; and the presence of strangers, or most friends, however close, put a stop to nearly every subject. Clementine consequently had to plan any social entertaining with particular care, and, indeed, it was reduced to the bare minimum, with 'outsiders' grouped together. There was the small 'golden circle' of trusted colleagues, and nearest relations known to be 'padlock', to whom, of course, that trust was sacred.

Our domestic life, perched on the top two floors of Admiralty House during the opening months of the war, was very close and happy. Diana was serving in London as an officer with the Women's Royal Naval Service (WRNS), and her husband, Duncan Sandys, a Territorial Officer, was stationed with his Anti-Aircraft Regiment in the London area. Sarah and Vic had a flat in Westminster Gardens, and were busy and happy acting. I was seventeen, and having just left school was revelling in my first taste of London and 'grown-up' life. While continuing my education part of the time, I also worked in a canteen and a Red Cross workroom. Living at home, I was able to help with the entertaining, and to be a companion and confidante to my mother.

In late September Randolph, who was in his father's old regiment, the 4th Hussars, became engaged to Pamela Digby, the elder daughter of Lord and Lady Digby. Since his regiment might be ordered abroad at any time, the marriage took place very quickly, on 4 October, in London. The wedding reception was held at Admiralty House, where the state rooms were opened up for this joyous family occasion.

During the eight months Winston was at the Admiralty Clementine took a close interest in matters concerning naval welfare, and shared in much of the work encouraged and organized by the Navy wives. In January 1940 she made a public appeal for contributions to a Minesweepers and Coastal Craft Fund. These small ships, hurriedly commandeered to meet the emergency, and collected in large numbers in small ports, had few of the comforts on board or the amenities ashore that were available to bigger, regular service ships. Clementine's appeal touched the particularly warm spot in the British public heart reserved for the Navy: considerable sums flowed in, and she kept a close personal interest in the work of the Fund. Diana Cooper recorded, after a visit to her at Admiralty House in the early months of 1940, 'She makes us all knit jerseys as thick as sheep's fleeces, for which the minesweepers must bless her.'[7] In November 1939 Clementine also began her work for the Fulmer Chase Maternity Home for the wives of officers of all three services, with which she became deeply involved.

During these autumn and early winter months, Britain had been slowly adjusting to the new conditions of life. People had expected action, violence and suffering: instead, they had to stumble around in the darkness of the blackout, and were asked to conform every day to more and more instructions from Whitehall. They complied with considerable goodwill and an astonishing degree of self-discipline. No one doubted that sterner days would come, but in the meantime people tried to remember to carry their gas-masks and identity cards, and to 'Talk Victory', as recommended from the hoardings. The news from the only real scene of action – the war at sea – was almost unrelievedly disastrous and depressing. But as the first year of the war drew to its close, the anxious dreariness was dramatically interrupted by the Battle of the River Plate*, which ended with the scuttling of the German battleship *Graf Spee* after a fierce encounter with three British cruisers. The suspense at Admiralty House was intense, as we followed the battle message by message. It was a glorious victory, and brought a gleam of light into a dark December.

We spent this first Christmas of the war in London, gathering around us quite a number of the family, as well as two old and valued friends and 'Chartwell regulars', Brendan Bracken and 'The Prof'. The New Year of 1940 opened quietly, and towards the end of March Winston accompanied Clementine when she went up to Barrow-in-Furness to launch an aircraft carrier, the *Indomitable*. There are few more moving, thrilling sights than the launching of a great ship. One of the best photographs ever taken of my mother was of her waving the newly named ship away; it became my father's favourite, and after the war he painted a charming sketch-portrait from it.

* * *

The trance-like atmosphere of the 'twilight war' was abruptly shattered on 9 April 1940, when Germany leapt upon Denmark and Norway. Denmark

* Rio de la Plata, the South American river estuary lying between Argentina and Uruguay, both neutral countries.

collapsed at once, but the Norwegians fought the invader with desperate heroism, appealing to Britain for help. A naval and military force was hurriedly despatched, but they could make no impact on the German defences, and were exposed to the deadly harassment of the Luftwaffe. The consequences of the failure to heed Churchill's warnings about the might of Germany's air power were tragically clear: within three weeks all British troops had been evacuated except those carrying on a desperate but doomed struggle to seize the port of Narvik.

The disastrous campaign in Norway brought home dramatically the consequences of irresolute leadership and inadequate planning. When on 7 May Chamberlain opened a two-day debate in the House of Commons on the Norwegian campaign, bitter and damaging speeches were made against the Government and the Prime Minister from all sides of the House. Churchill, who felt he bore a major responsibility for the events in Norway, asked to be allowed to wind up the debate, and made a spirited defence of the Government, although it included many who had turned a deaf ear to his warnings and had opposed him with determination and sometimes hostility in past years. The closing stages of the debate were passionate and noisy: 41 Conservative backbenchers came out in open revolt and, when the House divided, the Government's majority had slumped from 240 to 81.

Faced with such a loss of confidence even within his own party, Chamberlain sought to discover whether the Opposition parties would be prepared to serve in a National Government: the answer was that both the Labour and Liberal parties were ready to join a National administration – but only under a new Prime Minister. Chamberlain's first hope was that his Foreign Secretary, Lord Halifax, might succeed him, but Halifax demurred, feeling that in such a situation the Prime Minister must be in the House of Commons. It became manifestly clear as the hours wore on that Churchill was emerging as the only acceptable leader around whom all the major parties would rally in this dire crisis.

On the night of 9–10 May, German forces invaded the Netherlands and Belgium. These events clinched the political situation at home. At six o'clock on the evening of 10 May Chamberlain went to Buckingham Palace to tender his resignation: half an hour later the King sent for Churchill and asked him to form a Government. That same evening Churchill returned to the Palace to submit the first list of names of members of the new administration for the King's approval.

During these tense and anxious days Clementine was away from London; her brother-in-law, Bertram Romilly, had died at his home in Herefordshire on 6 May, and two days later she travelled north to be with Nellie, to whose sorrow great anxiety had been added by news of the capture at Narvik of her elder son, Giles, a war correspondent for the *Daily Express*. It was anguishing for Clementine not to be with Winston during these days; and he, sensing that events were moving towards a climax, telephoned, asking her to return as soon as possible. Accordingly, the day following her brother-in-law's funeral, she travelled back to London, arriving at Admiralty House shortly before Winston left to go to Buckingham Palace to receive the King's command to form a Government. And so they were together to share this strange and grim apotheosis after the years of exile from power and unheeded prophecies. Great events always found Clementine at her best, and at this moment Winston was as calm as he was resolute. He would later write: 'I felt as if I were walking with destiny, and that all my past life had been but a preparation for this hour and for this trial.'[8]

Churchill lost no time in constructing his Government, which included members of all the three principal parties, and whose scope was widened to include men of ability and stature from outside the political field. On 13 May he addressed the House of Commons for the first time as Prime Minister, ending his speech with a peroration which struck the keynote of his leadership:

> I would say to the House, as I said to those who have joined this Government: 'I have nothing to offer but blood, toil, tears and sweat.'
>
> We have before us an ordeal of the most grievous kind. We have before us many, many long months of struggle and of suffering. You ask, what is our policy? I will say: It is to wage war, by sea, land and air, with all our might and with all the strength that God can give us; to wage war against a monstrous tyranny, never surpassed in the dark, lamentable catalogue of human crime.[9]

That afternoon Germany attacked France across the Meuse, and the next day came the ominous tidings that German tanks had broken through the French line at Sedan, which lay in the gap between the northern end of the Maginot Line and the British-fortified front along the Belgian border. This thickly wooded region of the Ardennes had been considered an impassable natural barrier against the advance of large bodies of troops or vehicles. Unhappily this comfortable theory was abruptly disproved as the Panzer units poured through the fatal gap, driving a wedge between the Allied armies.

In the early morning of 15 May, Paul Reynaud, the French Prime Minister, telephoned Churchill and told him: 'We have been defeated.'[10] Churchill was not inclined to accept so early the idea of defeat, and he flew to Paris the following afternoon. This was the first of five flights he was to make to France that May and June: he made three visits to Paris, and after the French Government left the capital, he flew once to Briare, near Orléans, and finally to Tours. On each occasion he tried to encourage and rally our hard-pressed and increasingly demoralized ally and to offer the French such practical help as lay within our power. He always stressed that Britain would fight on 'whatever happened or whoever fell out'.[11]

Clementine shared to the full Winston's hopes and fears for the French army and nation, and grieved at the growing realization that not only the military might of France, but also the country's spirit, was being crushed on this anvil. His journeys to France, decided always on the spur of the moment, caused her sharp anxiety, for they involved a considerable degree of risk: usually escorted by guardian Spitfires, on one occasion he had to leave France and return without his escort; during this flight his aircraft had to dive down to about 100 feet above sea level to avoid attracting the attention of two German planes which were strafing fishing boats. On his last visit, on 13 June, it was found that a severe bombing raid at Tours airport during the previous night had left large craters in the runways, making landing difficult and dangerous. Although Clementine knew the dangers of these flights, she never tried to dissuade Winston, nor did she give way to her inner fears. In a favourite phrase of his, she knew how to 'rise to the level of events', and her strongly disciplined nature enabled her at times like these to carry on with all the seemingly petty, but necessary, concerns which compose so much of life.

Less than five days after breaking through the French defences the German forces reached the north-east coast of France, thus cutting off the Allied armies in Flanders. On 27 May the order was given to evacuate the maximum force possible. The surrender of the Belgian army on 28 May left the British flank exposed, with no alternative but to make a dash for the sea, through a narrowing corridor held open with desperate heroism. Churchill told the House of Commons on 28 May that it should prepare itself for 'hard and heavy tidings',[12] as the Allied troops fell back fighting all the way to the cruelly exposed beaches of Dunkirk. In the course of the next few days over 338,000 British and Allied troops were embarked and ferried across a benignly smooth Channel in vessels that ranged from destroyers and sloops to merchant vessels, pleasure launches, yachts and fishing boats. The whole of Britain and the Commonwealth watched and waited and prayed as the saving miracle was accomplished.

On 4 June, my mother and I went to the House of Commons and listened while my father told of this great epic. His speech contained the first warning to the country that we might soon be left alone to face the storm, and he spoke of our determination and ability to 'defend our island home . . . if necessary for years, if necessary alone'. He ended his speech with words that have passed into history:

> [W]e shall not flag or fail. We shall go on to the end. We shall fight in France, we shall fight in the seas and oceans, we shall fight with growing confidence and growing strength in the air; we shall defend our Island, whatever the cost may be. We shall fight on the beaches, we shall fight on the landing-grounds, we shall fight in the fields and in the streets, we shall fight in the hills; we shall never surrender.[13]

Events continued to move relentlessly and swiftly: on 10 June Italy declared war on France and Britain; on 14 June the Germans entered Paris; and eight days later France capitulated. Winston and Clementine, with their long attachment to France and its people, witnessed with intense grief and dismay the downfall of that great country. But if most of the French leaders were at this time lacking in the will to endure, and failed to galvanize their divided and harassed countrymen into a firm and united resolve to continue the fight, the true spirit of France lived on in the person of General Charles de Gaulle who, on the day following Reynaud's resignation and the assumption of power by the ancient and defeatist Marshal Pétain, flew to England in a British plane. He raised the standard of 'La France Libre', and in his now famous broadcast of 18 June called on all French men and women to rally to him.

It is difficult to recapture the atmosphere of those days and weeks; there were indeed to be many more anxious times, and many hard tidings to bear in the course of the next five years; but never again, I think, did one feel one could scarcely breathe. We got through the days living from news bulletin to news bulletin, and dreading what each one might bring.

* * *

As soon as Winston became Prime Minister, Clementine began the arrangements for their move to No. 10 Downing Street. Throughout the succeeding weeks of oppressive anxiety and unrelentingly disastrous news, she carried on with all the necessary plans, and they were installed there by 17 June. 'No. 10' is a charming house, where the Prime Minister both lives and works. On the ground floor are the Cabinet Room and the Prime Minister's Private Office; on the first floor is a series of dignified communicating rooms, including a large state dining-room, and a delightful (then) white-panelled passage room, where we always lunched or dined when we were not more than eight people. Upstairs, the bedroom floor, with its eggshell-blue passages, cheerful red carpets and pleasantly sized rooms with sash windows looking out over the garden and Horse Guards Parade, gave the impression of a country house.

During the spring months of this year, Winston and Clementine had been able to spend quite a number of weekends picnicking in the new Orchard Cottage at Chartwell; but the visits there were to become fewer and fewer, partly on the grounds of safety once the bombing of London started, for Chartwell was on the direct target route, and easily identifiable from the air. Also, Winston now had at his disposal the Prime Minister's official country residence, Chequers, near Aylesbury in Buckinghamshire – a large Elizabethan house, fully staffed and run by a charming and efficient Scottish curator, Miss Grace Lamont. Winston and Clementine spent their first weekend at Chequers on 1 June, while the evacuation from Dunkirk was reaching its climax.

The fact that Winston might be in the country made little difference to his programme of work or his contact with events. Chequers was large enough to accommodate in comfort a skeleton Private Office, two or three telephone operators, the detectives, chauffeurs and despatch riders, as well as family members and up to eight guests. Quite soon the pattern for the weekends took shape: there was nearly always a nucleus of the family, and occasionally a close friend or two, and superimposed on these intimates would be a succession of service chiefs, Cabinet colleagues or specialists in various subjects, who were bidden to 'dine and sleep'.

Since all weekends were, as far as Winston was concerned, working days, most guests were invited on a business basis, and it was rare for wives to be included. Many an overworked general, civil servant or Government colleague was reft away from a weekend in the bosom of his own family, to spend perhaps a gruelling twenty-four hours at Chequers. Nevertheless, their sacrifice was not in vain, for it would be quite impossible to overestimate the value of these working weekends, which often resulted in far-reaching and vital decisions, or in a closer understanding between Churchill and those in whose minds and hands lay the waging of the war. Chequers was also invaluable for entertaining important guests from overseas, who particularly appreciated the country-house atmosphere and the lovely surrounding countryside.

At first the original civilian domestic staff grappled with the almost continuous weekend parties, but as the war went on, it became impossible to find servants, and, moreover, the requirements of security became increasingly stringent. It was therefore decided that it would be a proper use of womanpower for Chequers to be staffed by members of the women's services, and so volunteers from the Auxiliary Territorial Service (ATS) and the Women's Auxiliary Air Force (WAAF) presently took over the running of the house.*

* * *

After the fall of France, Britain stood alone. The greater part of our Army had been, almost miraculously, snatched from destruction or captivity, but it had lost nearly all its arms and equipment in the flight to the sea and the evacuation from the beaches. From the moment the conquest of France was complete, we knew that this island was in daily increasing danger of invasion by the Germans. The fate of Poland, and now more recently of the Low Countries and of France, had left people in no doubt as to what they might expect if our turn came. But the British accepted this dark prospect with characteristic phlegm; they bent themselves now, day and night, to remedy the neglect of the purblind years, and everywhere preparations forged ahead to repel an enemy onslaught.

There occurred in early July a sequence of events which, more than any other, impressed upon the world at large that Britain was in deadly earnest, and would stop at nothing in its fight for survival. Following upon the collapse of France, the vital question arose as to the fate of the French fleet. It was a matter of the utmost importance that these powerful ships should not pass into enemy control. Had this come to pass, the balance of power at sea would have been disastrously – perhaps fatally – altered. When all hope

* The practice has continued to this day.

had faded that a large part, if not all, of the French fleet would sail for Allied or neutral ports, the War Cabinet took a drastic decision: on 3 July, the French ships lying in British harbours were taken over by swift, determined action, and at Oran in North Africa – after fraught parleyings – the Royal Navy opened fire on the French fleet: three battleships were destroyed, with the loss of 1,300 lives, and the remaining French ships at Oran and in other North African ports were either destroyed or immobilized.

It must have been just at the time of these searing events – the painfulness of which no one felt more keenly than Winston himself – that General de Gaulle lunched at Downing Street. The conversation turned to the future of the French fleet, and Clementine said how ardently she hoped that many of its ships and crews would carry on the fight with us. To this the General curtly replied that, in his view, what would really give the French fleet satisfaction would be to turn their guns 'on you!' (meaning the British). Clementine from the first had liked and respected this dour man – but she found this remark too much to bear and, rounding on him, rebuked him, in her perfect, rather formal French, for uttering words and sentiments that ill became either an ally or a guest in this country. From the other side of the table Winston sensed that something had gone amiss and, in a conciliatory tone, said to the General: 'You must forgive my wife. *Elle parle trop bien le français*.' Clementine interrupted him, and said in French: 'No, Winston, it is because there are certain things that a woman can say to a man which a man cannot say, and I am saying them to you – General de Gaulle!'[14] After this verbal fracas, the General was much upset, and apologized profusely. The next day he sent her a huge basket of flowers.

Throughout the ups and downs of Winston's relationship with this awe-inspiring, brilliant and difficult man, Clementine never wavered in her esteem and liking for him; her sentiments were obviously returned, and their mutual regard for one another lasted on long after the war. Clementine also made friends with Mme de Gaulle; she greatly respected the dedication and dignity of this woman, the circumstances of whose life during the war years were particularly lonely and difficult.

Clementine was keenly aware of the demoralized state and unhappy situation of the many French servicemen who now found themselves in this country, many of them having been evacuated with our troops from the Dunkirk beaches. Some joined the newly forming Free French forces straight away; for others the choice was more complicated, and they felt a clash of loyalties. They were offered the alternative of repatriation, and while waiting for arrangements to be made they were accommodated in the White City Stadium, which had been turned into a temporary camp for them. Urged by friends who were trying to help these bewildered and downcast people, Clementine visited White City in August. It made a painful impression on her to see the looks of doubt or despair, and even glances of downright hostility, on so many – for the most part young – faces. She talked to some of the men, and came home sad and troubled. For the British, life had become very simple: we meant to fight; we thought we would win;

but we would fight anyway – and we were spared the agonies of divided loyalties or complicated issues.

It is easy, now that we know the ending of the story, to under-estimate the grim burden of anxiety and responsibility that lay on Churchill's shoulders in those days of June 1940. During these bright summer months the world watched, while our fate hung poised in an agonizing balance. He never doubted the determination and courage of the British people to endure whatever might come, but he knew the deadly nature of the onslaught which was being prepared just across the Channel; and, above all, he knew our nakedness. It is idle to pretend that the knowledge of all this did not weigh on him. He drove himself, and he drove others, with a flail in his desire to prepare the country for the assault which seemed certain to be unloosed. In his determination to drive on through all difficulties he must at this time have become extremely overbearing and tyrannical to many of those who served him, as the following letter from Clementine shows:

10 Downing Street,
Whitehall
June 27, 1940

My Darling,

I hope you will forgive me if I tell you something I feel you ought to know.

One of the men in your entourage (a devoted friend) has been to me & told me that there is a danger of your being generally disliked by your colleagues and subordinates because of your rough sarcastic & overbearing manner – It seems your Private Secretaries have agreed to behave like school boys & 'take what's coming to them' & then escape out of your presence shrugging their shoulders – Higher up, if an idea is suggested (say at a conference) you are supposed to be so contemptuous that presently no ideas, good or bad, will be forthcoming. I was astonished & upset because in all these years I have been accustomed to all those who have worked with & under you, loving you – I said this, & I was told 'No doubt it's the strain' –

My Darling Winston. I must confess that I have noticed a deterioration in your manner; & you are not as kind as you used to be.

It is for you to give the Orders & if they are bungled – except for the King, the Archbishop of Canterbury & the Speaker, you can sack anyone & everyone. Therefore with this terrific power you must combine urbanity, kindness and if possible Olympic [*sic*] calm. You used to quote: – 'On ne règne sur les âmes que par le calme –' I cannot bear that those who serve the Country & yourself should not love you as well as admire and respect you –

Besides you won't get the best results by irascibility & rudeness. They will breed either dislike or a slave mentality. (Rebellion in War time being out of the question!)

Please forgive your loving devoted & watchful
Clemmie
[drawing of cat]

This brave letter must have cost Clementine dear to write; indeed, the post-script is most revealing: 'I wrote this at Chequers last Sunday tore it up, but here it is now.' No written answer from Winston exists. But he cannot have been enraged, for the letter has survived. Many wives might never have had the perception or the courage to write in such a way: and the letter surely bore fruit, for although Winston could be formidable and, indeed, over-bearing, many people who served him at all levels during those hard years have put on record not only their respect and admiration for him as a chief, but also their love for a warm and endearing human being.

But the strain of all the events of the past months was not weighing only on Winston – Clementine also was in a state of tension, as is shown in the account of an extraordinary outburst to Cecil Beaton, who came to No. 10 to photograph both Winston and her on separate occasions in the summer. The afternoon's photography with Clementine, as described by Beaton in his diary,[15] passed 'breezily and easily'. His observant eye noticed that 'the rooms were a delight with the sun streaming in from beneath the blinds onto bowls of sweet peas from Chartwell . . . Mrs Churchill, a bright, unspoilt and girlish woman, is full of amusing and shrewd observations about people.' He photographed her in evening dress, sitting on a sofa, 'her hair set for the occasion like Pallas Athene', and also in a simple summer dress pouring out tea for them both. But the picture in evening dress eventually published by *Picture Post* caused her dismay; an obliging friend had told her it made her look 'like a hard-bitten virago who takes drugs'![16] And parts of the accompanying article (for which Beaton had no responsi-bility) greatly upset her. When he came again to Downing Street to show Winston the photographs he had taken of him and his grandson 'little' Winston, he also saw Clementine – who took him to task in an astonishing and painfully embarrassing scene, which he described in his diary:

> She was working herself into an even greater state of nervous hysteria. Her face flushed, her eyes poured with tears, 'Really it's too damnable. It isn't as if my life has been too easy. It hasn't but when I married Winston he loved me,' she blubbed and I held her hand and comforted. 'But he still <u>does</u>. We all know that!' She let herself go and for me it is so awful a sight to see some-one weeping that I kissed her on the forehead and held tight her hand. But instead of 'coming round' she wept more uncontrollably and confided, 'I don't know why it is, but I suppose my friends are not exactly jealous but they think that other people could do the job better and that I shouldn't have been married to Winston. After all he is one of the most important people in the world. In fact he and Hitler and President Roosevelt are the most important people in the world today.' This was embarrassing for me, but it was no use to argue or split hairs, but now I felt the moment had come for her to behave with more dignity. She was entirely abandoning herself to a complete stranger. It was really rather reprehensible.[17]

One of Clementine's characteristics was that after one of her 'electric' storms, she could recover herself amazingly quickly; and she did so on this occasion. But looking back as we now can on what she would achieve in the war in the area of her public work, and how widely and genuinely she would become respected and admired as the wife of the great wartime Prime Minister, this episode reveals how painfully lacking in self-confidence she was at this point.

Her composure restored, Clementine largely retrieved the embarrassing situation by admiring the photographs of her husband and grandson. Cecil Beaton 'left exhausted. I took a taxi to a flower shop where I bought her a wonderful bunch of dark red and white roses, violets, yellow orchids and dove carnations. Such a bouquet to touch any heartstrings. I wrote an affectionate note, and got an ecstatic telegram reply.'[18]

* * *

On 10 July, the full force of the Luftwaffe began to be directed against Britain. In the first phase the enemy drew our air forces into battle over the Channel and south coast. Then the southern counties, and particularly the airfields, were the object of daily raids by large numbers of aircraft; and finally London itself became the principal target. The Battle of Britain has passed into history – perhaps it will be the last example of war by individual combat. People stood in the summer meadows and watched, while overhead desperate aerial fights raged, fought out by a few hundred young men, many of whom had joined the Air Force for fun, and found themselves the champions of all that free men and women hold dear.

During August and September, the invasion of these islands was expected by the mass of the population with the same sort of grumbling acquiescence with which farmers anticipate rain and storms to spoil the hay harvest. The daily papers abounded with helpful hints for disastrous contingencies; the cartoonists were never funnier. But what had been a virtually undefended island in mid-June presented a much more formidable proposition even by the end of July, and every day our strength increased and our organization improved. On 7 September, the German air force began its series of raids on London; the 'Blitz' had begun, and for fifty-seven nights in succession an average of 200 German bombers attacked London. Despite the toll of death and injury, the destruction of buildings and the general dislocation of daily life, the German objective of making this vast city uninhabitable and of breaking the spirit of its inhabitants remained unachieved. 'London can take it' was the popular slogan. It also became historical fact.

Soon after we were installed in No. 10, changes had to be made again owing to air raids and the extremely unsafe nature of the Downing Street buildings. Below the ground floor of No. 10, on the garden level, were the kitchens, the air-raid shelter and several offices, normally inhabited by secretaries known as 'the Garden Room girls'. These offices were now converted into a sitting-room and dining-room, after being suitably strengthened by great beams used in the fashion of pit-props, and with steel shutters on the windows. Thus fortified, these rooms were not without their own charm, giving the impression of a ship's wardroom, and they certainly made it possible for social life to continue in a seemly way, despite constant alarms and not infrequent bombs. These arrangements, however, were soon deemed to be inadequate: the whole fabric of No. 10 was too frail, and the air-raid shelter itself was considered too small and unsafe for those living and working there. This opinion was reinforced by an incident which occurred in mid-October, while Winston was entertaining several colleagues to dinner. An air raid was in progress, and a bomb fell quite close, out on Horse Guards Parade. Winston, on a sudden impulse, left the table and went into the kitchen, where the cook, Mrs Landemare, and her kitchen-maid were working, and ordered them both to go, with the other servants, to the shelter; this they all somewhat reluctantly did. A few minutes later, a bomb fell very much closer, shattering the great glass window of the kitchen and turning the kitchen into a shambles of broken wood, fallen plaster and splintered glass. Answering a letter at this time from Violet Bonham Carter, Clementine wrote:

10 Downing Street
Whitehall
26th October 1940

Dearest Violet

The bomb on the Treasury was very bad and, alas, killed two Civil Servants and two messengers. Their bodies were retrieved only last night.

We have a very comfortable shelter at No. 10. We do not hear much noise, but I fear it is not safe as it is on a level with the garden. But it makes one feel safe and that is a great thing.

I think it is very courageous of you to be a Warden, and what you tell me about Bongie [her husband, Sir Maurice Bonham Carter] fills me with admiration. Please give him my love. I should very much like to see you soon, but we have been 'blown' out of Downing Street and are living in two rooms – one of them your former sitting-room looking on to the garden. We have no gas or hot water and are cooking on an oil stove. But, as a man called out to Winston out of the darkness the other night, 'It's a grand life if we don't weaken!'

Yours very affect.ly
Clemmie

A secure underground complex – the Central War Rooms* – situated under the modern and extremely solid block of government offices next to Storey's Gate in Westminster, had already been prepared in 1938 in anticipation of heavy aerial bombardment. Further strengthened, by July 1940 this vast underground warren included meeting rooms, map rooms, cypher offices and a telephone switchboard. Here the War Cabinet and the Joint

* Now the Cabinet War Rooms, part of the Imperial War Museum, and open to the public.

Planning Committee would hold their meetings; additionally the Home Command had its HQ there. A limited number of emergency bedrooms were available for key personnel, and a simple suite for the Churchills. (Winston, in fact, slept there on only three occasions.)

The Central War Rooms thus formed a subterranean headquarters, and would have been a 'last ditch' redoubt in the event of invasion; but they were no solution to the day-to-day living and sleeping arrangements for Winston and Clementine once No. 10 became unfit for habitation. However, an admirable solution to this problem was achieved by the swift conversion into living apartments of offices overlooking St James's Park, immediately above the War Rooms, with which they communicated directly by an internal staircase. At first the Churchills and their immediate staff only slept there, continuing to entertain in the fortified Garden Rooms at No. 10, while Winston continued to do business as much as possible in the Cabinet Room. But towards the end of 1940, conditions were such that it was both safer and easier to transfer the whole of their life to the No. 10 'Annexe' (as the new flat had come to be called) until the air raids eased up.

Typically, in a short time, Clementine managed to impress her own taste on the gaunt and unprepossessing rooms, having them painted in pale colours and hung with well-lit pictures, and with much of their own furniture, they became almost attractive. The flat had a somewhat public quality about it, since the rooms gave off the main passage which connected various government offices, and embarrassed officials would often encounter Winston, robed like a Roman emperor in his bath towel, proceeding dripping from his bathroom across the main highway to his bedroom. However, 'The Annexe' served very well, and there they lived until July 1945, snug and secure.

Anxieties were also felt at this time for Winston's weekend arrangements. The very fact that he proceeded every weekend to Chequers was thought to hold dangers; moreover, Chequers was easily identifiable from the air. So an alternative base was sought, and found, in Mr and Mrs Ronald Tree's beautiful house Ditchley, in Oxfordshire, quite close to Blenheim.

Ronnie Tree was American by birth, but had been brought up wholly in Britain and was a British subject; he had been for many years a Conservative Member of Parliament. Although not until now an intimate friend of Winston and Clementine, he had long supported Winston's campaign for rearmament. His beautiful and gifted Virginian wife, Nancy, was a daughter of the one of the famous Langhorne sisters, and thus a niece of Nancy Astor. The Trees welcomed not only Winston and Clementine and their essential entourage, but also their official guests and members of their family. Whenever, therefore, the moon was high, we all repaired to Ditchley. We were becoming so accustomed to the drabness and ugliness of war – khaki, mountains of sandbags, the blackout, and the dust and desolation of ruined buildings – that we gazed with keener appreciation on elegance and beauty, and glowing, lighted interiors.

* * *

As soon as Winston became Prime Minister, Clementine's already busy life became busier still. Her field of public activity was no longer mainly concerned with his departmental office, but ranged widely over many aspects of life in wartime England. Her mailbag doubled; the amount of entertaining increased considerably; and her care and vigilance for Winston, now on this awe-inspiring pinnacle of power, were even more necessary and demanding.

Nor was her vigilance merely for his physical well-being. Clementine's 'antennae' were sensitive to everything that bore relation to him – or, through him, to a wider sphere. An example of her watchfulness and the speed with which she was wont to act arose when she learnt that Winston's great-niece – Sally, the five-year-old daughter of Jack Churchill's elder son, Johnnie – was about to leave the country with a group of other children for the United States. In the early months of the war quite a number of children were sent to America and Canada to be sheltered from the impending storm by transatlantic friends or relations; and in June 1940, the Government had reluctantly introduced an official evacuation scheme. Churchill disapproved of it, describing it as a 'stampede from the country'*. Clementine, on hearing of Sally's impending departure, realized at once the effect it might have if it became known that a 'Churchill child' had left the country at such a time. Although the child was on the point of embarkation and the changing of her plans caused great inconvenience and some distress to her and to those who had made the arrangements, Clementine insisted (to the point of having her passport withheld) that in no circumstances could she be allowed to depart.

Since Winston trusted her completely, Clementine had an extra responsibility to bear: she often knew not only what had befallen, but what was about to befall, or what might befall. This burden of secret knowledge placed a barrier between her and other people; and while she was not, in any case, a woman of many friends or prone to chattering, she tended now to shun even the few close and trusted friends she had. Sylvia Henley told me how often, during the course of the war, she felt that Clementine literally 'pushed her away'. She did not want to talk banalities with Sylvia, and she was duty-bound not to talk anything else; and she perhaps feared that her tongue might betray her with so intimate a friend. All these pressures and considerations tended to make the small family circle draw even closer together.

Despite the separations and dislocations of war, our family life still revolved around the lodestar of our parents. They always rejoiced to see any one of us, whenever we could turn up, but of course our father was more than ever preoccupied. Since Goonie Churchill was not well and was living

* The scheme came to a tragic end in September 1940 after the *City of Benares* was torpedoed, with the loss of seventy-three children (Longmate, *How We Lived Then*, p. 74).

permanently in the country, Jack came to live in Downing Street, and later somewhere was found for him in that veritable warren in the Storey's Gate building: self-effacing and discreet, he was the most loyal of brothers and was always there when he was needed; but he made himself scarce otherwise.

Randolph, temporarily home from abroad, was returned unopposed as a Member of Parliament for Preston in a by-election at the end of September 1940; my mother went to watch him take his seat in the House, sponsored by his father. His wife Pamela was expecting her first child, and she came to Chequers in the middle of September to await the birth in peace and quiet. There, on 10 October, 'little' Winston was born.

Sarah's life at this time was full of problems. Her marriage had run into difficulties which were sharpened by the fact that her husband, Vic Oliver, felt he should return to the United States; he was now an American citizen, but Austrian by birth, and a date-line was due to be set shortly with reference to his nationality status. To Sarah, the idea of leaving England and her own family at this moment was an agony, and she found herself torn by a cruel dilemma. During the autumn and winter of this year, she wrestled with this personal and painful problem. She confided her difficulties to her mother, and found support and understanding – if not a solution.

At the end of July, I had gone to Norfolk to stay with Venetia Montagu, whose daughter Judy and I were contemporaries; but my holiday was prolonged into many weeks, for when the Blitz started my mother arranged for me to stay on with my hospitable cousins. I expected to return to London at any time, but my parents were adamant that they did not wish me to live in London while the raids persisted – they had quite enough worries already. So presently I was packed off to Chequers; I was somewhat disconsolate at first, but soon was very busy working for the Women's Voluntary Service (WVS) in nearby Aylesbury. I used to look forward to the weekends when the great gloomy house came to life with the arrival of my parents and their guests.

During this first year of the war, two people came to help Clementine: Grace Hamblin, her Private Secretary, and Mrs Georgina Landemare, her cook. Neither was new to her: Grace, having been with us at Chartwell since 1933, now left her parents, with whom she lived in the country, and came to work for Clementine in London. She soon was a built-in part of my mother's private and official life, and her devotion, tact, efficiency and charm soon became a byword among all those with whom she had dealings.

Mrs Landemare was a superb cook, combining the best of French and English skills. She had learnt her craft the hard way, starting as No. 6 in the kitchen, over which reigned the French chef, Monsieur Landemare, whom she eventually married. Clementine had come to know and appreciate her talents and her delightful personality during the thirties, when she used to come to Chartwell for special parties or busy weekends to boost and teach the rather inexperienced cooks or promoted kitchen-maids whom Clementine could then afford. When they moved into Downing Street, Mrs Landemare came to cook for Winston and Clementine on a permanent basis.

Through all the difficulties of wartime rationing, she managed to produce delicious food. After the war she stayed with us until 1953 when she retired, aged seventy.

One aspect of Clementine's new role as wife of the Prime Minister was the great number of letters from the general public she received every day, all of which she read herself. Each one, whether complimentary, sensible, abusive or futile, received an answer, either personally dictated and signed by her, or written by Grace Hamblin on her behalf, after they had together discussed what action should be taken or what enquiries made. Clementine set particular store by these letters, and never grudged the time spent in dealing with them. Through them she felt she had a contact with ordinary people's everyday lives, at a time when harassments and aggravations were often caused by a host of rules and regulations, which were often the cause of more grievances than the more horrific or tragic aspects of war. Sometimes nothing could be done, but frequently Clementine was able to sort out some difficulty or injustice by making a personal enquiry to the right ministry or authority. All this involved a good many hours' work each week, and often several letters were necessary before a final answer could be given. Clementine was greatly helped by Grace Hamblin, who was herself deeply interested in the human problems thus brought to light, and threw herself into the sometimes tedious enquiries with zest and tenacity.

Clementine dealt with most of the problems raised herself, but from time to time she would edit the facts of a case, or of a series of similar cases, and put a businesslike memorandum in front of Winston; she was able in this way to draw his attention to conditions or regulations which seemed to be causing dissatisfaction or unnecessary hardship. Winston nearly always took up the cases she put before him; he knew her judgement was sound, and that she did not enlist his help without good reason. But while she was ready to seek Winston's aid over genuine grievances or valid requests, she was also a stout shield between him and tiresome people (usually 'fringe' friends, or distant relations), who wanted special (and usually unsuitable) favours, and who thought she would be their advocate and channel. Politely, but quite firmly, she would regret her inability to help; she could even on occasion be quite fierce.

It was through her postbag that Clementine was first alerted to the appalling conditions existing in many of the London air-raid shelters. She received a number of letters on this subject, and the complaints were borne out by one or two people she knew personally who were working in the shelters, who begged her to go and see for herself what the conditions were really like. As early as the end of June 1940, when air raids had just started sporadically in the provinces, Clementine had sent a minute to Winston's Principal Private Secretary asking him to find out for her what the various arrangements were to shelter people whose homes had been wrecked or damaged in raids, what welfare arrangements existed, whether there was any scale of compensation for injuries, and answers to other pertinent questions. In December she determined to investigate for herself the conditions under which so many people were spending their nights: accordingly, she made a series of visits to various London boroughs, calling in unannounced on as many of the shelters as she could. Her chief guide on these expeditions was Mrs May Tennant of the Red Cross; she was also often accompanied by Jock Colville*, a young diplomat and member of the Private Office, who as time went on was to become a close family friend, but who now was often irritated by, and critical of, Clementine, whom he described in his diary (30 October 1940) as being 'ridiculously overdressed in a leopard-skin coat' on one of these tours; still, he admitted that she 'was loudly acclaimed'.[19] The following year, again visiting shelters with her, in Stepney, he noted: 'Mrs C., who looked beautiful, was followed by an admiring crowd of women and made quite a good speech standing on a chair in the shelter.'[20]

Winston highly approved of Clementine's decision to make these surprise visits, and was eager to hear the results of her investigations. On one occasion when Mrs Tennant called for her, he insisted that they should go in an armoured car for protection should a raid start while they were on their tour. Both Clementine and Mrs Tennant were indignant, but bowed (albeit rebelliously) to 'orders'. However, as soon as they reached the East End, which was their target for that particular night, they abandoned the vehicle, which embarrassed them, and made their way on foot to the various shelters.

Clementine was horrified by the conditions she saw in most of the places she visited, and she formed the opinion (reinforced by the views of responsible people who worked in the shelters and by many local authority officials) that, while life in the shelters could never be Ritz-like, there were nevertheless many things that needed to be done, and that could be done, to improve the standards of hygiene and comfort. When she had inspected a representative cross-section of shelters, she submitted a series of memoranda to Winston.

Her suggestions give a vivid and horrifying picture of the conditions which many thousands of Londoners endured for nights on end, in addition to the risk and fear of death, injury and the loss of their homes; and they also show her practical approach to problems:

> Making Waterproof. Every shelter should be water-proof. In several instances the water was dripping through the roof and seeping through the walls and floor . . .
>
> Lice. Lice must spread . . . if the present overcrowding continues and if the bedding in the shelters is not regularly stoved . . . No soldier in the trenches in the last war was offended at being asked to go to the de-lousing station . . .
>
> Stoving. All the bedding in the shelters should be stoved two or three times a week. This should be compulsory even though the bedding is private property . . .

* John Colville (1915–87). Assistant Private Secretary to the Prime Minister, 1939–41 and 1944–5. As a pilot in the RAF (1941–4) he took part in D-Day sorties. Private Secretary to the Princess Elizabeth, 1947–9, he returned to WSC's Private Office, 1951–5. A principal motivator behind the foundation of Churchill College, Cambridge. He was knighted in 1974.

> Sanitary accommodation. Each shelter should be provided with adequate latrines . . . if possible not in the dormitory portion of the shelter itself.

In a subsequent memorandum, Clementine came out strongly against the overcrowded sleeping arrangements and poor sanitation:

> The more one sees of the 3-tier bunks the worse one feels them to be. They are, of course, much too narrow; . . . They are also too short . . . Where mothers have their babies sleeping with them it must be quite intolerable, as the baby has to sleep on top of the mother as the bunk is too narrow for it to sleep by her side . . . Would it be possible to stop any further orders on the existing model and have it redesigned?

On the subject of latrines, she urged that

> The latrines should be doubled or trebled in number. This is easy as they are mostly only buckets . . .
>
> The buckets, instead of standing on porous ground, (as they mostly do) could be placed on big sheets of tin with turned-up edges like trays . . .
>
> Light should be provided over the latrines. The prevailing darkness merely hides and, of course, encourages the dirty conditions . . .
>
> There should be separate latrines for little children, with low buckets and chambers [potties].

It was, of course, inevitable that in the first weeks of the Blitz living and sleeping arrangements in the shelters had to be improvised, but these early and rudimentary arrangements were, in too many cases, allowed to continue. Clementine's opinion was 'that the main difficulty in getting things right lies in the number of Authorities concerned, namely Local Authorities, Home Security, Ministry of Health, Ministry of Labour, Ministry of Supply; and the fear of these Authorities of using compulsion where persuasion fails'. She also had something to say about Government policy (or lack of it):

> There seems to be general uncertainty as to the Government policy re bad shelters. The reason (or excuse?) given for doing nothing is that the particular shelter is unsafe and that it is not worth while to spend money on it. Meanwhile people continue to live there perhaps fourteen hours out of the twenty-four in really horrible conditions of cold, wet, dirt, darkness and stench.

The sustained and close interest which Clementine took in the shelters at this time was largely instrumental in bringing about rapid and considerable improvements. The chief qualities which characterized her approach to these, as to other, problems were tenacity and an unwillingness automatically to accept the official answer or explanation, even if it happened to be given by the minister concerned. Of course, she had the priceless advantage of being able to approach ministers personally; but most of her considered reports were addressed in the first place to Winston, and they went out to the appropriate departments with all the authority and priority which papers emanating from the Prime Minister's Office possessed. Nevertheless, had Clementine's representations been ill-founded, trivial or impractical, even Winston's backing would not have got them very far. As it was, her interest in these matters became well known, and soon it was not only long-suffering members of the general public who drew bad cases to her attention: Members of Parliament, social workers, doctors and officials, exasperated at the inertia, indifference or inefficiency of their own superior authorities, appealed to Clementine to exert her influence.

* * *

At the beginning of November 1940, Neville Chamberlain died. Earlier in the year he had undergone a major operation for cancer; since then he had continued his work with Spartan courage. Winston and Mrs Chamberlain together had finally persuaded him to leave London, where the almost continual bombing added to the difficulties of nursing him. Winston and Clementine were on friendly terms with the Chamberlains, and admired the spirit of patriotism and loyalty which he had shown by serving under Winston after so many bitter passages had passed between them. Also, Clementine felt a special compassion for Anne Chamberlain, who had seen her husband the object of fawning adulation at the time of the Munich Agreement, only for him to be cast aside and decried by many who in the palmy days of appeasement had hastened to join the chorus of praise.

On Friday, 8 November 1940, Mrs Chamberlain issued a statement to the press, saying that her husband's strength had been failing and that he was now gravely ill. Clementine wrote to her at once:

November the 9th 1940

My dear Anne,

I was saddened last night when on the wireless I heard your message.

Winston has told me that you were very anxious – But the announcement is a shock to me, as I know it must be to the whole Nation.

I feel for you so much & think of you continually and I pray that the Peace of God is with you both.

Yours affectionately,
Clementine S. Churchill

Neville Chamberlain died on 9 November, and later Anne Chamberlain wrote Clementine a letter full of affection and gratitude.

When Chamberlain had resigned from the Government in early October he had also relinquished the Leadership of the Conservative Party, and Churchill had been invited to take his place. There were, and maybe still are, two views as to whether he was right to do so. Churchill himself had no doubt whatsoever; all his experience as a politician made him quite sure that he should grasp with a firm hand the Leadership of the Party which held a commanding majority in both Houses of Parliament. The opposing view, which Clementine herself held with passionate conviction, was that Winston had been called by the voice of the whole nation, irrespective of party, at a time of grave national emergency, to head a National Government, and that, although he might feel his position to be stronger in Parliament by accepting the Conservative Leadership, he would affront a large body of opinion in the country. Clementine expressed her view with vehemence, and all her latent hostility towards the Tory Party boiled over; and there were several good ding-dong arguments between them on the subject. But Winston's view prevailed: he accepted the Party's Leadership, and felt sustained through some tough times by the knowledge that he could command its loyalty.

True to her custom, Clementine accepted his decision without recriminations. But she never altered her opinion that this step was a mistake, and that it alienated much of the support which Winston received from the working classes through the vindication of his pre-war prophecies and his record as a war leader.

* * *

A joyful family occasion lightened the grey winter days of 1940: 'little' Winston was christened at Ellesborough Parish Church near Chequers on Sunday, 1 December. Our family party was remarkably complete, and Pamela's parents and her brother came from Dorset. There was a complete muster of godparents on parade: Max Beaverbrook, Lord Brownlow and Brendan Bracken; and Virginia Cowles, the American journalist and author. The christening was held after Matins, and most of the congregation stayed on to join with us in the service. Afterwards we returned to Chequers for luncheon, when Winston proposed the toast to 'Christ's new faithful soldier and servant'.

In just over a week, however, there was cause for more than family rejoicing: the Eighth Army launched an attack in North Africa on 9 December, and by the fifteenth the Italian forces had been driven from Egypt, with the destruction of five enemy divisions and the capture of 38,000 prisoners. And so, as we gathered for the second Christmas of the war, our hearts were buoyed up with not ill-founded hope and natural jubilation.

Diana and Duncan, Randolph and Pamela, Sarah and Vic, and myself made the tally of children present complete. In addition there was John Martin, one of my father's Private Secretaries. 'Everyone in good spirits . . . No reports of any air, land or sea activity,' I wrote in my diary on Christmas Eve. And on Christmas Day I recorded that 'This was one of the happiest Christmasses I can remember. Despite all the terrible events going on around us . . . I've never before seen the family look so happy – so united.'

Perhaps it is because human beings have this capacity to draw down the blinds, and to shut out the wild winter weather, and to take comfort in the familiar faces in the lamplight, that mankind has survived at all. At any rate, it was fortunate for us that the Christmas of 1940 proved to be such a happy and peaceful one, for never again during the war were so many of us able to gather together as a family.

CHAPTER TWENTY

Taking up the Load

THE YEAR 1940 HAD SEEN THE OVER-RUNNING OF WESTERN EUROPE BY Hitler's hordes, leaving Britain dramatically isolated; but in 1941 the war was to become a worldwide conflict. The battle in the North African desert swung to and fro as Rommel's Afrika Korps pushed the Eighth Army back over the ground they had captured from the Italians; and the Germans invaded Greece and Yugoslavia.

We were still braced for an invasion when, on 22 June 1941, Hitler hurled his armies against Russia, on a front extending from the Baltic to the Black Sea. Within a month Britain and the Soviet Union had signed a treaty of mutual aid, and although the help we could give – or receive – was sorely limited, we were no longer alone in our struggle against Hitler.

Meanwhile, succour from a powerful source began to reach us: in March 1941 Congress passed the Lend-Lease Bill, and each month brought more help from the United States. On 7 December 1941 the Japanese air force bombed the American naval base at Pearl Harbor, and the United States entered the war. The days of our splendid isolation were at an end, and as the New Year of 1942 dawned we found ourselves partners in a Grand Alliance, pledged each to each to make both war and peace together.

But throughout the year, the depredations of the German U-boats on our merchant shipping threatened our slender lifeline; and in spring 1941 it was the turn of our ports and industrial cities to endure heavy bombing raids: Portsmouth, Merseyside, the Clyde, Hull and Plymouth all suffered grievously in the enemy's attempts to cripple our harbours. Manchester, Coventry and Bristol, too, sustained heavy attacks, though their ordeal was not so long-drawn-out as London's.

Through his frequent broadcasts on the BBC, Winston Churchill had become closer to the ordinary people of Britain than had any previous Prime Minister. His sombre tones, relieved by shafts of robust, sardonic humour, the total absence of cajolement by facile hopes, and the dogged courage all struck a note in tune with the mood of people of all classes. By this means Churchill forged a unique bond with his compatriots and was able to bring them hope and reassurance. He also sought every opportunity to visit places which had suffered severely, or were bearing the brunt of enemy bombardment, and Clementine always tried to go with him. They travelled, with Winston's essential entourage, on the special train arranged for this purpose, which not only provided comfortable and restful travelling conditions, but avoided all the difficulties of security which would have been involved by staying in hotels. In addition, secret, 'instant' communication with London was always available, as the train could be plugged in to the nearest telephone line, where a special 'top priority' number established contact with Downing Street.

These tours were often long and tiring. They usually involved trudging through ruined streets, visiting rest centres and canteens, sometimes even scrambling over piles of rubble, but Clementine, despite her not over-strong constitution, stood up to them wonderfully. The scenes they saw and the people they met had a profound effect on them both. Winston was time and again moved to tears by the bravery of the people he saw, and by the affection and regard they expressed for him. The effect of these visits to the bomb-battered cities was indescribably moving. He had indeed become the symbol of his country's hope and courage, and under the flail of war the usual reserve of the British character had melted away. Wherever Churchill went, crowds appeared, cheering and clapping; even from the ruins of their own houses people ran to greet him. And many, particularly the women, came to expect to see Clementine at his side.

Indeed, Clementine's own personality was now becoming widely known and appreciated, not only because of her habitual presence at Winston's side, but also through her postbag, her appeal broadcasts and her war work. These things combined in the war years to carve out for her a special place in people's minds.

It was not only her personality which made its mark: in the dreary, dusty shabbiness of wartime conditions, Clementine's lovely appearance and her own distinctive style of dressing added to people's pleasure at seeing her. An observer wrote of her visit with Winston to Plymouth in May 1941: 'There was much pleased feminine comment on Mrs Churchill's striking appearance: snow leopard coat with flounced sleeves, a coloured handkerchief over her head with phrases of her husband's speeches printed on it. Some were surprised Winnie had such a fine, fresh-looking wife.'[1] Her elegant, unaffected taste in clothes was perhaps seen at its best during the 'austerity' years of the war, when she managed always to look exceptionally well-dressed, without seeming in the least out of tune with the times. Clementine adopted and popularized a fashion which was a graceful compliment to all the thousands of women factory workers throughout the country: for safety reasons, and also to keep dirt and dust out of their hair, factory girls nearly all tied their heads up 'bandanna' style in headscarves. Clementine disliked hats, and she also thought they looked singularly unsuitable in the context of ruined streets and shattered houses; but, meticulously neat, she did not wish to become dishevelled during the course of a long day's programme. So she adopted the 'turban-bandanna' style, and, using every kind of material

– silk, cotton, crepe, tulle and chiffon – she assembled a 'library' of turbans. Taking trouble to arrange her turban, anchored by her earrings, carefully at the start of the day, she scarcely touched it again; it was impervious to wind and weather, and looked equally suitable whether she was visiting a hospital, clambering about in a ship, or lunching with a Lord Mayor in the Mansion House. These turbans, in which she was much photographed, acquired the nature of a distinctive and personal 'trademark'.

Although Clementine took great trouble about her appearance, she never, at any time in her life, allowed clothes to dominate her time or her interest; she gave thought to her dressing, organized it, and then forgot about it. Before the war, she had from time to time 'treated' herself to a dress from Schiaparelli, Molyneux or some other couturier; but most of her clothes were made by various small dressmakers, usually from paper patterns. Now, with the war, she bought many of her outfits from Molyneux, whose simple and supremely elegant clothes suited her to perfection. As to time spent at the hairdresser, it was practically nil. Clementine went about once every three months to her hairdresser for permanent waving; otherwise she managed her hair entirely by herself, cleaning it every few days (to the family's acute and constant alarm) with neat benzine!

* * *

During 1941 Clementine undertook two new and large responsibilities in addition to the work with which she was already concerned. In February she became President of the Young Women's Christian Association's Wartime Appeal, for which she was instrumental in raising many thousands of pounds. And in October she undertook the Chairmanship of the Red Cross Aid to Russia Fund, thus beginning by far the most formidable job she tackled in either war, and one which was to make enormous demands on her. Clementine began this new work at a moment when her time and energy were, one might have supposed, fully subscribed, but there were several reasons for her undertaking this further major commitment.

When the German onslaught was loosed upon Russia in June 1941 Churchill, despite his life-long hostility to communism, had at once declared Britain's determination to ally itself with the Soviet Union in its struggle against the Nazis, and he had spoken with great feeling of the sufferings of the Russian people in the face of the German advance. Spontaneously a warm wave of sympathy swept through this country, as people learnt with mounting horror of the sufferings of the Russian civilian population. The Soviet Embassy in London received hundreds of letters every day expressing friendship and sympathy, and individuals and organizations started to send money to help the Russian war effort. Working-class support was particularly strong; the Mineworkers' Federation sent the Russian Ambassador a cheque for £60,000 only a few days after the German attack.

These expressions of sympathy in the first few months found no clearly defined organization which could focus their offers of help and gifts into an effective form of aid to our new and struggling ally. In early October, however, Field Marshal Sir Philip Chetwode, who was the Chairman of the Executive Committee of the War Organization of the Red Cross and St John, reported that it had decided to form an Aid to Russia Sub-Committee, of which he would be Chairman, and that Clementine had consented to be the Vice-Chairman. A few days later, it was announced that she would also be Chairman of the Aid to Russia Appeal.

Deeply moved herself by the sufferings and heroism of the Russian people in the defence of their homeland, Clementine also understood, and accurately gauged, the strength and volume of the sympathy for the Russian cause among the mass of the population in this country. There was, moreover, a political aspect which did not escape her notice: from the moment that Russia was attacked by Germany, the Soviet Government appealed loudly and continuously to Great Britain for aid of all kinds, and made imperious demands for British landings in Europe, regardless of our own difficulties or the preparation involved in mounting a Second Front. The British Communist Party, which had hitherto condemned the 'capitalist and imperialist' war, now called for immediate and impossible measures to help the Soviet Union. All over the country 'Second Front Now' was scrawled on walls and hoardings. Churchill lost no opportunity of sending such aid as was possible but, as he commented wryly, 'there was little we could do, and I tried to fill the void by civilities.'[2]

Clementine understood these difficulties very well, and often discussed them with Winston, who later wrote: 'My wife felt very deeply that our inability to give Russia any military help disturbed and distressed the nation increasingly as the months went by and the German armies surged across the Steppes.'[3] And so when Clementine was invited by the Red Cross to head the Aid to Russia Appeal, she eagerly accepted the task. Here was the opportunity to demonstrate in a practical way the genuine desire of this country to render what help it could to Russia, and the fact that the Prime Minister's wife was heading the Appeal showed that the British Government was at one with the feeling in the country, and eager to give expression to it. Moreover, the Joint War Organization of the Red Cross and St John provided a machine capable of coping with the work involved, which Clementine from the first realized would be tremendous.

A more detailed account of Clementine's war work is given in the next chapter; here it is enough to note that shouldering during 1941 both the YWCA Wartime Appeal and the Aid to Russia Fund greatly increased her burden of work, and called forth all her reserves of energy and stamina. And, amid all these major preoccupations, time and thought had also to be found for domestic and family life. Although Clementine's days now were full to overflowing, she always managed somehow to be on hand when Winston wanted or needed her. A more personally ambitious woman, or a more conceited one, would perhaps (albeit subconsciously) have allowed her own personal role and importance to develop into a separate, almost competitive, entity of its own. But her career had always been simply to be Winston's wife, and the fact that she had now become a public figure in her own right did not alter the situation by one whit.

The only time it was normally feasible for Clementine and the family to see Winston was at mealtimes, and then there would almost certainly be outside guests. But we all made the most of these opportunities, and any of the children were always welcome to any meal they could manage. Winston's day started when breakfast and his red boxes were brought to him, usually around eight o'clock; unless it was a Cabinet day, or he had some other outside appointment, he would work in bed, receiving officials or service chiefs, until luncheon-time. In the afternoon he worked again until around six o'clock – then took the blessed, life-restoring rest for an hour to an hour and a half. When he awoke, marvellously refreshed, he would bath and change. Soon after the beginning of the war Winston abandoned the dinner jacket in favour of a zip-up 'siren' suit (or 'rompers' as the family disrespectfully dubbed them). Clementine took trouble and pleasure in the designing of these comfortable garments which, made in a variety of materials, remained a key part of his wardrobe to the end of his life. Sometimes as a birthday or Christmas present she would give him a sumptuous version in mulberry or midnight blue velvet; the family would often weigh in with slippers to match, embroidered with his monogram in gold thread. After dinner, at about ten or ten-thirty, he would say 'good night' and go off back to work, probably till one or two o'clock in the morning.

Clementine was most punctilious in not interrupting or disturbing Winston at his work. If she had an urgent message for him that could not wait until they met at a meal, or that it was not possible to discuss in front of guests, she sent him a typed 'memo' (often decorated, like her letters, with a 'kat'). Since Winston also worked solidly throughout the weekends, when anyhow there were nearly always guests staying at Chequers, the opportunities for them to be alone together were rare indeed, and precious to them both beyond price. It was unusual for them to be alone at luncheon, but they entertained less in the evenings, and Clementine never let anything come between her and the chance of a quiet, candle-lit evening alone with Winston. She always changed for dinner, and she usually wore one of her ravishing housecoat-cum-dressing gowns made in beautiful materials, and in glowing colours. She always tried to be, and to look, at her best for him.

On the family front, this year of 1941 brought its full ration of worries and concerns. Duncan Sandys was seriously injured in a car accident in April. For a while it was feared that one of his feet might have to be amputated; mercifully this was not the case, but the injury was to cause him pain and impediment ever after. Sarah's marriage finally broke up. From the beginning both her parents had feared it would not last, but Clementine, in fact, had become fond of Vic Oliver, admiring his taste and charming manners; Winston, however, had never got to know or like him; both now felt deeply for Sarah. During 1941 she and Vic parted, and shortly afterwards she joined the Women's Auxiliary Air Force (WAAF). Beginning in the ranks, she was soon commissioned, and was trained in the skilled work of photographic interpretation. In this exacting, secret labour she sought to forget the sadness and aridity of her life at this time. The unit which carried out this work was stationed at Medmenham in Buckinghamshire, and as this was not very far from Chequers, Sarah was quite often able to go home to visit her parents.

In July, Goonie Churchill died from cancer; she had been ill for some time. Her death brought back to Clementine memories of that other war, when she and her unusual, fascinating sister-in-law had shared a house, and Goonie had been her loyal and almost only confidante in the dark and troubled days of the Dardanelles crisis.

I managed to cause a diversion on the domestic front, when in May I somewhat precipitately became engaged, after a very short acquaintance, to Eric Duncannon, the son and heir of the Earl of Bessborough. He was charming and intelligent and our parents were long-time friends: it was an altogether suitable match – but from the very first my mother was convinced that I was not really in love. Strangely enough, she confided her fears to Max Beaverbrook:

> It has all happened with stunning rapidity.
>
> The engagement is to be made public next Wednesday; but I want you to know beforehand because you are fond of Mary –
>
> I have persuaded Winston to be firm & to say they must wait six months – She is only 18, is young for her age, has not seen many people & I think she was simply swept off her feet with excitement – They do not know each other at all. Please keep my doubts and fears to yourself.[4]

My mother, not unnaturally, did not relish bearing the entire responsibility for intervening in this delicate matter, and running the risk of being thereafter accused of 'wrecking my whole life'. But my father was totally preoccupied with events of national importance, and so she had to grapple with this emotional situation herself. Her views, and friendly advice from various quarters, combined with misgivings of my own, resulted in the indefinite (and, as it proved, permanent) postponement of the announcement of the engagement. In this matter, as in so many more, her judgement was right.

It is hardly surprising that the pressure of the work which Clementine had undertaken, combined with the strain of her social and domestic life, was now beginning to tell; earlier in the year she had suffered from quite a severe attack of bronchitis, and by the summer she was desperately tired, both physically and mentally. Since the beginning of the war she had had no holiday, merely an odd weekend here or there, and occasionally a day or an afternoon spent at Chartwell. It was not easy, however, for her to find a suitable moment to ease up, and she hated the idea of leaving Winston, even for a week, in these hard and anxious days. However, in August 1941 a great event took him far from home. For two years Winston had corresponded with President Roosevelt, and already a strong and intimate link had been forged between them, but for some time he had been hoping for an opportunity to meet the President face to face. Towards the end of July 1941, Harry Hopkins*, the President's personal envoy, told Winston that the President was equally anxious for a meeting.

Churchill's voyage to meet President Roosevelt in August 1941 was the first of many journeys across the world he was to make during the war by land, sea and air. The security arrangements protecting his movements and plans were of course elaborate, for the risks involved were many and great. Clementine was also somewhat concerned for Winston's health during this journey, and she tried to persuade him to take with him Sir Charles Wilson†, who had been appointed his doctor when he became Prime Minister. She wrote him the following note on 1 August: 'I feel very strongly that on this all-important journey you should have a Doctor with you . . . Please take Sir Charles Wilson.' This note is signed with a chubby 'Kat', and overleaf there is this comment: 'Brendan agrees with me.' On this occasion Winston was not to be persuaded; but on nearly all his subsequent journeys Sir Charles Wilson did accompany him.

Winston and his party sailed on 4 August in the battleship *Prince of Wales*, and arrived at the rendezvous – Placentia Bay, Newfoundland – on 9 August. The announcement of this first meeting of the two statesmen, which lasted three days, reverberated around the world. The outward and visible result of the meeting was the publication of the Atlantic Charter, which gave glimpses of a brighter world to struggling humanity. But the real importance of the occasion was the chance it gave Churchill and Roosevelt to establish the friendship already begun, and to come to understand each other's train of thought. There can have been few meetings which held so much significance for the history of the world.

Clementine took the opportunity of Winston's absence to spend ten days at Dr Lief's health establishment at Tring in Hertfordshire. She made the most of her rest there, and felt truly refreshed; but before she left his care, her doctor warned her that she was driving herself too hard, and he suggested that she should try to set aside one day in every week as a 'rest day', when not only should no public obligations be undertaken, but she should rest quietly at home, seeing the minimum of people and doing no work of any kind. Clementine agreed to this excellent plan, and for the first seven weeks after she left Tring, 'rest day' is marked down punctiliously once a week in her diary; but thereafter no mention of it appears again: it just simply was not practicable, and she abandoned it.

But it was a restored and rejoicing Clementine who greeted Winston at King's Cross on the morning of 19 August. The King sent Sir Alexander

* Harry L. Hopkins (1890–1946). US social worker and administrator. As a close confidant of President Roosevelt from New Deal days, he lived in the White House.

† Later Lord Moran. See note p. 77.

Hardinge to meet him; Mr Winant (the American Ambassador)*, Mr Fraser (the New Zealand Prime Minister), and many members of the Government were also on the platform. Although the time and place of Winston's arrival had not been announced, there was nevertheless a large crowd gathered both in and around the station. It was a wonderful welcome home, and it moved him deeply. As he walked down the platform and out to his car, with Clementine, he was cheered repeatedly.

In early September 1941 my cousin Judy Montagu and I joined the Auxiliary Territorial Service (ATS). We had become fired with the idea of serving with one of the new 'mixed' (men and women) anti-aircraft batteries, which were being formed about this time. Our parents understood our wish to go 'a-soldiering' and encouraged us to enlist. So Judy and I set out on what was, for us, a great adventure, our mothers waving us goodbye at Paddington Station.

We went to a training centre at Aldermaston, near Reading, for a month's basic army training, and then to a huge camp near Oswestry, in Shropshire, to learn the mysteries of anti-aircraft instruments. I duly regaled my parents with detailed accounts of my life as a private soldier (at 1s 8d a day). While we were at Aldermaston my mother and Venetia Montagu would come down at the weekends to see their soldier-daughters, bringing delicious picnic luncheons. Judy and I were very nervous lest our mothers should disgrace us by not conforming in the most minute details with the rules and regulations, or by failing to recognize the immense importance in our lives of the Guard Commander (usually a lance-corporal) or such exalted beings as the company sergeant-major.

After one of my mother's Sunday visits she wrote to me:

> I did love seeing you yesterday; but I came away with rather a heavy heart – driving away to luxury & comfort & leaving you both to rough it.
>
> My Darling I do love & honour you. I know you have done what is right & that even if it is uncomfortable & rather desolate now you will always all your life be glad you did it & didn't drift into a half & half makeshift job.[5]

She was indeed right: I never regretted for one moment my decision to join up.

In February of this year Diana had left the WRNS to grapple with her family life – two children under five and, from the spring, a badly injured husband who was considerably dependent on her in the early stages of his recovery.

Despite all the demands of her life, my mother was a wonderful correspondent, and wrote both her 'active service' daughters long and sparkling letters, of which the following are a few examples:

* John Gilbert Winant (1889–1947), 'Gil', a former Governor of New Hampshire, had succeeded the defeatist Joseph P. Kennedy as US Ambassador in April 1941. He was an immediate success at many levels, and became a great personal friend of our family. Soon after his return to America in 1947 he committed suicide.

10, Downing Street,
Whitehall
Sept. 16, 1941

My own Darling Mouse

Last night, Mr Harriman, freshly arrived from America in a huge Catalina flying boat came to dinner . . . He brought Kathleen [his daughter] with him, looking very pretty. They both asked after you, particularly Mr Harriman who sent you his love & salute. Next time you & Judy go to Reading, do be taken by the local photographer, together & alone. It will be such fun to have a picture to put up on my shelf & to stick in my War Scrap Book. And don't forget to ask for a pass for me, becos on Sunday 'I will be right there with bells' – at 2 o'clock.

Papa & I are going to St Paul's Cathedral this morning by special wish of The King. There is to be a Thanksgiving Service for King Peter of Yugoslavia's coming of age. They grow up quickly in the Balkans – I believe he is only 18; but he has shewn an old fashioned Kingly thirst for power by sacking his Prime Minister! I understand he is being told that this is unwise & premature! Then we are to lunch at Buckingham Palace.

10, Downing Street,
Whitehall
September 30, 1941

My own darling Mary

I am thinking of you constantly – I wonder if you & Judy will be a little sad at leaving Aldermaston or if you are full only of excitement & expectation? I shall miss our picnics. As soon as you have a breather do write me all your news . . .

Yesterday I went to my Maternity Hospital & tried to buy a house for a Convalescent Home for the Mothers. But it was too expensive, so I am scratching my head & wondering what to do next . . . To-day for 3 hours I have been trudging round the Borough of St Pancras looking at A.R.P. Canteens & decontamination centres till I thought I should drop. I had to make 3 speeches & grin & look gracious the whole time! Both my stockings laddered & a heel nearly came off my shoe!

Chequers
Butler's Cross – Aylesbury
November 2, 1941

My darling Mary

In my last letter describing to you the start & progress of the Russian Fund I forgot to tell you that the King & Queen lunched alone with Papa & me last Tuesday.

Mrs Landemare was in a flutter & produced a really delicious luncheon & it was really all most enjoyable, becos' The Queen is so gay & witty & very very pretty close up . . . Papa tried to interfere with the Menu but I was firm & had it my own way, & luckily it was good.

The Queen asked about you & admired a photograph of you scrubbing (the full face one in the open doorway) . . . The King did not say much – He looked thin & rather tired . . .

Princess Elizabeth & her sister are being 'educated at Eton'! that is, The Vice-Provost of Eton & one or two of the most agreeable & brilliant Masters go to Windsor Castle two or three times a week & instruct them & they enjoy it very much. I think they have the most amusing lives with lots of dogs (altho they are only those horrid 'corgies') & poneys [*sic*] & a delightful Mother.

10, Downing Street,
Whitehall
November 12, 1941

My own Mary,

It is three days since we parted, & it feels already like three weeks.

I do hope these coming weeks will not be so strenuous & rushed for you as the last fortnight . . .

To-day I went to see the King open Parliament – As you know the Lords very handsomely gave up their Chamber to the Commons when the House of Commons was destroyed; & now the Lords sit in the Robing Room which has been converted into a miniature, one might almost say toy House of Lords. There is room for only about 100 Peers instead of 800, the Woolsack has been reduced in size & the thrones are elegant but tiny. Altogether it is most attractive & cosy. The Queen looked lovely in plum coloured velvet, pearls & sables & the King in Naval Uniform instead of full Regalia led her in by the hand.

The Russian Fund is staggering – Yesterday £36,000 arrived by one post & to-day £46,000. We are well over Half a Million Pounds.

* * *

Churchill was at Chequers on the evening of Sunday, 7 December 1941, when he heard the report of the Japanese attack on Pearl Harbor on the nine o'clock news. It was a fitting coincidence that dining with him on that momentous evening were Gil Winant, the American Ambassador, and Averell Harriman, the President's adviser on Lend-Lease. Clementine too was at Chequers but, feeling exhausted, she had gone to bed for dinner, leaving the men on their own. As soon as he heard the news, Churchill put through a call to the White House, and within a few minutes he was speaking to the President. He recorded his passionate feelings of relief and sombre rejoicing at the realization that America, with all its might, was now committed with Britain to the struggle. There could be no more doubt about how the conflict would end: 'I went to bed,' he wrote, 'and slept the sleep of the saved and thankful.'[6]

The entry of the United States into the war made it imperative for Churchill to see the President again as soon as possible. Accordingly the intricate and highly secret arrangements were made, and on 13 December he and his colleagues sailed for America in the *Duke of York*. The weather was wild, and of all the many journeys he was destined to make during the war, this voyage set a record for sheer discomfort. Winston described to Clementine the almost unceasing and violent gales which delayed their progress. He was not afflicted by seasickness, but during the voyage he spent the greater part of the day in bed.

> I manage to get a great deal of sleep and have also done a great deal of work in my waking hours . . .
>
> We make a very friendly party at meal times, and everyone is now accustomed to the motion. The great stand-by is the cinema . . . I have seen some very good ones [films]. The one last night, 'Blood and Sand', about bull-fighters, is the best we have seen so far. The cinema is a wonderful form of entertainment, and takes the mind away from other things . . . Being in a ship in such weather as this is like being in prison with the extra chance of being drowned.

Then he went over, almost as in a conversation with her, the debit and credit account of the war at this time, and wrote also of his eagerness to arrive in Washington: 'You can imagine how anxious I am to arrive and put myself in relation to the fuller news and find out what is the American outlook and what they propose to do.' But his thoughts were not only of grand strategy: clothes rationing had been introduced in June, and he added: 'I wish particularly to know the length of your stockings, so that I can bring you a few pairs to take the edge off Oliver Lyttelton's coupons.' He finished dictating this letter after his arrival at the White House, and added in his own hand:

> I have not had a minute since I got here to tell you about it. All is vy good indeed . . . The Americans are magnificent in their breadth of view.
>
> Tender love to you and all. My thoughts will be with you this strange Christmas Eve.
>
> Your ever loving husband,
>
> W[7]

Back at home, Clementine sat down and wrote to him on Friday, 19 December:

> My Darling
>
> You have been gone a week & all the news of you is of heavy seas delaying your progress . . .
>
> I hope you are able to rest in spite of wind & weather and the anxiety in the Far East – How calm we all are – Hong Kong threatened immediately, Singapore ultimately? perhaps not so ultimately, Borneo invaded – Burmah? to say nothing of the blows to America in the Pacific.
>
> Here I am bound to my Russian Fund. We have passed the Million Pound

target & that without what will come in from the Flag Days, held not only in London but all over the country. I visited many depôts all over London from dawn till dusk. The people came running everywhere. They are so good & sweet especially the old & they all asked about you.

Yesterday Mary's leave came to an end; I took her & Judy in your car & deposited them as night was falling at their new camp near Enfield. In the gathering darkness it looked like a German concentration camp.

. . .Well my beloved Winston. May God keep you and inspire you to make good plans with the President. It's a horrible world at present, Europe overrun by the Nazi hogs & the Far East by yellow Japanese lice.

I am spending Christmas here at the Annexe & going to Chequers on Saturday the 27th.

Tender love and thoughts,
Clemmie

Since Judy and I had only just joined our new battery*, Christmas leave was of course out of the question; but we enjoyed the cheerful, noisy celebrations organized in our camp. My mother sent us, for the benefit of our 'barrack room', a vast paper cracker, fully four feet long and a foot in diameter; this was a huge success, and greatly added to the fun and merriment. On Christmas Eve she wrote to tell me that the previous afternoon my father had telephoned from the White House:

> He might have been speaking from the next room. But it was not very satisfactory as it was a public line & we were both warned by the Censors breaking in that we were being listened to!
>
> 'Nana' [Whyte] and I are all alone this Christmas, but we are not lonely because we feel Papa is saving the World & you & Sarah are also, & are happy in your work.

Many thousands of miles away from blacked-out London, Winston spent Christmas in Washington. On Christmas Eve, thirty thousand people gathered in the grounds of the White House to see and hear the President and his British guest, and to watch the traditional lighting of the White House Christmas tree. Clementine's thoughts were not only of Winston, but also of all he was hoping to achieve. 'I have been thinking constantly of you,' she wrote on 29 December, '& trying to picture & realise the drama in which you are playing the principal – or rather it seems – the only part. I pray that when you leave, that the fervour you have aroused may not die down but will consolidate into practical & far-reaching action.'

Her hopes were realized, for the 'Arcadia' Conference cemented the merger of the war efforts of Britain and the United States, and on New Year's Day 1942 the first declaration made in the name of the United Nations –

* After completing our initial training, Judy and I were posted to a mixed heavy anti-aircraft battery at Enfield, north of London.

the United States, Britain, the USSR and China – pledged that both war and peace would be made together. Churchill had addressed both Houses of Congress on Boxing Day, and then had flown to Ottawa for a three-day visit, including making a speech to the Canadian Parliament, returning to Washington to continue his talks with the President and his advisers.

The pace and pressure of these days were tremendous and telling, and on the night of 26 December Winston suffered a very slight heart attack: so slight was it that his programme remained unaltered, and not a word was said; but it was an indicative sign. Fortunately Sir Charles Wilson was with him, and his judgement and care were of prime importance. Kindly Mr Stettinius (at this time President Roosevelt's Lend-Lease administrator) offered Winston his house in Florida, and he was able to go there for five restful days. It was his first holiday for nearly three years.

Winston travelled home in a Boeing flying-boat, arriving on 17 January; he had been absent for nearly five weeks, and he was greeted with warmth and relief by his family, his colleagues and the country alike. Upon his return Churchill found a somewhat troubled political situation and a groundswell of public dissatisfaction. Now that as a nation we were no longer in mortal danger of invasion, the general sense of relief was accompanied by a reawakening of the faculty of criticism. There was a general demand for changes in the War Cabinet, and a feeling that in view of the importance of our relationship with the Soviet Union, the socialist element should be strengthened. Within the Government itself there were difficulties and personal tensions, particularly between Lord Beaverbrook (then Minister of Supply) and other ministers.

A week after Churchill's return there was a debate in the House of Commons; he insisted that the House should divide on a Vote of Confidence, so that the country and the world should see how the National Government stood. The vote brought an overwhelming majority for the Government of 464 to 1. But the discontent within the Cabinet was unappeased, and was indeed reinforced by suggestions from the outside that the time had come for the introduction of 'new blood' into the War Cabinet. An obvious candidate was Sir Stafford Cripps, who had just returned from being our Ambassador in Moscow: he was a man of outstanding intellectual achievements, and a socialist of unimpeachable integrity and strong views.

Meanwhile Winston was experiencing much difficulty with Max Beaverbrook, who was suffering acutely at this time from one of his recurring bouts of asthma, and who was in a highly tricky and unpredictable frame of mind. Already, while in Washington, Winston had brushed aside an impulsive resignation, and now Beaverbrook was alternating between a deep distaste for the constraints of office and the demand for wider powers as a minister. The complicated and delicate task of rearranging his team at the same time as placating Beaverbrook, whose gifted services he hoped to retain, superimposed on the burden of the daily running of the war, added greatly to the strain of Winston's life.

The war news at this time was bad. In North Africa, Rommel had launched a new offensive towards the end of January, and the British Army was once more driven out of Cyrenaica. In the Far East, the Japanese were advancing through Malaya, and on 15 February the country learnt of the surrender of sixty thousand of our troops at Singapore: it was the 'largest capitulation in British history'.[8] These were dark and perplexing days indeed. Clementine hated to see Winston so beset on all sides, and resented the extra fret and worry caused by a temperamental colleague. She was often present at meals when all these difficulties were discussed, and she had her own opinions about some of the proposed appointments and re-appointments in the Government.

Since the beginning of the war, Clementine's long-held antipathy to Max Beaverbrook had softened. She was second to none in her admiration for the amazing feats he performed at the Ministry of Aircraft Production; and the exploits as a fighter pilot of his gallant son, Max Aitken, and his father's pride in them, warmed her towards 'big' Max, who also had been charming to her in these dramatic years, and helped her generously with the causes for which she worked. But through this period of 'truce', deep down there slumbered her strong conviction that he was a baleful influence on Winston. And now, as she saw Winston struggling with tensions and difficulties in his team caused in large part by the contradictory humours of this curious man, she boiled over: during a discussion with Winston about this problem, she became very heated and expressed herself in immoderate terms. Afterwards she wrote him this letter (sent by 'House-post'*):

Thursday [probably 12 February 1942]

My Own Darling,

I am ashamed that by my violent attitude I should just now have added to your agonising anxieties – Please forgive me. I do beg of you to reflect, whether it would not be best to leave Lord B entirely out of your Reconstruction.

It is true that if you do he may (& will) work against you – at first covertly & then openly. But is not hostility without, better than intrigue & treachery & rattledom [*sic*] within? You should have peace <u>inside</u> your Government – for a few months at any rate – & you must have that with what you have to face and do for us all – Now that you have (as I understand) invited Sir Stafford, why not put your money on him.

The temper & behaviour you describe (in Lord B) is caused I think by the prospect of a new personality equal perhaps in power to him & certainly in intellect.

My Darling – Try ridding yourself of this microbe which some people fear is in your blood. Exorcise this bottle Imp & see if the air is not clearer & purer – You will miss his drive & genius, but in Cripps you may have new accessions of strength. And you don't mind 'that you don't mean the same

* The term used for their communications when under the same roof.

thing'. You both do in War & when Peace comes – we can see. But it's a long way off.

Your devoted
Clemmie

In the event, after much cajoling, Lord Beaverbrook became Minister of War Production, but suddenly resigned a fortnight later, giving the state of his health as the reason. The reconstructed War Cabinet was announced on 19 February, Sir Stafford Cripps taking the offices of Lord Privy Seal and Leader of the House of Commons, and Clement Attlee becoming Deputy Prime Minister (a new office). The strengthening of the left wing in the War Cabinet made the National Government more truly representative, and helped to secure public approval for the stringent controls which the nation was to endure as the war dragged on.

That the combination of bad news from the war fronts and domestic disagreements within the close circle of his own colleagues had a wearing effect on Winston is borne out by an entry in my diary for 27 February 1942:

Papa is at a very low ebb.

He is not too well physically – and he is worn down by the continuous crushing pressure of events.

And in a letter to her sister Nellie Romilly on 28 February 1942, Clementine wrote:

These are as you say days of anguish for Winston, so full of strength & yet so impotent to stem this terrible tide in the Far East.

We must pray that the Country will show patience and constancy & then All may be Well.

Clementine was, of course, deeply concerned for Winston: she saw every day the toll the weight of the war exacted from him, and he confided to her all his worries and tribulations. With her brave and stoical outlook, she accepted that the strain and burden of the work could scarcely be otherwise; but however and whenever she saw a chance to ease his burden or to spare him extra worry, she did so, and she could be fiercely protective if she thought unnecessary anxieties or problems were added to those it was inevitable he had to bear.

Among personal problems which weighed heavily on both Winston and Clementine at this time was the increase in tension between Randolph and Pamela, whose marriage was on the rocks. Since the spring of 1941, Pamela had been having an affair with Averell Harriman, which was soon common knowledge in social circles: Randolph was convinced his parents knew about this situation, and he was embittered by his belief that they condoned the affair and that they tended to take Pamela's side in their mutual disagreements. There was a major scene between Randolph and his father in the spring of 1942, which so upset Winston that Clementine feared he might have a seizure. It so happened that I too was at home on leave at the time, and was so outraged that Randolph should assail his father with such verbal violence that I wrote my brother a furious letter suggesting he should rejoin his unit forthwith; this was not helpful – indeed, it caused further ructions. All this bore heavily on my mother.

In April Randolph, who at this time had a staff job in Cairo, volunteered to join a parachute unit. Strangely enough, when Clementine heard this news, she reacted strongly against Randolph's decision; it was a most uncharacteristic reflex, for she greatly admired spontaneous and brave actions. But the reason for her initial reaction to Randolph's decision is clearly revealed in a House-post letter she wrote to Winston, after an obviously not very happy conversation on this subject:

> My Darling,
>
> Please don't think I am indifferent because I was silent when you told me of Randolph's cable to Pamela saying he was joining a parachute unit . . . but I grieve that he has done this because I know it will cause you harrowing anxiety, indeed, even agony of mind –
>
> I feel this impulse of Randolph's . . . is sincere but sensational. Surely there is a half-way house between being a Staff-Officer and a Parachute Jumper. He could have quietly & sensibly rejoined his Regiment & considering he has a very young wife with a baby to say nothing of a Father who is bearing not only the burden of his own country but for the moment of an un-prepared America it would in my view have been his dignified & reasonable duty.
>
> I think his action is selfish & unjust to you both, & as regards Pamela one might imagine she had betrayed or left him – [This last sentence indicates that at this time Clementine did not know of Pamela's affair.] . . .
>
> My Darling – Do you think it would be any use my sending an affectionate cable begging him on <u>your</u> account to re-join his Regiment & give up this scheme in which if he begins one feels he must perhaps persevere?[9]

In the event Clementine took no action, and Randolph pursued his intention of joining the Special Air Service. One cannot help feeling that in this matter her judgement went a little astray; but at this time of intense strain her first thought was for Winston, and it is an example of how she sought to protect his peace of mind.

Not long after this, in May, Randolph suffered fairly serious injuries when the truck in which he was returning from a long-range raid on Benghazi overturned. He was several weeks in a Cairo hospital before being invalided back to England until October.

* * *

By June 1942 differing views on grand strategy between the British and American high commands made Churchill anxious to see the President again. Visiting Britain in April, Harry Hopkins and General George Marshall* had argued in favour of landing an Allied force in Western Europe that very year – a course which Churchill and the British Chiefs of Staff were convinced would be calamitous. Churchill saw a danger that America, if deterred from attacking the enemy in the West, might lend its ear to the powerful 'Pacific lobby' and turn its interest and its strength too much towards the Far East. To counter this possibility, and to advocate an alternative plan of action for 1942, in the form of Allied landings along the French North African coast (Operation Torch), Churchill arranged in mid-June to fly once more to America.

During his ten-day visit to Washington he received the news of the fall of Tobruk. It was a grievous military blow, but worse, it was a humiliation. Ever afterwards Churchill remembered with emotion and gratitude the attitude of the President and Harry Hopkins: 'Nothing could exceed the sympathy and chivalry of my two friends. There were no reproaches; not an unkind word was spoken. "What can we do to help?" said Roosevelt.'[10]

After concluding satisfactory talks with the Americans, and advancing plans for the execution of Operation Torch later in the year, Churchill felt he himself must impart to the Russians the unpalatable tidings that there could be no Second Front in the immediate future, and he therefore decided that after his return from Washington he would set off on his journeys once more and combine visits to Cairo and to Moscow.

On Friday, 31 July Winston and Clementine paid a fleeting visit to Chartwell. It must have been a precious day for them both – they had so little time together, and now he was on the brink of another long, fatiguing journey; she did not doubt its necessity, but she knew, all too well, the risks he must run. Chartwell had a charm and appeal all of its own during the war. The big house, with shuttered windows, like blind eyes, stared out gaunt and silent over the beautiful valley. The grounds, virtually untended, had become like the garden of the Sleeping Beauty. But the fish were there still, waiting to be fed, and the Canadian goose (which had earned the name of the 'Naval Aide de Camp', because it would always accompany Winston on his tours of the lakes) was there to greet them. A few black swans, too, remained, having survived the depredations of the foxes.

Two days later, Winston flew off to Egypt for a week of conferences and inspections, during which important changes were decided upon in the military hierarchy: General Alexander was to assume command of the forces in the Middle East, and General Montgomery was to take over the command of the Eighth Army. Churchill then journeyed on to Russia, bearing his unwelcome news. 'It was like carrying a large lump of ice to the North Pole,'[11] he later wrote. The tidings were indeed ill-received, and the talks threatened to break down: but finally disagreeable reality

* George Marshall (1880–1959), US general and diplomat. At this time Chief of US army. He would be the prime mover in the European Recovery Program, which became known as the 'Marshall Plan'. He was awarded the Nobel Peace Prize in 1953.

was accepted, and comradeship preserved.

Back in Cairo for a few days on his way home, Churchill already sensed a new atmosphere permeating all levels in the Army. On 24 August, his Liberator aircraft touched down at Lyneham, where Clementine and Randolph were waiting to greet him.

The fruits of Winston's visit to Cairo were not long in ripening. On 23 October, the Eighth Army attacked the enemy at El Alamein, launching the heaviest artillery barrage recorded in the history of war, and on 4 November, after twelve days of heavy fighting, it inflicted a severe defeat on the German and Italian forces under Rommel's command; thirty thousand prisoners were taken. Here at last was victory – real, solid and overwhelming. This was the turning-point of the war. 'It may almost be said,' wrote Churchill, '"Before Alamein we never had a victory. After Alamein we never had a defeat."'[12]

In his diary, Sir Harold Nicolson gives a detailed description of a luncheon party he attended on Friday, 6 November, two days after these tremendous tidings:

> At 1.15 I stroll across to Downing Street where I am to lunch . . . I go downstairs to the basement where the Churchills are living, since the upper floors* have been knocked about. They have made it very pretty with chintz and flowers and good furniture and excellent French pictures – not only the moderns, but Ingres and David [lent by the National Gallery].
>
> I find Lady Kitty Lambton and Lady Furness and Clemmie Churchill. We are given sherry. Eddy Marsh comes in, and then the Private Secretary, Martin . . . He tells us not to wait for Winston, as he is late. We go into luncheon: sea-kale, jugged hare and cherry tart. Not well done. In a few minutes Winston comes in. He is dressed in his romper suit of Air Force blue and he carries a letter in his hand . . . He gives the letter to Clemmie. It is a long letter from the King written in his own handwriting, and saying how much he and the Queen have been thinking of Winston in these glorious days. Winston is evidently pleased. 'Every word,' he mutters, 'in his own hand.'[13]

Sir Harold then goes on to give an account of the conversation. Lady Kitty and Lady Furness had just escaped from the South of France; they were old 'fringe-friends', obviously asked for old times' sake. 'Winston talks to Lady Kitty and I talk hard to Lady Furness, as she is frightened of the gigantic figure on her right, and Winston is bad at putting people at their ease. Nor does Clemmie Churchill help much.'[14]

This is a clear example of a luncheon party where the topics of conversation were strictly confined. Both Winston and Clementine knew that, while they were all rejoicing at the wonderful feat of arms in the Egyptian desert, another tremendous operation was impending: Torch, the invasion of French North Africa, was planned to take place in two days' time.

* They were actually 'living' at the Storey's Gate Annexe, but when the bombing diminished they still used the basement rooms at No. 10 as much as possible for entertaining.

Sir Harold continued his description of this occasion by recounting how Brendan Bracken (then Minister of Information) had appeared, and how Churchill told him to arrange for the church bells to be rung the following Sunday. 'Some hesitation is expressed by all of us. "Not at all," says Winston, "not at all. We are not celebrating final victory. The war will still be long. When we have beaten Germany, it will take us two more years to beat Japan."'[15]

By a strange coincidence, I also recorded the conversation about the church bells in my own diary, because towards the end of this luncheon I arrived from my battery at the beginning of a short leave. I noted that it was a 'very sticky' occasion, and that when 'the party broke up, Mummie [was] being violent (quite rightly I thought) with Papa who wanted to have all the bells rung on Sunday'.[16]

Since the beginning of the war the church bells had been silent – they were to be the warning that the invasion had started: now those dark days had passed, and Winston wanted the bells to ring out in celebration of the resounding victory at El Alamein. Clementine was more cautious, and fearful lest some reverse should occur which would make nonsense of such a mark of triumph and elation, and she vehemently begged Winston to bide a while – and on this occasion he heeded her arguments. But soon the bells did ring out – joyously and triumphantly – on Sunday, 15 November. By then, the enemy forces in French North Africa had laid down their arms, and the victorious British Army had once more entered Tobruk.

* * *

Clementine was particularly busy just now because, at the invitation of the Queen, Mrs Eleanor Roosevelt had come on a three-week visit to Britain; for the first few days of her visit she stayed at Buckingham Palace. During her time in Britain she carried out a full and, to any other woman, totally daunting programme, visiting places as far apart as Glasgow and Londonderry. The main purpose of her visit was to acquaint herself with what the war really meant to the British – and particularly to women. In addition, she visited United States troops stationed in this country. Winston and Clementine were, of course, most anxious to entertain her, and for the first weekend of her visit Mrs Roosevelt stayed at Chequers. The following Tuesday, they gave a dinner party for her at No. 10, and Clementine asked some distinguished British women to meet her.

During the course of her visit, Clementine went with Mrs Roosevelt on a good many of her expeditions. The 'First Lady's' pace was too much for her – visibly, on one occasion, when they were touring a WVS clothing distribution centre: Clementine simply sat down on a marble staircase, and joined up with her later! She was not the only one to feel the strain: toughened and younger journalists were worn out by trying to keep up the marathon pace set by the President's wife.[17]

Halfway through Mrs Roosevelt's visit, Clementine wrote a long account of her activities to the President, telling him the effect his wife's presence had 'on our women and girls. When she appears their faces light up with gladness and welcome.' Clementine was much struck by the way she spoke on several occasions to groups of young servicewomen: 'Each time she said something significant, fresh and true, and she gave all who heard her a sense of being in the presence of a remarkable and benevolent personality.' Clementine also admired her handling of a press conference: 'I was struck by the ease, friendliness and dignity with which she talked with the reporters, and by the esteem and affection with which they evidently regard her.'[18]

At dinner one night, Mrs Roosevelt and Winston had a 'slight difference of opinion' over Loyalist Spain:[19] they never really got on – but they esteemed each other for what they were. Clementine played a neutral, conciliating role, and brought the issue to an end by leading the ladies out of dinner. She liked and admired Eleanor Roosevelt, but evidently did not drop her guard, leaving her guest to wonder what sort of a woman lay behind the attractive, surprisingly youthful exterior and charming personality observed by Mrs Roosevelt, who commented in her diary: 'One feels that she has had to assume a role because of being in public life and that the role is now part of her, but one wonders what she is like underneath.'[20] She knew about Clementine's involvement with aid to Russia, and noticed that she 'is very careful not to voice any opinions publicly or to associate with any political organizations'.[21] This was, of course, in strong contrast to herself: she gave many press interviews and wrote a regular newspaper feature, 'My Day', which was syndicated throughout the United States. Eleanor Roosevelt's political views and involvements were at times an inconvenience, if not an embarrassment, to the President.

But if at times the British Tory and the New World radical took a tilt at each other, there was no reserve in Winston's praise and gratitude for the effect of Mrs Roosevelt's visit. Clementine went to say goodbye to her on the day of her departure; Winston was at Chequers, but she brought Mrs Roosevelt a letter from him in his own hand: 'You certainly have left golden footprints behind you,'[22] he wrote, with grace and sincerity.

Every year, whatever the news, Clementine gathered together a party of the family and closest friends to celebrate Winston's birthday. It was immensely touching to him – and to all of us – how since the beginning of the war the whole country, and indeed much of the wide world, seemed to want to wish him 'Happy Birthday'. Greetings and presents used to pour in, and despite the rationing and austerity people would send delicacies of every kind. This year of 1942 – Winston's sixty-eighth birthday – we were a small party of only seven apart from my father and mother who gathered at The Annexe to celebrate: Diana, Sarah, Pamela, Jack Churchill, Venetia Montagu, Brendan Bracken and myself. My mother arranged a lovely party, as usual, and the rooms looked particularly pretty with so many birthday flowers. We were all conscious of the thousands of people who joined with us in spirit as we drank Winston's health – and this year the bright gleam of victory added lustre and joy to these family rejoicings.

CHAPTER TWENTY-ONE

All in the Day's Work

WHATEVER NEW PREOCCUPATIONS AND COMMITMENTS CAME HER WAY AS THE war ground on, Clementine remained faithful and efficacious as a member of the committees for the Fulmer Chase Maternity Home, for which she had been working since November 1939. There was a real and pressing need for this home: many officers in all three services made unusually early marriages, many of them were of small means and few had established homes. Their wives for the most part lived with their own families, or 'followed the drum', living in lodgings or rented accommodation; many of them had been serving themselves in one of the women's services. When the time came for them to have their babies, their husbands might be abroad or at sea; and some of them were already widows. It was felt that it would be a source of reassurance to their husbands to know that there was somewhere where their wives would be looked after skilfully and kindly, and where they could be in comfort and, if possible, safety, when they gave birth.

A house had already been made available for this purpose by Mrs Edward Baron, who lent her large and comfortable mansion (free of rent and rates) – Fulmer Chase, in Buckinghamshire. As a result of the original appeal letter to *The Times*, it was possible to open the home for thirty patients early in 1940.

From the start, Clementine took the closest interest in the maternity home; her particular care was the food, over which she took endless trouble to ensure that despite the difficulties of staffing and rationing, the patients' and staff's meals were very good, and appetizingly served.

As the pressure on accommodation at Fulmer Chase increased, the period for which mothers could remain in hospital had to be reduced from three to two weeks, and many of them had to take their babies back to inadequate accommodation. To improve this situation, an anonymous benefactor gave Clementine a sum large enough to buy a house, and in February 1942 Fircroft Post-Natal Home (at nearby Gerrards Cross) was opened. It served as an annexe to Fulmer Chase, and here mothers could spend a further fortnight after leaving the maternity home. Fulmer Chase was not very far from Chequers, and Clementine was therefore able the more easily regularly to attend the frequent meetings of the committees; and her association with the Home continued until it was wound up some years after the war.

When I appealed through the press for material for the first edition of this book, a large number of the letters containing personal reminiscences of my mother came from women who had borne their babies at Fulmer Chase, and who spoke in glowing terms of the care and comfort they had experienced there, and of the pleasure her visits gave them.

* * *

In February 1941 Clementine became President of the YWCA's Wartime Appeal, in succession to Lady Halifax, who accompanied her husband to Washington on his appointment as our Ambassador there. During the war, the YWCA's principal task lay in providing hostels, clubs, huts and canteens for women war-workers, and particularly for the ever-increasing numbers of young servicewomen, nearly all of whom were far from their homes, and many of them serving overseas. In order to provide adequately for their needs it was essential for the YWCA's appeal to be headed by a considerable public personality.

Fund-raising on the scale on which she was now to be engaged was a new venture for Clementine. In nearly every case the first salvo fired was a letter to *The Times*, to be followed by a broadcast appeal. Clementine was not an experienced broadcaster, but she was soon to become one; and if her script was of necessity in the main constructed for her so that all the pertinent points were made, she always wrestled with the text herself, to try to make it come to life. As time went on and she acquired more confidence, and more detailed knowledge of the cause in question, her appeal letters and broadcasts became more natural in manner and even more effective. Unaccustomed as she was to making appeals to individuals for money, and although at first very shy, she soon became very good at it, and was delighted by the gratifying response she nearly always received. She took great pains with her individual appeal letters, which were very much 'her own', and her thanks were always gracefully and warmly expressed.

Surprisingly, Clementine was not a 'natural' as a public speaker, although on occasion, at political meetings, she could, when aroused (as in the old Dundee days), forget her shyness and launch forth upon what Winston was wont to call 'the unpinioned wing'. Now for her wartime causes she overcame her natural reticence in public, and would welcome any opportunity, and face any audience, in order to further the purpose in hand. One such occasion was in December 1943, when she visited the London Stock Exchange and appealed for funds to build a YWCA hut. The money was duly forthcoming, and when in March of the following year she officially opened the hut, Sir Robert Pearson, the Chairman of the Stock Exchange, said in his speech that he had mentioned Clementine's visit to this essentially and tenaciously male preserve to the Queen, who had commented that she 'was a brave woman'. Her 'audacity' had been generously rewarded.

But Clementine did not regard her role as President of the Wartime Appeal as limiting her simply to the task of raising large sums of money; she became deeply interested in their work, and wanted to have her say in the general policy and standards of the hostels and clubs. Although she in no way ignored the high Christian mission of the YWCA movement, she did not see at all why high principles need go hand in hand with low standards of physical comfort. It was in this field that she saw a real opportunity to make a contribution, and she bent all her powers to improving the running and the standards of the Association's hostels.

Her approach to this new job was characteristic: find out the facts; grasp the problem; grapple with the difficulties. As with her investigations into conditions in the shelters, she would never allow official appraisal of what could be done to discourage her or to deflect her from aiming at the highest standards. It was, of course, very difficult to get alterations or repairs done on the home front, and bomb-damaged housing had priority. Everything was in short supply: labour; materials; hot water; food – in fact, almost all the things that contribute to comfort. Nevertheless, Clementine soon formed a clear view of what should be the YWCA's policy in regard to its hostels: it should try to produce the highest standards possible despite all the difficulties.

As soon as she had launched her first appeal for the 'YW', Clementine lost no time in visiting as many of their clubs and hostels as possible. Her eagle eye went straight to the essentials: what was the ratio of baths to the number of girls in the hostel? Was the hot-water supply adequate and efficient? What were the mattresses like? (She never hesitated to have the bedclothes whisked off so that she could judge for herself.) Were there proper facilities for washing and ironing? Were the rooms too crowded? (Unfortunately they nearly always were, owing to the great number of girls who wanted accommodation.) Were there bedside lamps? But Clementine was not solely obsessed with physical comforts, important though she thought them to be: she never under-estimated the vital role played in each hostel by its Warden; and she appreciated that many a hostel had a reputation for being a happy one because of its Warden's personality, even if the living standards left much to be desired.

But all this was not achieved by merely looking gracious and smiling. Clementine liked to forge ahead, and was intolerant of delays – nor could she stand complacency. She cannot always have been easy to work with, as she found it difficult to conceal or dissimulate her opinions, and was often impatient or sceptical of official explanations. But her colleagues and the officials of the YWCA who worked with her closely throughout those years have paid generous tribute to her work.

* * *

The response to the Appeal for the Red Cross Aid to Russia Fund, headed by Clementine and launched in mid-October 1941, was immediate, and contributions large and small flowed in from all over the country. On 28 October, when the appeal was only twelve days old, Clementine in a broadcast was able to announce that the total raised stood at £370,000. For the first three months of its existence, the Aid to Russia Fund received all of the Red Cross Penny-a-Week Fund, and thereafter, for the rest of the war, a quarter. Three days before Christmas, Clementine again broadcast, and she was able to give the almost incredible news that in less than three months the Fund had passed the £1 million mark.

Nor was this a momentary wave of passing enthusiasm: all over the country organizations, bands of schoolchildren, and groups of factory and office workers set themselves to collect money for this cause on a regular basis. Individual donations also poured in: millionaires and old-age pensioners alike gave, and gave again. Throughout the country, towns and cities held flag days and organized Anglo-Soviet Weeks from which the Aid to Russia Fund reaped a rich harvest. Auctions, exhibitions and theatrical galas took place; artists donated pictures; actors and musicians gave generously of their time and talent – the great pianist Moiseiwitsch, for instance, giving a series of concerts. The proceeds of the Wembley International in January 1942 were devoted to the Fund, and the England v. Wales football match in February 1943 raised £12,500 for it, at that time the largest sum ever raised for charity by one sporting event.

But the great difficulty of making the aid really effective was the vast scale of Russia's needs. In 1942, the British Red Cross sent a representative to Moscow to find out exactly what their requirements were: he was given a schedule of gigantic proportions. Some of the items demanded were at first unobtainable but, nothing daunted, the Red Cross set about finding ways and means to procure as many as possible of the things that had been asked for. Several British factories undertook the manufacture of instruments and medical equipment never before produced in this country.

While it was through Clementine's efforts and influence that the nation's sympathy for Russia was translated into a vast and continuing flow of money, the Aid to Russia Committee of distinguished and knowledgeable men and women wrestled with the technical problems and the practical difficulties of delivering the goods, once procured.

The Russians were at times extremely difficult to deal with, chiefly on account of their often unreasonable demands, and their pathological suspicions of our motives or actions. However, despite all difficulties and discouragements the work of the Fund forged ahead and, by the end of December 1942, close on £2,250,000 had been subscribed. Owing to the scale of Russian demands, and the necessity of placing orders sometimes six months ahead, the Appeal Committee had deliberately allowed the Fund to become heavily overdrawn. On New Year's Eve, Clementine made a broadcast: she gave an account of all that had been hitherto subscribed, and she told her listeners of the shipments that had already been despatched; she then announced that the Fund was in debt, and gave the reason for it; she added that she knew from experience that she could rely on the public to rise to this urgent need. Her optimism was fully justified: on 20 February, she was able to announce that in response to her New Year appeal she had received £170,000 in donations and that, combined with the contributions from the Red Cross Penny-a-Week Fund, had not only cleared the debt but would also enable the supplies sent to Russia to be increased.

By the end of October 1943, two years after the launching of the Aid to Russia Fund, the £4 million mark was reached. In a statement issued by Clementine, she said:

> This sum of money is small compared with the size of Russia, the number of her wrecked cities and of her heroic unconquerable population. But it is a token from the hearts of all the British people all over the world. The rich and well-to-do contribute generously, but the larger part comes from the workers – hundreds of thousands of them give week by week. It is indeed the people's fund, and through it they express to the Russian nation their respect, wonder and admiration for unsurpassed courage, patriotism, and military achievement.[1]

By December 1944 over six million pounds had been raised, and although active collection for the Fund ceased in 1945, contributions continued to be received, and by June 1947 the grand total stood at seven and a half million. During the Fund's first fourteen months nearly 300,000 tons of supplies were despatched to Russia.

Clementine's work for the Aid to Russia Fund added greatly to her labours: hours had to be spent in committee, wrestling with the lists of requirements and how to set about supplying them; then all sorts of difficulties arose over shipping the goods to Russia. In all these matters she took an active part. She also made many expeditions to plants and factories which were turning out special orders for the Red Cross Fund, and she was invited to an endless succession of functions and events organized all over the country in support of her Fund; she tried to go to as many as possible, and this often entailed long, tiring journeys.

Apart from other correspondence connected with the Fund, all the gifts and donations were of course officially acknowledged, but many of them were sent especially and personally to her; these she always tried to acknowledge herself, and it was not only large and splendid contributions which received personal replies: for instance, Clementine used to take especial trouble to thank schoolchildren who had made several small collections for her Fund. The same direct and personal touch which characterized all her public work did not desert her now, when faced with such a mammoth undertaking.

In all this work her chief link with the Soviet Embassy in London was the Ambassador's wife, Mme Agnes Maisky. She was not an easy woman to deal with, and Clementine had to spend much time and exercise great patience in listening to her long lists of imperious demands or complaints. Fortunately, however, Clementine was neither small-minded nor easily offended – and she often gave quite as good as she got! But she was soon accustomed to Mme Maisky, and came to like this brusque, intelligent woman, whose fiery patriotism was so clearly evident. When the Ambassador was recalled to Moscow in September 1943 Agnes Maisky wrote her a warm letter of farewell: 'During my stay here I greatly appreciated our close relations, and your friendliness and kindness to me; but most of all I am grateful to you for all that you have done to send medical aid to my countrymen. This is a great work, and we shall never forget it.'

There were other organizations for sending help to Russia: a number of communist and left-wing bodies started funds, including the Joint Aid to Russia Committee, under the leadership of Dr Hewlett Johnson, the Dean of Canterbury (known as the 'Red Dean'); various trade union groups also raised money for Russia independently; and sums were contributed direct to the Soviet Red Cross. But none of these agencies caught the confidence or the imagination of the British public as greatly as 'Mrs Churchill's Fund', which had three signal advantages. First, it was backed by the organizational power and prestige of the Red Cross. Second, there is no doubt that the name Churchill, combined with the respect and confidence Clementine herself inspired among all classes in Britain, lent the Fund a lustre, and fired people's enthusiasm. Third, all the other funds had a political bias; through the Red Cross Fund, people with no sympathy for communism could pay their testimony of admiration and gratitude to the people of Russia, whose heroic fight for their native land, and whose terrible sufferings, moved all but clay hearts.

The crowning moment of Clementine's long and arduous work for the Russian Fund came in 1945, when, in response to the invitation of the Soviet Red Cross, she and Miss Mabel Johnson (Secretary to the Aid to Russia Fund) went to Russia for six weeks, visiting mainly those centres where the goods and equipment provided by the Fund had been distributed. The story of this interesting journey – one of the most thrilling and moving events in Clementine's already not uneventful life – will be told in a later chapter.

In the course of this memorable visit both Clementine and Miss Johnson were to receive Russian decorations – Clementine, the Order of the Red Banner of Labour, and Mabel Johnson, the Medal of Labour Distinction – in recognition of their work for Aid to Russia. It is therefore strange, and rather sad, that thirty-three years later, one of the few countries which did not send a representative to Clementine's Memorial Service in Westminster Abbey was the Soviet Union.

In June 1946 Clementine received from the King at Buckingham Palace the Grand Cross of the Order of the British Empire, the citation being 'for public services' (it pleased her greatly that this honour was bestowed on the recommendation of the Labour Prime Minister, Clement Attlee).*

*Gratified though she was by this accolade, which carries the title 'Dame', she would never allow herself to be called 'Dame Clementine' – preferring to remain (until Winston received the Garter) Mrs Churchill, GBE.

The work undertaken in the war by women of all ages and classes was truly remarkable in its scope and variety, and Clementine's own efforts can take an honourable place in the records of women's war work. The name she bore, and her position as wife of a powerful Prime Minister, lent strength to her arm; but to these undeniable advantages she brought the unstinted, unwearying offering of her own talents and personality, which reaped a rich return for those causes for which she worked, and was in itself a shining example.

CHAPTER TWENTY-TWO

Conference and Crisis

THE YEAR 1943 WAS ONE OF DEARLY BOUGHT ADVANCE ON ALL FRONTS. The Russians began the reconquest of their homeland, the Germans were driven from North Africa and the Allies invaded Italy; on the other side of the world, the American forces were locked in a deadly grip with the Japanese in the Pacific, and the British battled to defend the Burma Road. It might well have been named the 'Year of Conferences', for it saw four meetings of great consequence between Churchill and Roosevelt, and the first meeting of the Big Three at Teheran in November. Wide divergencies of view arose between the British and American Chiefs of Staff over the strategy to be adopted once the North African campaign was over. The main argument revolved around the respective priority to be given to the planning and timing of Operation Overlord (the invasion of Normandy), and the campaign in Italy and the Mediterranean. In order that a united policy might be forged, meetings were vital between Churchill and Roosevelt and their colleagues. Winston also set great store by personal contact with commanders on the spot; he therefore undertook many long, exhausting and dangerous journeys. Nowadays, when air travel is a commonplace, safe and comfortable means of transport*, it is difficult to grasp just how different were the conditions of flying then. And to the physical discomforts, and the much greater technical limitations and hazards, was added the ever-present possibility of attack by the enemy.

For my mother, 1943 was a year of almost continual concern for Winston's health. That this was already a cause of worry when the New Year opened is borne out by an entry in my diary for Sunday, 3 January, when I spent twenty-four hours' leave at Chequers:

> Went for a long & lovely walk alone with Mummie after lunch.
>
> We talked entirely of the family – & especially of Papa. It appears that he might get a coronary thrombosis – & it might be brought on by anything like

* I wrote this before hijacking, suicide bombers and the scandalously cramped conditions in tourist class could be deemed to have put in question this broad statement.

a long &/or high flight –

The question is whether he should be warned or not. Mummie thinks he should not – I agree with her.

For my mother, the knowledge of this additional and agonizing possibility shadowed all the journeys (and particularly the flights) which my father undertook hereafter.

Apart from this continuing concern for Winston's health, there were also continuing anxieties and tensions within our family arising from the break-up of Randolph's and Pamela's marriage. The main burden of these difficulties, which inevitably involved his parents, had to be borne by Clementine: she did her best, but any intervention in such situations nearly always results in bitter reproaches being levelled, and this case proved no exception to the rule. In moments of reflection and calm Randolph appreciated his mother's difficult situation. 'I hope you don't think me unappreciative of your position in all this disagreeable business,' he later wrote to her. 'I know how difficult things have been for you and I do know that you have tried to understand my point of view.'[1]

For much of the time my mother could not discuss these intimate and painful family matters with anyone, for my father was utterly engrossed by the war, and Sarah and I were for most of the time away with our units. Diana, although living in London, was now expecting her third child, and was greatly preoccupied with caring for her family in wartime conditions, and with her duties as an Air Raid Warden.

The first of the meetings between Churchill and Roosevelt in 1943 took place in Casablanca in January, before the conquest of Tunisia was complete. Winston left England in a Commando plane on 12 January. The travelling conditions were austere and uncomfortable: a newly installed heating apparatus was found to be in danger of overheating and causing an explosion, so that it had to be turned off halfway through the night, and the passengers suffered acutely from the cold.

The President and Prime Minister conferred for about ten days at Anfa, on the outskirts of Casablanca, and then Winston prevailed upon his friend to stay on for a further day or two traversing the desert so that he might show him Marrakech and the glories of the sunset on the snows of the Atlas Mountains. After the President's departure, Winston lingered for a further two days in this place which so beguiled him, and while maintaining a constant flow of minutes to the War Cabinet, painted the only picture to come from his brush during the war. Later he would give this picture to the President.

On 14 January Clementine had written to Winston from London:

My Darling,

The 'Annexe' and 'No. 10' are dead & empty without you – Smoky [the cat, a great favourite with Winston] wanders about disconsolate . . .

Everything is quiet – so far at this end 'the secret' is water-tight.[2]

Indeed, security arrangements for the meeting had been amazingly successful, and the journalists summoned to a news conference at Casablanca on 24 January were flabbergasted to see the 'Big Two' – and even more astonished to learn that they had been there in conclave for nearly a fortnight.

Meanwhile, learning that air raids on London had started again (18 January), Winston had concern for those at home, sending the message to his Private Office: 'Air Cde Frankland* wishes you to ensure that Mrs Frankland and the servants go down to the shelter in event of air raids warning'.[3]

Winston went on to Cairo, and then (after overcoming considerable opposition to this plan from the War Cabinet) to Adana, on the border of Turkey and Syria, for talks with the Turkish President: the object of the meeting was to try to persuade President Inönü to assist the Allies, and to facilitate in particular the establishment of British bases in Turkey. Returning to Africa, at Tripoli he witnessed with emotion and pride a march-past of two divisions of General Montgomery's victorious Eighth Army; and finally he called in at Algiers for talks with General Eisenhower before flying home. His movements were decidedly ad hoc and, necessarily, subject to the greatest secrecy. Clementine was apprised by the Private Office of all the changes in his programme, and received two long letters from Winston himself, but although she was accustomed to sudden changes of plan, she was particularly anxious at this time because she knew of the fears for Winston's safety during these extended travels.

There was a charming exchange of messages during the last part of his journeyings, when Winston's code-name was 'Mr Bullfinch':

From Clementine to Winston, on 2 February:

> I am following your movements with intense interest. The cage is swept and garnished fresh water and hemp seed are temptingly displayed, the door is open and it is hoped that soon Mr Bullfinch will fly home[4]

From Winston to Clementine, the following day:

> Keep cage open for Saturday or Sunday. Much love[5]

And on 5 February:

> The Bullfinch hopes to make a long hop home tonight[6]

But her concern shows through the note she sent by hand to greet him at the airfield on his return on 5 February:

* 'Air Commodore Frankland' was WSC's code-name for this journey.

My Darling,

Welcome Home. The anxiety & tension has been severe. What an inspiration was the visit to Turkey. And how glad I am you did not allow yourself to be deviated from that extra lap of your journey. I'm thinking of you flying thro' the tenebrous dark & pray you make a good land-fall.

Your loving & expectant
Clemmie
[Picture of lovely fat Kat]

Although Winston returned elated by the success of his mission, his journeyings – which might well have wearied a much younger man – had taken their toll. He had travelled thousands of miles in varying degrees of discomfort and experienced sharp changes of temperature; he had slept in luxury villas, trains, aeroplanes and caravans; and wherever he went, day and night, he carried the burden of the conduct of the war. When, three days after his return home, he addressed the House of Commons, he was already suffering from a cold and a sore throat. On 12 February he developed a temperature and spent a day or two in bed. On the evening of the sixteenth, while dining alone with Clementine, he felt ill and found his temperature had soared. Pneumonia was diagnosed the following day, and bulletins were issued.

I came home for the day on Sunday, 21 February, and found

M was ready for church. In her lovely dark cloth coat trimmed with beaver & the flame-coloured scarf bursting out – beaver muff & hat . . .

We two went to service at the R[oyal] Military Chapel [at Wellington Barracks] – It was rather lovely & comforting.

M is not seriously worried about Papa – but he is pretty ill.

I was shocked when I saw him. He looked so old & tired – lying back in bed.[7]

The next day I left for an anti-aircraft practice camp at Whitby, in Yorkshire, and my mother wrote to me on the twenty-fourth:

First of all I write to tell you that Papa is I think really better. The Doctors still give out the 'no change' bulletin because the temperature has not completely gone down, but I can see for myself that he is better. His face looks quite different. He has lost that weary look. I know, my darling, this will relieve you. I have really been very worried about him.

I am longing to hear how you are, and all you are doing, and how you are settling down. I love detail about everything if you have time.

The entry in my diary, and my mother's subsequent letter, show to what extent she could maintain a calm exterior and conceal her inmost fears even from those nearest to her.

As soon as Winston was well enough the doctors insisted that he should have at least a ten-day convalescence; so he and Clementine retreated to Chequers. Clementine also took the opportunity to have a holiday from her work, and all but the most pressing problems were fended off. During this period she spent one weekend with the Trees, from where she wrote to me: 'As you see I am at Ditchley for the week-end. Papa has got "Prof" & Uncle Jack who is also having a rest cure – & masses of films – War & Hollywood.'[8]

Two days later Clementine was back at Chequers, and she gave me this account of her 'quiet' week in the country:

> My Darling,
>
> Here are Papa & I living for a whole week at Chequers! Most unusual – By the time it is over I shall have settled down & be quite enjoying it! Papa is progressing very slowly but (I hope & believe) safely through his convalescence to his normal strong state of health.
>
> Yesterday morning, as Papa could not go to his weekly luncheon with the King – the King came here & paid a morning visit in the White Parlour à la Jane Austen. I did not see him as I was motoring back from my week-end at Ditchley . . . Tomorrow night Anthony & Beatrice Eden are coming for the night – Thursday Mr Amery [Secretary of State for India] arrives & Mr Butler [Minister for Education]. Friday I go to Nottingham, so you see it's pretty well non stop . . . Sir Stafford & Lady Cripps are bearing down & David Montgomery [aged fifteen] from Winchester. Papa is going to give him a wonderful portrait of his Father painted by Neville Lewis the same artist who did you when you were a Baby & would not sit properly! We are having the Desert Film for him.
>
> On Monday we return refreshed (!) to London . . .
>
> Your loving but somewhat exhausted
> Mummie[9]

At the end of March I was posted to 481 Heavy Mixed Anti-Aircraft Battery, which was stationed in the middle of Hyde Park, where my father would quite often 'drop in' on us during the course of an air raid, sometimes bringing his dinner guests with him.

By now Winston was quite restored in health, but Clementine was far from well. Once again the pace and strain of her life and work were telling on her, and the anxiety she had felt over my father's illness was now taking its toll. She developed a painful boil, and had to undergo a series of inoculations and X-ray treatment, which were most exhausting, and for a short period she was virtually laid up. Although the treatments soon resulted in improvement, Clementine was so obviously in an exhausted and run-down state that her doctor advised her to go away to the seaside.

Much though she hated leaving Winston, and had conscientious qualms about her work, Clementine was feeling so low that she acquiesced and, with Jack Churchill as a companion, she went to Weymouth, where they stayed at the Royal Hotel for about ten days. From there she wrote to me on 24 April:

> This must be a charming old-fashioned sea-side 'resort' in peace time; but now the lovely golden sands are a thick tangle of barbed wire & there are two rows of iron barricades half under water at high tide – all to keep off an invasion. Last night we had an Alert – planes flying over to Birmingham I'm told. It's rather curious & delicious to be quite out of the world & inside knowledge for a bit & very restful tho' dull. I'm horribly tired & can't yet shake it off. The hotel is comfy, but the food dreary, Bird's Custard at every meal! The 'Times' says London had an alarm last night so I expect you were standing to.

When Clementine returned, refreshed and revived, from Weymouth, she found that preparations were afoot for Winston's imminent departure for Washington; he had decided that he must see the President again now that the North African campaign was nearing its close. His doctors had taken the definite view that, on account of his illness in February, it would be unwise for him to undertake any long flights just for the present: Winston and his party therefore travelled by sea, in the *Queen Mary* this time, leaving on the night of 4 May.

Less than a year before, Winston had been in Washington when he received the grim news of the fall of Tobruk; now, on 13 May 1943, it can be imagined with what deep feelings he read this triumphant message from General Alexander:

> Sir:
> It is my duty to report that the Tunisian campaign is over. All enemy resistance has ceased. We are masters of the North African shores.

On that same day, many thousands of miles away, Clementine's thoughts were with him: 'My darling Winston,' she wrote, 'How I wish that in this hour of Victory I were with you – so that we could rejoice together & so that I could tell you what I feel about your North African Campaign. You must be deeply moved by these events although you planned them & knew beforehand that they could be achieved.'[10]

But in her gratification at the victory, Clementine's watchful eye had detected an aspect of the Conference in Washington that disturbed her, and in the same letter she wrote: 'I'm worried at the importance given by the Press (notably The Times) to the presence of Wavell & his East Indian Naval & Air Colleagues in your Party. I'm so afraid the Americans will think that a Pacific slant is to be given to the next phase of the War – I have cut out the piece of The Times which disturbs me. Surely the liberation of Europe must come first.' And in an emphatic footnote she added: 'Do re-assure me that the European Front will take 1st place all the time.'

On 13 May the King had sent this message to Winston:

> Now that the campaign in Africa has reached a glorious conclusion, I wish to tell you how profoundly I appreciate the fact that its initial conception and

successful prosecution are largely due to your vision and to your unflinching determination in the face of early difficulties. The African campaign has immeasurably increased the debt that this country, and indeed all the United Nations, owe to you.

George R.I.

Winston was much moved by this message, and Clementine was so overjoyed to see his part in the victory recognized that on a sudden impulse she wrote to the Queen.

10, Downing Street, S.W.1.
14th May, 1943

Madam,

I venture to write to Your Majesty to say how moved and touched I am by the wonderful message which the King has sent Winston. It made me cry with joy and I should be so grateful if Your Majesty would tell the King how I feel about it.

I wish that at this time of thanksgiving & rejoicing that I were not separated from Winston; but I know that it is important & his duty that he should be with the President now that new plans are taking shape.

I hope it is not presumptuous in me to have written this letter.

I am Your Majesty's devoted & obedient Servant,
Clementine S. Churchill

Two days later, she received the following reply from the Queen, written in her own hand:

Buckingham Palace
May 15th 1943

My dear Mrs Churchill

I have just received your most charming letter, and hasten to send you a line of thanks for what you have said in it. I know the King sincerely feels all that he wrote in his message to the Prime Minister, and we both feel more grateful than words can say for your husband's unfaltering courage and over-the-mountains vision in the anxious days behind us. We look forward to the future battles with the comfortable feeling that we've got the right man as Prime Minister, and the best and most civilized people in the world to fight those battles – tolerant & brave & humorous. May God give us true victory to build a well governed world for all peoples.

I am,
Yours very sincerely,
Elizabeth R.

On Sunday, 16 May the church bells rang out again. Clementine did indeed miss Winston's presence at this triumphant hour, writing to me in a letter on 14 May: 'These glorious victories are uplifting but I have been alone – no-one to rejoice & give thanks with – No Papa – No Mouse [Mary] – "nothink"! So I've been rather low which is very wrong.'

While Winston was away, she took the opportunity of going to Chartwell for nearly a week, staying in Orchard Cottage. In Washington, Winston addressed Congress for the second time on 19 May, and the following day Clementine wrote to him:

My Darling

Your address to Congress was grand & a masterpiece of 'walking delicately' –

It warmed me to hear your voice so strong, resonant & resolute.

Yesterday I went with the Attlees to St. Paul's to give thanks for Tunisia – The Service was short & restrained & the Cathedral was cold as Charity, but when we emerged into brilliant sunlight & clanging bells & whirling pigeons & sweet clapping city stenographers who, sensing a ceremony, had stayed on after their work, one's blood began to flow & one's pulses to beat in time with great events.

Diana had her baby [Celia] in one swift hour. Doctor & Nurse only just in time. She looks lovely & well but such a crashing labour is a bit of a shock & she must be well nursed. She & Duncan are over the moon with joy. I have just now come to the cottage for a week to enjoy the heavenly spring weather – Chartwell has on her bridal dress – She is a lovely untidy bride – 'A sweet disorder in the dress kindles in clothes a wantonness'*.

We have had four sleepless nights [in London] caused by nuisance raids – Alerts at Midnight, 2 & 4, with the Parks' guns & Prof [Lindemann]'s rockets barking & shattering.

Last night down here 2 big bombs, I don't know where, shook the little cottage – They say it is the Army down the valley who shew lights & attract them – Here, guns not crashing but muttering in the distance, but a constant drone of 1 enemy plane at a time <u>seemingly</u> circling round & round!

Tender Love Darling
from
Clemmie[11]

Clementine's time at Chartwell (despite these interruptions) did her good, and she returned to her committee work, entertaining and a fund-raising tour for 'Wings for Victory' week.

Winston flew from Washington to Algiers – this time in a comfortable flying-boat – and was able to write to Clementine from Gibraltar, where they paused en route.

* Robert Herrick, 'Delight in Disorder', 1648.

28th May 1943

My darling Clemmie,

You really have been splendid in writing to me. Hardly a day has passed that I have not had a letter to give me so much pleasure and delight. I, on the other hand, have been most remiss but I really have been hunted altogether beyond the ordinary. Not only have I had all the big business about which I came, but I naturally took advantage of all the time the President could give me, which was a very great deal . . .

Harry Hopkins and his wife* were most agreeable and friendly, and evidently in the highest favour. Mrs Roosevelt however was away practically all the time, and I think she was offended at the President not telling her until a few hours before I arrived of what was pouring down on her [i.e. WSC and his entourage]. He does not tell her the secrets because she is always making speeches and writing articles and he is afraid she might forget what was secret and what was not. No-one could have been more friendly than she was during the two or three nights she turned up.

They all made great complaint I had not brought you with me, and made me promise that next time you must surely come . . .

My friendship with the President was vastly stimulated. We could not have been on easier terms. There is no doubt that the speech I made [to Congress] showing the success which has attended our joint efforts, and his part in it, strengthened his position.

The next day Winston and his colleagues continued their journey to Algiers, using for the first time the York aircraft – a transport plane taken over by the RAF and specially converted for these long journeys. Winston had a private cabin, but there were also comfortable bunks for the other passengers, and a small galley where proper meals could be prepared.

In Algiers, Winston stayed in Admiral Cunningham's† villa on the hills above the harbour; he decided to stay on for a few days, resting and working in this pleasant place. Randolph spent a few days with his father, who was able to report to our mother how well and sunburnt he looked.[12] Clementine thoroughly approved Winston's plan to remain for a little longer 'in the sunshine and cool breezes'.[13] He arrived home on 5 June; he had been away a month.

Although 1943 was a year of heavy burden and stress, there was no more doubt about which way the war was going, and the truly agonizing strain had been, albeit almost imperceptibly, eased. It was in this year that Winston once more painted a picture; cards, too, laid by since the beginning of the war, were brought out again, and Winston and Clementine from time to time enjoyed a little bezique. Another great pleasure – the theatre – was also rediscovered, and during this spring and summer they saw several plays:

* Mrs Louise Macy (his third wife), whom he had married in July 1942.

† Admiral Andrew Cunningham, at this time Naval Commander-in-Chief, Expeditionary Force, North Africa; later Admiral of the Fleet 1st Viscount Cunningham of Hyndhope.

Flare Path by Terence Rattigan; *Arsenic and Old Lace*, a glorious comedy by Joseph Kesselring, which gave great pleasure; and two plays by Noël Coward, a friend whose work Winston had always admired, *This Happy Breed* and *Present Laughter*. Whenever Winston and Clementine took their seats only a few minutes before 'curtain-up', they would be recognized: a ripple of applause, which grew to a thunder, would fill the whole theatre – everyone was really glad that they too were having a 'night off'.

It seemed like the apotheosis of the great North African victory when, on 30 June, Winston received the Freedom of the City of London. Sarah and I were able to be with our parents, and drove with them in an open landau from Temple Bar, past shattered buildings, to the Guildhall, where with time-honoured ceremony the Lord Mayor of London conferred on Winston the highest mark of honour the City of London can bestow.

* * *

By the late summer Winston once more felt the need for personal talks with the President to concert the Allied attack on the mainland of Italy and continue preparations for Overlord. The city of Quebec was chosen for this meeting, and the King offered the use of the Citadel there, which is a royal residence. On this occasion Clementine was to accompany Winston, and I was given special leave to act as my father's ADC. The whole party totalled about two hundred people, and we all embarked on the *Queen Mary* on 5 August, reaching Quebec five days later. The Citadel stands imposingly on the cliffs above the St Lawrence river, overlooking the city; we had arrived late in the evening, and my mother and I stood for quite a long time gazing at the twinkling lights below – it seemed a most marvellous sight after four years of blacked-out Britain.

Despite her holiday at Weymouth in the spring and later some days of rest at Chartwell, the nervous strain under which Clementine had lived for the last four years was exacting its price, and when she embarked for Quebec she was already in a state of profound physical and nervous exhaustion; but she hoped that the four days at sea would revive her. Unfortunately, however, during the voyage she suffered from a bout of sleeplessness, and she arrived in Quebec still jaded and fatigued. President Roosevelt had invited us all to spend the few days before the opening of the Conference at Hyde Park, his family home on the shores of the Hudson river. Clementine, however, decided to stay quietly in seclusion at the Citadel, in the hope that she would recover her energy and spirits, and be ready for all the activity and demands of the following week when the Conference would be in full swing. She had Grace Hamblin with her, and Mr Mackenzie King*, with whom she had made friends on his visits to England, promised to look after her.

Winston was greatly disappointed that Clementine could not accompany

* William Lyon Mackenzie King (1874–1950), Liberal Prime Minister of Canada, 1921–6, 1926–30 and 1935–48.

him to Hyde Park, but he realized that she was really at the end of her tether, and that this was the wiser decision; my mother insisted that I should go with my father. On our return to Quebec three days later we found her looking better, but although she had had no public duties, 'quiet little lunches' and long sightseeing drives with kindly strangers are not restful and relaxing to someone like Clementine, whose strict sense of obligation always meant she would appear at her best, and be at her best, however tired she might feel. I realized that the beneficial effect of the last few days had been only superficial.

On Tuesday, 17 August, the Governor-General of Canada, the Earl of Athlone, and HRH Princess Alice Countess of Athlone* came from Ottawa, and shortly after them the President and his retinue arrived; and the next day the Conference got going in earnest. My mother and I were kept extremely busy with a mixture of entertaining arising from the Conference itself, and the receiving and returning of hospitality from numerous charming and welcoming Canadians; we also had some sightseeing and shopping expeditions. Clementine's programme was full and without respite, but she saw it through with grace and goodwill. In addition she made two broadcasts, one for England, where the Aid to Russia Fund was holding a flag day, and the second an appeal on behalf of the Canadian Young Women's Christian Association. As always she took great trouble with her script, delivering part of the broadcast in French.

When the Conference was over on 24 August our party went into 'retreat' for four or five days as the guests of Colonel Frank Clarke, at his fishing lodge on the Lac des Neiges in the Laurentian Mountains, about sixty miles from Quebec. Our host's log house was marvellously comfortable, and Winston thoroughly enjoyed these days in the beautiful wild woods, and indulged in some fishing. But my mother was by now too overtired for enjoyment, so she and I came back to Quebec ahead of my father. By this time I was greatly concerned by her nervous state, the slightest things causing her perplexity and worry out of all proportion to their importance. Our time in Quebec was nearly at an end, but a visit of more than ten days lay ahead of us all in Washington, where once more much would be demanded of her.

In Washington, we stayed at the White House as the guests of the President and Mrs Roosevelt. Outside in the city, the atmosphere resembled a Turkish bath, but the White House was deliciously air-conditioned. Clementine braced herself again and, in fact, enjoyed much of her visit, which included some sightseeing, as well as various obligations such as visiting the Headquarters of the British War Relief Organization. And on one evening, Lady Halifax (wife of the British Ambassador) allowed her to give a party in the Embassy for the wives of all the British officials in Washington.

The highlights of our sightseeing were a personally conducted tour by the

* A granddaughter of Queen Victoria.

President himself of Mount Vernon, George Washington's family home on the shores of the Potomac; and a flying visit to Williamsburg, the wonderfully reconstructed and preserved colonial capital of Virginia. It was there, on the afternoon of Wednesday, 8 September, while we were in one of the lovely houses, that a strange man suddenly appeared and told us that the unconditional surrender of Italy had just been announced on the radio.

On our last day in Washington, while in a bookshop, Clementine missed her footing and fell down some steps, cracking her elbow; the injury was not serious, but very painful and hampering, as she had to carry her arm in a sling. We all left Washington by train on the evening of Saturday, 11 September, travelling once more to Hyde Park to see the President, with whom we spent the following day.

During our visit we had seen a great deal of Mr Roosevelt, and received much kind attention and hospitality from him. One could hardly fail to be dazzled by the magnetic quality of his charm, but Clementine never really fell under 'FDR's' powerful spell, spotting very quickly that his personal vanity was inordinate. However, they got on very well, although she complained to me after an early encounter that she thought it 'great cheek of him calling me Clemmie'! It was surely meant as a compliment; but my mother always regarded the use of her own, or other people's, Christian names as a privilege marking close friendship or long association. The President was not the only one to earn a straight look over this personal foible of hers. Margaret Suckley, a relation and close companion of the President, who had accompanied my mother and me on several of our sightseeing expeditions, and was present at the *en famille* dinner, observed in her diary on Monday, 13 September: 'With all her charm of manner, Mrs C. is very English & reserved.'[14]

That Sunday, 12 September, was my parents' thirty-fifth wedding anniversary, and the day passed very happily, my mother confiding to me that my father had told her that 'he loved her more and more every year'.[15] At dinner, the President proposed their health, and afterwards drove us all down to the little railway station near Hyde Park, where we took our leave of him.

The return voyage was made in HMS *Renown*, and our journey home was a happy one. My father was in relaxed and genial form, and my mother managed to get some good nights' sleep. The voyage for me personally was memorable, for I celebrated my twenty-first birthday; and I was also very nearly drowned, when I was swept the full length of the quarter-deck by a vast wave. However, we all eventually arrived safely home on Sunday, 19 September.

* * *

For some time past both Churchill and Roosevelt had sought to persuade Stalin to join them in their confabulations; and he eventually agreed to meet them in Teheran towards the end of November 1943.

Winston left Plymouth, once more aboard *Renown*, on 12 November, this time taking Sarah with him as his ADC; also included in the party (most fortunately, as it turned out), was Lord Moran. In fact, Winston was far from well when he started out on this journey; he had a heavy cold and was suffering from the effects of several inoculations, and during the ten-day voyage he remained for most of the time in his cabin.

They disembarked at Alexandria on 21 November, and flew on to Cairo, where preliminary talks were to be held with the President. These meetings were most frustrating to the British party, for Roosevelt was almost entirely occupied with Generalissimo Chiang Kai-Shek, whom he had invited, and Chinese and Far Eastern affairs dominated the sessions.

On 27 November Winston flew on to Teheran, and there the 'Big Three' met at last. This Conference was an example of the homely proverb, 'Two's company, three's none', for Roosevelt and Stalin hived off together in a most marked manner. This was naturally wounding to Churchill, whose relationship with the President was so close; but much more than any personal feelings of hurt, he feared for British interests. In a telegram to Clementine he commented: 'The atmosphere at mealtimes is genial but the triangular conferences are grim and baffling'.[16] However, there was one occasion of open amity and comradeship: on 30 November – Winston's sixty-ninth birthday – Roosevelt and Stalin dined as his guests at the British Legation. Sitting between the President of the United States of America and 'the Master of Russia', Winston reflected upon how times had changed: 'I could not help rejoicing,' he later wrote, 'at the long way we had come on the road to victory since the summer of 1940, when we had been alone . . . against the triumphant and unbroken might of Germany and Italy.'[17]

The meeting at Teheran was, however, for Winston a most exhausting conference. He had not succeeded in throwing off his cold, and he was still suffering from a sore throat, prevented from losing his voice only by constant spraying. On 2 December, he and his party returned to Cairo, where he had four more days of discussion with the President. Everyone could see that he was exhausted, and Lord Moran tried to persuade him to abandon his intention of visiting the battlefront in Italy on the way home. But Winston would not be deflected from his course.

Clementine was kept in touch by telegrams of Winston's movements and the general trend of events. On 2 December she wrote to him:

> My Darling
>
> I'm overjoyed to get your message that 'things have taken a very good turn'. Allelulia! Allelulia!
>
> Yesterday F[ield] M[arshal] Smuts* came to luncheon with me & was in the gayest spirits . . . [He] brings you this letter – Sarah has written me the most delightful & colourful accounts of your doings & hers. She writes beautifully

* Jan Christian Smuts (1870–1950), commander of Boer forces in Anglo-Boer War. Prime Minister of South Africa, 1919–24 and 1939–48. A supporter of the Allies in both world wars, he was appointed an Honorary Field Marshal in 1941 and sat with British War Cabinet, 1943.

> & unusually. Perhaps she will write seriously one day. Give her my love & kiss her for me in case I have no time to write to her as well.
>
> Your loving but 'blackout & winter wearied'
>
> Pussy [Kat]

Four days later she wrote again: 'Your telegram just came saying you have gyppie tummy & absolute whirl of business. I do hope the tummy will yield to treatment; I fear it can be rather obstinate.'[18]

On 11 December Winston, ignoring the advice of those around him, left Cairo for Tunis, planning to spend one night on the way with General Eisenhower. From there he intended to go on to see both 'Alex' and 'Monty'. But on arrival at 'The White House' (General Eisenhower's villa near ancient Carthage) he was feeling so ill and weary that he went straight to bed. The next morning, his temperature was 101°F; Lord Moran judged that this was no trivial, passing indisposition, and sent at once for nurses and a pathologist: pneumonia was diagnosed.

Clementine and the Cabinet were informed of the course of events in a series of telegrams which swiftly succeeded one another, often giving contradictory impressions, which was confusing and alarming. But once she learnt that Winston had pneumonia Clementine, strongly urged by his Cabinet colleagues, decided that she must fly out to be with him, and arrangements were put in hand at once. The next morning, 16 December, the bulletin was not reassuring; the whole day was filled with packing, and constantly changing times for her departure; Grace Hamblin was to go with her, and also Jock Colville.

During the day a thick blanket of fog crept up, and by early afternoon all the airfields near London were out of action; so at teatime Clementine and Grace Hamblin, Jock Colville and I set off for Lyneham in Wiltshire, where the airfield was still clear. It was a long, slow, disagreeable drive through swirling fog, but after about four hours we arrived at Lyneham, where we were given dinner in the RAF Mess. Clementine was to have travelled in a heated aircraft, but an hour or so before take-off a fault was discovered, and the only suitable plane available was an unheated Liberator, from which '[t]he bomb-racks had been removed . . .; there were no seats; and the furnishings consisted of some rugs and RAF blankets on the floor.'[19]

After dinner, Clementine and Grace were zipped into padded flying suits, and I went to see them stowed aboard the plane, where I wished them Godspeed; then an officer took me up to the control tower to watch them take off. Jock Colville described the journey:

> We took off into the blackness and Mrs Churchill, who in spite of a gay and apparently unconcerned exterior was deeply worried about her husband, announced that she could not possibly sleep and that she had brought a backgammon board with her. So crouched on one rug and swathed in two others we played backgammon and drank black coffee . . . As the first grey light of dawn appeared we touched down on the Gibraltar runway, the first

and most dangerous leg of our journey safely completed. We must have played at least thirty games of backgammon and I was £2 10s 0d to the good.[20]

It may be imagined with what relief and thankfulness we at home received news of my mother's safe arrival. On 18 December I received a special message: 'Your Mother is here. All is joyful. No need to worry. Tender love. Papa.'

The news of Winston's illness and the fact that Clementine had flown out to be with him were made public on 17 December, thereafter daily bulletins being issued; and soon we were able to rejoice in improving news of my father's health. My mother knew how thirsty we in the family would all be for an inside account, and her letters, written to us individually or as 'circulars', tell the story of Winston's illness and his convalescence.

18th December 1943

My darling Mary

On arrival at this Villa yesterday afternoon at about 3.30 I was taken straight in to see Papa. The joy of seeing him was overcast by the change I saw in him. But when later in the evening I mentioned this to Sarah and Lord Moran they both said that if I had seen him 48 hours earlier I would indeed have been shocked, and they thought that now he was looking remarkably better. This is accounted for by the fact that it is nearly six weeks since I had seen your father. During the whole of this time he has had one little ailment after another, culminating in the illness which I have now ascertained started a week ago . . . I think he is very happy that I am here. After my arrival I stayed with him till six, when he slept for two hours and I rested too. I then had dinner with him. After dinner Sarah, and Randolph [who had flown from Cairo] came in and we all talked. At eleven o'clock Lord Moran came in to say 'Good-night' and I then saw that he wanted us all to go; so we did . . . But Papa showed no signs of fatigue, and once or twice when I got up to go to bed, he would not let me go. At one in the morning I saw a light under his door, and went into the passage and saw the night nurse who said: 'He is awake and feeling very cheerful and having beef tea – come in and see him.' So I did, and sat with him for a quarter of an hour, after which he went to sleep and I returned to my bed.

This morning, Sawyers [Winston's valet] came in early and said would I breakfast with Papa at 8.30? So I went in with him at that time and found Lord Moran in his room, and that at 3.15 in the morning he had had another attack of auricular fibrillation.* The doctors are disappointed but not surprised. Papa is very upset about it as he is beginning to see that he cannot get well in a few days and that he will have to lead what to him is a dreary monotonous life with no emotions or excitements. Yesterday morning, before my arrival, because he felt well again, he was as happy as a lark, and began to

* A slight one.

smoke again, which of course is wrong; but Lord Moran agrees that it is not so bad for him as it would be for most patients. The smoking has of course been quite cut out to-day . . .

I asked Lord Moran how I could help to carry out his wishes and he said that not more than one person should be in the room at a time. So as there has been a constant stream of people all day I have not done more than poke my nose round the corner of the door.

The Courier is going so there is no time for much more. I love you my Darling & wish you were here.

Health Bulletin by C.S.C. (not the doctors) for Sunday December the 19th
Papa much better to-day. Has consented not to smoke, and to drink only weak whisky and soda. In fairly good spirits. New cook arrived. Food much better and Papa enjoyed good luncheon and dinner, though appetite is not very good.

19th December 1943

This Villa is on a little promontory right on the sea with terraces leading down. It is only one storey high, and is like a white box . . . Here I should say that this house is the Headquarters of General Eisenhower who is at present I don't know where – I should think in Italy . . . so that here we are, an enormous party, all the guests of the U.S.A. Army.

. . . Sarah sleeps in the Villa, but I have not yet seen her room as I literally have not had time. Tommy [Thompson]* also lives here, but everyone else sleeps out, either at the guest house or in other villas. All these establishments are run by lieutenants – one a mere man, and the other called the 'White lieutenant' who is a W.A.C. [Women's Army Corps]. She is supposed to be a very high class dietician and was fetched in from Tunis to manage the Villa for us. The food, however, is 'vurry' American. Last night we had a partridge for dinner which Sawyers informed us was cooked for an hour and a half! The result was concrete! Sawyers rather rashly informed the American cook that Mrs. Landemare cooks partridges for only fifteen minutes. Your poor father literally cannot eat the food.

Later

Sarah has been and is a pillar of strength. She has been a great joy to your father; but indeed everybody loves her. She has taken so much trouble with everybody, and to smooth out any little crossnesses and difficulties which might have arisen. At Cairo, I am told that she was a perfect hostess at the Casey villa when, for instance, Papa gave a big dinner for the Chiang Kai Shek's. In Teheran she handled the rather difficult and delicate problem of being the only woman with perfect tact, and now here she looks after everybody at the Villa.

* Commander C. R. Thompson, RN, Flag Officer to Churchill when First Lord, remained with him as personal assistant throughout the war.

Before her mother's arrival Sarah had been her father's watchful companion, reading Jane Austen to him. On 16 December she and Randolph had cabled better news to Clementine: 'Papa had excellent night and his condition is much improved . . . We are glued to Pride and Prejudice. He says you are so like Elizabeth.'[21]

Health Bulletin for Monday December the 20th (C.S.C.)
Papa very refractory and naughty this morning and wants to leave this place at once. All doing our best to persuade him that complete recovery depends on rest and compliance with regulations. Progress continued.

Monday, December the 20th
Yesterday, Sunday, Papa was much better and in very good spirits. I breakfasted, lunched and dined with him in his bedroom and had lovely talks with him . . .

A little gun boat patrols up and down in front of the house in case a German submarine should pop up its nose and shoot up the Villa. Of course if the enemy knew we were here they could wipe out the place with dive bombers from the airfields near Rome. It is less than two hours' flight from here to the toe of Italy & one hour from Palermo in Sicily.

Tuesday, December the 21st
Your father has had a very good night so I am told by the night Nurse; but I have not seen him because quite early he sent for Mr. Kinna* to dictate some difficult paper.

In the afternoons, Clementine and other members of the party often went sightseeing. On one such expedition Lord Moran went with them, Winston being left in the care of Brigadier Bedford (the heart specialist) and Randolph. When they returned after darkness had fallen,

we found Papa sitting up in bed looking very pink and mischievous and announcing he had had a hot bath with Dr. Bedford's consent. Your father has taken a great deal of trouble to seduce Dr. Bedford, and Lord Moran is quite jealous. After the bath he played innumerable games of bezique with Randolph. In spite of all this he was not really exhausted, though when bedtime came he was looking a little pale and wan.[22]

That same afternoon they had visited Longstop Hill, the scene of a violent battle. With them was Bill Harris, a Coldstream officer, who had taken part in it. Clementine described how when he saw 'this terrible hill for the first time after he crawled up it with his hundred men, he could hardly speak'. She went on to describe how

* Patrick Kinna, MBE, a shorthand writer in WSC's Private Office, who was with him on many of his travels.

All along the road from Tunis to Longstop and on to Medjez-el-Bab, the fields on either side were littered with war debris – tanks, crashed planes, piles of ammunition, blown up guns, and so on. It is a horrible sight – the stale remains of war. We did not stop to look at the cemeteries, for we had to get back, but Bill says there are several – some British, some German and some where former foes are buried together.[23]

Health bulletin by C.S.C. Thursday, December 23rd
Very good progress. I think the only anxiety is impatience to be well too soon.
C.S.C.

26th December 1943
Yesterday we all of us thought of the party gathered at Chequers, and of Mary in her battery. We had a most extraordinary Christmas Day ourselves. All the, what Americans call 'high ranking Generals and other notabilities' converged here, and Christmas day was spent by them and Papa in a series of conferences . . .

We had a most interesting Christmas luncheon party with all these notabilities assembled, and your father at the end made a charming speech about his distinguished guests . . . This was his first meal outside his bedroom . . .

Early in the morning Sarah and I went to early Church with the Coldstream Guards. The service was held in a corrugated iron shed full of ammunition and stores. It had been cleared in the centre for the Service. Just as the Service was drawing to a close, there was a fluttering of wings, and a little white dove circled round and settled on a ledge just above the altar. I heard a hefty Guardsman behind me whisper to his neighbour: 'Look now, that means peace'. Your father declared that he was sure it was a conjuring trick, and that the padre had released the dove from under his surplice. General Alexander, who was there, hurried quickly back to the house to find out whether Hitler had indeed thrown in his hand. We were much disappointed when we found that this had not immediately occurred, and we settled down to eggs and bacon!

During the morning we went to the 'Parade' service of the Coldstream which was held out of doors. We all sang Christmas Hymns accompanied by a lovely little pipe and flute band . . .

This morning Randolph has gone back to Italy. General Alexander very kindly took him in his plane.

Meanwhile, back at home a depleted family party gathered at Chequers: Diana and Duncan with their three children, Aunt Nellie, and Uncle Jack.

Everyone combined to make the party go with a swing, and enthusiastic accounts from various pens were sent to Winston and Clementine. Someone had given oranges and lemons (great luxuries); the President had sent a Christmas tree from his own plantation at Hyde Park; there was turkey and plum pudding and presents for everyone – and through all the usual festivities ran the thread of thankfulness and relief.

After this serious illness, the doctors were adamant that Winston must have a proper period of convalescence before returning home. He himself felt so depleted and feeble that for once he did not oppose their wishes. Marrakech seemed the obvious place, and once more the lovely Villa Taylor, where he had stayed only a year before with the President after the Casablanca Conference, was put at his disposal by the ever-hospitable United States army. From there Clementine continued to give all of us at home lively accounts of their life:

Marrakech
New Years Day [1944]

. . . We are living in a glorious villa not far from the Mamounia Hotel where Papa and you [Diana] and Duncan all stayed some years ago. The villa is a mixture of Arabian Nights and Hollywood, and extremely comfortable as regards beds, hot baths, etc., but alas very sunless, as I suppose it is built for protection against the torrid heat of Summer.

First I must tell you about Papa. He is gaining strength every day, but very slowly indeed. He is disappointed that his recovery is slow; but when in the end he is quite well, the slowness of his convalescence may be a blessing in disguise, as it may make him a little more careful when he has to travel. For instance, when we came from Tunis here, the doctors did not wish him to fly above five thousand feet, but he insisted on going up to twelve thousand, to get above the clouds which of course made it safer as we were over mountainous country. He had oxygen and I think really it has turned out all right, but yesterday he said sadly, 'I am not strong enough to paint'. But this morning, New Year's Day, he came early into my room, and said 'I am so happy. I feel so much better'. So don't be anxious, but we must all realise that he really needs great care. He is not staying in bed now more than he does in his ordinary life. This is his day: He works in bed all the morning and gets up just in time to go for an expedition or for lunch in the garden, after which he goes for a drive. Then he goes back again to bed till dinner. We try to prevent him sitting up late, but I am afraid he does . . .

The sun is delicious, but the air is very sharp and cold. One needs one's warmest Winter clothes.

We had a great surprise one morning. Max [Beaverbrook] suddenly arrived! I knew he was coming, but we did not know the day, and he walked in at breakfast time, having flown in a Liberator . . .

Yesterday there was great excitement. General Eisenhower and General Montgomery arrived for a pow-wow. General Eisenhower has now gone . . . but General Montgomery is staying till to-night.

3rd January 1944

We all went for a picnic on New Year's Day, taking General Montgomery with us. On the way out he drove with Papa, and on the way back he came with me. He really is a thrilling and interesting personage – naive and sincere, good tempered, and with the same sort of conceit which we read Nelson had. He has piercing blue eyes and a wirey [*sic*] young figure . . .

After dinner he flew away home, and Sarah who for security reasons has been in mufti since her arrival here, put on her new Austin Reed uniform (22 coupons – not yet raised) to see him off . . .

Yesterday was bad temper day! Everyone was cross – me especially. Sarah was very sensible and said 'Let's have a bad day, and make a fresh start tomorrow'. So that is what we are doing. This morning everyone is smiling and we are going for another picnic.

There is a very nice British Consul-General here – Mr. Nairn. He and his wife are both Scotch. I like them both very much . . . She is an M.A. London University, and was a professional artist before she married. Tomorrow she is going to take Sarah and me to see the Glauie's [*sic*] 'household' (harem) so we shall have to try and make ourselves look tidy.

The following afternoon Margaret Nairn duly escorted Clementine and Sarah to visit the Glaoui, and on 5 January Clementine wrote us an amusing account:

In the afternoon we went for our carefully arranged visit to the Pasha's 'household' . . . To our great disappointment we did not see the Ladies, but only the Glauie or Pasha . . . who fed us with mint tea and all sorts of oriental sweetmeats . . . After tea the Pasha conducted us into every corner of the Palace, all except the harem, the doors of which we passed, but they were bolted . . . He conducted us to the door, and in the outer vestibule there was a row of myrmidons. He took from one of them a green velvet case which he presented to me for Papa with a little speech. It contained a most beautiful dagger in a gold scabbard. Three huge carpets, all rolled up, were then produced – one for Sarah, one for Mrs. Nairn, and one for me.

Although Clementine had been much taken by 'Monty' during his short visit (and they would over the next years all become true friends), she had had cause to rebuke him; Jock Colville was much amused and later recounted the story. One evening,

when it was time to have a bath before dinner, she turned to the A.D.C., Noel Chavasse, and said she looked forward to seeing him in half an hour.

'My A.D.C.s don't dine with the Prime Minister,' said Monty tartly.

Mrs Churchill gave him a withering look. 'In my house, General Montgomery, I invite who I wish and I don't require your advice.'

Noel Chavasse dined.[24]

Clementine continued her letter home on 10 January:

Since my last letter there has been a great deal of coming and going. All the 'big shots' have been here for a Conference . . . They all arrived one evening . . . Next morning they met again at 9.30, always of course with Papa. All was over at 11.30 and they were airborne back to Algiers, Italy etc. by noon. It was all over in a flash, and I believe everything 'clicked'. They all went away happy having reached complete agreement.

Meanwhile Randolph came back with Fitzroy Maclean*. They were brought by General Alexander from Bari. They both stayed an extra day and have now gone back and will soon be operating in their special sphere with Tito in Yugoslavia . . . I think Randolph is well fitted for this work, and I hope it will be a great success. They went off in good spirits.

. . .We have been for several more picnics; in fact we go out every day when it is fine, but we can rarely go any distance because your Father works all the morning. So except on one or two occasions we have not been able to penetrate deep into the Atlas, but just establish ourselves by the side of a stream in the foothills . . .

I think Papa is now really quite himself again. Every day sees an improvement. He simply loves the picnics. He has been reading Jane Austen – Pride and Prejudice first, which Sarah read aloud to him practically non-stop during the two days that he was really ill, and then to himself Northanger Abbey which he did not like nearly so well, but still I noticed he could not put it down. Now I am re-reading Mansfield Park.

Meanwhile, a visit from 'the General' impended: 'General de Gaulle has been by way of coming several times,' wrote Clementine, 'and his visit has each time been put off generally by himself. I think he really and truly is coming one day soon. I hope there will be no explosions!'[25]

Two days before de Gaulle's arrival Duff and Diana Cooper joined the party; he was at this time the British representative with the French Committee of Liberation in Algiers. On 12 January the several-times-anticipated visit materialized. Clementine wrote:

To-day we are rather on tenter-hooks because at last General de Gaulle really is arriving by plane from Algiers in time for luncheon. This part of the visit is hush-hush and security, but afterwards he repairs to his local palace and bursts into publicity and military reviews . . . It is expected that he will have a rapturous welcome both from the French and the native population. I am rather nervous and de G. has again perpetrated one of his chronic incivilities by sending Papa (who really has worked for the French at all these Conferences and also to keep the President sweet, who simply loathes de G.) a boorish message. I am trying to smooth Papa down so I hope this function will pass off without further unpleasantness.[26]

And Diana Cooper told how 'Clemmie has given him [Winston] a Caudle

* Diplomat, traveller, author and soldier. At this time WSC's personal envoy to Tito and his Yugoslav partisans.

curtain lecture* on the importance of not quarrelling with Wormwood [her nickname for de Gaulle]. She thinks it will bear fruit.'[27]

But if Clementine was 'lecturing' Winston, and smoothing him down, she also had some plain words for the General (with whom she had from their first encounter got on very well). Sarah told me that she was witness to a conversation between her mother and the General while they were walking in the garden, during the course of which Clementine said, bravely: 'Mon Général, you must take care not to hate your allies more than your enemies.' His reply – if any – is not known. But, at any rate, the luncheon party and subsequent 'pow-wow', which had been the subject of so much nervous anticipation, both went off much better than had been feared.

The guests were General and Mme de Gaulle, Duff and Diana Cooper, Max Beaverbrook, and Mr and Mrs Nairn. The General was in the best of humour and insisted on speaking English throughout the meal. 'To make things equal,' wrote Winston, 'I spoke French.'[28] During the afternoon the ladies diverted themselves in the bazaars and the men had a long and serious talk. Despite the fact that many thorny subjects were broached, the conversations ended on a pleasant note, with the General inviting Winston to attend a review of French and Moroccan troops the next day.

Following the review, the party went for their last picnic (the eighth). Diana Cooper described how Winston ventured down a steep gorge, and tried to climb on top of the biggest boulder: 'Clemmie said nothing, but watched him with me like a lenient mother who does not wish to spoil her child's fun nor yet his daring.'[29]

This long and ill-starred journey was now drawing to its end. The party flew from Marrakech to Gibraltar, where they went aboard the *King George V*; they disembarked at Plymouth and arrived back in London by train early on the morning of 18 January 1944.

Clementine had been a marvellous correspondent, keeping us all at home fed with news of Winston's steadily improving health, and lively, amusing accounts of their doings. Once the intense anxiety about Winston had subsided, she had been able to enjoy the pleasures of sightseeing and the numerous picnics; and she had relished (as she always did) the tinge of the exotic. But underneath her courage and buoyancy there lurked a built-in weariness – the fruit of the emotional and physical strains with which she had now lived for four years or more.

Diana Cooper recounted a 'curious calm and sad conversation' she had with Clementine one evening in Marrakech before dinner:

> I was talking about post-war days and proposed that instead of a grateful country building Winston another Blenheim, they should give him an endowed manor house with acres for a farm and gardens to build and paint in. Clemmie very calmly said: 'I never think of after the war. You see, I think

* Referring to the lectures given behind the bed-curtains to Jacob Caudle by his wife, in the nineteenth-century *Punch* feature.

Winston will die when it's over.' She said this so objectively that I could not bring myself to say the usual 'What nonsense!' but tried something about it was no use relying on death; people lived to ninety or might easily, in our lives, die that day . . . But she seemed quite certain and quite resigned to his not surviving long into peace. 'You see, he's seventy and I'm sixty and we're putting all we have into this war, and it will take all we have.' It was touching and noble.[30]

CHAPTER TWENTY-THREE

The Last Lap

EARLY IN 1944 THE ENEMY ONCE MORE TURNED HIS ATTENTION UPON THE cities of this country, London being the chief target. Londoners accepted this resumption of the air raids stolidly, but people were just that much wearier; three years of the sheer slog of wartime life since the first Blitz had inevitably taken their toll. During the 'Little Blitz' the noise was truly appalling, most of it being caused by our own, now much more formidable defences, and even a quiet night brought little rest to many thousands of men and women who, after their ordinary day's work, went home to do their stint as Air Raid Wardens or firewatchers. As well as using the now extensive and greatly improved shelter accommodation, people once more took to sleeping in the Underground stations, alongside those already rendered homeless in previous attacks. Westminster was no more immune than other parts of London: on the night of 20 February 1944, Downing Street and Whitehall once again suffered bomb damage.

During the previous year Clementine had made several extensive tours of air-raid shelters in various areas of London with officials from the Ministry of Health or the Red Cross and other concerned organizations. Since 1940 the entire organization and administration of air-raid shelters had been continually improved, and in most cases the standards of hygiene and (relative) comfort were by now as good as could reasonably be expected. Shelter life had become a feature of wartime London, and had developed its own code and folk culture.

Clementine's programme was as full as ever of outside obligations and activities, over and above the ordinary business of keeping her domestic life running to her own meticulous standard, answering her large postbag, and entertaining the constant stream of official visitors either in London or at Chequers. Luncheon was the principal occasion when visiting notabilities were entertained; Winston also frequently invited the Chiefs of Staff, the Planners, and all sorts of people who were in close contact with him in his work; thus 'business' was conducted before, during and after meals. The evenings were more often the occasion for the last-minute invitation – for the family, 'The Prof' or Brendan Bracken, or some other close crony. But there were 'arranged' dinners too. During 1944 the King dined at No. 10 six times.* This was a singular and totally unusual sign of honour, and marks the closeness of the relationship that had grown up during these hard years between the Sovereign and his Prime Minister.

The following figures give some idea of the volume of official entertaining undertaken by Winston and Clementine at this period. Between mid-January and early September 1944, they gave seventy-five luncheon parties and nineteen dinner parties. (These figures exclude any occasions which could be described as 'purely family'.) Of the other evenings during that period, Clementine's diary rather sadly marks only four of them: 'Dinner alone with Winston.' At Chequers, too, despite the charming and efficient housekeeper and a permanent staff of cheerful servicewomen, the weekends were far from rest cures for Clementine: the war never let up – and nor did Winston. Between January and September 1944, out of a total of thirty-three weekends, twenty-four were all, or partly, taken up with entertaining official guests from abroad, or commanders home for a short while from their theatres of operations, in addition to the many men Winston had constantly around him for the purpose of conducting affairs.

Sometimes in one weekend there would be two, or even three, 'shifts' of guests: some lunching, others 'dining and sleeping'. This constant coming and going put a considerable strain on Clementine, whose sense of hospitality and code of good manners meant that she was almost constantly 'on duty'. Very often she returned to London to face a full and exacting week, as tired as when, the previous Friday, she had arrived at Chequers.

The year 1944 was not a happy one in personal terms for some members of our family. Sarah was still wrestling with the problems and unhappiness caused by the breakdown of her marriage; although she and Vic had separated in 1941, they were not in fact divorced until four years later. Meanwhile she gave herself unsparingly to her work in the WAAF dealing with highly secret photographic intelligence: a job which was as arduous as it was interesting.

Randolph also had difficult personal problems to resolve between himself and Pamela, and distance from all the people involved made for misunderstandings. Throughout his life an easily aroused and explosive person, he was at this time under considerable strain and, sadly, the brief meetings with either or both of his parents were often marked by painful scenes, which left their mark on all concerned.

Randolph had for some time been leading a hectic and hazardous existence. Having in the spring of 1942 given up a staff job in Cairo to join the Special Air Service, he had been quite severely injured in May of that year, returning from a dangerous raid on Benghazi with the Long Range Desert Group. When he had visited his father in Marrakech at the end of 1943 he had been awaiting his orders to be parachuted into Yugoslavia to

* The King dined here fourteen times in all during Churchill's tenure: on two occasions the company was forced to withdraw to the nearby shelter on account of aerial bombardment. There is a plaque commemorating these dinners in the downstairs garden rooms at No. 10.

join the British Military Mission which, headed by Brigadier Fitzroy Maclean, was attached to Marshal Tito and his forces. Early in February 1944, Randolph was dropped into Yugoslavia, where he shared the life and vicissitudes of the partisans.

In mid-July 1944 he had another brush with death when the transport plane in which he was travelling in Yugoslavia crashed on landing and burst into flames; he was among the ten people who escaped alive from the wreckage. Although Randolph, amazingly, was not very seriously injured, he took some time to recover fully from damage to his knees and spine. Ever kind and hospitable, Duff and Diana Cooper* invited him to convalesce with them in Algiers after his discharge from hospital in Bari, and he was still staying with them when his father 'dropped in' for a few hours on his way to Italy in August. This meeting was all affection and rejoicing, and Winston wrote to Clementine: 'No reference was made by either of us to family matters. He is a lonely figure by no means recovered as far as walking is concerned. Our talk was about politics, French and English, about wh there was plenty of friendly badinage & argument.'[1]

Amid all these disturbances, activities and anxieties, Clementine had little time left over for herself. Always a reserved person, she was isolated more and more by the demands of her wartime life and the burden of the secret knowledge which she so often bore, even from her closest and most trusted friends. She leaned, I think, mainly on Sarah and myself, as people to whom she could unburden herself and with whom she could talk freely. But of course, we were often not there because of our jobs; telephone conversations were couched in carefully guarded language, even letters could not really be regarded as safe or sure, and we were all meticulously security-minded. Fortunately she had in Grace Hamblin a constant, steady and 'padlock' companion.

Among Winston's colleagues, Clementine had some true and 'cosy' friends – such as Sir Peter Portal, Chief of the Air Staff (whose wife, Joan, was also a friend and a colleague of hers at Fulmer Chase), and 'Pug' (Sir Hastings) Ismay, Chief of Staff to Winston in his capacity as Minister of Defence. And in the members of Winston's Private Office she found a continual source of loyal help, understanding and support in the many problems and vicissitudes of their everyday life in these years.

But the tide of war brought to Winston and Clementine – and indeed, to all our family – one very special friend: Gil Winant, the American Ambassador. A man of quiet, intensely concentrated charm, Gil had very quickly become a dear friend of us all, entering into our joys and sorrows, jokes and rows (in these last, always as a peace-maker). He was deeply fond of Clementine, and understood intuitively her character, and the strains and difficulties in her life. He fell in love with Sarah, and they had a close relationship, which could not be a simple one because neither of them was

* Duff Cooper (1890–1954), later 1st Viscount Norwich, was at that time British Minister in Algiers and the Government's Representative with the French Committee of Liberation.

free, and both were ever-conscious of his true loyalty to his family. Gil's life was to end tragically in 1947 when, shortly after his return home to Concord, he shot himself.

But for all these loyal friends, in a life crowded with people, Clementine remained a somewhat lonely woman. One of the former WAAF officers described to me how, from the windows in the Cypher Office in Storey's Gate, she used to watch her 'in her smart London clothes, going for a walk in St James's Park, quite by herself. One would imagine, just to think quietly among the peace of the lake and the trees, very often after a night of bombing. It used to impress me greatly.'[2]

* * *

As 1944 wore on, the whole country was gripped by the consciousness that some time soon the long-awaited, awesome attack on the Continent of Europe must be launched. Passing visitors, and Britons returning home during these spring months, all had the same impression – the whole country had become a vast assault-craft. England was splitting at the seams with trained men, and every road or free space (especially in the south-east) was filled with equipment and vehicles of every type. Very few knew the day or the hour, but the whole nation shared the knowledge that the event was imminent. Both those under military discipline and the population as a whole were equally conscious of the necessity for the strictest general and personal security measures. In all history there can surely be no other example of such a vast secret being so miraculously preserved. From 28 May, when subordinate commanders were informed of the date of 'D-Day' – 5 June – all the camps holding assault troops and all ports of embarkation were sealed; telephone communications were forbidden; all mail was impounded; and foreign embassies could no longer communicate with the outside world. The whole country held its breath.

During these last days Winston was visiting the camps and embarkation ports; he earnestly longed to view at closer quarters the assault from a British warship, to watch over and in some way to share in this tremendous enterprise in whose conception and growth he had played so close a part. Opposition to this idea was strong and almost unanimous from the King downwards. Clementine understood his wish, but thought it totally impractical: Winston waged a stubborn fight, but was finally compelled to abandon his plan at the direct and strongly expressed wish of the King.*

Two days before the planned hour, the whole vast project had to be postponed for twenty-four hours, due to increasingly bad weather conditions. This decision was right, but agonizing. All those who held the knowledge and responsibility for the attack knew that if it could not be launched within the following forty-eight hours, a whole fortnight would have to elapse before moon and tide conditions would once more be suitable. The problem

* In the event Churchill visited the D-Day beaches and General Montgomery's HQ in the field on 12 June.

of keeping vast numbers of men cooped up in their embarkation 'cages', and the risk run at every day's delay of the enemy accurately estimating the place and other details of the attack, made the prospect of further postponement a grim one, and was perhaps decisive. However, the weather improved, although much discomfort and some damage and loss were caused among the smaller troop-carrying craft, and the conditions on the approaches to the beaches were difficult.

On 4 June the Allied forces entered Rome after hard months of costly fighting, but even this glorious piece of news was overshadowed by impending events at home. On Monday, 5 June, Winston was in his official train after making a final visit to Hampshire; the train had been shunted into a siding near General Eisenhower's Headquarters in the woods behind Portsmouth. Clementine had returned to London from a visit to Chartwell on the Sunday evening. That Monday morning, before leaving for her usual committee meeting at the Fulmer Chase Maternity Home, she wrote a note to Winston to go down in the official box which was always despatched to him wherever he was.

10 Downing Street
Whitehall
Monday Morning

My Darling,

I feel so much for you at this agonising moment – so full of suspense, which prevents one rejoicing over Rome.

I look forward to seeing you at dinner . . .

Tender love from:
Clemmie [lovely picture of a cat]
Just off to her Hospital

On his return from D-Day Headquarters, Winston and Clementine had one of their few dinners alone together. Afterwards he went to the Map Room to study the final dispositions and plan of attack. Already great convoys were moving in the darkness to their appointed stations off the coast of Normandy. Just before going to bed, Clementine joined him in the Map Room; Winston said to her: 'Do you realise that by the time you wake up in the morning twenty thousand men may have been killed?'*

While the attention of the whole country was fixed in absorbed and agonized interest on the progress of the Allied armies in France, in the second week of June those who lived in London and south-east England once more had cause to feel that they too were part of the battle: for in the early hours of 13 June the Germans launched the first batch of their long-awaited 'Secret Weapon' – a pilotless aircraft carrying a high-explosive warhead. The

* Incident recounted in Pawle, *The War and Colonel Warden*, p. 302. Mercifully, the casualties on D-Day were not as horrific as had been feared: about 4,300 British and Canadian, and 6,000 American (*The Oxford Companion to the Second World War*).

'Doodle-bugs' or 'Buzz Bombs', as they were quickly dubbed, were exceedingly frightening and disagreeable, chiefly because one could hear them for quite a long time before the engine suddenly cut out, when one knew that the brutish thing would plunge earthwards; there were then a few agonizing moments of silence between the cut-out and the sickening explosion, and the slow rising of a huge cloud of dust. The blast damage was particularly vicious because the bomb usually exploded before penetrating the ground. The flying bombs were fired in batches at intervals throughout the day and night, and the anti-aircraft defences and civil defence services practically never stood down, meals and sleep being snatched when and as possible. Winston and Clementine were anxious about Diana, who was in the thick of it as an Air Raid Warden, and myself with my battery in the middle of Hyde Park.

During the early days of the flying-bomb raids the south of England was blessed with 'glorious June' weather, and despite the new terror, the streets and parks of London were full of people enjoying the sunshine. Indeed, one evening I telephoned my mother to tell her how our guns in Hyde Park had been in action that afternoon, witnessed by an interested crowd of people who had assembled to watch an American baseball match on a pitch adjoining the gun-site. The *sangfroid* with which people continued to carry on with their normal lives, however, resulted in some calamitous incidents, such as the occasion when a flying bomb scored a direct hit on the Guards' Chapel in Wellington Barracks on Sunday, 18 June. Morning Service was in progress in the packed church, and among the congregation were many people whose relations were with the Brigade of Guards fighting abroad. The casualties numbered 121 civilians and soldiers killed. My mother had called at my gun-site to leave a note for me, together with some roses and strawberries from Chequers; my section was on a brief 'stand-down', but within a few minutes of her arrival the guns were firing, and together we watched the fatal plunge to earth of the flying bomb which caused this ghastly incident.

The first and most severe phase of these attacks came to its end in the early part of September 1944, with the over-running of the launching areas in north-east France by the Allied armies; but there were further spasmodic attacks from more remote sites, and the ordeal from the air was not yet over: Hitler's second 'Secret Weapon', the V2 rocket, was hurled upon us for the first time on 8 September, and October and November saw a daily routine delivery of two to four rockets a day. The 'shelling' lasted for seven months and caused heavy casualties.

Between January and mid-March 1945, Clementine visited the sites of rocket-bomb damage in Wanstead and Woodford (Winston's constituency), Westerham, and Fircroft, near Gerrards Cross (the convalescent annexe for Fulmer Chase) – miraculously, in this last place no one was either killed or injured. The rockets put people, already wearied, under further great strain. But stolid courage held out, and the advance of the armies and the sure knowledge that we really were now treading a visibly victorious road enabled people in the target areas to grit their teeth to the end. Like everyone else, Clementine got on with her everyday life as best she could, accepting untoward interruptions, noise and consequent fatigue as all part of the day's or night's work.

At the end of August there was another, short but real, moment of panic about Winston's health. He had been in Italy since 11 August, visiting General Alexander; he had also met, for the first time, the heroic and redoubtable Yugoslav leader, Marshal Tito; he had watched from afar (too far, so far as he was concerned), in a British destroyer, the Allied landings on the Riviera coast; he had had talks with Italian politicians and the Greek Prime Minister; and in Rome he had been received in audience by the Pope. On Tuesday, 29 August he flew home to England.

My battery, in the general redeployment of the anti-aircraft defences, had moved several times during that eventful summer and was now encamped at Fairlight Cove, near Hastings, under canvas. On Friday morning, 1 September, I received a letter from my mother which told me of my father's sudden illness, which had developed in the last lap of his Italian journey:

> I had a great shock on Tuesday. I went to Northolt to meet Papa. The 'York' made a lovely landing & taxied right up to where everyone was waiting for him. Lord Moran emerged & ran across the tarmac to the car where I was sitting & said: – 'He has a temperature of 103. We must get him back quickly & get him to bed.'
>
> Then Papa emerged looking crumpled & feverish. I got him straight into the car & we rushed away leaving everyone stunned & astonished including the special correspondents. But all is well I hope. The two beautiful Nurses from St Mary's Hospital appeared as tho' by magic – Doctor Geoffrey Marshall the lung specialist took blood tests & X rays & gave M. & B.* It is a slight attack – there is a small shadow on one lung, but in himself he is well & this morning, now 7 a.m. the temperature is normal. I was sick with fright Tuesday night & yesterday. We hope it will not be necessary to publish bulletins & strange to say Lord Moran says that in about 5 days he can go to Canada† by sea. I shall go with him now, Sarah too. I do wish you could but your job is essential.
>
> Darling – he says he must see you before he goes, so try to come up, ringing me first.[3]

Needless to say, I was dismayed by my mother's letter, but a telephone call conducted in veiled language indicated to me that my father was improving. I was allowed a 24-hour leave pass, and I bolted up to London. On my arrival at The Annexe I found my parents having dinner together in my father's room – he was sitting up in bed attired in his glorious many-hued bedjacket, with my mother by his side in one of her lovely housecoats:

* The sulphonamide drug which was the first effective treatment for pneumonia.
† For the Second Quebec Conference.

it was a most reassuring sight, and I enjoyed the loving welcome they gave me.[4]

This very brief, but none the less worrying, bout of pneumonia had been disconcerting, but no bulletins were issued, and only the smallest circle of people knew he was ill. Gil Winant, writing on 1 September to Harry Hopkins, rightly judged it: 'Tonight his temperature is back to normal and he seems on the way to a quick recovery. But each journey has taken its toll and the interval between illnesses has been constantly shortened.'[5]

His forthcoming journey was a somewhat anxious prospect, and we were all thankful that my mother could go with him on this trip. Sarah, who was originally intended to accompany them, stayed behind, as her unit was under great pressure of work. As usual my mother was an excellent correspondent, and we could all follow the events of the next few weeks through the 'Diary' account she sent to us.

The party embarked at Greenock on the *Queen Mary* on 5 September. During the voyage the weather became very hot and sticky – they had taken a southerly course following the Gulf Stream, partly to get good weather, and partly to avoid U-boat packs. Winston, whom Clementine reported as being 'in low spirits and not very well',[6] worked as usual in his cabin most of the morning. On 7 September, after reading the early news received by ship's radio, Clementine sent this note from her cabin to his:

> Darling – How are you this morning?
>
> What a rousing news Bulletin this morning! Calais, Boulogne, Dunkirk, Le Havre, more & more closely invested – 19,000 Prisoners to the poor unnoticed British! Is the Moselle the frontier between France & Germany? because if so we are in the Reich.*
>
> Then side by side with the announcement about the Home Guard† the Americans (with an eye I suppose to the Election) announce demobilisation! Wurrup & Allelulia!
>
> [Signed with a fat cat with very alert ears]

Lord Moran had brought with him on the voyage a pathologist (Brigadier Whitby) and a nurse; and in the event he was proved to have acted with foresight, for Winston sprang a temperature during the last part of the sea journey. Fortunately, however, it disappeared quickly, and there was no return of the dreaded pneumonia. On 10 September Clementine was able to report to us at home that

> This morning he woke up well, and by then we were out of the Gulf Stream, in glorious normal summer weather, with an oily sea, lovely sunshine and a light breeze . . .

* The German frontier was actually crossed on 11 September.
† On 5 September the War Office had announced that although the Home Guard were still necessary, plans had been prepared for standing them down when appropriate.

> The news had got around in Halifax and Papa had the most marvellous welcome. We had a great send-off from the ship which is going on to New York. The taffrails were garlanded with cheering, shouting, waving men, and then crowds came running and gathered round the waiting train. Now we are but two hours on our journey. All along the line people waved in the fields and at a place called Truro there was a tremendous crowd waiting for us and again at Moncton. At Halifax, while there was a prolonged delay before the train started, the crowd all sang songs, 'Tipperary', 'Pack up your Troubles', 'Roll out the Barrel' and 'Canada', ending up with 'God Save the King'. It is thrilling and moving and, oh, how I wish that Sarah and Mary were here . . . The train is so luxurious, all done up in peach-coloured paint and different coloured chintzes, that I cannot bear the idea of getting off it to-morrow.

The Conference party were once more staying in the Citadel, and when Winston and Clementine arrived they found the Governor-General and Princess Alice, Mr Mackenzie King and the Roosevelts already there. While the men conferred, Princess Alice looked after Clementine and Mrs Roosevelt. One afternoon they all went for a picnic and a four-mile walk up in the hills along the Jacques Cartier river, getting back to the Citadel just in time for a dinner party for the Chiefs of Staff. Clementine wrote on 13 September: 'There is a gay almost hilarious atmosphere as I understand the military conversations are going very well.'

Clementine was less 'public relations' conscious than Mrs Roosevelt, and was somewhat put about when the latter announced her intention of broadcasting to the Canadian people, and told Clementine that she had been invited to do so too. 'I was staggered and reluctant,' she wrote on 14 September. 'First of all I said I could not possibly do it at such short notice and thought I had nothing to say. But finally I was hounded into doing it.' She was helped by a British journalist, Mr Crookshank, and between them they composed a suitable text, part of it being in French. The joint broadcast was delivered that same evening, and pronounced successful; 'but in my opinion,' added Clementine, 'the whole thing was a work of supererogation.' The final straw for Clementine was that after they had finished broadcasting, 'we were expected to do bits of it again as a News Reel. It was mortifying and frightful. Mrs Roosevelt throughout was calm and determined to do it all.'[7]

Clementine's struggles with the preparation of the broadcast had been severely interrupted by the necessity for her to attend, with Mrs Roosevelt, a luncheon party given by the wife of the Lieutenant-Governor of Quebec. 'I am sorry to confess', Clementine wrote,

> that I was in a filthy temper . . .
>
> The luncheon party at Lady Fiset's was just like the one that Mary and I attended a year ago, only this time, instead of 24 women round the table, there were 65. I just don't know how they were all got into the dining room. The blinds were drawn, all the electric lights turned on, and there were seven

courses:– Pâté de Foie Gras, Chicken Consommé, Oyster Patties, Galantine of Veal, Cheese Fondu, a very elaborate Ice, and then finally Brandy Cherries covered with pink icing. I ate it all! & enjoyed it. There were four wines handed round, and the ladies ended up with several liqueurs, after which we reeled back into the drawing room. The weather was dull, and here again curtains were drawn and full electric light as though it were a dinner party. Mrs. Roosevelt and I then had a few words of conversation with each of the 65 guests, after which Lady Fiset said Mrs. Roosevelt would like to say a few words – which she did. I tried to hide behind a palm tree because I saw what was coming next. But it was no use. When Mrs. Roosevelt had ceased, Lady Fiset said Mrs. Churchill would now like to reply to Mrs. Roosevelt. Mrs. Churchill was fished out from behind the palm tree – and I won't repeat to you what I said because I have forgotten, being under the influence of the luncheon.[8]

That evening there was an official reception and supper given by Mr Mackenzie King, who with Mrs Roosevelt and Clementine shook hands with more than seven hundred people. Clementine's account of this hectic day ended on a rueful note of admiration: 'Besides doing all these things which I have described, on top of it all Mrs. Roosevelt wrote her daily column.'[9]

Back in England Clementine's daughters eagerly awaited these entertaining instalments of our mother's Quebec journal. Sarah commented: 'Poor Mummie they are certainly making you "work your passage"! But I am sure all the speeches are a great success even if you don't remember them afterwards!'[10]

The Second Quebec Conference ploughed on its way, each day bringing for Clementine its ample ration of visits to organizations such as the Navy League, the Imperial Order of the Daughters of the Empire, and the Red Cross, who, in addition to cherishing their own Canadian forces, generously gave to British funds, military and civilian alike. Every day there was a luncheon, tea or dinner party – or all three – and in the intervals Clementine combed the shops for presents for the stay-at-homes – such rareties as face tissues, nylon stockings, soap and hairnets, quite apart from the odd dress or glamorous négligée.

At the end of the Conference on 17 September Winston and Clementine, as in the previous year, were invited by the Roosevelts to spend a day or two with them at Hyde Park before returning home.

18th September, 1944

The President and Mrs Roosevelt and their charming daughter, Anna Boettiger and her husband, John, came down to meet us . . . She is the eldest child, aged 38, and the only daughter. She is charming, gay and high-spirited. So far as I have been able to observe she seems much the nicest of the large Roosevelt brood . . .

[The Duke of] Windsor came to luncheon and had a long talk with Papa.

I had not seen him for several years. He is now 51 or 52 and, in the distance, still has a boyish appearance which is rudely dispelled when one sees him close to.

Harry Hopkins also appeared and stayed till the next morning. He seems to have quite dropped out of the picture. I found it sad and rather embarrassing . . . Almost exactly a year ago we were in Washington, and the President seemed hardly to be able to draw breath without him. Harry was then in hospital beginning his long, grim illness from which he has but just recovered. Two or three times during our fortnight's stay at the White House, when things were stuck, he dragged himself out of his bed and came tottering along more dead than alive, and in half an hour the difficulty was resolved. Now it seems that the intimacy has ended, and I cannot but feel that this is a disadvantage to Anglo-American relationship.

We cannot quite make out whether Harry's old place in the President's confidence is vacant, or whether Admiral Leahy [the President's Chief of Staff] is gradually moulding into it. One must hope that this is so, because the President, with all his genius does not – indeed cannot (partly because of his health and partly because of his make-up) – function round the clock, like your Father. I should not think that his mind was pinpointed on the war for more than four hours a day, which is not really enough when one is a supreme war lord.

Last year I was at Hyde Park only for the inside of a day, and Mrs. Roosevelt had flown away on her journey to the Pacific. She loves her meals out of doors and so life at Hyde Park is a succession of picnics. It is rather fun really, and clever, because, when you have a lot of foreign guests whom you do not know planked [*sic*] down on the top of you, it all fills in the time.

I went for two terrific walks with Mrs. Roosevelt, who has very long legs and out-walks me easily.

The Churchills travelled home again in the *Queen Mary*, arriving in London on 26 September. But it was only for a brief 'turn-around' period of ten days, before Winston was off on his travels again – this time flying on 7 October with Anthony Eden* to Moscow. Clementine at first thought she might well have to accompany him to Russia, but his health had improved so much that she decided to remain at home, where her family and many other considerations were calling for her attention.

During my parents' absence in Quebec my battery had returned to its original base in Hyde Park, and I was now most opportunely due for a week's leave. My mother, reprieved from having to go to Moscow, gave me a wonderful time, and we enjoyed an orgy of theatre-going: *Scandal at Barchester*; *Richard III*; *Pink String and Sealing-wax*; and *The Last of Mrs Cheyney*. We may have gone short of many things in the war, but the quality, variety and number of good plays running at any given moment in London remains clearly in my memory.

* Anthony Eden, later 1st Earl of Avon, KG (1897–1977), at this time Foreign Secretary.

Winston's 'alias' for this journey was 'Colonel Kent', and in the two weeks of his absence 'Mrs Kent' received, as usual, a series of cables and a letter. On 13 October Colonel Kent reported by cable: 'Everything is going well here and there is great cordiality.' During this visit Winston was able to renew the friendship begun earlier in the war in England with Averell Harriman, who was now United States Ambassador to the Soviet Union. On the evening of 13 October Winston had dined with Averell and his daughter Kathleen, and wrote to Clementine from their house. They had played some bezique, and Winston reported with satisfaction that he had won two 'Rubicon' games. He went on:

> This is just a line to tell you how I love you & how sorry I am you are not here . . . I do hope that you are happy w yr dear & my Maria. Give her my dearest love. It is wonderful to get the London papers the same day at about 6 p.m. But the couriers also bring heavy bags & in between meals of 12 or 14 courses & conferences of various kinds I am hard at it. I am very well except for a little Indy [indigestion].
>
> The affairs go well. We have settled a lot of things about the Balkans & prevented hosts of squabbles that were maturing. The two sets of Poles* have arrived & are being kept for the night in two separate cages & tomorrow we see them in succession. It is their best chance for a settlement. We shall try our utmost. I have had very nice talks with the Old Bear [Stalin]. I like him the more I see him. Now they respect us & I am sure they wish to work w us. I have to keep the President in constant touch & this is the delicate side.
>
> Darling, you can write anything but war secrets & it reaches me in a few hours. So send me a letter from yr dear hand.

Clementine spent the weekend of 13 October at Chequers, and had all three of her daughters with her, Diana bringing her baby Celia. Uncle Jack, Cousin Moppet and Gil Winant completed a cosy party. On Sunday, 15 October, Clementine heard a description on the news of the rapturous ovation Stalin and Winston had received from the audience at the Moscow Opera, and she wrote to him the same day: 'I wish I could have witnessed this hard-won recognition of your untiring and persistent work.'

While Winston was away, Clementine got on with her own very busy life, but the pressure from official entertaining was eased. She must, however, have been feeling the effect of the 'back-log' of fatigue just now, because two separate days in her engagement book have simply 'in bed' written across them; she was just taking this interlude to 'recharge her batteries'.

In early November she felt shock and sadness when the news came of the assassination by the Jewish extremist Stern Gang of her old friend Lord Moyne, who was at that time Deputy Minister of State in Cairo. Both she and Winston, who greatly esteemed and liked this distinguished and

* The pro-communist 'Lublin Committee' and the London-based Government-in-Exile.

delightful man, went to his Memorial Service. Clementine's mind must have gone back to those sunlit cruises in the *Rosaura* in the years before the war. She had enjoyed them so much: now her host and friend was brutally dead. Recalling those journeys, in a bleak November in the fourth winter of a gruelling war, must have seemed like dreaming of another life on another planet.

After all the hard, dark times one could not but feel in these last months, with mounting excitement, the tide of victory, which now was flowing strongly. There were still hard losses to bear, and reverses to sustain, but nothing now could long impede the inexorable, victorious progress of the Allied armies. One great milestone on the road had been marked by the liberation of Paris by General Leclerc's Division at the end of August. Now, in November, General de Gaulle invited Winston and Clementine to the French capital for the commemoration of Armistice Day on the 11th. I was 'in attendance', and so witnessed the unforgettable scenes of emotion, joy and thanksgiving, as the people of Paris applauded their two great heroes of the hour – de Gaulle and Churchill.

My father had spent his previous birthday far from the family in Teheran; this year, which marked his seventieth, we had him for ourselves. The passing of a decade always seems 'special', and this year I think our minds were full of all the last twelve months had held: of his serious illness in North Africa; of all his journeyings; of all the moments of historic drama we had lived to witness and wonder at. There was a joyous dinner party at The Annexe: the family present were Diana and Duncan, Sarah, myself, Nellie Romilly, Jack Churchill, Nana Whyte and Venetia Montagu; the only 'outsiders' were Max Beaverbrook and Brendan Bracken.

* * *

In the last month of 1944, this year of such tremendous events, two crises erupted. One was the violent offensive launched by the Germans in mid-December in the Ardennes which, while it could not stop the now inevitable advance of the Allied armies, caused a serious hold-up, and was costly in men, materials and time. The other was the Greek civil war, which culminated over Christmas in a bitter struggle for Athens and the political future of Greece.

The guerrilla forces controlled by the Greek Communist Party had played a pre-eminent part in the long struggle against Germany. As the Germans withdrew from Greece, the Communists prepared to seize political control in the wake of liberation. Churchill was determined to prevent this, and to secure democratic elections. A British force was despatched to Greece in October 1944 to collaborate with the liberating forces and secure the authority of an all-party provisional Government. Just before Christmas violent fighting broke out in the streets of Athens between the Communist and pro-monarchist groups. Churchill thereupon decided that his personal intervention was the only hope of resolving the deteriorating situation and avoiding full-scale civil war. Accordingly, he arranged to fly off to Athens on Christmas Eve, taking Anthony Eden with him.

Clementine had arranged a large gathering of relations and friends for the Christmas weekend at Chequers: Uncle Jack and his daughter, Clarissa; and Aunt Nellie (a sad but gallant figure, with one son, Esmond, killed flying in an RAF bombing raid, and Giles a prisoner in the sinister castle-prison of Colditz). The three Sandys children arrived a day or two ahead of their parents, Diana and Duncan. Sarah and I both had to be with our units over Christmas Day, but were to join the party on Boxing Day. Family apart, there were 'The Prof' and Jock Colville (doubling as a welcome friend and the Private Secretary on duty); Gil Winant was expected on Boxing Day.

This Christmas seemed to have a special atmosphere; the dark war years were drawing to their end like a long and bitter winter; even the least optimistic could reasonably feel that this, the sixth, would be the last wartime Christmas. Everywhere people were making their arrangements, despite gaps in the family, despite the blackout and despite rationing. Among them, Clementine had taken infinite plans to prepare a glowing Christmas-tide. Nellie Romilly arrived at Chequers at about teatime on Saturday the 23rd to find Winston sitting in the Great Hall; he welcomed her, and then, under the seal of secrecy, told her that he was off the next day to Athens; he begged her to go and find Clementine, who was upstairs and 'very upset'. Nellie went to her sister's bedroom and found her in floods of tears. It was so rare for Clementine to give way – she was accustomed to sudden changes of plan, and had, in these last years especially, developed a strict sense of priorities; but somehow this sudden departure of Winston laid her low. On the following evening Winston and Jock Colville left the party and flew off to war-torn Athens.

For two days meetings took place under conditions of siege. Churchill established his headquarters in the British cruiser *Ajax*; while he was ashore, conferring in the British Embassy, the building was shelled and machine-gunned. He met leaders of all parties, including the Communists, and persuaded them to agree to the establishment of a regency pending a plebiscite on the maintenance of the monarchy. His personal intervention on the spot was undoubtedly the only way in which agreement could have been reached so rapidly.

By the end of January 1945 the German armies were being slowly compressed within their own frontiers, and the end could not be far off; but with victory so near it became imperative for the 'Big Three' to decide on the immediate shape and manner of things when Germany finally surrendered. Once more, therefore, Winston was off on his travels, this time to Yalta on the Black Sea. With his usual colleagues and staff, and with Sarah in attendance as ADC, he set off on 29 January 1945 to meet Roosevelt and Stalin.

During the flight to Malta he produced a sudden high temperature, which fussed everyone considerably; however, he managed to throw it off and revived amazingly quickly. During these days Winston was evidently brooding deeply on the horror and suffering in the world, for Sarah wrote in a letter to her mother on 6 February: 'Last night just before he went to sleep Papa said: "I do not suppose that at any moment in history has the agony of the world been so great or widespread. Tonight the sun goes down on more suffering than ever before in the world." ' And Winston dwelt on the same theme in his letter to Clementine on 1 February:

> I am free to confess to you that my heart is saddened by the tales of the masses of German women and children flying along the roads everywhere in 40-mile long columns to the West before the advancing Armies. I am clearly convinced that they deserve it; but that does not remove it from one's gaze. The misery of the whole world appals me and I fear increasingly that new struggles may arise out of those we are successfully ending.

His long letter, which was dictated, ends with this *envoi*, in his own writing:

> Tender love my darling.
> I miss you very much. I am lonely amid this throng.
> Your ever-loving husband
> W

Ending a letter to Winston on 3 February, Clementine wrote: 'Tender love & grapple close to the President.' This last piece of advice was easier said than done: Churchill's personal relationship with Roosevelt had been, and was still, close, but his role at Yalta was delicate and difficult, and Britain's bargaining power was weak compared with the strength in wealth, men and materials of either the United States or Soviet Russia. Far more even than at Teheran, Churchill found himself isolated at this last meeting of the 'Big Three'. Roosevelt was already in frail health, and with hindsight one can now see from the photographs taken at the time that he indeed looked a stricken man. The President's inclination to conciliate the Russians on almost every point was cleverly exploited by Stalin. Churchill had to wage a lonely campaign, and came away from Yalta with deep reservations and anxieties about the future.

On his way home, Winston 'dropped in' on Athens, which seven weeks before had been in the grip of civil war; now he was acclaimed by a vast crowd in Constitution Square. From there he flew on to Egypt, and took leave of the President on board the USS *Quincy*. Winston described Roosevelt as seeming 'placid and frail. I felt that he had a slender contact with life . . . We bade affectionate farewells.'[11] It was to be their last meeting.

On 19 February, the day of Winston's return to England, there was a thick fog, and his plane was diverted to Lyneham, to where Clementine set out to greet him. After various changes and counter-changes of plan, they actually met in the Manager's Office of the Station Hotel at Reading, where my mother described to me how she found Winston ensconced 'imbibing whisky and soda. He is marvellously well – much, much better than when he went off for this most trying and difficult of Conferences.'[12]

Meanwhile, I was on the verge of what was for me personally, and indeed for all of us 'battery girls', a great adventure. Our battery was one of those chosen to go to Europe to form part of the defence of Brussels which, with other cities, was still vulnerable to aerial attack from bases in the interior of Germany. We set out at the end of January via Dover and Ostend, for a snowy (and subsequently extremely muddy) hillock about fifteen miles from Brussels. I wrote my mother long accounts of our daily life, and as usual she kept me supplied with all the news it was discreet to pass on.

She was somewhat frustrated by not knowing my exact whereabouts, my postal address being merely a series of letters and numbers; and I was far too well-trained in security even to indicate whether I was encamped outside Brussels or Timbuctoo. However, on 17 February she informed me triumphantly:

> I know exactly where you are. Who do you think told me? Your 'Boss' [Field Marshal Montgomery] who sent Captain Chavasse to look you up. He says 'I sent an ADC (Noel Chavasse) to visit Mary & to ensure that all was well.
>
> Her battery is at . . . about 10 miles south west of . . . , & they all seem to be well housed & to be enjoying themselves; morale generally is very high, so that seems very satisfactory.
>
> I have not had time to visit her myself; just at the present I am very occupied: on German soil I am glad to say!!'

Winston and Sarah arrived back from their travels on 19 February, and the next day he and Clementine dined at Buckingham Palace. On 21 February my mother gave me this account of the evening:

> Last night Papa and I and Anthony Eden dined with the King and Queen. It really was great fun. The King kept Papa and Anthony a long time in the dining-room, so I had more than an hour alone with the Queen, but it passed like a flash because she is gay and amusing and has pith and point. When the King came in with Papa and Anthony we had a knock-about turn over de Gaulle and his last fit of sulks – the King and Queen and Anthony and myself against Papa who routed the lot of us from the fender where he occupied a commanding post.

* * *

Clementine was herself on the threshold of a long and exciting journey: she had been invited by the Russian Red Cross to go to Russia and inspect the use that had been made of the materials sent over the last years by her Aid to Russia Fund. Her departure for Moscow, where she was to spend a week before beginning a tour which would take her many thousands of miles through Russia, was planned for the end of March, and she was to be away for more than a month.

Although Clementine had worked continuously for the Red Cross, she had not up to this moment been a 'uniformed member'; now it seemed best that for the purpose of this journey she should wear uniform – which was also the sensible solution to the otherwise vexing and time-consuming question of 'what to wear'. She was accordingly granted the rank of a Vice-President of County of London Branch, British Red Cross, and had early in the New Year ordered her uniform. I offered some seasoned advice: '[A]re you having your shirt collars attached – or like a man's – separate? My experience is the latter look better and are in the end more comfortable.'[13]

On 11 March she wrote to me: 'I'm getting very near my flight to Russia – another 16 days & I can't get my uniform done. I expect everything will be ready at the last moment. Perhaps when I get back you will be sitting in the ruins of Cologne.' The uniform continued to be troublesome: 'My uniform does not fit, & makes me look like an elephant!' she complained to me on 20 March. But finally, three days later, she reported triumphantly: 'My uniform will really fit & I've got a Battle top & a Monty beret!'[14]

The Queen invited Clementine to tea on the eve of her departure: 'It was nice of her to send for me just before my great adventure. The King was there. The Queen looked very sweet & soignée like a plump turtle-dove.'[15]

Although she looked forward to this great experience, Clementine had some preoccupations: she hated leaving Winston for so long – and they would be so very far apart. She really hated flying, and was as aware as anyone else of the element of risk. That this played on her mind is shown in a letter she wrote me on the day of her departure: 'Darling supposing anything happened to me (e.g. air crash) Do you think you could be released from the A.T.S. on Compassionate grounds to look after Papa? Because he would need it.'[16] And to Sarah she wrote: '[L]ook after Papa as much as you can & say a little prayer now & then for your devoted Mother who altho' she keeps up a brave front sometimes feels like a nervous old lady!'[17]

Accompanied by Grace Hamblin and Miss Mabel Johnson (Secretary to the Aid to Russia Fund), Clementine flew off on 27 March. Their first stop was Cairo, where they were looked after by the Minister Resident in the Middle East and his wife, Sir Edward and Lady Grigg. Here, bad weather delayed their onward flight for several days; but she was happy and comfortable with the Griggs, and was by no means idle, visiting the principal YWCA and three other clubs and hostels, as well as the Military Hospital.

During this enforced pause in her programme, Clementine was assailed by pangs of homesickness; she ended a letter to Winston on 28 March:

> My beloved Winston This is a long separation. Think of your Pussy now & then with indulgence & love.
>
> Your own
> Clemmie

The day before Clementine set out Lloyd George had died, at the age of eighty-two; a few days later Churchill led the tributes in the House of Commons to his former chief and colleague. Clementine's mind went back, not to the difficult and sometimes embittered moments in the First World War, but to those glorious, crusading days of radical reform, which had left such an undying impression on her mind: 'I loved your speech about Ll. G. It recalled forgotten blessings which he showered upon the meek & lowly.'[18]

During her unexpected stay in Cairo, Clementine took the opportunity to write to Field Marshal Montgomery. The crossing of the Rhine by British troops on 23 March had seemed, as indeed it was, a dramatic milestone on the way to victory. Winston had made a two-day visit to the Field Marshal's Headquarters, traversing the Rhine twice and visiting British troops in advanced positions. She wrote, on 28 March:

> My dear Sir Bernard
>
> I am perched here for 24 hours on my way to Moscow.
>
> I want to send you my heartfelt & grateful congratulations & rejoicings upon the Forcing of the Rhine – a glorious Victory which will for ever live in history when we have departed from this earthly scene.
>
> Winston loved his visit to you. He said he felt quite a reformed character & that if in earlier days he had been about with you I should have had a much easier life! referring I suppose to his chronic unpunctuality & to his habit of changing his mind (in little things) every minute! I was much touched & said I had been able to bear it very well as things are. So then he said perhaps he need not bother to improve? But I said 'please improve becos we have not finished our lives yet'.

In his reply, Field Marshal Montgomery wrote: 'It was a great pleasure to have the P.M. at my H.Q. during the Battle of the Rhine; he was very amenable to discipline and never argued about what was to be done!'[19]

Clementine spent her sixtieth birthday on 1 April in Cairo. It was Easter Sunday, and while she was in church the British Ambassador, Lord Killearn, handed her a telegram from Winston. She was so pleased, and wrote to tell him so later in the day.

After the delay caused by unsuitable weather conditions, she and her companions finally got off on the last lap of their journey, arriving at Moscow Airport on 2 April. Among the government and Russian Red Cross officials there to meet her were the Maiskys, whom she had known in London; the British Ambassador, Sir Archibald Clark Kerr; and Averell Harriman, the American Ambassador. 'This is one of the most inspiring and most interesting moments of my life,' Clementine said, with perfect sincerity, in a short speech on her arrival; she had also managed to 'work up' a few apt sentences in Russian, which duly gave pleasure. Since the Yalta Conference, the danger of Soviet domination in Eastern Europe had become an ever more visible threat; suspicions of Stalin's intentions had led to an increasingly strained relationship between Britain and Russia; and there had been doubts about the wisdom of Clementine's journey. But in the event her visit provided a welcome occasion for agreeabilities. Winston cabled her immediately after her arrival: 'Lovely accounts of your speech and reception received here. At the moment you are the one bright spot in Anglo-Russian relations.'[20]

During the next few crowded and, for Clementine, peripatetic weeks, cables through the British Embassy rather than letters were their mutual means of communication, and hers gave a graphic description of her thrilling and moving journey, while half a world away Winston kept her as up to date as he could with events which followed in rapid succession as the tide of victory flowed swiftly forward. Although Clementine was totally absorbed, physically and mentally, by the demands of her formidable programme, she found it frustrating to be away from the source of instant news, and she minded above all being parted from Winston in these last tremendous weeks of the struggle they had lived through together, moment by moment, from its beginning.

During her week-long visit to Moscow, Clementine and her companions stayed in the State Guest House. Her programme was a full one: visits to two hospitals, a factory and a children's home were interspersed with official luncheons and dinners, one being given for her by the Molotovs (he was Soviet Commissar for Foreign Affairs); and on two evenings she went to the ballet. Clementine was also received 'most amiably' in private by Mr Molotov in the Kremlin. In a cable to Winston she reported that 'He referred to present difficulties, but said that they would pass and Anglo-Soviet friendship remain.'[21]

Winston, in a cable to her, told her that he had asked the Ambassador to show her Foreign Office telegrams, and with reference to her contacts with the Russians he added: 'Please always speak of my earnest desire for continuing friendship of British and Russian peoples and of my resolve to work for it perseveringly.'[22]

On 6 April Clementine attended a Meeting of the Praesidium of the Executive Committee of the Soviet Red Cross. It was during the course of this meeting that she was awarded, 'amidst stormy applause' (in the words of the official report), the Distinguished Red Cross Service Badge.

However, the highlight of these days in Moscow was her visit to Marshal Stalin, who received both her and Miss Johnson at the Kremlin on the afternoon of 8 April. They were shown into a vast room, at the far end of which Stalin was sitting at a writing desk. Speaking through an interpreter, the Marshal expressed himself graciously about the work of the Fund. Clementine then presented him with a gold fountain-pen from Winston, as a souvenir of their meetings during the war. 'My husband wishes me to express the hope that you will write him many friendly messages with it,' she said.[23] The graceful acceptance of presents was evidently not a prime trait in Marshal Stalin's character, for although he took the pen with a genial smile, he put it on one side saying, 'But I only write with a pencil.' He also added: 'I will repay him.'[24] Clementine, who was now quite inured to Russian ungraciousness, was unabashed and much amused.

Between 6 and 9 April, Winston dictated a twelve-page letter to Clementine. In it he ranged over a wide field of events and thoughts: the war, the political scene and likely future developments there, and domestic details from their 'home front'. Indeed, the letter reads more like a conversation than a literary exercise:

> My darling Clemmie,
>
> Since you were swirled away into the night, I have had the most exacting time. What with looking after Bernie Baruch and all the Dominion Premiers, as well as overwhelming toil, I have not found a minute to write. It is now Friday, and I have just finished my sleep and am going down to Chequers, where Smuts is spending the week-end with me . . . I arranged for Sarah to attend on me for week-ends on duty while you are away, to help me with our various official guests.

With the advance of the Allied armies, the danger from air attack was diminishing daily; in reporting to Clementine that there had been no 'bombs or bangs' for a whole week, Winston told her:

> I have moved the Cabinets back to No.10, and have also had one or two meals there. I am giving orders for the rehabilitation of that dwelling, otherwise we shall not be able to use it this year. If you have any strong views about this, pray write to me, so that I can have them carried out. The garden has become very nice now. The lawn is in good order, the herbaceous border is fully stocked and the little magnolia tree is at its best.

He had received messages from Roosevelt through Bernard Baruch, and he was already aware that the President's strength was failing:

> [N]ow that Harry [Hopkins] is ill and Byrnes [US Director of War Mobilization] has resigned, my poor friend [the President] is very much alone, and according to all accounts I receive, is bereft of much of his vigour. Many of the telegrams I get from him are clearly the work of others around him.

Winston was gravely disturbed about the difficulties over Poland and Romania, for Russia had already violated agreements made only three months before at Yalta.

On the political front at home, with the approach of victory party politics were beginning to revive. Some colleagues were already making speeches in a hostile vein, and Winston told Clementine that

> there is very little doubt that the Government will break up shortly . . . I expect the General Election about the middle of June.

He had some gratifying news for her about the success of her visit:

> There is no doubt that your visit is giving sincere pleasure. Gousev [Gusev, the Soviet Ambassador to London] called at the Foreign Office yesterday, as Anthony thought, to begin a long attack, but instead he spent a long time in conveying a message from his Government in praise of you and your work, and asking whether they might offer you the Order of the Red Banner of Labour.

He had also heard from the British Ambassador that her visit was doing 'the utmost good at a most difficult time'. But he was anxious lest she should become exhausted:

> I do hope you will be sensible and not overdo it. Insist upon days of leisure. They will understand. Otherwise you may be killed with kindness.

Clementine was due shortly to leave Moscow for Leningrad and far distant parts of the Soviet Union. Winston knew well how much she set store by knowing the news, but he warned her that once she left Moscow their telegrams to each other would be *en clair*, and that all private information would perforce cease:

> You will not have a cypher, and much of this stuff is dynamite. The same is true of letters. I do not feel able to write freely because I do not know how letters will be forwarded from Moscow.

Although this letter was dictated and typed, he added the ending in his own hand:

> My darling one I think always of you . . . Yr personality reaches the gt masses & touches their hearts. With all my love & constant kisses
>
> I remain ever your devoted husband
>
> W
>
> [charming drawing of a pig]

On the night of 8 April Clementine and her party boarded the special train provided for them by the Soviet Government, which was most roomy and comfortable, and, in order that the visitors should be able to enjoy the scenery, travelled at sightseer's speeds. Throughout their journeyings Clementine and her companions were accompanied by, among others, an interpreter and the head of the Press Department of the Russian Foreign Office. They reached Leningrad late on the afternoon of 9 April, and were installed in a comfortable villa in the suburbs. 'Think Leningrad is the most beautiful city I have ever seen,' she cabled Winston the next day.

Her visit lasted three days; she genuinely enjoyed it and was greatly touched by the warmth and spontaneity of the welcome she was accorded. Leningrad had been blockaded for nearly two and a half years, and much of the population had suffered from the effects of prolonged starvation; one of Clementine's visits was to a scientific institute and hospital where there were many children being nursed back to health and strength.

The Chairman of the City Council was most attentive, calling upon Clementine and later escorting her to the ballet. On the last night of their visit, she and her party were entertained again by the Chairman and his wife, who gave a dinner and musical entertainment, which sounded very jolly: 'We ended the evening with community singing. I think our Russian hosts were astonished that we were so familiar with the Song of the Volga Boatmen!'[25] It was at this dinner that it was announced that Clementine had been awarded the Order of the Red Banner of Labour, and Miss Johnson the Medal of Labour.

Their next destination was Stalingrad, and on the way the train had to pass through Moscow. Only a very short stop there had been scheduled on the afternoon of 13 April, but as the train drew into the station Clementine saw that a large silent crowd was waiting on the platform, and in the front were the Molotovs; they boarded the train. 'We come with bad news,' said Mr Molotov. 'President Roosevelt is dead.' Clementine was stunned by these grievous tidings; she at once understood their import to the course of events, and knew also what a deep and personal loss the President's death would be to Winston. But she collected herself instantly, and suggested to the Molotovs that they should sit in silent recollection for a few minutes in the railway carriage; this they did. Clementine remained a few hours in Moscow, and was able to speak to Winston from the British Embassy. She also, in some haste, wrote a letter to Averell Harriman:

> My dear Averell
>
> Upon my arrival this afternoon I was told the tragic news. It has upset me very much & I am deeply grieved –
>
> The President has truly died a Warrior's death – No one fought more valiantly than he to save the world. It is cruel that he will not see the Victory which he did so much to achieve – But he saw it coming.
>
> I know you will be very unhappy dear Averell & I send you my affectionate sympathy.[26]

Then they headed south for Stalingrad. The siege of this great city had lasted for six months (13 September 1942 to 2 February 1943), and every house had been fought for: even amid all the other horrors of the war, the eyes of the world had been fixed upon this savage and searing episode, and now, two years later, the ravaged city made an indelible impression on Clementine, who was later to write: 'What an appalling scene of destruction met our eyes. My first thought was, how like the centre of Coventry or the devastation around St Paul's, except that here the havoc and obliteration seems to spread out endlessly . . . The imagination is baffled by the attempt to encompass calamity on so vast a scale.'[27] Clementine was shown the rebuilding operations in progress: factories, hospitals and schools (the latter working in two shifts), and a prefabricated wooden village which was being built to house twenty thousand people. She was deeply impressed.

Between 17 and 19 April they visited three spa towns in the Caucasus: Kislovodsk, Essentuki and Pyatigorsk. From Kislovodsk she cabled Winston on 18 April, describing how they had been to the local theatre: 'The audience were rapturous in their welcome and threw bunches of violets from the gallery. This morning we spent visiting sanatoriums full of severely wounded Red Army soldiers. The whole town turns out to greet us every time we go out and I am continually amazed and moved by so much enthusiasm.'

During these travels Clementine realized how far away – how very far away indeed – Europe was to the Russians. The immensity of their own sufferings as a nation, the progress and losses of their own army, and the immediate problems of reconstruction absorbed all their thoughts, emotions and energies. They were genuinely grateful for the solid help provided through the British Red Cross Aid to Russia Fund but, as Clementine herself wrote: 'I could not honestly say that I found a great deal of curiosity about British life or affairs.'[28]

On 21 April the party arrived at Rostov-on-Don, which was the focal point of their visit, for it was in this city that the Aid to Russia Fund was undertaking the complete re-equipping of two large hospitals, both of which had suffered severe damage at the hands of the Germans. It had been proposed to equip the two hospitals with five hundred beds each, but when Clementine and her party saw the need and potentialities there, it was decided to increase the provision to fifteen hundred beds.[29] In a cable to Winston en route from Rostov to Sebastopol she said:

> I am happy that our hospitals are in this town [Rostov] in which . . . thousands of citizens were put to death, many other thousands of women and children deported to Germany and the remainder stood the brunt. It must have been a lovely city once and stands high overlooking the Don which is stiff with delicious fish. We were mobbed by friendly crowds. Everywhere we see smiling faces and even in the hospitals grievously sick people sit up and say greetings to British women.[30]

Thousands of miles away at home, Winston tried to keep Clementine up to date with major events, but things were moving at hectic speed, and now that she was so far from Moscow their cables took longer to arrive and often crossed with each other. As the liberating armies in Europe swept forward they discovered the terrible concentration and death camps, the stark reality of which now became apparent to all. As Winston wrote to her on 21 April, 'Intense horror has been caused by the revelations of German brutalities in the concentration camps. They did not have time to cover up their traces.'

The next stage of Clementine's journey was four days spent in the Crimea, where Sebastopol made a particularly sad impression: '[I]t presented so melancholy a spectacle . . . Before the Nazis destroyed it in their blind rage, it must have been a dream of beauty, as lovely as a poem, with its many pillared and frescoed houses.'[31] In Yalta Clementine and her party stayed in the Vorontzov Palace, and she slept in the room Winston had occupied during the Yalta Conference. Here they all had a 'day off'.

On 20 April the Russians had entered Berlin, and three days later the United States and Russian armies met at Torgau, south of the city. This splendid news percolated to Simferopol on 28 April, where Clementine was spending the last day of her Crimean visit; she at once cabled to Winston:

> I have just heard the glorious news of the meeting of the Red Army with the Allies and by your statement on the wireless it seems the end is near. I long to be with you in these tremendous days and I think of you constantly.
>
> All my love
> Clemmie

The party's next stop was a two-day visit to Odessa, the great port on the Black Sea, where they toured the city, visited two hospitals, went to the opera and witnessed a May Day parade. Of especial interest to her and her companions were the British prisoners-of-war (approximately 250 of them) who had been liberated by the Red Army and were in a camp awaiting repatriation. Clementine was taken to visit them, and also to another camp in which were French, Belgian and Dutch ex-prisoners, many of them having been deported to Germany for slave labour, who now were on the threshold of their return home.

Meanwhile dramatic news crowded in almost hourly: Mussolini had been executed by the Italian partisans on 28 April; the German forces in Italy had surrendered on 29 April; and on 30 April Hitler had shot himself in the air-raid bunker of the Chancellery in Berlin as the sounds of the Russian guns came ever nearer.

As she heard the news of these events in disjointed telegrams and press reports, triumphant joy and relief were mingled for Clementine with an almost overpowering desire to get home to England and to Winston.

But her journey was not finished yet. Although not originally in the programme, Clementine was invited to visit Kursk. She was later to write: 'I was stirred to witness that in this city whose civic chief is a woman, thousands of women have volunteered to help in its rebuilding. I saw them working with a will among the ruins. I am sorry they had only spades with which to clear away the rubble – I wished they could have had bulldozers.'[32]

On 5 May, Clementine and her companions arrived back in Moscow; Winston had sent a cable to greet her: 'Delighted with all your telegrams. Skymaster [the aeroplane used for all Winston's long flights] starts for you tonight with many papers and messages . . . Several crises are coming to a head, and as you see both our great enemies are dead . . . My hours are shocking but I am very well.'[33]

On 5 May Winston had sent her a second, much longer telegram – now that she could receive news safely through the Embassy he felt free to confide to her his gnawing anxieties:

> You seem to have had a triumphant tour and I only wish matters would be settled between you and the Russian common people. However there are many other aspects of this problem than those you have seen on the spot . . . It is astonishing one is not in a more buoyant frame of mind in public matters. During the last three days we have heard of the death of Mussolini and Hitler; Alexander has taken a million prisoners of war; Montgomery took 500,000 additional yesterday and far more than a million to-day; all north-west Germany, Holland and Denmark are to be surrendered early tomorrow morning . . . and we are all occupied here with preparations for Victory-Europe Day . . . I need scarcely tell you that beneath these triumphs lie poisonous politics and deadly international rivalries. Therefore I should come home after rendering the fullest compliments to your hospitable hosts.

In reply Clementine cabled to Winston:

> I am full of joy at the overwhelming victories of Alexander and Montgomery. I long to be with you but have some necessary engagements to fulfil and some loose ends to tidy up after which I shall joyfully fly home by Malta leaving at dawn on the 11th. In the midst of all this military glory I know you are having harassing and sometimes sad experiences.[34]

These last few days in Russia were full of engagements: Clementine visited a school, several hospitals and the Moscow Metro; she attended receptions given in her honour; she gave a press conference, and went to a concert and the ballet. She and Miss Johnson received their decorations at the Kremlin from the hands of the First Vice-Chairman of the Supreme Soviet of the USSR and, at a luncheon given two days before their departure by the Soviet Government in honour of the visiting British party, Mme Molotov presented Clementine with a most beautiful diamond ring as a token of 'eternal friendship'.

Her journey had been a tremendous success: it had been 'laid on' officially at the highest level wherever she had been, but there is no doubt that her own personality and open, intelligent approach had played a major part in turning an official exercise in public relations into a genuine two-way expression of gratitude, appreciation and friendship between the British and Russian peoples.

But it must not be supposed that the wonderful welcome and gilded hospitality she had received had entirely blinded Clementine's eyes to the true state of things, or that, when the need arose, she was not able to make firm representations. During her journey one of the few letters forwarded to her was from Miss Eleanor Rathbone, the Independent Member of

Parliament for the Combined English Universities and a woman of great distinction. She was on the Parliamentary Committee for Refugees, and the previous winter had made a speech in the House of Commons drawing attention to the fate of the vast numbers of Poles who were being deported by the Russians to labour camps in distant parts of the USSR. When Miss Rathbone heard of Clementine's visit she asked her to make some kind of representation about these wretched Poles. After her return home Clementine wrote the following letter, which shows so well her directness, her instinctive reaction in favour of some immediate action, and her clearsightedness:

> 28th May 1945
>
> My dear Miss Rathbone,
>
> I was glad to have your letter with me in Russia, but alas, I was not able to do any good. I saw Mr Maisky several times during the two periods when I was in Moscow, and I did let him know that I was acquainted with some of the facts concerning the Polish deportees. Actually he started the subject in rather a school-masterish lecturing manner! After listening as patiently as I could I am afraid I rather flew at him and said that I was kept informed by our Ambassador of the telegrams which were passing between the Foreign Office and the Kremlin, and that I would not discuss the matter with him as I was afraid it would make me very angry. So he beat a diplomatic retreat.
>
> From my small and limited experience I am convinced that the only hope is that the English-speaking world should act together and go so far as to inform the Kremlin that if they do not mend their ways we may have to break off Diplomatic relations with them.
>
> I am terribly grieved at the progressive deterioration in the relationship between our countries. When one has been in Russia, only for a few short weeks, one cannot help loving the people, and one must always separate them from their Government which is mysterious and sinister, and terribly strong.
>
> Yours sincerely
> Clementine Churchill

She added this postscript in her own handwriting, referring to her suggestion about breaking off diplomatic relations: 'Winston would disapprove! I am sending you an article I have written on my journey.'

* * *

Tuesday, 8 May 1945 was Victory-in-Europe Day. Early in the morning Clementine sent Winston a short cable:

> All my thoughts are with you on this supreme day my darling. It could not have happened without you.

Arrangements were quickly made by the British Embassy in Moscow for a service of thanksgiving; there was no official Chaplain, but an RNVR officer who was a Methodist minister conducted the service.

At luncheon at the British Embassy that day, where Mr (later Sir) Frank Roberts, the Chargé d'Affaires, acted as host in the absence of the Ambassador, the guests included General Catroux, the French Ambassador, and M. and Mme Herriot. He had been President of the Chambre des Députés at the fall of France, he and his wife having later been deported to Germany; they had both just been released by the Russians. Later in the afternoon they all listened to Winston's broadcast from London. M. Herriot wept. He said to Clementine:

> I am afraid you may think it unmanly of me to weep. But I have just heard Mr Churchill's voice. The last time I heard his voice was on that day in Tours in 1940 when he implored the French Government to hold firm and continue the struggle. His noble words of leadership that day were unavailing. When we heard the French Government's answer, and knew that they meant to give up the fight, tears streamed down Mr Churchill's face. So you will understand if I weep to-day, I do not feel unmanned.[35]

The next evening Clementine broadcast over Moscow Radio a message from her husband to Marshal Stalin: 'It is my firm belief that on the friendship and understanding between the British and Russian peoples depends the future of mankind.'[36]

On her last night in Moscow, Clementine was taken to the ballet to see *Swan Lake*. She described the scene at the end of the performance: 'At the fall of the curtain the prima ballerina, with exquisite grace, turned the applause from herself to our box. She wheeled towards us, clapping and smiling, the whole company followed her, and then the great audience took up the applause.'[37] This graceful and heart-warming scene made a perfect ending to her memorable journey.

In his cable to Clementine of 2 May, Winston had told her: 'You should express to Stalin personally my cordial feelings and my resolve and confidence that a complete understanding between the English-speaking world and Russia will be achieved and maintained for many years, as this is the only hope of the world.' Now, on 11 May, the morning of her departure from Moscow, Clementine put this message in her own words in a handwritten letter:

> My dear Marshal Stalin,
>
> I am leaving your great country after a wonderful & unforgettable visit.
>
> I have seen with sorrow some of the ravages caused by a wicked & ruthless Enemy & observed the dignity, courage & patience of your people.
>
> I have enjoyed the most warm hearted hospitality & everywhere I have been welcomed with the greatest kindness & enthusiasm. And my happiness has been crowned by being received by you & by the decorations which you have bestowed upon Miss Johnson & myself.

> I know of the international difficulties which have not been surmounted, but I know also of my Husband's resolve & confidence that a complete understanding between the English Speaking World & the Soviet Union will be achieved and maintained as this is the only hope of the World.
>
> Yours sincerely,
> Clementine S. Churchill
>
> I count myself fortunate to have been in your country in these days of Victory, and to have seen the Sun of Peace rise in Moscow.

Soon after her return, Clementine wrote a small booklet entitled *My Visit to Russia*, which was sold for the benefit of the Aid to Russia Fund. She described the welcome she had received, and the impression made upon her by all she saw, and the people she met. The oft-repeated expressions of friendship on either side, and the pious hopes for a new dawn of amity and understanding between the Russian and the British people now of course have a somewhat hollow ring.* Even at the time we see that Winston and, sharing his knowledge, Clementine, had strong reservations and anxieties which were to be only too dismally fulfilled. But in reading the account one must try to recapture the atmosphere of that time. The British people had been awestruck by the heroism and sufferings of the Russians, our allies in a tremendous and agonizing struggle. The extent of that admiration was expressed in the sustained and amazing support for the Aid to Russia Fund from people of all classes. Much that soon became common knowledge was then unrevealed, and the journey from admiration and hopeful aspirations to grey disillusion and suspicion was a long-drawn-out process.

Leaving Moscow in Skymaster on the Friday, Clementine, Mabel Johnson and Grace Hamblin had a swift flight home, arriving at Northolt early on the morning of 12 May. Winston was determined to go and meet her himself, but (predictably) he did not leave Storey's Gate quite in time, and Skymaster had to make a few extra tours round the airfield to allow a loving but tardy Winston to be on the tarmac to welcome home his Clemmie.

* * *

Although Winston had tried to keep her abreast of events (private and national), some of his cables had, in fact, only met Clementine on her return to Moscow, and there was much of course that she could only learn from him when she got home.

In the family circle, Winston had been deeply anxious for his brother Jack, who had suffered a heart attack and had for a number of days been seriously ill; now, happily, he was on the road to recovery. Clementine's sister, Nellie Romilly, had been in agony for the safety of her son Giles who, with Lord Lascelles (son of the Princess Royal), Earl Haig and other *prominenti*, had been removed from their camp a few hours before the liberating

* This was first written in the Cold War climate of 1978.

Americans arrived and taken off to an unknown destination, no doubt as potential hostages: after a period of acute anxiety, however, she had received the joyful news that he and his companions had been freed. Sarah was well, but during her mother's absence her divorce from Vic Oliver had been granted: it was the final milestone on what had been an increasingly unhappy road.

My battery was now encamped outside Hamburg, and I had some leave just at the time of my mother's return. My parents took me with them when they went on 13 May to St Paul's Cathedral, to a moving and wonderful service attended by the King and Queen. 'This is such a national thanksgiving as never was before,' said the Archbishop of Canterbury in his address, 'for everyone from greatest to least has borne his share of suffering and toil.'

Only those who lived through those years can understand the emotional relief, joy and gratitude which the proclamation of victory in Europe inspired. Now we know the disillusions of the years that came after. Yet a vile tyranny had been laid low, and proud countries liberated from a hideous yoke by the years of sustained endeavour and sacrifice. All were entitled to their hour of spontaneous rejoicing.

111. Clementine, looking really splendid, leaves No.10 for Westminster Abbey on Coronation Day, 2 June 1953, wearing the petunia mantle of a Dame Grand Cross of the Order of the British Empire. She was beginning to suffer from neuritis, and her arm is in a sling.

112. Surrounded by his colleagues at the Conservative Party Conference in Margate, on 10 October 1953, Winston makes his triumphant return to the fray after his stroke the previous June. On his right are Anthony Eden and Clementine.

113. Diana and Duncan Sandys (at this time Minister of Supply) and Celia, aged ten.

114. Clementine stands between King Gustav Adolf and Queen Louise in Stockholm, where she had received the Nobel Prize for Literature on Winston's behalf, December 1953.

115. The night before Winston tendered his resignation as Prime Minister, 4 April 1955, the Queen and the Duke of Edinburgh dined at No.10. Here Winston and Clementine are greeting Her Majesty on her arrival.

116. Winston and Clementine celebrated their Golden Wedding on 12 September 1958 at Max Beaverbrook's villa, La Capponcina, at Cap d'Ail. With them are Randolph and his daughter, Arabella, aged nearly nine.

117. In May 1960 General and Mme de Gaulle made a State Visit to London. Here they are guests at the Royal Hospital, Chelsea, on 8 May. Left to right: Mrs John Profumo (Valerie Hobson, wife of Minister of State for Foreign Affairs); Clementine; Winston; General de Gaulle; Prince Philip; Mme de Gaulle; Lady Dorothy Macmillan (wife of the Prime Minister).

118. Off to the races! Winston had flown back from France to see one of his horses run at Ascot on 15 June 1961.

119. Sarah with Henry Audley, 23rd Baron Audley, whom she married on 26 April 1962. Their visible happiness was short-lived: Henry died of a massive heart attack in Granada on 3 July 1963.

120. Clementine playing croquet at Chartwell, summer 1964. Taken by Viscount Montgomery ('Monty').

121. Clementine and Grace Hamblin, secretary to the Churchills since 1932, with 'Robbie' the poodle, walking in Kensington High Street, summer 1964.

122. Sylvia Henley, Clementine's cousin and friend since childhood, in November 1969: truly a woman for all seasons; she died in 1980.

123. During the dark chill days of January 1965 when my father lay dying, my mother, Christopher and I paced the parks – killing time, while time killed him.

124 Winston Churchill's State Funeral, 30 January 1965. Our family, led by Clementine and Randolph, watch the Bearer Party found by the 2nd Battalion the Grenadier Guards bear the coffin into St Paul's Cathedral, where the Queen awaited her greatest commoner.

125. The Baroness Spencer-Churchill of Chartwell, GBE, with her sponsors, Lord Ismay and Lord Normanbrook, House of Lords, 15 June 1965.

126. 'Monty' and Clementine, at the Passing Out Parade at Sandhurst in August 1967, when her grandson, Nicholas Soames, was being commissioned.

127. At the launching of the Winston Churchill Centenary Appeal, at the Banqueting House, Whitehall, 1 April 1974. The three party leaders raise their glasses to Clementine: it was also her eighty-ninth birthday.

128. After a commemorative service in St Martin's Church, Bladon, Clementine at her husband's graveside, with Minnie and Winston Churchill, 30 November 1974.

129. Christmastime 1975, staying with Winston and Minnie. Clementine with Jack and Marina Churchill.

130. Charlotte Hambro (Soames) with the 'new' Clementine after her christening in London, with her great-grandmother, Clementine, 12 January 1977.

131. Clementine at ninety, photographed by Lord Snowdon. 'And calm of mind, all passion spent.' (John Milton, *Samson Agonistes*, 1671)

CHAPTER TWENTY-FOUR

The Two Impostors

If you can meet with Triumph and Disaster
And treat those two impostors just the same
Rudyard Kipling

'AFTER THE WAR' HAD SEEMED A FAR-OFF, GOLDEN PERIOD – THE RECURRING theme of many contemporary songs – but few people had had time or energy to dwell upon it in detail; sufficient to the day was very much the sweeping up of last night's shattered glass, or queueing for too few oranges. Now peace was upon us, and the end of the war illuminated deep disagreements which had been obscured while our survival was at stake. It is true that political discords had never been far below the surface, but the three main parties had honoured the electoral truce. The Coalition Government had remained united, and had been sustained by an overwhelming and unfaltering parliamentary majority. But by the end of 1944 unity within the Government and in the House of Commons (which still remained unshakeable on matters concerning the defeat of our enemies) was very fragile on all future domestic issues.

In moving the annual Bill to prolong the life of Parliament in October 1944, Churchill made it clear that he foresaw victory in Europe as marking the end of the Coalition Government. If he perceived this situation with realistic clarity, he accepted it with reluctance; he had hoped that national and political unity could be maintained until all our enemies had been defeated. But the present Parliament was ten years old, and Churchill was adamant that the coming of peace must be followed at the earliest opportunity by a General Election in which the people's voice could be heard.

It was not only politicians and their parties who felt they could now afford the luxury of considering their own interests and of casting their minds forward to the return of normal times; families and individuals also started making plans for the future, and among them, Clementine took prudent thought for the morrow. During the winter of 1944–5 she started quietly looking around for a new London house. Her thought (to which her deepest wishes were certainly the father) was that at the end of the war and the breaking up of the Coalition Government, Winston would retire. This idea never formed part of his own personal post-war plan: his mind was totally dominated by the war and the problems which would soon confront the victorious Allies and a prostrate Europe. Nevertheless, although Clementine did not involve him in her discreet house-hunting, he was in favour of their again acquiring a London home.

Early in the New Year of 1945 Clementine saw a charming house in Kensington – No. 28 Hyde Park Gate – in a secluded cul-de-sac and with a delightful garden; she thought this would be just right for them. My mother had told me about her discovery in one of her first letters to me in Belgium: 'Yesterday afternoon I took Papa to see the little house I covet. He is mad about it, so now I must be careful not to run him into something which is more than he can afford.'[1]

But Clementine's mind was not solely preoccupied with thoughts for their own domestic arrangements in the early months of 1945. Her letters to Winston while he was away at the Yalta Conference show that she was most attentive to conditions both in this country and on the Continent. In February, Joan Portal* had been staying with a French service family in France, and had been shocked by the want and cold afflicting their civilian friends and relations. On her return to London, she and her husband had lunched with Clementine, who wrote to Winston on 12 February, recounting what Lady Portal had told her and adding: 'Both "Houses" are working up steam & the Press is writing some moderate & some exaggerated articles. At least I hope they are exaggerated . . . There is, it seems, a great deterioration since November the 11th [1944] when we saw Paris exalted.' She enclosed with her letter a 'snip' from the same day's *Daily Mail* headed 'CATS – 30s Each in Paris', in which it was stated that thirty thousand cats had 'disappeared' in Paris since the Liberation and that 'Acute food supply difficulties have produced a systematic cat hunt in the city. A cat is now worth 30s.' Clementine begged Winston to 'influence General Marshall† or the President to release enough lorries to distribute the food & coal already in France'. The cold and misery were to be widespread throughout Europe that icy winter. The wonderful, saving 'Marshall Plan' was not to get going for over two years.

Before Winston had left for Yalta on 29 January, a crisis had arisen in Britain over the distribution of coal, due to lack of manpower combined with some very severe weather. From their letters it would seem that Clementine had the simple, straightforward idea that soldiers could be used to help out with coal deliveries in this critical situation. On 30 January, she ended a letter to Winston: 'The radio announced this morning that coal distribution by Army lorries begins today! It didn't mention who thought of this.' And in Winston's letter of 1 February, which must have crossed with hers, he wrote: 'Keep your eye on the coal shortage as long as

* Wife of Sir Charles Portal, Chief of the Air Staff, and a great friend of Clementine's.
† See note p. 354.

it lasts, and keep prodding at them to give the necessary military lorries etc.'

* * *

A fortnight after Victory Day in Europe, on 23 May 1945, Churchill tendered his resignation to the King, the Labour Party having refused to continue the coalition until after the defeat of Japan. (In all fairness it must be remembered that the swift collapse of the Japanese after the bombing of Hiroshima in August could not then have been foreseen.) The King charged Churchill with the formation of an interim administration, and the Caretaker Government, as it was called, held the fort for the few weeks remaining until the dissolution of Parliament.

It was widely thought in this country, and certainly generally assumed abroad, that the Conservative Party would win the election – if only on the prestige and record of its leader. Yet there were signs to the contrary for those few who read them aright. Several important by-elections since 1940 had been won by Independent candidates, and since 1942 the Gallup Poll (a then fairly new device in this country and not widely published) had shown that Labour consistently led the Conservatives in public popularity. The Conservatives' opponents were always quick to emphasize that in the years before the war Tory Governments had failed to prepare the country for the coming conflict – a failure which of course Churchill had continually condemned. The Conservatives' lack of sympathy for the victims of the Depression, and their methods of dealing with the economic situation, were also constantly held up to contumely.

Electioneering began as soon as polling day was fixed for 5 July. Winston and Clementine lost no time in making a tour of his constituency, which had changed in character owing to the growth and movement of population from the near-country seat of Epping into the populous borough of Woodford. By chivalrous agreement, none of the opposition parties nominated candidates to stand against him; he was, however, opposed by a middle-aged eccentric, Mr Alexander Hancock, who stood as an Independent.

Due to his still being Prime Minister, and the need also for him to undertake speaking tours covering the whole country, Winston was of necessity somewhat of an absentee candidate. Clementine, however, did several tours in the constituency on his behalf, making short speeches and visiting supporters and campaign workers.

The *Woodford Guardian* of 6 July recalled how, 'Dressed in a black wool costume, with a bright red scarf and a chiffon turban, and standing on a low wooden platform', Clementine had addressed an open-air gathering:

> A cross against his [Churchill's] name, she said, would mean they were glad he had been head of the Coalition Government and had approved of its conduct of the war and that they had confidence in him for the future.
>
> She thought that the election had come as a grief and disappointment to

many. After the cooperation of the war years they were flying at each other's throats. 'But we must not allow ourselves to be depressed, for after the election I am sure we will all work harmoniously together for the good of our country.'

Winston did not relish the return to party political electioneering: it was, for him, a traumatic change of gear from being the nationally accepted leader of a united nation, working with tried and trusted colleagues regardless of party affiliations, to find himself overnight the leader of a faction, and of necessity cast in a contentious and divisive role. But although it was a very real effort to plunge once more into the maelstrom of party politics, he was a pugnacious animal, and a doughty champion of his party. As for Clementine, she had never made any secret of the fact that, in her view, Winston should, at the victorious conclusion of the war, resign from office and not seek re-election at all: she felt very strongly that, having led a Coalition Government and a united nation, he should retire rather than become the leader of one-half of the nation against the other. To this view Winston was wont to retort that he was not yet ready 'to be put on a pedestal'.[2]

Churchill launched his national campaign in his first party broadcast on 4 June. He reproached the Labour and Liberal parties for withdrawing from the Coalition, and for putting party before country, and he also delivered a forceful attack on socialism, in which he suggested that, to bring about a socialist order of things, the Labour Party 'would have to fall back on some kind of Gestapo, no doubt very humanely directed in the first place'. These words were ill-received by thousands of his listeners. Clementine, to whom he had shown the script of his broadcast, had spotted this unfortunate sentence at once, and had begged him to delete the odious and invidious reference to the Gestapo. But he did not heed her.

During the campaign, Clementine was once more to deplore the political influence of Lord Beaverbrook, who, both as a platform speaker and through his newspapers, did vigorous battle for Winston and the Tory Party (although, as some wit was to remark, he greatly preferred 'the jockey to the horse'). But the jocular and overbearing tone adopted by the Beaverbrook press was out of tune with the serious mood of the electorate. Clementine sensed this, and she was also anxious about the advice Winston received privately from Max Beaverbrook and from Brendan Bracken, both of whom had constant access to him. She thought much of their counsel was ill-judged and their outlook rashly optimistic. Her view was shared by several of Winston's colleagues, and the Beaverbrook–Bracken influence was much commented upon in hostile criticism from the Opposition.

There is a revealing sentence in a letter my mother wrote me just before my father was to make the third of his four election broadcasts; it shows that he himself was not happy with his own performance or judgement: 'Papa broadcasts tonight. He is very low, poor Darling. He thinks he has lost his "touch" & he grieves about it.'[3]

The socialists cleverly and effectively presented two images of Winston to the public: one was 'Churchill the Great Wartime Leader', who had led the nation to victory; the other was 'Churchill the Party Leader': irresponsible, out of touch with ordinary people, subject to the malign influence of Lord Beaverbrook, and not to be trusted in peacetime. The result was that thousands of people who turned out to cheer him in the streets silently voted against him in the polling booths.

The opinion polls pointed a trend to the left, but none forecast the coming landslide. The Services vote was one unknown quantity: three million men and women in the Forces were entitled to vote by post or proxy, and as they were scattered far and wide throughout the world and the votes had to be gathered in, a space of twenty-one days had been arranged between polling day and the counting of votes. It is now generally assumed that the Forces vote (which could not be represented in the opinion polls) was overwhelmingly Labour.

The Conservatives' early predictions of the outcome of the election were cautious, but after Winston's tour of the Midlands, the North of England and Scotland at the end of June, the Tory pundits took a more sanguine view of their prospects, and Winston's own spirits revived. His tour was a triumphal progress: in four days, he made about forty speeches and travelled over a thousand miles. Clementine was with him for most of the time and, as Winston was still directing the war against Japan, he was able to use his official train with its mobile communications centre. Everywhere he went he was received by dense, cheering crowds, who had often waited many hours for his cavalcade to arrive. The tour was an overwhelming personal triumph, and at its end, gruelling although it had been, Winston was revivified and full of hope. But Clementine, although she had rejoiced to see him so tumultuously received, had her own private doubts: she had noticed from time to time, as she sat beside him in the open car, that, four or five rows back in the crowds, there were watchful and even hostile faces.

The election campaign was now entering its last stages, and in the run-up to polling day Winston visited many constituencies in London. Here for the first time in the campaign he encountered active and vigorous opposition. At Walthamstow Stadium, on 3 July, before an audience of twenty thousand people, he was almost prevented from making his speech because the heckling and booing were so intense, while at Tooting Bec, on the eve of the poll, a squib was thrown at him, all but exploding in his face. However, he was too seasoned a campaigner to take much notice of such incidents, and most of his critics at these meetings appeared to be youths below voting age.

On the eve of polling day, while Winston toured the South London area, Clementine addressed no fewer than six meetings in Woodford, and on the day itself she accompanied Winston on his traditional tour of the constituency.

The fight over, there ensued an uncanny and tantalizing three-week pause between polling day and the declaration of the poll. Winston and

Clementine took advantage of this space to go to France for a brief holiday. Through the good offices of Bryce Nairn (the newly appointed British Consul at Bordeaux, and a friend from the days of Winston's convalescence at Marrakech in January 1944), they were offered hospitality by General Raymond Brutinel at the Château de Bordaberry, near Hendaye, a seaside resort close to the Spanish frontier. The General was an agreeable and unusual man: of Scottish–French descent, he was a naturalized Frenchman who had served in the Canadian army; he and Winston got on famously. I had saved up my leave, and so was able to join my parents for the first holiday we had shared since just before the war. Our party also included Lord Moran, 'Tommy' Thompson and Jock Colville.

Winston at first was low and tired, but the magic of painting soon laid hold of him, absorbing him for hours on end, and banishing disturbing thoughts of either the present or the future; he also enjoyed the sea bathing enormously. Although Clementine unfortunately stubbed and cracked her toe during the course of the visit, she still enjoyed this much-needed respite.

Their holiday could last only a week, for grave matters awaited Winston, and he flew off from Bordeaux on Sunday, 15 July, to Potsdam, where the three Allied victors were to determine the future of Europe, and much else. As always he wanted Clementine to go with him, but she was committed to report to the Women's British Soviet Committee on her journey to Russia, and she was also anxious to press on with her plans for opening up Chartwell against Winston's return; so she returned to London, confiding him to my care.

From the Conference area on the outskirts of Potsdam (an area soon earning the title 'Babelsburg') I wrote her almost daily accounts of our life there. My father had been allocated a handsome house of rose-pink stone, with a lawn sloping down to a romantic-looking lake. My mother was eager to hear about his first meeting with, and impression of, President Truman, up to now a new and unknown quantity. I was able to send her a 'news-flash' on 16 July:

> This morning Papa paid his first visit to President Truman, who is installed in a monstrously ugly house about 400 yds down the road . . . Papa remained closeted for about two hours – during which time Anthony [Eden] joined them . . .
>
> When Papa at length emerged we decided to walk home. He told me he liked the President immensely – they talk the same language. He says he is sure he can work with him. I nearly wept for joy and thankfulness, it seemed like divine providence. Perhaps it is F.D.R.'s legacy. I can see Papa is relieved and confident.

The Conference was due to continue beyond the date of the result of the election, and Winston had therefore invited Mr Attlee to accompany him to Potsdam, on the basis that, in the event of a change of Government, there would be continuity of knowledge and thought. While grappling with all the work of the Conference, there was not much time to dwell on the secret hidden in the ballot boxes; but none of us could banish the cloud of uncertainty over the future. On 17 July I reported to my mother:

> Papa is well, although experiencing tiresome bouts of indigestion every now & then. Last night he was in bed quite fairly early – nor do I think he will stay up late tonight.
>
> Isn't it agony not knowing the results of the election? & it will be awful when we do.

However, this mood of pessimism did not persist; Max Beaverbrook's forecasts were mainly optimistic; both Conservative and Labour quarters estimated that the Tories looked like having the lead; the Gallup Poll showed a swing to the left, but it was thought (no doubt wishfully in our circle) that this would only diminish, not destroy, the anticipated Conservative majority. Even Stalin entered the 'Forecasting Stakes', and told Winston that his sources confirmed his own belief that the Conservatives would be returned by a majority of about eighty.[4]

Two nights before the British contingent were to leave for home, Winston gave a dinner party for the President and Stalin. I did what I could to help with the arrangements; and the following day I wrote an account of the evening to my mother, describing how the house had, about an hour before the party, been surrounded and laid siege to by what appeared to be half the Red Army.

> At last the guests started to arrive . . . then a fleet of cars swooped up & out skipped Uncle Joe attended by a cloud of minions . . .
>
> Uncle J. wore the most fetching white cloth mess jacket blazing with insignia. Close on his heels the President arrived, having walked from his house.[5]

I did not dine with them but, at my father's suggestion, returned later to the dining-room and, from the vantage point of a stool behind his chair, witnessed the end of the dinner:

> The party appeared to be a wild success – and presently Uncle Joe leapt up and went round the room autograph hunting! The evening broke up about midnight, in a general atmosphere of whoopee and goodwill.
>
> Tomorrow I shall be home. I long to see you, darling Mummie.

We flew back to London on Wednesday, 25 July. In the mood of confidence which now prevailed in our party, Winston and his colleagues made farewells at Gatow airport of a distinctly 'be-seeing-you-again' nature. Clementine met us at Northolt on our return, and that evening we dined *en famille* at The Annexe. Randolph was full of confidence, and left soon after dinner to go back to Preston* for the end of the count there. Diana and

* For which seat he had been returned, unopposed, in a by-election in 1940.

Duncan, however, who came in after dinner, were gloomy about Duncan's fate in Norwood. Next morning, when I breakfasted in my mother's bedroom, as I always did when I was at home, she told me that during the night my father's own confidence had vanished – as he later described in his own words: '[J]ust before dawn I woke suddenly with a sharp stab of almost physical pain. A hitherto subconscious conviction that we were beaten broke forth and dominated my mind.'[6]

Since he had to be at the centre of events throughout the day, Winston did not attend the count in his own constituency; Clementine went down to represent him, and took me with her. We arrived quite early, and as the count proceeded my mother spotted that Winston's somewhat 'crack-pot' opponent was amassing a considerable number of votes, which disquieted her; and although Winston was returned with a majority of over seventeen thousand, she realized this could not bode well for the way things would go in the rest of the country. As soon as she was able to, she telephoned the Private Office, and was told that early reports showed forty-four Labour gains to one for the Conservatives.

We were both rather silent on our drive back from Woodford, and as soon as we arrived at The Annexe we went to the Map Room, where all had been prepared to display the results. On our way we met Jock Colville, looking grave. ' "It's a complete debacle," he said, "like 1906."*'[7] In the Map Room we found my father, David Margesson (former Chief Government Whip and minister) and Brendan Bracken; the results were rolling in, showing Labour gain after Labour gain. It was now one o'clock, and it was already quite clear that the Conservatives were defeated. Every minute brought news of the defeat of friends, relations and colleagues; Randolph and Duncan were both out; everyone looked dazed and grave.

> We lunched in Stygian gloom – David Margesson – Brendan (himself just defeated) – Uncle Jack . . .
> Sarah arrived looked beautiful & distressed –
> We both choked our way through lunch.
> M. maintained an inflexible morale . . .
> Papa struggled to accept this terrible blow – this unforeseen landslide . . .
> But not for one moment in this awful day did Papa flinch or waver.
> 'It is the will of the people' – robust – controlled –
> The day wore very slowly on – with more resounding defeats.
> All the Staff and all our other friends looked stunned & miserable.[8]

But even in these bleak hours my father's sense of irony and humour did not desert him. During the day my mother had said: 'It may be a blessing in disguise,' to which my father replied: 'Well, at the moment it's certainly very well disguised.'

* The great Liberal landslide victory, when the Liberals won 400 seats and had a majority of 130 over all other parties (*Chronology of the 20th Century*).

During the afternoon of this day, which seemed interminable, Winston sat on in the Map Room, where he had so often studied the ebb and flow of our fortunes on the seas and battlefields; now he watched the tale of resounding political defeat. Clementine had accepted the situation, and did not particularly want to watch the death throes; so, feeling she could be more useful later, she retired to rest.

The final count showed that, with 393 members in the new House, the Labour Party had a majority of 146 over all other parties and groups.[9] It was a tremendous victory. At seven o'clock that evening Churchill drove to Buckingham Palace to submit his resignation to the King, and to lay down the burden he had shouldered in darker times. Shortly afterwards Clement Attlee was received by the King, who invited him to form a Government.

That night we were ten for dinner – all family and faithful friends. Diana and Duncan, Sarah and myself; Uncle Jack, still looking thin and frail after his heart attack; Venetia Montagu, ill, battered by cancer, but stalwart to the end; Brendan; Anthony Eden. The meal was a somewhat muted affair – understandably so, with everyone trying to help and say the right thing. My father still maintained his courageous spirit, and I described my mother as 'riding the storm with unflinching demeanour'.[10]

The next few days were, if anything, worse than that dreadful Thursday. After years of intense activity, for Winston now there was a yawning hiatus. The whole focus of power, action and news had been transferred (with lightning speed, as it always is) to the new Prime Minister: the Map Room was deserted; the Private Office empty; no official telegrams; no 'red boxes'. True, letters and messages from friends and from countless members of the general public started pouring in: sweet and consoling, expressing love, indignation and loyalty. But nothing and nobody could really soften the bitterness and humiliation of the blow.

My mother, I noted in my diary on 27 July, 'for the first time looks tired & shattered'. But at least she had something to do – namely, pack up and move out. She did not delay, and with Grace Hamblin started in at once on the organizing. The ensuing days were made more painful too, for both Winston and Clementine, by many leave-takings: the Chiefs of Staff and the Private Secretaries, who were off back to Potsdam with the new Prime Minister to take up the threads of the Conference; and Winston's personal detectives. His Private Office had been more than usually close to their chief and his family, with all the crises, and the journeyings, and weekends on duty at Chequers or elsewhere. Saying goodbye and thanking them was not easy.

Although they were seriously considering the house in Hyde Park Gate, no firm decision had been taken as yet, and Clementine was pushing on with the plans to make Chartwell habitable once more; she knew Winston would now want to move back into the 'big' house as soon as possible, instead of perching as they had done in Orchard Cottage. Realizing that domestic help would be very difficult to get, she had already decided to abandon the ground floor and basement rooms, and to organize their life on the top two floors of the house; but at this moment, the work entailed in carrying out this very practical plan had only just begun.

Winston and Clementine were therefore at this moment on the brink of homelessness. They wanted to get out of No. 10 and The Annexe at Storey's Gate with all possible speed, both for the convenience of the Attlees, and because these places had now become hateful to them. This urgent problem was solved for them by the swift and warm-hearted gesture of Diana and Duncan, who within forty-eight hours of the election results offered to lend them their lovely flat in Westminster Gardens until they found somewhere to live.

Winston had been to see No. 28 Hyde Park Gate in January, but it was only now that he concentrated his mind upon it; another visit confirmed his first opinion, and impressed him with its charms and possibilities. During the last few months they had visited various other houses, but now the decision was taken in favour of No. 28, and negotiations put in hand forthwith.

With the Attlees' kind and courteous permission, they spent the weekend immediately following the election result at Chequers, to take leave of the staff there and to collect their personal belongings. They gathered round them for this last visit all the family, and in addition Lord Cherwell ('The Prof'), Gil Winant, Brendan Bracken, Tommy Thompson and Jock Colville came for all or part of the weekend. It would, in other circumstances, have been a very jolly party; but it was certainly quieter than usual with such a gathering of the family, and although everyone tried to be gay, we were all still rather stunned by the events of the previous week. Winston made valiant efforts to be cheerful; he played cards, and there was croquet (which he liked watching, especially admiring Clementine's expertise and strategy); neighbours called in to say their goodbyes, and after dinner there were films. Clementine quite firmly went to bed before the cinema; she was exhausted, and always found the film sessions over-long at the best of times. The hardest moments were when, after the film was over, Winston came downstairs: it was normally at this point that he would get all the latest news from the Private Secretary on duty; there might even be a 'box', brought down by despatch rider with some urgent and secret communication; now there was nothing. We saw with near-desperation a cloud of black gloom descend. Tommy and Sarah and I played gramophone records for him: Gilbert and Sullivan (usually top favourites) were unavailing now, but French and American marches struck a helpful note, and finally 'Run, Rabbit, Run' and 'The Wizard of Oz' had a cheering effect. Eventually, very late, he felt sleepy, and we all escorted him upstairs to bed.

On Sunday, 29 July, we sat down fifteen to dinner, and before we went to bed we all signed the Chequers Visitors' Book. My father signed last of all, and beneath his signature he wrote: 'Finis'. We know now that it was no such thing, but that is how he and my mother and all of us felt then.

* * *

While Clementine addressed all her energy to moving out of The Annexe and No. 10, and while Diana and Duncan were making haste to move into a furnished flat, Winston divided his time between the penthouse suite at Claridges and Chartwell. Orchard Cottage was once more a haven, and our devoted Cousin Moppet took on the cooking. The gardens were a wilderness, the roses all a-tangle and the grass knee-high, but Winston roamed through them, and found them peaceful and beautiful.

The night before I left to rejoin my unit, my parents took me to see a revival of Noël Coward's *Private Lives*. It was a complete contrast to the events and emotions through which we were passing; my father simply loved it, and we all roared with laughter. To our moved amazement, on his entering and leaving the theatre, the whole audience cheered and clapped; and, after the final curtain, the principal actor, John Clements, came forward and made a most charming speech. During the following days and weeks there were many such heart-warming incidents.

It was a beautiful, pearly morning on 3 August, when my mother walked with me very early across the park to the assembly point from where I was to start my journey back to Germany, after a leave I was unlikely to forget. Soon after my return to my battery we were visited by Field Marshal Montgomery, and my first letter home described how,

> just before he climbed back into his minute 'plane he talked to me. He was charming and kind and wanted to know how Papa was, & he asked specially about you. He really was kind & said that should you need me it could be arranged for me to be flown back. So I thanked him and said I would remember what he said in case of need.
>
> And then he flew off.[11]

During these difficult weeks everyone they knew seemed to want to help. A warm-hearted friend, Audrey Pleydell-Bouverie, who years before, in 1931, after Winston's serious accident in New York, had lent them her house over there, once more came forward with the generous offer of her house in the country, pending Chartwell's rehabilitation. Clementine's reply illustrates her feelings at this time: 'Now that again we have suffered another unexpected shock, once more you offer to take us in. It is sweet of you, dear Audrey, and I shall never forget it. But actually I think it is best to struggle into Chartwell and re-adjust our lives.'[12]

But that it *was* a struggle is amply shown in the long letter my mother wrote me from Chartwell on 14 August:

> My darling Mary,
>
> Time crawls wearily along. It's impossible to realize that it is not yet three weeks since your Father was hurled from power. He is lion-hearted about it . . .
>
> We have settled into Westminster Gardens & I shall never forget Duncan & Diana's prompt & generous action in lending it to us. Now Nana & I are

struggling to get the flat down here ready. Our 'married couple' on whom we pin great hopes arrive tomorrow, & all yesterday Nana, Miss Hamblin & me scrubbed and polished their rooms – they are still not ready. But I must be truthful & say that I was not allowed to scrub & polish as it seems I am too old & inefficient – I merely heaved furniture about which you know is one of my favourite hobbies – And I opened hampers stuffed full of old curtains, rejoicing when I found some long-lost treasure & despairing when I discovered that mice (I repeat mice) have nibbled holes in some lovely old rose damask covers & made nests everywhere. As for the moth!!! 'Lay not up for yourselves treasures upon this earth . . .'

But these are not treasures but necessities becos you can't buy 'nothink' [clothing and household fabrics were still rationed] . . .

I hope I shall be able to organise a comfortable & happy home for Papa; but the truth is that it is silly to have two homes just now (London & Country) because of the rations which get all eaten up in one place & becos' of servants – Rich people now-a-days have a small house in the country & when in London have permanent rooms in an hotel . . . I blush to think that I who organised the Russian Fund, the kitchen in Fulmer Maternity Hospital & who complained about the organisation of the Y.W.C.A. am stumped by my own private life. But I think I shall learn to do it. The difficulty is that Papa is unpredictable in his movements & altho' I'm used to that, I still find that very tiring to cope with . . . Nana has been cooking for us & has been sweet to Papa, who feels her kindness very much just now.

'Now mind during these last months to keep your conduct & morale as high as it has been during the whole war,' says your old Mother who is dropping to pieces herself!

Your devoted Mummie

'Please,' I begged in my reply 'on no account push/heave or lift grand pianos, thereby straining yourself. I am appalled by your account of the moth/mouse Fifth Column activities. It's too much to think humans have been rationed, while they have gorged themselves on our curtains.'[13]

But, if Clementine opened her heart to me about her domestic trials, she was well aware that hundreds of thousands of people were facing greater difficulties in finding homes and adjusting their lives, and only those close to her knew of her harassments. Both she and Winston were too dignified and too proud to present anything other than a stoical and urbane aspect to the curious, albeit sympathetic, eyes of the world at large.

* * *

Churchill did not lurk long licking his wounds; when Parliament re-assembled on 1 August, less than a week after the election results, he took his new place on the Opposition front bench. The Conservatives started singing 'For he's a jolly good fellow' as he walked to his seat; the mass of the Labour members rejoined by singing the 'Red Flag', which was on the whole thought to be ungracious considering the magnitude of their victory. But not all parliamentary exchanges were to be on this level: Japan surrendered unconditionally on 14 August 1945, and two days later Attlee, the new Prime Minister, paid a generous tribute to Churchill's leadership in the House of Commons: 'In the darkest and most dangerous hour of our history this nation found in my Right Honourable friend the man who expressed supremely the courage and determination never to yield which animated all the men and women of this country. In undying phrases he crystallized the unspoken feeling of all . . . His place in history is secure.'[14]

My father also spoke in the same debate, and on 18 August my mother wrote to me from Westminster Gardens:

> Papa made a brilliant moving gallant speech on the opening of Parliament. He was right back in his 1940–41 calibre. The new house full of rather awe-struck shy nervous members was rivetted & fascinated – On V.J. Day many people gathered round this block of flats to see Papa & cheer him & he got mobbed in Whitehall by a frenzied crowd. These friendly manifestations have reassured & comforted him a little. He says all he misses is the Work & being able to give orders.
>
> The crowds shout 'Churchill for ever' & 'We want Churchill'. But all the King's horses & all the King's men can't put Humpty Dumpty together again.

This letter, with its trace of bitterness, was followed on 26 August by a sad and despondent one written from Chartwell. After saying she had received a letter from me, and commenting on various aspects of my Army news, my mother continued:

> But what really excited & relieved me was that you say you [my regiment] may soon be disbanded & that you will apply for a post in England. Now my Darling please ask for a job at the War Office, so that you can live at home in your lovely bed-sitting room at Hyde Park Gate. Because I am very unhappy & need your help with Papa.
>
> I cannot explain how it is but in our misery we seem, instead of clinging to each other to be always having scenes. I'm sure it's all my fault, but I'm finding life more than I can bear. He is so unhappy & that makes him very difficult. He hates his food, (hardly any meat) has taken it into his head that Nana tries to thwart him at every turn. He wants to have land girls & chickens & cows here & she thinks it won't work & of course she is gruff & bearish. But look what she does for us. I can't see any future. But Papa is going to Italy & then perhaps Nana & I can get this place straight. It looks impossible & one doesn't know where to start . . .
>
> Then in a few days we shan't have a car. We are being lent one now. We are learning how rough & stony the World is.

I was much disturbed by this letter, for I realized that the adjustment from the wartime tension and tempo of their life was bound to be painful and difficult for both of them. For although they had lived for over five years on an exhaustingly heroic level, they had not suffered the physical shortages and domestic difficulties which had been the lot of ordinary people. Because they had to entertain so many people officially, they received 'Diplomatic Rations'; Chequers was staffed by the Women's Services, and civilian staff had been available for London; cars and petrol were always at hand. Now, of course, all these comforts and facilities disappeared. I understood how, in their emotionally and physically exhausted state, the combination of petty problems with the hurt and bewilderment of Winston's political downfall produced tensions and flare-ups between them. Although I did not want to leave my unit before the regiment was actually disbanded, I felt a new priority had appeared and that, in the circumstances, I could ask for a home posting. A short time later, on 31 August 1945, I was able to write with some good news:

> My darling Mummie,
>
> I was so grieved to read in your letter, which arrived yesterday, of all your worries and desolations. I feel for you and Papa more deeply than I can say; and I know only too well, how much of the burden weighs on you.
>
> I asked at once for an interview with the 'Queen AT' over here, and today I drove down to Headquarters . . . (we are about 120 miles away from them) and saw her . . . and explained everything. She was kind and very helpful. A posting to England (London) is going to be arranged for me. So darling Mummie, I hope so much to be with you perhaps within a month.

True to their word, the Army authorities arranged for me to be posted as an administrative officer to the War Office Holding Unit in London, in Radnor Place. I did not live at home, but as my parents' new house was just the other side of Hyde Park from my billet, even a few hours off meant I could visit them easily and often.

On 7 September, just before I left my unit in Germany, my mother had written me a much more buoyant letter which also contained some sagely maternal advice: 'Tomorrow week you will be twenty-three, a perfect age, & a perfect springboard from which to leap into married life! But I hope you won't think of it for a year after you get home! because I am sure it will take you a little while to adjust yourself to civilian life after being a soldier for so long. (I'm finding it damned hard to adjust myself!)'

Meanwhile, in early September, Winston went to Italy for a holiday, where he stayed in a commandeered house on the shores of Lake Como lent to him by Field Marshal Alexander. It would have done Clementine good too, for she was exhausted, and of course Winston wanted her to go with him; but with all she had to do at Chartwell, and with the new London house to be arranged, she preferred to stay at home. Also, it is quite clear from her letters to me and to Sarah that she was in a tense and unhappy state, and one has to be in the right mood even to enjoy blue skies and sunshine.

'Cast care aside,' Winston was wont to say when Clementine was wrought up – but with her nature, that was easier said than done. Fortunately, Sarah was able to get leave to accompany him, and Lord Moran went too. The thoughtful Field Marshal assigned two young officers from Winston's old regiment the 4th Hussars, Major John Ogier and Lieutenant Tim Rogers, to look after Winston and to see that all went well with the arrangements. Not only did they do that, but they also proved to be delightful and cheering companions – brave and young and full of fun, they appealed to him in every way. The day after his arrival, 3 September, Winston wrote to Clementine:

> This is really one of the most pleasant and delectable places I have ever struck. It is a small palace almost entirely constructed of marble inside. It abuts on the lake, with bathing steps reached by a lift . . . Every conceivable arrangement has been made for our pleasure and convenience.
>
> . . . Yesterday we motored over the mountains to Lake Lugano, where I found quite a good subject for a picture . . . I have spotted another place for this afternoon . . . These lakeshore subjects run a great risk of degenerating into 'chocolate box', even if successfully executed.

The letter was dictated, but at the end he added in his own writing:

> I have been thinking a lot about you. I do hope you will not let the work of moving into these 2 houses wear you down. Please take plenty of rest. With fondest love,
>
> Your devoted husband,
>
> W
>
> [there is a little pig at the bottom of the letter]

The same day Sarah wrote a most understanding letter to her mother:

> I wish you were here with us. I was so distressed to see you so unhappy and tired when we left, and so was he. We never see a lovely sight that he doesn't say 'I wish your mother were here'. Wow! You know, I expect you will feel a little low and tired and oppressed by the million domesticities that will now sink down upon you, for a bit. Six years is a long time to live at such a high tempo, knowing as fully as you did all the moments of anxiety and worry, and decisions. You are bound to feel a reaction – as he does, and will for some time.

Sarah then went on to give a lively and vivid description of their surroundings and daily activities: 'I really think he is settling down – he said last night – "I've had a happy day"! I haven't heard that for I don't know how long!' In a P.S. Sarah added: 'He really is distressed about you – wow – He has just prowled round my room, and tried to read my letter! He says I'm to say – he's good. He really is!'

At home, as well as grappling with Chartwell, Clementine also paid a visit to Winston's constituency, and on the way she observed and was depressed by the general squalor and dilapidation – the legacy of six years of war.

4 September 1945

My darling Winston,

On Saturday the day you flew away, I went to Woodford to open a Vegetable Show held by the local Allotment Holders . . . Driving along the road so familiar for more than twenty years, I thought how the War had changed it – Many of the humble but neat little homes were shattered, all were battered & squalid – In every space, where before had been busy shops & houses, huge menacing Venereal Disease posters were erected – After passing half a dozen of these – suddenly I saw a new design – the picture of an insect (upside down so that you could see his mandibles & count his crawling feet) magnified to 12 feet and across it written 'Beware the Common Bed Bug'.

Winston wrote to her from his very different scene on 5 September:

We have had three lovely sunshine days, and I have two large canvasses under way, one of a scene on the Lake of Lugano and the other here at Como. The design is I think good in both cases, and it has been great fun painting them . . .

Alex [Field Marshal Alexander] arrives tomorrow, and I am looking forward very much to seeing him. I cannot say too much for the care and authority which he has bestowed on making my visit pleasant . . .

It has done me no end of good to come out here and resume my painting. I am much better in myself, and am not worrying about anything. We have had no newspapers since I left England, and I no longer feel any keen desire to turn their pages. This is the first time for very many years that I have been completely out of the world. The Japanese War being finished and complete peace and victory achieved, I feel a great sense of relief which grows steadily, others having to face the hideous problems of the aftermath. On their shoulders and consciences weighs the responsibility for what is happening in Germany and Central Europe. It may all indeed be 'a blessing in disguise'.

But although these were soothing, sunlit days, Winston continued to miss Clementine, and that she was often in his thoughts is borne out in a long epilogue, in his own hand, to one of his letters:

My darling, I think a great deal of you & last night when I was driving the speed-boat back there came into my mind your singing to me 'In the gloaming' years ago. What a sweet song & tune & how beautifully you sang it in all its pathos. My heart thrilled in love to feel you near me in thought. I feel so tenderly towards you my darling & the more pleasant & agreeable the scenes & days, the more I wish you were here to share them & give me a kiss.

You see I have not forgotten how to write with a pen. Isn't it awful my scribbles?[15]

The only inconvenience Winston had to report was that he was suffering from a hernia, writing (in his own hand) in the same letter:

Darling a tiresome thing has happened to me. When I was vy young I ruptured myself & had to wear a truss. I left it off before I went to Harrow & have managed 60 years of rough & tumble. Now however in the last 10 days it has come back. There is no pain, but I have had to be fitted with a truss wh I shall have to wear when not in bed for the rest of my life* – Charles [Moran] got a military surgeon from Rome who flew & has been w us for the last 3 days.

The original plan had been that the whole party should return about 20 September, but Winston was so relaxed, and so beguiled by the sunshine days, that he decided to drive along the Riviera to Monte Carlo, and spend a little time in that part of France where he had enjoyed so many happy prewar holidays. Sarah had to return to her unit, and Lord Moran to his other patients, so with his two ADCs, now the firmest of friends, his valet Sawyers and his detective Sergeant Davies, he set off. He told Clementine of his intentions in a letter on 18 September:

I hope you will not mind my change of plans. The weather has been so good and the prospects seem so favourable that the opportunity of having another four or five days in the sunshine was too tempting to miss . . .

I really have enjoyed these 18 days enormously. I have been completely absorbed by the painting, and have thrown myself into it till I was quite tired. I have therefore not had time to fret or worry, and it has been good to view things from a distance. I think you will be pleased with the series of pictures . . . which I have painted. I am sending them home by Sarah, who will give you all our news . . .

Sarah has been a joy. She is so thoughtful, tactful, amusing and gay. The stay here wd have been wrecked without her.

When Winston and his party arrived at Monte Carlo, they stayed at the Hôtel de Paris for a day or two, until Ike (General Eisenhower) most hospitably pressed them all to go and stay in his commandeered villa at Cap d'Antibes, although he himself was not there. Winston hoped that he could prevail on Clementine to join him during this last lap of his holiday – Alex's plane would have wafted her to him – but she preferred to wait for him at home. As always, when he could not persuade her, he was a little rueful: 'Naturally I am vy sorry you cd not come out by Alex's plane,' he wrote (in his own hand) from Eisenhower's villa on 24 September. 'It would have done you good to bask in this mellow sunshine for a space.' On the same day he

* In the event he was successfully operated for this condition in 1947.

also dictated a long letter to her, full of his painting expeditions, plans for his return, and (having read a whole batch of English newspapers) his comments on world and home events. He ended this last 'holiday' letter:

> There will be no lack of topics to discuss when we all come together again. Meanwhile, this rest and change of interest is doing me no end of good and I never sleep now in the middle of the day. Even when the nights are no longer than 5, 6 or 7 hours, I do not seem to require it. This shows more than anything else what a load has been lifted off my shoulders.

While Winston was away, Clementine had been able to get to grips with the alterations and rearrangements necessary to make Chartwell liveable. Fortunately, little needed to be done at Hyde Park Gate, other than some redecoration. In the immediate post-war period there were shortages of everything, and with so many houses to be repaired as a result of bomb damage, and the requirements of the new house-building programme, materials and labour were both scarce and restricted. However, when Winston arrived back early in October, Clementine had just managed the move in time to welcome and install him there.

Gradually the new arrangements at Chartwell took shape. The large kitchen, scullery and pantry in the basement were abandoned for ever as such, and much smaller and labour-saving kitchen arrangements were organized in a former staff bedroom on the top floor, with a service-lift to the floor below, where the 'best' visitors' room took over as the dining-room. Clementine's beautiful azure-coloured bedroom with the 'barrel' ceiling made a lovely sitting-room, with views over the gardens and lakes. Clementine herself ascended to the top floor, and made her bedroom in the Tower Room, which was a charming self-contained suite. Winston's study-bedroom remained unchanged, and the large drawing-room on the ground floor made a most handsome studio. The 'Office' on the ground floor remained as before, and the secretaries (still two of them) lived in Orchard Cottage. In 1946 Sir Alexander Korda gave Winston the most modern film projector and screen, and the old dining-room was turned into a cinema. This rearrangement of Chartwell lasted for several years, and made it possible for Winston and Clementine to live there despite all the difficulties of the post-war period.

But in spite of all these practical rearrangements a major question mark still hung over the long-term future for Chartwell, and Winston was reluctantly considering the necessity of selling it. Lord Camrose*, so long a staunch friend and counsellor, was horrified when he learnt this, and assembled a group of friends and admirers who anonymously (at the time) bought Chartwell from Winston and presented it to the National Trust, with the proviso that Winston and Clementine could remain there for their lifetimes. This generous and far-sighted act meant that not only could Winston enjoy

* William Berry, 1st Viscount Camrose (1879–1954), Chairman and Editor-in-Chief of the *Daily Telegraph*.

without care his beloved Chartwell to nearly the end of his days, but that this house and grounds, with all its memories and links with him and the years so fateful for our country, would be for ever part of Britain's heritage. And Clementine crowned this gift, after Winston's death, by giving to the house, in perpetuity, not only a large number of his pictures, but also the greater part of the furniture which had always been there.

After the neglect of the war years, the gardens, fields and woods at Chartwell were of course in a shaggy and neglected state, but gradually their condition was restored. Clementine had to fight some brisk battles with Winston, who, as in years gone by, would distract the gardeners from their proper work by taking them off to help with forestry, brushwood clearance or wall and fence construction, to all of which activities he attached a high priority, and the organizing and supervising of which afforded him hours of carefree pleasure. She, however, attached much more importance to velvet lawns, the remaking of the flower garden and the production of vegetables. Winston therefore recruited his own personal workforce, which consisted of a part-time old-age pensioner, Kurn (who in days gone by had helped him with his bricklaying), and another less venerable assistant. A little later two German prisoners-of-war came to work full-time. This team did valuable works of reclamation, excavation and reconstruction, and the garden staff were left to their proper tasks, much to Clementine's and the Head Gardener's relief.

The Churchills' pre-war attempts to farm on a small scale, it will be remembered, had not been a signal success, and Clementine dreaded starting once more on schemes similar to those which had been such a constant source of vexation and expense in the past. Winston, however, was determined to live in a pastoral and patriarchal way, and of course it was highly advantageous to produce one's own food in the bleak years of rationing, which were to be prolonged for several years after the end of the war. Max Beaverbrook started them off on their post-war 'smallholding' economy by sending them as a present a flock of hens. These arrived while Winston was away in Italy, and Clementine wrote gratefully and enthusiastically to Max about his handsome present on 14 September 1945:

> My dear Max,
>
> I must tell you about your lovely hens & their companion, that glorious cock. They are the admiration of all beholders – they are looked after by my cousin Marryot Whyte [Cousin Moppet] who is thrilled by them. They are laying, not very freely yet, but the numbers of eggs are increasing & they have a most delicate flavour.

Clementine gradually warmed to the idea of 'home-producing', and the old beehives (occupied by unsupervised and dozy bees) were galvanized into renewed life, and increased output. There was an exciting development when the Royal Jersey Agricultural Society presented Winston with a beautiful Jersey cow (called May Belle of the Isles), who arrived at Chartwell in the autumn of 1946. To wait upon May Belle (and a companion cow) the services of a Land Girl were procured, and Chartwell truly became a land flowing with milk, honey and eggs.

Although the alterations to the house, the outside operations and the advent of livestock caused Clementine some worry, in the longer term she would not be harassed by anxieties as to the financial consequences of these undertakings, as she would have been in earlier years: now – for the first time in their lives – Winston would be rich. He had always kept his family by his pen, and in the years he was to be in opposition he would write six volumes of war memoirs*; moreover, his writings, both past and present, now commanded sums far exceeding anything they could have dreamed of in times past. Chartwell would always have its problems, but it was never again to be the monster worry of former times. In 1946, the farm adjoining their land, but lower down the valley, came up for sale; Winston bought it, with its herd of Shorthorn cows. Farms and farming were wise investments just then, and these activities suited Winston's inclinations. In the following year he also bought a market garden at French Street, near Westerham, and another neighbouring farm, Bardogs, at Toy's Hill. Presently, at Bardogs, he established a Jersey herd, and later Landrace pigs. In all he now farmed about 500 acres.

Winston had, if not a veritable kingdom, at least a principality, the governing and welfare of which were a constant interest and pleasure to him. The Grand Tour by land rover, accompanied by family and guests, became a regular feature of Chartwell life. When he was away he liked to have detailed reports about everything, and he, in turn, kept Clementine posted in her absence. A typical Chartwell 'bulletin' in August 1947 reads:

> The harvest is proceeding with tremendous vigour and in perfect weather. Most of the fields are already cut and stooked and some have been put up on tripods . . . The lettuces in the walled garden were sold for £200, though they cost only £50 to grow . . . The hot-houses are dripping with long cucumbers. The grapes are turning black and a continuous stream of peaches and nectarines go to London. I have one a day myself – 'le droit du seigneur'.[16]

But for Winston, all these new acquisitions and interests were, of course, only pleasurable sidelines. As he took up the challenge as Leader of the Opposition, politics involved him as much as ever. Soon the writing of his memoirs also occupied many hours, and 'in the holidays' travelling and painting were resumed. Although the strains and exertions of the war years had undoubtedly left their mark, yet he revived astonishingly rapidly from his immediate post-war fatigue; and as he became more and more engrossed in his new and former occupations, he ceased to brood at any length on the hurt and humiliation of his political overthrow.

But it took Clementine much longer to 'get her wind', both physically

* *The Second World War* (vol. I published 1948; vol. VI, 1953).

and mentally. The domestic problems with which she had chiefly to grapple were essentially dreary as well as difficult, and she must have missed the challenge and interest of the various aspects of her war work: nor had she got the outlet that Winston's many 'toys' gave him. Moreover, she was not really reconciled to the fact that he was determined to battle on in politics: she had longed – and she still longed – for him to retire. All her life her energy had been in excess of her stamina, but like many others, during the crisis years Clementine had 'kept going': now accumulated exhaustion was catching up on her. Throughout the autumn and winter of 1945 her health was only middling, and she suffered from depression and nervousness. The build-up of her worries, which her fatigue served to enlarge, made her often impatient and irascible with Winston; and he, for his part, could be demanding and unrealistic. These months saw a series of scenes between them. After any quarrel both suffered pangs of remorse, and both were always anxious to make it all up; but these were difficult days for them both.

One unexpected problem arose when Clementine realized that No. 28 Hyde Park Gate, while being a delightful house for private life, was inadequate when it came to providing the necessary office room to accommodate a secretariat. Subconsciously, I suppose, while house-hunting she had had in her mind a home for an old, retired couple, happily and honourably released from long years of active service. The real facts were different: the public part of their lives seemed scarcely abated; hardly a potentate or distinguished visitor came to London without wishing to call; old wartime comrades and colleagues were faithful; and there was the constant come-and-go of active political life. The amount of 'general public' mail was still amazing, and it used to grow to floodtide proportions at Winston's birthdays, when extra staff had to be enlisted. If the press gave out that he had a cough or cold (let alone a real malady), kind letters advising (and often accompanied by) remedies would pour in. And in addition to all this was Winston's literary work. Consequently four full-time Private Secretaries were needed – three for Winston's work, and one for Clementine's – and this secretariat required considerable room-space.

The room at No. 28 destined for the Office was not at all adequate to the actual situation, and took up a much-needed extra sitting-room. Clementine was at a loss to know how to solve this urgent problem when – fortunately – at the crucial moment, the next-door house came on the market. Although buying a complete house and garden seemed an extravagant way to acquire more office space, it really was the only solution, and No. 27 was duly bought. Eventually Clementine's Scottish prudence was appeased when it was found possible to create a charming maisonette on the top two floors, which was let.

Now that the war was over, many people wished to express their gratitude and admiration to Winston: here, in Great Britain, towns and cities proffered him their highest honour – the freedom of their boroughs; foreign countries, and particularly those which had suffered enemy occupation, sent pressing invitations. France had received Winston and Clementine officially in November 1944, only a short time after the liberation of Paris; now, in the winter of 1945, the Regent of Belgium and his Government invited them both to Brussels. Sadly, when the time came, Clementine did not feel up to going, and so to me fell the chance to see my father received with overwhelming love and enthusiasm in Brussels and Antwerp, and given many honours at the hands of the Belgian Government and people.

A few weeks before the Belgian visit, however, on 20 October 1945, Clementine had shared in an honour which was accorded to them both, when they received together the Freedom of the Borough of Wanstead and Woodford. That she was made a Freeman in her own right was a true and graceful tribute to the part Clementine had played in the affairs of Winston's constituency, and the place she had won in their affections in the twenty-odd years he had represented them in Parliament. In her brief speech she said:

> My dear Friends,
>
> I feel this to be a significant and most honourable day in my life. It has been my privilege often to visit you when my Husband would have wished to come but has been prevented by the fact that his duties have been, not only to his own constituency, but to the whole country. On these occasions you have always been very kind to me and never allowed me to feel I was a mere substitute.
>
> When we first came to you in 1924, we had been through rough political waters, and I remember that, at the first meeting of women I attended, in replying to a very kind welcome, I said I hoped we would be with you 'for keeps'. It has turned out to be so, and when the tale is told it will be seen that our association with you is woven into the pattern of our lives in rich and happy colours.

But although this was for her a gratifying and happy day, and although she looked beautiful and smiling – yet, on this occasion, as on many others, her appearance belied her true state. Her morale at this moment was very low, and I must have been particularly worried about it, for on the day after their sunshine day of cheers and prizes I wrote my mother a long letter:

> My dearly beloved Mummie,
>
> I felt I had to write to tell you how much my thoughts are with you. I hate to see you so hedged round with so many tedious vexations – and I grieve that you feel so low and exhausted in spirit . . .
>
> It is difficult to put down on paper all I want to say. But I do understand that these days for you are not smiling ones . . .
>
> You spoke yesterday in your lovely speech about 'the pattern of our life'. Turn your eyes from the intricate, tiring, vexing piece you have got to now, and rest them by dwelling on all you have accomplished already. I gaze at it with so much love and admiration.
>
> It seems to me such a triumph that after so many events – which have all of them left their marks on your own private life and experience you and Papa

should have come through still loving each other and still together.

To me – it is one of the most wonderful and admirable rocks to which I cling amid the daily evidences of ship-wrecked marriages – & so many of them not even ship-wrecked – but just deserted.

For despite all his difficultness – his overbearing – exhausting temperament – he does love you and needs you so much.

I know you often feel you would gladly exchange the splendours and miseries of a meteor's train for the quieter more banal happiness of being married to an ordinary man . . .

But one of the ingredients of your long life with Papa is the equality of your temperaments. You are both 'noble beasts'.

Your triumph is that you really have been and are – everything to Papa. Many, many great men have had wives who ran their houses beautifully and lavished care and attention on them – But they looked for love and amusement and repose elsewhere. And vice versa. You have supplied him with all these things – without surrendering your own soul or mind.[17]

So as not to tax any further her nerves and energy, Clementine wisely took her doctor's advice, and severely restricted the number of engagements or obligations she accepted for some time. And her spirits revived as, slowly but surely, her domestic life assumed organized shape: No. 27 took the strain off No. 28, and the new arrangements at Chartwell were finally completed during November.

This year saw the first of Winston's peacetime birthday parties, one of many such glowing and glorious occasions for us all. And in order to gather as many of the family together as possible, Winston and Clementine decided to spend Christmas in London. It was quietish, but very pleasant, with Diana and Duncan, Sarah, myself, Uncle Jack and Aunt Nellie all mustered. It made a quiet and peaceful ending to a year that had seen such cataclysmic events for the world, and had been full of drama in our own personal lives as well.

CHAPTER TWENTY-FIVE

Swords into Ploughshares

WINSTON AND CLEMENTINE SPENT MOST OF THE FIRST THREE MONTHS OF 1946 in the United States. Lord Moran had told Winston that it would be wise for him to be out of Britain during the bleakest months, and that a long rest would do him nothing but good. With 'resting', Winston combined a number of public and private engagements: he was officially received in Cuba; he conferred in Washington with President Truman; he addressed the State Legislature in Richmond, Virginia; he received honorary degrees from the universities of Miami and Columbia. Making a detour to visit Hyde Park, the family home of President Roosevelt, he laid a wreath on the grave of his old friend and comrade. In New York he was accorded a civic welcome and the city's Gold Medal, and was received by cheering crowds and tickertape.

But the most important event of his transatlantic visit took place at Westminster College, Fulton, Missouri, on 5 March 1946 when, at the invitation of President Truman, and in his presence, Winston made a speech which in retrospect has assumed historical importance. In the course of his address he used a phrase that has passed into the bleak phraseology of the political history of our time, when he referred to the 'iron curtain' which had descended 'from Stettin in the Baltic to Trieste in the Adriatic', behind which now lay 'all the capitals of the ancient states of Central and Eastern Europe'. As with his warnings between the wars, this realistic, forward-looking speech was received with much disapproval in some quarters, and he was considerably scolded for pointing out the fact that Soviet Russia was successfully establishing an iron grip on all the countries within its orbit.

On nearly all these varied occasions Winston was accompanied by Clementine; and Sarah, who had been demobilized from the WAAF shortly before Christmas, flew out to join them at Miami. Apart from these numerous visits and functions, they also had several weeks of sea and sunshine, staying with Colonel Frank Clarke (their Canadian friend from the First Quebec Conference days) at his house at Miami Beach. From there on 18 January 1946 my mother wrote to me:

> We arrived here 48 hours ago in tropical heat, rather much but delicious – And lo & behold in the same night it changed & we are shivering among rustling palm trees & grey skies . . . Papa has not yet settled down to painting, and is a little sad and restless, poor darling. I hope he is going to begin writing something.
>
> Later in the day. The weather has slightly improved and Papa has now started a picture of palms reflecting in the water. I visited him, and draped a knitted Afghan round his shoulders as he was sitting under a gloomy pine tree in a particularly chilly spot.

Very likely as a result of sitting and painting in the shade, Winston caught a slight chill, which caused everyone some anxiety – the fear of pneumonia always lurked in the background – but happily this was only a passing indisposition, and on 22 January my mother wrote again:

> Papa, thank God, has recovered or nearly so. We had a wretched 36 hours when we telephoned to Lord Moran & the temperature, tho' not very high simply would not go down & poor Papa was very nervous about himself, & yet very obstinate & would either take no remedies at all or several conflicting ones at the same time.
>
> But today he bathed! & loved it. The weather is now perfect but tropical. Crickets chirp all night – there are lovely flowering hedges of hibiscus pink lemon & apricot . . . The sea is heavenly – water about 70° . . .
>
> Papa has learnt a new card game 'Gin Rummy' & plays all day & all night in bed & out of bed. He has started 2 not very good pictures.
>
> Tender love (I must fly Papa is calling)
> from your devoted
> Mummie

That letter made all of us at home feel much relieved. The family security service had been efficient, and this health scare went unobserved by the press.

From my unit in Radnor Place I was able to keep an eye on the home fronts, and in a series of long gossipy letters to my parents I kept them in touch. At Chartwell things were going ahead despite hard winter weather; on 23 January I reported: '[T]he German prisoners are hard at work. Great wood chopping operations . . . the wood shed is three-quarters full.' At No. 28 the secretaries and staff had all been laid low with influenza, but were recovering. 'I miss you and Papa so much; I often go home – but the little Victory House seems very deserted and solitary.'[1]

Just before they were due to return home I had some personally exciting news: 'I'm longing to see you all again – and joy oh joy – your birthday and my demob[ilization] will both be on April the First! I feel quite light headed with the dual excitements of your return and my "liberation".'[2]

The *Queen Mary* docked on the evening of 26 March, and I went to Southampton to greet them. And I was not the only one – 'Right through S'hampton & way out into the suburbs there were groups of people waiting to welcome Papa home. And there was a little crowd waiting in the chilly darkness round the door of No. 28.'[3]

April the first, this year, was, as I had written to my mother, rather special for us both. I went through the official process of demobilization in the morning, and later drove home to Chartwell with Sarah and Mrs Landemare, bearing my mother's birthday cake.

> Arrived at Chartwell we fêted Mummie, and she seemed pleased with our presents.
>
> The 'flat' is looking lovely, and it was full of flowers – branches of forsythia and pots of magnolia stellata.[4]

We were a small, but loving company: my parents, Sarah, Nana and myself. After dinner, when we all 'drank her health & ate her cake',[5] the family fell to playing 'Oklahoma' – another new card game of the moment.

Now that I was an ex-servicewoman, I turned my mind and energies to civilized and civilian pleasures and occupations – first among them being the task of trying to catch up on a raggedly finished education. During this spring and summer my mother and I went together to many galleries, museums and exhibitions. At the weekends we nearly always headed for Chartwell. Life there was beginning to pick up in an enjoyable way; the alterations which had been made were proving a great success, and consequently Clementine could take pleasure in being there. The tennis court had been rehabilitated, and her interest in the garden was reviving.

During these next few years Clementine nearly always accompanied Winston, both at home and abroad, on the occasions when he received honours and distinctions, and invariably she received 'an honourable mention': it was clear that her role in his life, and her own wartime activities, were widely recognized. But Clementine also was the recipient of several outstanding honours in her own right. This was gratifying to her, and made Winston beam: 'A little bit of sugar for the bird', he would say with pleasure and pride. The first, and the greatest, was heralded by a letter from the Prime Minister, marked 'Confidential', and dated 20 May 1946.

> My dear Mrs Churchill,
>
> I feel very sincerely that it would not be fitting if the Victory Honours Lists, of which the forthcoming List is the last, did not include your name. I hope, therefore, that you will allow me to submit your name to His Majesty for appointment as a Dame Grand Cross of the Order of the British Empire in recognition, not only of your work for the Aid to Russia Fund, and for the promotion of Anglo-Russian understanding, but also of those other many services which made so marked and brave a contribution during the years of

> the war. I hope this will be agreeable to you, for I am sure it would be an Honour which would be widely acclaimed.
>
> Yours sincerely,
> C. R. Attlee

It could not have been more gracefully or more handsomely put, and it touched Clementine deeply. In writing to thank Mr Attlee for his proposal to submit her name for this honour, she added: 'I also thank you for the terms of your letter and it makes me happy that you feel that I was able to help a little in these last terrific years when we all fought together in heart and mind.' Her name duly appeared in the Birthday Honours List on 13 June 1946, and she received her decoration from the King at an Investiture in July, proudly watched by Sarah and myself.

During this same summer, two universities offered her honorary degrees: Glasgow and Oxford. In presenting her to the Principal of Glasgow University on 19 June 1946, for a degree of Doctor of Laws (*honoris causa*), the Professor of Scottish History, touching on the great services performed by women in the war, said the university wanted to honour her 'not only as a symbol of the great debt which we owe to our women, but also and chiefly, for her own great merits'. He then went on to refer to her work for the Aid to Russia Fund, and finally to her role as a wife:

> There are times when the fate of the world seems to depend on the life of one man. Such a time we have known. And we can but remember with gratitude what it meant to Mr Churchill that there stood beside him in the evil days one who added womanly grace and womanly wisdom, a power to achieve, a faith to persevere, and a full measure of the courage which, as we like to think, reflects the ancient valour of a Scottish ancestry.

In her reply, Clementine took this opportunity to speak of her thoughts and feelings on Anglo-Russian relations at the present time:

> Of my journey to Russia I would here say just one thing. During that to me ever-memorable visit I learned how different an event seems when you view it from Moscow instead of in London or Washington . . .
>
> I mention this because now that there are widening rifts and cooling estrangements and dark suspicions we must sometimes try to transport ourselves in thought across the crashed continent of Europe and imagine what a Russian would feel, isolated after the 1917 Revolution for twenty-five years; then drawn into partnership by terrible necessity and, now that the Nazi monster is laid low, reverting to his loneliness. But we do not want to be cut off from one-sixth of the human race, and however queerly the Kremlin may behave, I am sure the Russian people – if we could only get into human touch with them – want to be friends. The life-line which the British Red Cross threw to the Soviet Red Cross in 1941 still holds fast. If it is to be severed, the fault must not lie with us.

> Finally, Sir, you have mentioned me as the wife of my husband. This is really the reason why I am here among you. But if, as in your courtesy you say, I have contributed to any of his achievements, any faith to persevere, or any of his courage – then I am proud and glad, and deeply moved by your appreciation and the honour you do me today.

At Oxford, where later in the summer of 1946 Clementine was made a Doctor of Civil Law, Winston (who had himself received an honorary degree from the University many years before) walked in the procession headed by the Chancellor of the University, Lord Halifax, and witnessed the conferment of her degree. He heard the Public Orator, who had praised her war work, end his Latin oration by reminding himself, and the assembled company, that their Honorary Graduand was married:

> And how great a man her husband is! But I'm not concerned with him now as the Pilot of our Empire's destinies, or as the 'Father of his Country' [*pater Patriae*]. I'm concerned with him as the man about the house . . . he's a perfect volcano, scattering cigar-ash all over the house [*totas aedes Coronarum favillis conspergi*]!
>
> I present to you the very Soul of Persuasion, Guardian Angel of our country's guardian, Mrs Clementine Churchill . . . Let her assume the same silks as her husband and take her customary place – at his side.

Between 1944 and 1951, Winston was received officially in their capital cities by the Heads of State and Government of seven European countries: France, Belgium, the Netherlands, Luxembourg, Switzerland, Norway and Denmark. Great and famous cities also welcomed him – among them Antwerp, Rotterdam, Metz, Strasbourg and Nancy, Geneva and Zürich. On many of these occasions Clementine was with him, and shared in his triumph. Each country or city bestowed on Winston its highest honours and presented them with beautiful presents. But above all gifts and honours was the tribute paid by the great, cheering crowds of ordinary people, who greeted him wherever he went in all these places: crowds that stood for hours, often in the dark, cold and wet, to acclaim the man – and the country and cause which he had come to embody. These scenes have left an indelible memory in the minds of those of us, his family, who were with him on any of these occasions. It brought home to us all that Winston Churchill had signified – above all in the occupied countries; how much his voice alone had meant, to thousands of human beings. In the cold and dangerous darkness of their imprisonment, his words had been like a beacon of hope.

* * *

On 20 November 1947 Princess Elizabeth married Lieutenant Philip Mountbatten, RN, newly created HRH the Duke of Edinburgh. Winston and Clementine were invited to the ceremony in Westminster Abbey, and to other functions celebrating the marriage. In a letter Clementine wrote on 27 November to Horatia Seymour, who was spending that winter in South Africa, we have a sparkling account of this event, which shook us all out of the doldrums of post-war dreariness:

> [W]e are now entering the long tunnel of Winter & it's too soon to see the gleam of spring at the end of it.
>
> Still we have had the Royal Wedding which lifted our hearts & our feet off the ground. It really seems too good to be true – but true it is which is a shining miracle – The Ceremony itself was beautiful; and we went to an Evening Party at Buckingham Palace which was really gay & brilliant – The beauty of the Queens is like a fairy story. Our Queen's dazzling & magnetic charm the Queen of Denmark as beautiful as Princess Patricia* in her youth and with a gentle, thoughtful & intelligent face – & the Queen of the Hellenes lovely gay & sprightly like a tortoiseshell kitten. Then they were most beautifully dressed, our Queen & the Greek Queen shimmering gleaming & sparkling, the Queen of Denmark in pale lacey [*sic*] romantic clothes. The most beautiful & sumptuous of all was the Duchess of Kent. Her gown for the Wedding was pale silvery damask with a pattern of soft pink feathers.
>
> The bridesmaids are not really very pretty girls but they were beautifully turned out & their wreaths were romantic & unusual. (This letter is getting like the 'Tatler'!)

In London for the royal wedding were many notabilities. Clementine knew Horatia would be interested:

> Smuts is here & we have seen him several times. He really cares for Winston & is a source of strength & encouragement to him. Then there is your Friend Mackenzie King as unchanging as a Chinese image & then General Marshall, the hope of Mankind.[6]

* * *

We get a charming account of Winston's post-war life from a letter Clementine wrote in April 1948 to an old friend whom they had not seen for some time – Ettie Desborough†:

> We are both rather excited because three of Winston's pictures have been accepted by the Royal Academy!‡ And to-day we are going like professional artists to a private 'preview'. Winston spends nearly all his life now down at Chartwell writing and painting – He comes to London only for his political duties & then with many groans and sighs. When he was young his sole

*A granddaughter of Queen Victoria.

† Widow of 1st Baron Desborough (d. 1945). Two of their sons were killed in the First World War (the elder son was the poet Julian Grenfell). WSC/CSC often stayed with them at Taplow Court, Buckinghamshire.

‡ This spring WSC was elected an Honorary Academician Extraordinary by the Royal Academy.

acquaintance with country life was week-end parties among which yours were his happy favourites, but now he hates London!

I had forgotten to tell you that Winston has also though past seventy become a practical farmer – at least we hope we are practical – We have 30 cows & Mary's husband Christopher* manages it all including a Commercial Kitchen garden. I am writing all this, becos it is now so long since I have seen you & Winston and I often talk of you and the happy past.

We both send you our dear love

Your very affectionate
Clemmie

P.S. Winston says he has 40 cows! not a mere 30[7]

During these years, when Winston was in Parliament but out of office, he and Clementine (usually with one or other of the family) spent some agreeable holidays abroad, where they always sought sunshine, and good painting grounds. During this 'interim' period Winston, on the recommendation of his doctor, tried to be in a warmer climate during the worst of the British winter months. He was mainly faithful to the places he knew, Marrakech and the Côte d'Azur – with a diversion in January 1950, when they went to Madeira, from where Winston had to return ahead of schedule because of the dissolution of Parliament prior to the General Election. These winter holidays were productive not only of pictures, but also of chapters, for during these weeks he devoted many hours to writing.

Clementine set great store by being at home at Christmastime; moreover, she had exhausted the tourist delights of Marrakech and the Riviera: so, on the occasions when Winston longed to be away to the sunshine with his paint-box and his chapters as soon as possible after Parliament had risen for the Christmas recess, she used to remain at home, usually joining him later on. Although he missed her, he was never lonely, as a change-of-guard of pleasant companions had always been arranged. Every few days he would write her an account of how he passed his time. On 18 December 1947 he wrote from the Hôtel de la Mamounia, Marrakech:

I have been working very hard, rather too hard, in fact. My routine is: Wake about 8 a.m., work at Book† till 12.30, lunch at one, paint from 2.30 till 5, when it is cold and dusk, sleep from 6 p.m. till 7.30, dine at 8, Oklahoma with the Mule [Sarah] . . . At 10 or 11 p.m. again work on Book. Here I have been rather naughty; the hours of going to bed have been one o'clock, two, three, three, three, two, but an immense amount has been done and Book II is practically finished. I am not going to sit up so late in the future.

Clementine was not dull in Winston's absence, and on 16 December 1947

* I had married Christopher Soames in February 1947. A Coldstream Guards Officer, he had served with them in the desert; he was later attached to the Special Operations Executive, and operated in northern Italy and France.

† WSC's war memoirs, *The Second World War*, 6 vols, published 1948–53.

had written about an interesting encounter at a London dinner party the previous evening:

> I dined with the young Birkenheads* to meet Gen. Marshall. It was a really delightful party. I sat between 'General' Marshall [at this time US Secretary of State] & Lord Camrose . . . The Conference† had ended in dismal failure half-an-hour before but Mr Marshall did not refer to it once. He talked much about you & President Roosevelt with whom it seems he often disagreed & whom he sometimes did not consult – He said that the President would direct his mind like a shaft of light over one Section of the Whole subject to be considered, leaving everything else in outer darkness –
>
> He did not like his attention being called to aspects which he had not mastered or which from lack of time or indolence or disinclination he had disregarded. Mind you he did not actually use these words, but the gist & I thought much more were implied.

Clementine wrote that she had plans to see various grandchildren, and she spent Christmas with Christopher and me at Chartwell Farm, our first home: 'We spent a happy & peaceful Christmas Day & we drank your health & Sarah's before we fell to on the fat turkey.'[8]

Winston had been interested by Clementine's account of her conversation. 'I am so glad you had such an interesting dinner to meet General Marshall,' he wrote on 24 December, 'I think we have made good friends with him. I have long had a great respect for his really outstanding qualities, if not as a strategist, as an organiser of armies, a statesman, and above all a man.'

Clementine was always fearful on Winston's account of the treacherous winter climate of Morocco, because she knew that he was loath to break off from painting, and used to stay out after it was wise to do so. On Christmas Eve 1947, he wrote to reassure her of his prudence in this respect: 'The weather continues to be cloudless and lovely. The air is cold, and in the shade or when the sun goes down it is biting. I am very careful to wrap up warmly and never paint after 5 o'clock.' This was too late by an hour at least, as she was quick to point out – and he did catch a cold in the New Year, which developed into bronchitis. Fortunately the attack was not long or serious, but Clementine, accompanied by Lord and Lady Moran, arrived on 3 January to supervise, pet and scold him.

The annual long parliamentary summer recess gave ample opportunities for some lovely sunlit holidays, in places which were a pleasure also to Clementine, and as they nearly always invited some of us children for part or all of the time, she had companions, while Winston painted to his heart's content. One most enjoyable holiday was made possible by a group of generous Swiss admirers, who placed at their disposal a beautiful and

* Freddie, 2nd Earl of Birkenhead, son of F. E. Smith, WSC's great friend, and his wife Sheila (b. Berry), second daughter of 1st Viscount Camrose.

† The London Conference of major powers on Germany, held 25 November to 16 December 1947; it broke down due to the USSR's demands for reparations.

secluded house on the shore of Lac Léman, halfway between Geneva and Lausanne, where they stayed for nearly three weeks in the summer of 1946; Diana and Duncan and I stayed there with them too. During this visit Clementine sought out her old governess Mlle Elise Aeschimann, who lived in Lausanne, and whose kindness and tenderness to her as a small child she had never forgotten. My mother had always kept in touch with her, and a few years after this was able to help Mlle Elise financially when she fell on straitened times.

This restful holiday would have been perfect for Clementine, whose health had been far from good this year, except for a most unfortunate accident in a speedboat. She was standing up watching my most inexpert attempts at water-skiing when the speedboat made a sudden turn, causing her to fall; she struck herself violently against part of the boat's wooden structure, bruising herself badly and breaking several ribs. For at least a week she was in great pain, and under heavy sedation. At the end of their lakeside holiday, Winston and Clementine were due to be received in Berne by the Swiss Government, and in Geneva and Zürich by the cities' authorities; sadly, Clementine had to miss all these occasions, and it was many weeks before she was fully recovered. It was at Zürich that Churchill, on 19 September, had made his famous speech, calling on Europe to unite, and laying the foundation of the European Movement.

In the summer of 1948, Winston and Clementine stayed for several weeks in Aix-en-Provence; Christopher and I were included in this lovely holiday. We all stayed in the Hôtel du Roi René, the weather was perfect, and almost every day we used to set out, equipped with a delicious picnic, to spend the day in some lovely and 'paintatious' place. Les Calanques, Les Baux, the Roman arch at St Rémy, the Fontaine de Vaucluse (where we cooled our wine by hanging the bottles with string round their necks in the icy, aquamarine river that wells from a gulf in the earth) – these were some of the places we visited. While Winston painted under his great umbrella, the rest of us went for leisurely walks, or visited the local museum, or just sat and gazed at the beautiful summer scene.

In the summer of 1949 my parents went for their summer holiday to Italy, staying at a hotel on the shores of Lake Garda; the pleasant memories of this place, from his visit just after his defeat in 1945, had remained. They stayed here for nearly a month, Winston, as usual, combining hours of painting with hours of writing. On 10 August they both went to Strasbourg, where Winston was to attend the Inaugural Session of the Consultative Assembly of the Council of Europe; he took part in several debates, and addressed a mass meeting of the European Movement. On 15 August, applauded by large crowds, he was made a citizen of the city of Strasbourg. After this, Clementine flew home, while Winston went to stay with Max Beaverbrook at his villa, La Capponcina, near Monte Carlo.

It was during this visit that Winston suffered his first stroke: it occurred while he was playing cards in the early hours of 24 August. The arrival of Lord Moran the next day was spotted at the airport, and the press were immediately on the alert. Since, happily, the stroke was slight, and the effects began to wear off quite quickly, a bulletin was put out to the effect that Winston had caught a chill while bathing, and would have to spend a few days in retirement resting. Clementine, who had of course been kept informed about his condition, although anxious, remained at home, as her sudden arrival would certainly have aroused suspicions that he was suffering from something more than a chill. He made a wonderfully quick recovery and was able to fly home on 1 September.

Apart from the journeys and holidays Clementine shared with Winston, she also made some independent expeditions on her own, taking with her for company either one of us children, or a close friend or two. These jaunts were usually rather short, as Winston minded very much if she absented herself for long. She usually chose periods such as a parliamentary recess, when he would be installed contentedly at Chartwell, with writing, waterworks and painting to occupy him; and she always made sure there would be a succession of children and friends to keep him company. One such occasion was in August 1947, when she went for a motoring holiday in France with Sylvia Henley and one of her daughters, Rosalind Pitt-Rivers (a brilliant research scientist), and Rosalind's fifteen-year-old son Anthony. France was only slowly recovering from wartime conditions. From Auray in southern Brittany Clementine wrote:

> The weather is perfect, the beds comfortable, the food delicious, the sanitary arrangements deplorable! & no hot water except a trickle at 7 in the morning . . . Yesterday we went to Lorient & in the distance saw the German submarine pens – 15 of them visible from where we stood on a bridge –
>
> Havre has been knocked end-ways & on our road to Rennes we saw Lisieux, Falaise & numerous villages much destroyed.[9]

At Eastertide in 1950 Clementine went to Venice for a week, taking with her as a companion Penelope Hampden-Wall, her delightful young secretary, who was looking after the London side of life while Grace Hamblin concentrated on Chartwell. Clementine greatly enjoyed this Venetian jaunt; she had some agreeable friends there, and she was able to indulge in some intensive sightseeing. Winston wrote to her (in his own hand) from Hyde Park Gate, whither he had returned from Chartwell after the Easter recess on 18 April:

> I am so sorry that you have had disappointing weather. I do hope that you have enjoyed the change of scene & the relief of household cares, & that you will come back refreshed . . .
>
> I have thought much about you my sweet darling, and it will be a joy to have you back. Your flowers are growing beautifully on the Chartwell balcony, & here [Hyde Park Gate] the cherry tree is a mass of bloom. All yr arrangements have worked perfectly in yr absence, and no one cd have been more comfortable than
>
> yr P [pig].

Attached to this letter was the 'Chartwell Bulletin' – an edition of four double-sided pages – which contained the most detailed account of every aspect, animal, vegetable and mineral, of the situation there. He also reported on book progress: 'I have completely turned off politics these last ten days in a struggle to deliver Volume IV [*The Second World War: The Hinge of Fate*] in good condition on May 1.'

On her way back, Clementine visited Verona, Mantua and Milan, where she spent a few days, with the intention of visiting the Brerà Picture Gallery. On her arrival she discovered that the gallery had not been re-opened since the war, but that Milan was crowded with people from nearly every European country who had come for the annual Milan Fair. Disappointed in her hope of seeing the Brerà Gallery, Clementine decided to visit the Fair, where she was disgusted by the low standard of the British exhibit. In fact, when she arrived home four days later she was still boiling over with indignation and, on the day after her return, she dictated a fiery memorandum on the subject which Winston sent on to the President of the Board of Trade (at that time Harold Wilson, who would become Prime Minister in 1964).

* * *

During the spring and summer of 1950 Winston was concentrating deeply on getting ahead with his memoirs. The General Election in February of that year had left the Labour Government with their majority sliced to five, and it was obvious that their days in power were running out; he therefore decided to stay at home to concentrate on both politics and the book. There was no combined family summer holiday this year, and Clementine made another expedition abroad in June, this time to Spain, taking me with her. We stayed at the Ritz in Madrid, and indulged in a veritable orgy of sightseeing. Johnny and Peregrine Churchill (Winston's nephews) and Mary, Johnny's wife, were in Madrid at the same time, and we joined forces on several occasions. Although it was an entirely private visit, my mother was treated with great consideration and courtesy by the Spanish Government, who provided cars, detectives (apparently deemed necessary) and other facilities.

We had one very grand Spanish contact, who showed us much kindness and hospitality – the Duke of Alba. In a remote, romantic and historical way he was Winston's kinsman, being descended from James Fitzjames, Duke of Berwick, who was the natural son of James, Duke of York (afterwards James II of England) and Arabella Churchill, the sister of John Churchill, 1st Duke of Marlborough.

It was on this visit that Clementine made friends with a delightful woman, the Marquesa de Casa Valdes, her husband and her family. Much of the pleasure of this visit was due to them: an affectionate and enduring bond was formed between us all, and Clementine subsequently stayed with them in Paris. One of our most enjoyable days was spent visiting their beautiful patriarchal estate at Guadalajara. Another expedition was to the Escorial, and we launched several attacks on the Prado, from which we emerged exhausted, though not defeated. Unfortunately we rather overdid the programme of sightseeings and expeditions, and my mother became over-tired; but the visit on the whole was a great success, and had whetted her appetite for the beauties and splendours of Spain. The following Easter she went to Seville, where she stayed throughout Holy Week to watch the world-famous processions. As delightful companions she had Bill and 'Pussy' Deakin. In the thirties Bill Deakin had helped Winston with *Marlborough*; now, after a brave and brilliant war, in which he led the first British Military Mission to Yugoslavia and was awarded the DSO, he had returned to Oxford. At this time Warden of St Antony's College, he also directed all the research for Winston's war memoirs. Both he and his wife were great personal friends.

Throughout this visit to Seville my mother wrote me long and detailed accounts of all they did and saw. Reading her letters one feels the tug between the puritan/Protestant (and slightly anti-clerical) element in her nature, and her sensitivity to the great beauty and emotional appeal of much she saw; her radical instincts made her suspicious of official Falangist explanations. Here are some extracts from her letters:

Monday, 19th March 1951

Yesterday, being Palm Sunday, the great Religious Processions began . . .

The whole population of Seville is in the streets & chairs for hire are provided everywhere. Each church has its procession which wends its way by circuitous routes to the Cathedral. The procession is formed round the particular treasure of each Church which is carried in triumph on a decorated (very baroque & splendid) platform . . .

We saw 3 of the 7 processions . . . The figures are life-size in plaster & very well modelled . . . Magnificently clad Virgins with beautiful sad pensive faces . . .

The crowds chattered gaily & noisily, but here & there you saw people moved with deep emotion. It's a great pagan carnival but it shews Holy Church triumphant over pain, corruption, tyranny & unbelief.

Watching processions of penitents, where the acolytes were in black, bearing great black candles or crosses, and the series of religious tableaux, Clementine herself was deeply moved. She also described the carnival atmosphere in Seville:

I have been hailed in the streets as 'Clementina' which is a welcome change from 'Oh my Darling Clementine'!

The Hotel is packed with Americans & there are one or two 'croyants' French families who have brought their children . . .

So far, no sign of 'our cousin' the Duke of Alba . . .

Meanwhile 10 bulls are on their way here for Easter Sunday. A queer reaction to the Resurrection of our Lord.

And on 23 March:

> During Holy Week the population of Seville seem to live on shrimps & rusks. These are sold from hand-barrows & no-one goes home for a meal, except vulgar tourists like us. The shrimps are delicious (when fresh!).

These engaging accounts of Clementine's travels were duly circulated throughout the family.

* * *

At the end of the war Diana, Randolph and Sarah were respectively thirty-six, thirty-four and thirty-one years old, and, like everyone else of their generation, they set about picking up the threads of their interrupted lives. Diana was much occupied by her three children, the youngest of whom was two. Although Duncan Sandys' parliamentary career had been temporarily suspended by his defeat in 1945, he was never out of political life, concentrating on the European Movement in which he was, in this country, one of the prime movers. Winston esteemed and liked him very much, and their family bond was reinforced by their mutual concern for the cause of European unity. It was not long before Duncan was adopted for another constituency, Streatham, which he fought and won in 1950, and which he was to hold until he resigned in 1974 on becoming a peer. He held ministerial office continuously from 1951 to 1964; Diana, therefore, with her growing family and a husband in active politics, had a very full life. She and Duncan came often to Chartwell and later on to Chequers, and her relationship with her father was always warm and affectionate. But between her and her mother there seemed too often to be an atmosphere of watchfulness and carefulness, rather than ease or cosiness.

Randolph led his own rampaging existence: lecturer, journalist, author and politician – he always had lances to break, and hares to start. Although blazingly loyal and demonstratively affectionate, he was not a 'comfortable' person to have around, and too often was in the mood when he would – all else failing – pick a quarrel with a chair. Unhappily, the emotional and angry scenes which had so often marked his brief times at home in the war had left lasting scars on his relationship with both his parents. At no time, however, was the bond of loyalty severed, and letters between Randolph and his mother (mostly about arrangements for visits, plans for houses or thank-yous) show a persevering desire to maintain or restore affectionate and agreeable relations between them. But hurts and misunderstandings were, alas, frequent, and Clementine could not bear the noise and commotion of the arguments which erupted, particularly between Randolph and Winston, arising mostly from the former's bully-ragging of his father about some of his political colleagues – Anthony Eden being a chief target for his scorn and vituperation. Except on the rare occasions when she actually 'blew up' herself, she more often simply retreated into a chill silence, or removed herself from the scene altogether. Yet whenever Randolph did try to please his mother it made her truly happy, and he greatly appreciated all the very real practical help she was able to bring to his financial difficulties, as Chairman of the Trust set up by Winston after the war for his children and grandchildren. Yet so often neither of them seemed to be able to 'get through' to the other. A letter from me to my mother (undated, but written probably in September 1948) is revealing about the situation between them.

> My Darling Mama,
>
> When I dined with Randolph I sat next to him, and he said how very sweet you had been to him about all his difficulties . . .
>
> He spoke in warm, loving and grateful terms of you. He said you had spared him much humiliation. And I can see he is full of admiration and true gratitude of the way you are treating him in these painful affairs.
>
> I then said that you had told me how happy you were made by his gentleness & affection to you.
>
> I thought you would like to know this – because I know a little of the grief and perplexity you have gone through on his account, & because of his manners towards you.
>
> Please burn this letter.

I am glad now that my mother ignored my injunction, because so many years later this letter has cast a softer light on the so often strained situation that existed between her and Randolph.

In his relationship with his father – which he prized above all else – Randolph failed to perceive that, after the passage of the war years, Winston had lost much of his own relish for polemics in private. In days gone by he had enjoyed verbal knockabout turns, giving as good as he got (although he hated real rows). Now he was so many years – and a whole exhausting war – older, it upset him to hear present or erstwhile colleagues berated or vilified by Randolph. What stores of energy and strength remained to him were precious and were not to be squandered in mock battles of words, but reserved for the tasks that still lay ahead. Randolph's insatiable appetite for controversy, unabated by time, was undoubtedly the main reason why his father sought less of his company than in days gone by, and was more reserved in his confidences. Randolph was conscious of this change, and it grieved and embittered him. But both he and his father were too warm-hearted to maintain distances or hostile positions for long periods, and through all the ups and downs of their relationship Winston's loyalty to Randolph was equalled only by Randolph's loyalty to his father.

When Randolph's marriage with Pamela Digby had ended in divorce in December 1945, both Winston and Clementine had been saddened, as they had become very fond of their daughter-in-law, and they were devoted to 'little Winston', who had been born in 1940. As in the war, Clementine used to incur Randolph's displeasure at times because she maintained affectionate relations with Pamela (which, incidentally, greatly facilitated the making of holiday arrangements and so forth for their child). Many people will recognize this as a 'copy-book' situation, which is beset with difficulties and is a breeding ground for misunderstandings. However, as time went on Randolph grew to be less suspicious and touchy, and 'little Winston' very often visited and stayed with his Churchill, as well as his Digby, grandparents.

In November 1948 Randolph married June Osborne. Both Winston and Clementine welcomed her warmly into the family and earnestly hoped that now Randolph might find some measure of domestic peace and happiness. There was in his life an unassuaged loneliness, combined with a touchingly patriarchal sense of family; yet, notwithstanding his many gifts and great heart, he does not seem to have possessed the aptitudes for marriage. A beautiful child was born, Arabella, in 1949, and largely because of her, Randolph and June persevered with a relationship that was both unhappy and tempestuous. They soldiered on for thirteen years, finally divorcing in 1961.

Winston and Clementine always grieved very much about the shipwrecks of their children's marriages. Winston could never quite understand why 'the young people couldn't make it up and have another try'. He knew what he owed to his own rock-like marriage, and he took the simple view that you fell in love, married and lived happily ever after. Clementine was more comprehending of the emotional tangles and entrenched attitudes that grow up in unhappy relationships. Although her own standards for married life were very strict, I was often struck by how uncensorious she was, and how fair-minded, when it came to the marital mishaps and difficulties of her own children or friends. The draft of a gentle and understanding letter to June survives; it must have been sent in reply to a long and very sad letter from her daughter-in-law written in March 1950. Clementine wrote:

> Dearest June,
>
> I was very sad after I had read your letter and seen Randolph. Marriage strikes deep roots, and I pray that with you both in spite of storms these roots may hold. Do not hesitate to come and see me if you think I could help.
>
> Your affectionate
>
> Clementine S.C.

Soon after her demobilization Sarah once more took up her stage career, about which she had been constant in mind ever since her debut as one of 'Mr Cochran's Young Ladies' in 1935. Now it was not vaudeville, but the straight stage to which she bent her energy and talents. In the summer of 1946 she played the lead in a Victorian thriller, *Gaslight*, and both Winston and Clementine went to see her in it. That autumn she went to Rome, having signed a contract with an Italian film company, but soon after she started working there she fell ill, and when it became evident that it was more than a passing indisposition, my parents despatched me to Italy to keep Sarah company, and to see that she was taking proper care of herself.

I was naturally delighted, both at the prospect of an unexpected jaunt and at being with Sarah. However, this journey coincided with a crucial moment in my own private life. On a recent and very brief visit to Paris with my father I had met Christopher Soames, an officer in the Coldstream Guards, and an Assistant Military Attaché at the British Embassy. Despite distance and duty, our friendship was budding, and at the very moment of my departure for Rome to join Sarah, Christopher was due to come to London for some leave, when we had arranged to meet. He hastily changed all his plans and accompanied me to Rome instead. Fortunately my sister Sarah had an iron constitution, for although I went to Rome with the sole intention of nursing her tenderly back to health, I arrived in a state of high emotion with a good-looking young man in tow. Despite all this, she made a complete though slow recovery. During the short time Christopher could stay in Rome before his leave expired we became engaged. My letters home were, in consequence, a jumble of bulletins on Sarah's health combined with explanations of my own heart's state, which must have been confusing and unsatisfactory to all concerned – more particularly so since neither Christopher nor I had met each other's parents.

Like most people in love, we were blissfully and egotistically oblivious to all reactions which might contain an element of doubt or reserve on our families' part. On the whole, both sides took the sudden news of our engagement with stoicism laced with hopefulness, and Sarah from her sick-bed wrote moving pleas for confidence and understanding to my mother. Both my parents had many misgivings because of my precipitate action, and my mother in particular was extremely anxious. But they both made only welcoming and affectionate noises when Christopher and I arrived home.

From their very first meeting my father and Christopher took to each other. Thus began a warm relationship which grew in affection and trust; it had an enormous effect on Christopher's personality and life, and their real friendship added to the happiness I found in my marriage. And if my father gave much to Christopher, Christopher was able to bring the breath of fresh air from outside, from a younger generation, to my father. His ebullient, cheerful personality also brought fun back into my father's life, revived old enjoyments, and added a new one – the Turf. And once launched in politics himself, Christopher was able to perform a very real service in being a link between Winston and a younger generation of politicians on both sides of the House of Commons.

My mother was not such an easy conquest, and although in her more formal way she was welcoming and affectionate to Christopher, and expressed no open opposition to our engagement, she nursed deep doubts about this 'stranger' who had suddenly arrived on our family scene. She also, not unreasonably, in view of my (up-to-now) changeable affections, wondered if I really was settled in my mind and my heart. Six years earlier, when I had become seriously romantically entangled, my mother had not hesitated to use all her influence with me to break off my engagement, for she had been convinced that I was not really in love, that I was too young, and that I had had very little experience in these matters. Now the situation was different – I was twenty-four, and I had been out and about in the world; and when she saw that I had the 'bit between my teeth', and was determined to marry Christopher, she acquiesced gracefully, and did everything to help us. But we were both aware that her attitude was less enthusiastic than my father's.

During the latter part of this autumn Clementine was not well; the accident on Lac Léman during the summer holidays had thoroughly shaken her, and her ribs were painful for a long time afterwards. She became progressively more exhausted, and finally in early December, on her doctor's advice, she cancelled all her engagements and had a complete rest: that is, as much rest as it is ever possible to have in one's own house, with, moreover, a daughter's wedding in the offing. We spent Christmas very quietly at Chartwell.

Christopher and I were married on 11 February 1947, at St Margaret's, Westminster, where my parents had started out on their long road together. It was a bitter winter, and the country was in the grip of a major fuel crisis. We left a gloomy, freezing London for our honeymoon in snow and sunshine in Switzerland. On the Sunday after our wedding, my mother wrote to me from London:

> My Darling and beloved Mary,
>
> I addressed the envelope first, and writing your new name for the first time gave me a sensation of anguish and satisfaction – anguish that your life at home is ended, satisfaction that you are founding with the man you love your young own home with all its hopes & joys and experiences.
>
> Your jubilant telegram thrilled us all & we are in imagination sharing the blue sky & dazzling sunshine & in reality your blissful happiness. It is agonisingly cold here & in this completely electric house we are feeling the 5 hour daily switch off. But I got a doctor's certificate for Papa's bedroom as he can really not work in the icy cold. We have pulled his bed near the window so he can see without a lamp . . .
>
> My darling Mary. I am quite numb & I have not begun to miss you yet. I went once into your deserted bed-room, but not again . . .
>
> Diana has just been in. She sends you her love. She was very sweet & relaxed & I enjoyed her visit . . .
>
> Papa did not go to Chartwell this week-end becos' of the biting cold & I think he has been quite happy here. He sends you his dear love & messages to Christopher . . .
>
> Goodbye Darling for now. Give Christopher one kiss from me if you can spare it from yourself!
>
> Your devoted Mummie[10]

During our honeymoon Christopher was taken suddenly very ill with a duodenal ulcer, and we were stranded somewhat forlornly at St Moritz. My mother flew out to be with us, thereby causing a slight sensation and some amused comment about mothers-in-law among the other hotel guests! As usual, she was ready with a practical suggestion: it was clear that Christopher would have to leave the Army, and my mother suggested that we should live in the charming farmhouse at Chartwell Farm, which my father had recently bought. This wonderful solution to our housing problem was warmly approved of by my father, who further suggested that Christopher (who was making a steady recovery) should take on the management of the farms. So began an idyllic, pastoral period in our lives; and during the next ten years we lived, with our growing brood, at the bottom of my parents' garden. Christopher and Clementine grew to know and appreciate each other, and 'The Chimp' (as my parents affectionately called him) was soon a built-in part of the family, and a trusted confidant in all matters.

Sarah, now almost completely immersed in her stage and film career, spent the summer and autumn of 1949 in America playing the lead in a touring production of *The Philadelphia Story*, in which she had a great success. Some time previously, in London, she had met Antony Beauchamp who, after serving as an official war artist with the 14th Army in Burma, was busy establishing himself as a photographer. As their relationship developed Sarah had brought him to meet her parents and they had all been in Monte Carlo together in early January 1949. During that spring, after Sarah had returned to America for her work, Antony came several times to see them; unfortunately, neither Winston nor Clementine took to him. Clementine did try – but she never came to like or trust him. And although she endeavoured to make things easier between them, Winston remained resolutely hostile.

Later that year Antony went to America to join Sarah, and laid strong siege to her. She had earlier written to her mother that she felt she was 'very near now to taking the plunge'.[11] So Winston and Clementine certainly knew that a decision was imminent: however, Sarah wavered for some little time more. But between her summer and autumn seasons with *The Philadelphia Story* she and Antony went for a short holiday to Sea Island, Georgia, and there Sarah made up her mind to marry him: she wrote at once to tell her parents of her final decision. 'Thank you, darling Mummie,' she wrote on 14 October, 'for your patience and forbearance through these difficult months while I had to sort my mind and heart out.'

Although when she wrote the date and place of their marriage had not been fixed, Sarah and Antony now decided to be married then and there – on 18 October 1949, on Sea Island – while Sarah had two or three days more of holiday left. The arrangements were made swiftly, and cables informing their families sent. Unfortunately, the news broke in the press before Winston and Clementine had received any word, by either letter or cable, from Sarah. The shock and hurt of hearing of the marriage first from the press greatly upset them both – particularly Clementine, who took it very hard indeed. And despite ruinously long telephone calls, and fully explanatory, distressed and loving letters from Sarah, she retreated into a hurt silence, broken only by a few telegrams – some of them about business affairs.

During that autumn Sarah wrote several letters home describing her hectic life on tour, and always begging her mother not to be 'hurt'. She wrote a special and loving letter to her father for his birthday. In mid-December Antony wrote a long, and in part severely reproachful, letter to Clementine, saying how deeply unhappy her long and chill silence was making Sarah, and that the exhaustion of her theatrical tour, combined with her distress at not hearing from home, was having an effect not only on Sarah's spirits but also on her health, and he begged her mother to write to her.

Clementine's long silence, and her unspoken reproach, are a little difficult to understand, although her original reaction to the way the news of the wedding had reached her and Winston is comprehensible. But this episode certainly bit deep: and her sense of foreboding that the marriage would not be a happy one may also have made writing to Sarah hard for her. However, Antony's letter had the salutary effect of goading her into action, and she wrote both to him and to Sarah almost immediately long letters couched in conciliatory terms. She assured Antony: 'There was no displeasure or coldness. But Absence and distance & circumstances have perhaps clogged natural impulses.'[12] And about her long silence she wrote to Sarah:

> I confess I felt numb. I felt able only to telegraph to you from time to time. I am glad Antony's letter arrived & galvanized me. Again I am sorry and you must forgive me and believe in my love and care for you . . .
>
> We have made friends with Antony's father and mother* and we had an agreeable luncheon all together.[13]

Thus the long and wounding 'freeze-up' was ended.

Sarah and Antony came back to London for a brief visit the following May, between working engagements. They were, of course, greeted with open arms. Clementine gave a large dinner party uniting both the families at Hyde Park Gate, and at the weekend Sarah and Antony came down to Chartwell just before leaving to return to America. It was all a great success, and everyone was overjoyed to see everyone else. Clementine was able to tell Sarah that she had been doing some house-hunting for her and Antony, and that a choice house was in the offing. When Sarah and Antony flew back to America a few days later, the dark clouds of estrangement and misunderstanding had been blown away, and the sunshine warmth of Sarah's relationship with her parents shone out once more.

* * *

One of Winston's chief satisfactions in his old age was the provision he was able to make for his family and descendants, through the Trust† he had set up. He made Clementine Chairman of the Trustees; two of these were

* She was 'Vivienne', the society photographer.
† Chartwell Literary Trust.

'The Prof' and Brendan Bracken, who between them combined financial expertise and worldly wisdom with a thorough knowledge of the family. It delighted Clementine to think up ways in which to help each child in the particular way he or she needed. But although sympathetic and swift in action to rescue any of us from the results of unwise budgeting, or even downright fecklessness, Clementine always felt a pang to see what was the fruit of Winston's genius and generosity being poured, as it were, down the drain. It was natural that she should think of their own early impecunious years, and what an alleviation to their constant anxieties about money such a beneficent Trust would have been to her and Winston in days gone by.

But helping any of us in a constructive way filled her with interest, enthusiasm and pleasure. If a house or flat was needed, she was indefatigable in helping to search for one, and her practical eye prevented the acquisition of some seductive property which had a fatal fault. She was ever ready to recommend to her fellow Trustees that a child should be helped with some basic domestic improvement such as a new kitchen floor, or a service lift, or a modern boiler; or perhaps just a wonderful windfall to help towards furnishing, curtaining and carpeting our home-sweet-homes. There is a file of grateful letters from various children and grandchildren thanking for divers blessings, of which Winston was the source, but in which Clementine's practical and imaginative thought is always apparent.

CHAPTER TWENTY-SIX

A Woman's Work Is Never Done

WITH THE END OF THE WAR, TWO OF THE CAUSES FOR WHICH CLEMENTINE had worked so unremittingly – the Fulmer Chase Maternity Home and the Aid to Russia Fund – had come to a natural conclusion, although there was some lengthy and intricate winding-up to do in both cases, which she, with characteristic thoroughness, saw through to the end.

Fulmer Chase had been a real success story. By 1945, when the home closed, over 2,500 babies had been born there, and there had been no maternal deaths. The standards of nursing, care and comfort had been such that they had been described in *The Midwife* as 'beyond praise'.[1] It had been indeed a satisfying and satisfactory bit of work, and Clementine had loved her share of it.

In March 1947 she resigned from the Presidency of the YWCA Wartime Appeal. In a farewell broadcast appeal on 23 March 1947, she asked for continued support and help for the YW's work in providing yet more hostels for 'girls released from war service, girls from factories and offices, government employees and students – a continuous stream from all parts of Britain of girls with jobs but nowhere to live'. As a result of this broadcast a good sum of money rolled in, but Clementine also received some letters she must have greatly valued from people she had worked with, either at the YWCA headquarters or in the regions, all of whom expressed appreciation for what she had done in those war years for the organization, both in raising money and in improving standards in the hostels. One colleague wrote: 'I am sure others are feeling as I do tonight – a real sadness that this will be your last official public act for us! You have been so good and understanding, so practical & stimulating that we shall miss you sorely.'[2]

That they did miss her, and had valued her work, was amply proved when nearly two years later the Association sought her help once more, and in January 1949 Clementine was elected Chairman of the National Hostels Committee. It was a compliment and a challenge she accepted with real pleasure and interest. Between the wars she had never relished committee work, nor had she been gripped emotionally or intellectually by any one particular cause or social theme: but the last war had changed all that. Her position as wife of the Prime Minister had, of course, given her a certain power and influence which, as we have seen, she learnt to wield with effect. But this was not the whole story: Clementine had acquired the taste for doing a real job of work, and grappling with policies and problems 'in depth'. Her wartime involvement in the YWCA's work, and in particular that part of it which dealt with hostels, had really gripped her interest, and when she accepted the Chairmanship of the National Hostels Committee she felt she knew something about the work and its problems already, and had a very clear idea of the goal towards which the committee should strive.

A very few weeks after becoming Chairman, Clementine found herself taking the chair at the yearly Hostel Wardens' Conference at Southsea. Afterwards, in her official report of the Conference,[3] she remarked upon the fact that the programme dealt almost exclusively with the religious and social aspect of the YWCA's work, and admitted that she was surprised that all the manifold practical difficulties of managing hostels had not been examined in detail; and she concluded with a firmly expressed hope that the programme for the next conference would allow 'several sessions to deal with the physical standards of well-being, and with the practical details of administration'.

When it came to a discussion on actual living standards in the hostels, she was bluntly outspoken, avowing that she had been 'disgusted at the conditions that I saw when I visted the hostels. The war was being made an excuse. Now that the war is over it is still being made an excuse . . . We must exert ourselves and get out of the state of squalor and sordidness . . . I have joined the National Hostels Committee to see if we can try and make the hostels more comfortable.'[4]

Clementine set about this task with energy and assiduity. She regularly attended the committee meetings at Headquarters, and during the three years of her Chairmanship she made a series of tours, visiting hostels in every part of the country. After every tour she made detailed notes on the various hostels which were for her own and Headquarters' guidance, and she always wrote to the chairmen of the local committees and the wardens of the hostels she had visited. The letters show an amazing grasp of practical detail and understanding of the problems. They contain praise (but only, obviously, where it was due), and understanding and encouragement where people were struggling with intractable houses, lack of staff and meagre funds. Where she had criticisms to make, these were invariably constructive. Judging from the answers to her letters, her remarks, albeit sometimes critical, were appreciated rather than resented, and a former resident in a Glasgow YWCA recounted how, at the time of a visit by Clementine 'I and many of the residents were ex-servicewomen and therefore rather cynical about visits by "high heid yins" [high-ups]. We were consequently terribly surprised, and pleased, when as a result of your mother pronouncing our beds too hard and the facilities for doing our washings [*sic*] not good enough, these matters were remedied.'[5]

Although fund-raising was not her province as Chairman of the National

Hostels Committee, Clementine was in fact a champion collector for the work that gripped her so much. People would quite often give her money 'for a favourite charity', and as she longed to have a source from which she could help hostels with equipment or special improvements which fell outside the scope of their normal budgeting, she started collecting donations which were given to her into a special fund – 'Mrs Churchill's Special Hostels' Fund'. In 1950 the McConnell Foundation sent her £3,000 specifically for hostels, and Mr Giraudier, a warm-hearted Cuban gentleman, was another regular contributor. In 1949 Winston had given one of his pictures to be auctioned in aid of the YWCA, and the proceeds of the sale swelled Clementine's Special Fund, as did successive sums from the yearly openings of the gardens at Chartwell. From this Fund Clementine rejoiced to be able to make substantial grants towards the improvements she had so much at heart.

Soon after Winston became Prime Minister again in 1951 Clementine realized that she would be unable to continue her regular work for the YWCA, and it was with real regret that she wrote resigning from the job to which she had really given the cream of her thought and energy for nearly three years, and which had given her great satisfaction. But she never severed her links with the 'YW', and indeed remained a Vice-President of the Association, as well as a member of the Executive and Finance Board (which for a long time she attended when matters concerning the hostels were discussed).

When the new YWCA National Headquarters (which included a hostel for 100 girls) was ready in March 1953, Clementine was invited to open it, which surely gave her great satisfaction, for she had been involved in the plans from the start, and the hostel accommodation incorporated so many of the features and facilities she had striven for in the past. It was not inappropriate that there was in this 'model' hostel a 'Clementine Churchill Wing'.

* * *

After the Second World War, as in the twenties and thirties, Clementine's political activities were centred on Winston's constituency. She was his chief 'liaison officer' with the Divisional Association in Woodford, and she kept in constant and close touch with them. She studied most carefully the programmes for Winston's visits to the constituency, to make sure that they were not overloaded or too demanding; she could wax quite fierce – but she evoked not only respect and loyalty but also true affection from the leading men and women in the local Conservative Association. Every year her engagement diary shows a regular series of visits to the Woodford area. Quite often on a Friday or Saturday she would attend some afternoon or evening meeting or social function there, driving to Chartwell afterwards to join Winston. It was quite a long and tiring expedition for her.

As wife of the Leader of the Conservative Party, she was of course much in demand all over the country to address meetings or open bazaars; but, with very rare exceptions, she resolutely declined to accept political engagements outside Winston's constituency. She had too many other obligations (not least her YWCA work), and she knew her stamina would not be equal to it. There were, of course, a few major exceptions to this rule – she nearly always accompanied Winston to the Annual Conference of the Conservative Party, and she always tried to attend some of the sessions of the Conference of the Conservative Central Women's Advisory Committee, which was held in London in the spring of each year. She always received a heart-warming welcome from the delegates, whom she once described as representing 'the greatest, grandest constituency in the world. You represent the women of Britain'.

Clementine was nearly always on these occasions, as at constituency functions, called upon to speak; and occasionally she had to make a 'full dress' speech, which she prepared most carefully, often submitting it to Winston to 'vet'. Although all her married life she had had to make speeches – one recalls the accounts of her fiery campaigning in the old Dundee days – yet, in these later years, political speaking was not really to her taste. She usually read her speeches, and indeed someone who often heard her speak in Woodford described her as a 'nervous speaker'. But where she excelled was when she had to make an impromptu speech, probably at the end of a meeting – for then her charm of manner and naturalness shone out, and she never forgot to thank people by name for what they had done. During elections Clementine was still a doughty campaigner, doing her share of speaking, and unprepared occasions often found her at her lively best. But she shouldered making speeches as a necessary duty, not as a pleasure. Her real contribution in the constituency was her genuine interest in local and Association affairs; her true appreciation of what individuals did in different ways for the organization; and her willingness to spend time and to take trouble with people.

As the Party Leader's wife, Clementine, when making a speech, had to wield official facts and figures, and follow the Party line. Comparing the drafts of some of her political speeches with those on other subjects, one is immediately struck by the difference: many of the political ones seem stilted, and the Central Office handout touch is easily recognizable (though the drafts are copiously corrected in her own hand, and no doubt 'came alive' as she spoke). But where she was speaking to her own brief, on subjects such as Aid to Russia, maternity care, and the problems, policies and goals for YWCA hostels – all causes which had caught her enthusiasm and in whose work she was personally absorbed – the difference in style is very striking. Even on paper they have life and appeal.

* * *

When Peace had returned to a war-shattered Europe, and a battered, rather weary Britain, Plenty – her traditional companion – had not accompanied her. Many wartime shortages continued: clothes were rationed until the spring of 1949, and some foodstuffs were still rationed as late as 1953. During a series of economic crises bread and potatoes were actually rationed for the first time, and petrol rationing was re-introduced for a period. But perhaps the worst single problem was the acute housing shortage, which weighed heavily on young families and brought with it a crop of heart-breaking social problems.

For a time these hardships and hassles were accepted as being the obvious legacy of the war; but gradually people's patience and acceptance wore thin, and while dire predictions of an egalitarian, doctrinaire socialist state proved to be fanciful exaggerations, equally the benefits of nationalization and a more centrally organized economy were not blazingly apparent. From the nadir it had reached in the rout of the 1945 election, the Conservative Party in these years passed through a period of renaissance in organization, thinking and morale. By February 1950, the credibility and popularity of the Government were seriously eroded, and in the General Election held in that month Labour's majority was reduced to five.

During this period, while ordinary people in every country struggled to remake their personal lives, the outlook on the international scene dashed the bright optimism of those who had expected the dawn of a new age of harmony among the nations. The critical voices that had been raised by Churchill's warning in his Fulton speech of Russia's intentions were soon silenced, as the grip of the Kremlin tightened on Eastern Europe. The re-action to this threat was to make the Western powers – both allies and erstwhile enemies – draw closer to each other, and to the United States. World events seemed to lurch from crisis to crisis, and over all there brooded the mushroom shadow of nuclear power: a new, sinister and terrifying element, to the presence and implications of which the whole world had to adjust its thinking and its actions.

Still in the midst of public life, and in close contact with the most informed circles, Winston and Clementine now, as always, lived with the reality of events. The feeling of contrast between the preoccupations of daily life and the natural relief of 'getting back to normal', and the tensions and uncertainties caused by recurring crises in world affairs, is reflected in a letter Clementine wrote to Queen Elizabeth on 7 April 1948, after she and Winston had stayed at Windsor Castle:

> I can truthfully say that I have not enjoyed a week-end Party so much for what seems an immeasurable space of time.
>
> After 'Clumps', which I had not played for forty years, (or more) & then in a much more sedate fashion, I felt nearly forty years younger.
>
> It was moving and stimulating to feel the pulse of Life beating strongly in Your Majesties' family, delightful to see the wonderful treasures all in their places again, and to feel 'here firm though all be drifting' –

During these five years when the Conservatives were in Opposition, life at Chartwell, which had taken a little time, and a great heave, to get going once more, truly blossomed and flourished anew. There was hardly a week-end when Winston and Clementine were in England that they did not spend there, with usually one or two people staying, and – particularly in the spring and summer – family, friends and colleagues coming down for the day.

One of the new features of life at Chartwell was the warm and sunny relationship which soon developed between the Churchills and the neighbourhood. Before the war, although from time to time one or other member of the family was in demand as a foundation-stone-layer or bazaar-opener, there had been little other involvement in parochial affairs. Winston never went to church except on 'State' occasions, and although Clementine usually did at the great festivals, church-going was never a feature of our family life. Many people in the thirties had regarded Winston with mixed feelings, and some had looked upon him as an outright *enfant terrible*, while as far as the local shops were concerned, the Churchills had a bad reputation when it came to paying bills. But now all that was changed: the neighbour-hood took a pride in having Winston Churchill in their midst, and Winston and Clementine themselves were more accessible and outgoing.

A great feature of summer life in these years was to be the annual open-ings of the gardens to the public, when thousands of people came to visit Chartwell. These began in 1948 with one opening; in succeeding years the gardens were opened always twice, and in 1951 there were four open days. The regular beneficiaries of the considerable proceeds were the YWCA and local charities. Since Chartwell is situated mid-way between Westerham and Crockham Hill, the money from the open days devoted to local causes was divided between the two communities. So too was their organization, which brought Clementine in particular into pleasant contact with her neighbours. These Garden Days, and Winston's and Clementine's attendance from time to time at such local functions as the Westerham Whit Monday Gala Day and the Edenbridge Agricultural Show, did much to cement the now warm and friendly relationship they enjoyed with the neighbourhood.

Now that the house was easier to run, and she was no longer haunted by financial nightmares, Clementine began to enjoy Chartwell as she had never done in the bygone years. Croquet replaced tennis as the summer game, and the tennis court made a quick transformation into a 'regulation' croquet lawn. Clementine played very well, and most of all she liked 'real' croquet, but nearly all her friends (and certainly her children) preferred the shorter, more immediately rewarding Golf Croquet to the scientific *longueurs* of the classic game; she was happy to acquiesce in this compromise, and many pleasant afternoons were passed in this way. New recruits were found in the growing grandchildren, and although Winston did not play himself, he was an attentive spectator, and particularly enjoyed seeing 'Monty' (a frequent and faithful visitor) out-manoeuvred as a strategist.

Both Winston and Clementine enjoyed their grandchildren, who quite often came, with or without their parents, to stay, and Christmastime once more saw real family gatherings. Because Christopher and I lived literally at the bottom of the garden in the Farm House, our children saw a great deal of their grandparents, and had an easy, uninhibited relationship with them both, which has remained a glowing memory. From quite an early age our children could trot up unescorted to the 'big house' to pay unannounced social calls. It was a ritual in the summer that, as the nursery party wended its way with Nannie, the pram and attendant dogs to the swimming-pool, a child would be despatched to tell Grandpapa and Grandmama, and one or both of them would almost invariably come down to the pool with any guests to watch or join in the aquatic fun, often staying to share nursery picnic tea. Winston was now always a spectator, as he had been persuaded not to swim except in really warm water, and since the war the pool (although filtered and crystal clear) was unheated. My mother described a typical afternoon to me when Christopher and I were away in Scotland in August 1952:

> After luncheon we converged with your babies [Nicholas aged 4, Emma nearly 3 and Jeremy 3 months] upon the swimming pool & although it was by no means very warm, Sarah plunged in with Nicholas & Emma who screamed with delight & excitement. I had on the beautiful taffeta bathing suit which you gave me, but at the critical moment I flinched & stayed on the verge warmly wrapped up in a bathing robe! Papa came down & afterwards we all went to your house to tea in the Day Nursery. Giuseppe [our Italian cook] made the most delicious miniature doughnuts & some twirley whirley things made of the same mixture. Nicholas & I devoured the whole plate between us.[6]

Whenever we were away, holidaying or electioneering, my mother kept a grandmotherly eye on our nursery, writing us lively and reassuring reports, such as this one: 'You will have heard from Nannie that your babies are well, beautiful & happy – They have been to tea here twice & are coming back tomorrow. It's such fun.'[7] It rejoiced my heart then, as now, to feel that life at the Farm House was a source of pleasure and joy to both my parents, because for us all those were golden, unforgettable years.

A totally new interest and diversion for Winston, inspired and largely contrived by Christopher, was racing. In 1949 he registered in his own name Lord Randolph's colours*, and soon acquired a grey French colt, Colonist II, which won thirteen races for him, and large sums in prize money, before he was sold to stud in 1951. He soon had several other horses in training, which afforded him great pleasure as well as being financially successful. He also acquired a small stud farm at Newchapel Green, near Lingfield, which, being quite near to Chartwell, he was often able to visit. Winston was much gratified to be made a member of the Jockey Club in October 1950.

Clementine at first greatly disapproved of this new enterprise – but

* Pink, chocolate sleeves and cap.

nothing succeeds like success – and when she saw what enormous pleasure Winston derived from his racing activities and that, moreover, they more than carried themselves financially, she became reconciled to this new pastime. But she never could work up any real interest in racing herself, and went only on 'star' occasions to the races, or on visits to the stud, in order to please Winston. She wrote to Ronnie Tree (who had become and remained a great friend since the wartime weekends at Ditchley) in May 1951: 'Have you seen about his horse Colonist II? He has won about 10 races in a year & is now entered for the Gold Cup. I do think this is a queer new facet in Winston's variegated life. Before he bought the horse (I can't think why) he had hardly been on a race-course in his life. I must say I don't find it madly amusing!'[8]

* * *

After the General Election in February 1950 it was clear to everyone that the Labour Government would not be able to hold out for long with such a fragile majority. Opposition hopes and spirits were high, and inter-party strife was marked by its keenness and acerbity. During the summer of 1951 the Conservatives forced several all-night sittings with constant divisions upon the House; Winston took his part in these marathons, and in every way demonstrated his vitality and capacity to continue to lead his party. He was now seventy-six, and his health had been comparatively good in these last years. He had made a satisfactory recovery from an operation for hernia in 1947, and the slight stroke of 1949 (which had passed un-publicized) had left no obvious effects on him. He was, however, increasingly troubled by deafness, and eventually had to resort to a hearing aid, which he found most tiresome.

Clementine's health had also been quite good and, as we have seen, she had undertaken a considerable load of work for the YWCA since 1949, on top of her other obligations. But one of the underlying causes of the bouts of exhaustion, which from time to time laid her low, was gynaecological; she had staved off the inevitable for quite a time, but in May 1951 she had to undergo a major 'repair' operation which kept her in St Mary's Hospital, Paddington, for three weeks.

She spent the greater part of her convalescence at Chartwell, and then she and I went off at the end of July to Hendaye, close to Biarritz, where we had all holidayed after the 1945 election. We spent a fortnight leading a quiet seaside life, with plenty of bathing, which did her a great deal of good. We both joined Winston and Christopher in Paris in the middle of August, and went all together to Annecy. From there our party went to Venice, returning home in early September, all prepared for the forthcoming fight. On 4 October Parliament was dissolved, and the second general election campaign within twenty months began.

Clementine held the fort in Woodford, as before, while Winston travelled to various key centres to speak. He also held several large meetings in his own constituency, but he did not undertake quite so intensive a programme as in previous campaigns. There were several new features in this election: it was, for instance, the first where television played a major role. And, in the interests of presenting a united anti-socialist front, Conservative candidates stood down in some seats to allow a free run to Liberals, while reciprocally Liberals gave way in neighbouring seats to Tory candidates. The *Daily Mirror* ran an odious campaign on the theme 'Whose finger do you want on the trigger?', implying that war was more likely under Churchill's leadership. It may have been as a result of this 'war-monger' campaign that the opinion polls showed a marked recovery in Labour support during the last few days of the election. When the results started to come in late on 25 October, early trends showed only a slight swing to the right, and it was not until the late afternoon of the twenty-sixth that a Conservative victory was certain, their final overall majority being seventeen. The King summoned Churchill to Buckingham Palace that evening and invited him once more to form a Government.

Winston addressed himself to the formation of his first and only all-Conservative and peacetime Government; he must have discussed various appointments with Clementine, including the office he proposed to offer his very able son-in law, Duncan Sandys: ever watchful, she sent him this (House-post) note:

> My Darling –
> Do not be angry with me – But first – do you not think it would be wiser to give Duncan a smaller post – Se^try^ of State for War is so very prominent – then do you think it wise to have him working immediately under your orders as Minister of Defence.* If anything were to go wrong it would be delicate & tricky – first of all having to defend your son-in-law, & later if by chance he made a mistake having to dismiss him –
> Forgive me I think only of your welfare, happiness and dignity.[9]

When the new Government was announced, Duncan's appointment was as Minister of Supply – an important post, but a junior one to that of Secretary of State for War.

As Clementine prepared to move back, once more, into No. 10 she must have felt – for Winston's sake alone – some sense of satisfaction after the bitter defeat of six years before: but of elation she felt none. Nothing that had happened had changed the conviction she held in 1945, namely, that Winston should have retired at the end of the war. Nevertheless – here they were again, and she set about reorganizing their life yet once more. Ronnie Tree had written to her after the election, and in her reply on 4 November she wrote: 'I do hope Winston will be able to help the country. It will be uphill work, but he has a willing eager heart.' By contrast, her own spirit was

* Until March 1952 WSC combined the office of Prime Minister with that of Minister of Defence (as he had done in the war).

one of dogged, somewhat weary determination, rather than of enthusiasm. Physically she was only six months from a major operation; and, as always when her morale was low, difficulties loomed large. One event, however, that caused her (and indeed everyone in our circle) great pleasure and reassurance was the return of Jock Colville to the Private Office as Joint Principal Private Secretary. After 1945 he had for two years been Private Secretary to Princess Elizabeth, and since 1949 he had returned to his diplomatic career. In 1948 he had married Lady Margaret Egerton (Meg) who became as dear and trusted a friend to Winston and Clementine as was Jock himself. During these years of Winston's last Prime Ministership, when there were many private anxieties and problems, it was a wonderful support to have someone at hand who knew both the public and the private side of it all. He understood Clementine very well, and he was a friend and confidant whose sense, loyalty and humour never failed.

Within a week of the election results, Clementine and Grace Hamblin (recalled from Chartwell to help with the 'second innings') went to see what the form was at No. 10. The Attlees, when they took over in 1945, had very sensibly arranged a self-contained flat on the second floor of the house, using the first floor State Rooms only for large parties, when the staffing and catering were all done by Government hospitality. It was into these pleasant, unpretentious apartments that Winston and Clementine moved on 19 November – marked by Clementine 'D-Day' in her engagement diary.

From the start Winston longed to move back into the whole house, but Clementine stalled – servants were difficult to get, and the kitchen arrangements for the downstairs dining-rooms were ill-devised and labour-consuming. However, as time went on, with the Coronation in the offing bringing the prospect of much entertaining, Clementine, after some prodding from Winston, made a plan with the Ministry of Works, which was submitted to him. He commented in a memorandum to Clementine, dated 'July 1952': 'I think it [the plan] is a brilliant conception and should be done in the public interest, as it is a great pity these rooms are not available for use. We really ought to have them for Coronation Year. Look at all the distinguished people I have had to entertain in our poor little attic. I am sure they are surprised at the difference between the accommodation and the menu.' Although Clementine had obviously been rather dragging her feet about moving from their 'perch' aloft, she soon was herself persuaded, and the plan was duly carried out to everyone's satisfaction.

One aspect of their translation back into prime ministerial life which presented no complications was Chequers. Still staffed by servicewomen, and now, since the retirement of Miss Lamont, under the direction of Mrs Hill (Winston's secretary for many years until 1946), Chequers was to be again both a haven and the perfect place for the dispensation of official hospitality. But Winston could not bear the thought of abandoning Chartwell: much though he appreciated Chequers, Chartwell was always home, and where he would rather be than anywhere else in the world. The house in Hyde Park Gate was let, and of course it would have been much simpler from every point of view if Chartwell could have been shut up and all their life concentrated in the two official houses. But throughout these last years of office they spent many more weekends at Chartwell than at Chequers: invariably they were at Chartwell for Easter and Whitsuntide, and for long spells during the summer holidays. Nevertheless, Chequers was the ideal place for entertaining larger numbers of people and for giving a 'rest' to Chartwell. From the family point of view Chequers will always be remembered for the four successive Christmases we all spent there. The great, rather gloomy house truly seemed to come to life and glow on those occasions, with blazing log fires, miraculous decorations made by Mrs Hill, a towering Christmas tree in the Great Hall – and above all the wonderful jumble of generations.

One of the major strains imposed on Clementine during this last lap in office was a constant, gnawing anxiety about Winston's health, which undoubtedly preyed on her own nerves. Towards the end of February 1952 he suffered a brief period of aphasia, the result of a spasm of the cerebral arteries which caused temporary confusion in his speech. This occurrence worried him very much, and of course raised for him, and for the strictly limited circle who came to know about it, the whole question of his ability to carry the burden of work required by a Prime Minister. This question was to recur during these last years of office, and it always posed the same painful conundrum: Winston felt he still had things to do for his own country and the world that only he, with his unique position in international affairs, could do – and he was not alone in this view. But he knew the risk he ran by continuing in office, and he dreaded, not dying in harness – that would have been his chosen way – but incapacity, sudden or gradual.

At his best Winston was still amazing, both in public and in private, but it was all achieved at a higher expenditure of his stock of energy, both mental and physical; and there were times now when he could not cope with the long detail of problems. Lord Moran understood him thoroughly, and Winston was indeed fortunate to have as his doctor a man who not only understood the medical considerations and risks to his patient, but also was fully aware of the implications, with regard to the office he held, of his health at any time. Lord Moran also understood the relationships between, and with, Winston's colleagues: and from where one could expect loyalty, understanding, and total discretion – and from where one could not.

For a few weeks after this minor but disquieting incident, the people (mainly his closest colleagues) who knew about it, apart from his immediate family, considered ways to lighten Winston's burden, without his giving up being Prime Minister. The consensus of their opinion was that in his unique position he could conceivably continue as Prime Minister from the House of Lords. Winston dismissed this idea as impractical; and soon, as he felt better, confidence returned, and no more was heard of the matter. Clementine, who always reacted calmly and with sense to the reality of events, understood the implications of the aphasia, but she had accepted (however reluctantly) Winston's determination to soldier on, and it was not at moments such as these that she battered him with admonitions: instead she surrounded him with tenderness and understanding.

In September 1951, just before the General Election, the King had undergone a major operation. The tour he and the Queen had planned to make to Australia and New Zealand later that autumn was cancelled; but Princess Elizabeth and the Duke of Edinburgh set off on a Commonwealth tour early in the New Year. Although the King was known to be in a fragile state of health it came as a great shock to everyone when he died in his sleep at Sandringham in the early hours of 6 February 1952. None among his subjects felt sadder than Winston, who not only cherished the memory of the close relationship which had developed between Sovereign and Prime Minister during the war years, but had been genuinely personally devoted to him. Now the accession of the young Queen Elizabeth II aroused in him every instinct of chivalry.

During the early summer of 1952 Clementine's health was not good, and on her doctors' advice towards the end of June she cancelled all her public engagements for nearly three months. In early July she went with Mary Marlborough to Italy, to do a cure at a well-known health spa at Montecatini (the Duchess was a regular visitor there and had much recommended it to Clementine). The fortnight's cure was most beneficial for her, and she flew on from there via Rome to Naples, to join Sarah and Antony, who had invited her to share a holiday with them on Capri. This Clementine enjoyed very much, and it was while she was there that she received a letter from her niece Clarissa telling her that she was engaged to be married to Anthony Eden*, and that she wanted her aunt to be the first person to hear of it. A few days later Clarissa and Anthony went down to Chartwell to tell Winston their news. Although they had known each other for some time, their intention to get married had been a well-kept secret: their engagement and marriage a short while later created a considerable stir both inside and outside our family. Winston and Clementine were delighted, and at one in wanting Clarissa to be married from No. 10, and to give the wedding breakfast there afterwards. Clementine had to bustle on her return from Italy to make the arrangements, for only a week later, on 14 August, on a glorious summer's day and amid considerable public interest, the marriage took place. Scarcely a year later, when Anthony was struck down by illness, Winston and Clementine were once more able to spread a sheltering wing, by inviting them both to stay at Chequers for nearly three weeks in quiet seclusion, so that Anthony could gather his strength after two operations, and before setting out to undergo a third in Boston.

In the New Year of 1953 Winston went to the United States to visit General Eisenhower, the new President-elect, and also Mr Truman, the outgoing President; it was planned to combine these visits with a holiday in Jamaica, where a house had been offered to him. Here was a prospect of a short breather before the crowded year ahead for both Winston and

* His first marriage had been dissolved in 1950.

Clementine; and they took Christopher and me with them. We all boarded the *Queen Mary* on New Year's Eve; apart from my parents and ourselves, the party was composed of Jock Colville, two secretaries, a detective, a valet and a maid and '100 pieces of luggage'![10] My parents both enjoyed Atlantic crossings in the 'Queens', the five days passing so agreeably and restfully, with a sense of isolation from the world.

Arrived in Jamaica, we stayed in the house, with its lovely garden and paradisal beach near Ocho Rios, which Sir Harold and Lady Mitchell had lent my parents. Apart from a visit to Kingston, where Winston was officially received by teeming, tumultuous crowds, the fortnight was private and off-duty. We bathed and sunbathed; my father painted three pictures; my mother and I made expeditions and visits to agreeable friends who had properties in the neighbourhood – Noël Coward, Edward Molyneux, the Brownlows and Max Beaverbrook. After this golden holiday we all headed for home on the *Queen Mary*, arriving at Southampton at the end of January.

Clementine was quickly into the collar again on her return, and almost her first engagement was a proud and glorious one – she launched the aircraft-carrier HMS *Hermes* at Barrow-in-Furness on 16 February. In stark contrast to this glamorous event was the tour she made shortly afterwards of areas in East Anglia that had suffered severely in the unprecedented floods which had inundated wide areas from the Humber to the North Foreland, causing several hundred deaths and rendering thousands homeless at the beginning of February. By this time, although the waters had subsided, people were struggling to make their homes habitable once more. She met many of the people who had shared in the rescue work, and visited rest centres, a clothing distribution centre and a feeding centre.

After the 1945 election the King had wished to make Winston a Knight of the Garter, one of the oldest orders of chivalry, which is in the personal gift of the Sovereign: raw, and mortified by his defeat, he had then declined the King's offer. Now, in the year of her Coronation, Queen Elizabeth II offered him once more this most special distinction, and Winston, moved and gratified by her gesture, accepted. Letters of congratulation from people in all walks of life poured in – many of them addressed to Clementine. The Princess Royal wrote charmingly: 'I am very delighted about this particularly as I know how much the King wished him to have it.'[11] A fishmonger from Blackpool wrote:

> Permit me please to add my warmest of congratulations . . .
>
> I am not able to express what we all feel towards this wonderful person.
>
> Have, however shown a little material appreciation by sending you today the best halibut that I could find.[12]

Both Winston and Clementine were invited to dine and sleep at Windsor Castle on 24 April, when the Queen invested him with the Order. He was installed as a Knight the following year at the beautiful traditional service always held in St George's Chapel, Windsor.

Even before the many functions arranged around the Coronation had begun, Winston and Clementine had been involved in a good deal of entertaining, for Marshal Tito of Yugoslavia (in March) and Dr Adenauer, Chancellor of the German Federal Republic (in May), both made official visits to London. Now Churchill's burden of work was made the heavier by having to shoulder responsibility for foreign affairs for several months, Anthony Eden having become seriously ill. The Foreign Secretary had been due to give a dinner at Lancaster House on 5 June for the Queen and all the Heads of State and Government attending the Coronation. In Anthony Eden's absence Winston and Clementine presided, and Clementine assumed a measure of responsibility for the arrangements for this very large and important party, in addition to her own considerable load of entertaining. Luncheons, dinners and receptions succeeded each other. Government business still had to be done: and Winston attended all the plenary sessions of the Commonwealth Conference which took place also at this time.

Coronation Day, 2 June, was a splendour: despite cold and wet weather, nothing could douse the enthusiasm of the vast drenched crowds. Winston and Clementine made a splendid pair – he in the Garter Robes worn over the uniform of the Lord Warden of the Cinque Ports, and Clementine looking outstandingly distinguished and beautiful in the petunia satin robe of the Order of the British Empire; a friend had lent her a tiara. They rode in a coach in the procession, with a mounted escort of Winston's first regiment, the 4th Queen's Own Hussars. At No. 10 a large party of the grandchildren, nannies and various relations spent the day watching the procession from windows in the Treasury buildings overlooking Whitehall, and the service in the Abbey on television.

But the combined strain of junketings with an overload of work and responsibility was not slow to take its toll. It happened towards the end of a large dinner at No. 10 on Tuesday, 23 June, given in honour of Signor de Gasperi, the Prime Minister of Italy, and his wife: there were about thirty-eight people to dine, including Christopher and myself. In proposing his guest's health, Winston made a singularly delightful and witty speech, but as the company was leaving the dining-room Christopher suddenly noticed that his father-in-law was having difficulty in standing up. Christopher helped him out of his chair, and with Clementine, who had seen from across the room that something was amiss, escorted him to a chair in the State drawing-room. I only realized my father's plight a short time later, as I was busy entertaining the guests. When alerted, I hastened to his side; Christopher told me to try to guard him from the people, as he was having difficulty with his speech. I did my best, but it was not very easy – my father looked unhappy and uncertain and was very incoherent. Christopher managed to convey to Signor de Gasperi that Winston was very much over-tired, and the Italian Prime Minister, with kind understanding, soon took his leave, the other guests following his example. A few had noticed the slur in Winston's speech and his unsteadiness, but attributed it to his having had a little too much to drink; nobody guessed the real reason – that he had sustained a stroke. Meanwhile Jock Colville tried, unsuccessfully, to contact Lord Moran, but finally had to leave a message asking him to come to No. 10 in the morning. We escorted my father upstairs to his bedroom, and there he seemed to feel, and to be, much better.

Lord Moran came early the next morning and diagnosed a stroke: and he called into consultation Sir Russell Brain, the distinguished neurologist. Although there were visible signs of the stroke, Winston seemed better, and, amazing as it seems, he presided at Cabinet that morning, where none of his colleagues noticed that anything was amiss – although Harold Macmillan later said that 'Churchill looked very white and spoke little.'[13] There were no guests for luncheon – only my mother, Christopher and myself – which was fortunate, as by that time Winston was extremely tired, and once more had difficulty in getting up from his chair. Clementine was quiet calmness personified, but she was of course deeply worried. Winston had intended to answer Questions in the House that afternoon, but she and Christopher persuaded him not to do so, adding their weight to the doctors' warning that he might risk breaking down completely if he did.

My mother and I had been going to the Wimbledon tennis championships, but we cancelled our visit, and we all awaited developments with anxiety. Meanwhile the closest secrecy was maintained. It was of paramount importance that no rumours circulated, one of the chief reasons being that plans had already been announced for a meeting to take place in early July in Bermuda, between President Eisenhower and the French and British Prime Ministers. There was much comment and interest throughout the world at the prospect of this top-level conference, and pending a decision to the contrary, arrangements for it went ahead. Winston was due to leave for Bermuda on HMS *Vanguard* in a week's time – on 30 June.

On Thursday, 25 June, the effects of the stroke were rather more pronounced. When I went round to see him that morning I was painfully struck by his appearance, and found him very despondent. In the circumstances, it was thought better that he went down to Chartwell, where privacy could be more easily achieved, and he duly drove there later in the morning, accompanied by Jock Colville. Fate was kind – Winston had been able to walk unaided to his car at Downing Street*, but by the time he reached Chartwell he needed considerable help getting out of the car; there, fortunately, there were no outside or curious eyes to witness this.

On the way down to Chartwell, Winston had asked Jock to tell nobody what had happened: this was a request which it would have been wrong – and indeed impossible – to obey. Colville would be in touch with Sir Alan Lascelles at Buckingham Palace; the Secretary to the Cabinet, Sir Norman Brook; and – in the absence of Anthony Eden – Rab Butler, who had taken charge of the Cabinet. He also wrote at once in manuscript to Lord Camrose, Lord Beaverbrook and Brendan Bracken: these three 'achieved the all but incredible, and in peace-time possibly unique, success of gagging Fleet Street'.[14]

* In those days Downing Street was still a public thoroughfare.

During the next two days the paralysis increased, and it became quite clear that Bermuda was out of the question. Indeed on Friday, 26 June, Lord Moran told Jock Colville that he thought it unlikely that Winston would live through the weekend.[15] On Saturday afternoon the following announcement was made:

> The Prime Minister has had no respite for a long time from his very arduous duties and is in need of complete rest. We have therefore advised him to abandon his journey to Bermuda and to lighten his burdens for at least a month.
>
> (Signed) Moran
> Russell Brain[16]

The quiet, private agony of those succeeding days had faded even from my memory. Reading my diaries, written in the anxiety and sadness of the moment, has revived it:

> Saturday, 27th June 1953
>
> Saw Papa – felt wretchedly gloomy. There are nurses now, and he cannot walk, or use his right hand much. In the afternoon he had a fall – but beyond the jolt – no damage.

> Sunday, 28th June 1953
>
> Today he is gayer . . . Lord M. says there is a distinct improvement. It's so difficult to tell people when they ring up – Because when he's down – we're down – And when he's cheerful our spirits, too, revive . . . Mama is truly marvellous – tender, considerate, thinking of everyone's comfort. Unblinded by hope or fear – she teaches us all.

During these days Winston's closest family and friends, and some colleagues, came in succession to visit him: Max Beaverbrook; Brendan Bracken; 'The Prof'; Lord Camrose; Sir Norman Brook; Harold Macmillan; 'Monty'; the Salisburys. On Saturday, 27 June, Randolph and June came to luncheon, and a day or two later he wrote to his mother: 'I do want you to know how much I am thinking of you at this sad and difficult time. I thought you were magnificent on Saturday & doing everything possible to maintain Papa's morale. So long as that persists no miracle is impossible.'[17]

Sarah flew back from America – she told the press that her current television series was over, which was quite true. When she and Antony came to Chartwell on Monday, 29 June, Winston had had a good day, and Lord Moran had found him distinctly improved, but I wrote in my diary: 'She [Sarah] was distressed. It made me realise how used to it we have become, and how low we were – for today we feel almost gay – he seems so improved. But of course it is only comparative.'

The press and the public seemed to accept the explanation for the 'need to rest'; even Sarah's arrival, and the cancelling of a Chartwell 'open day', which was to have been on 1 July, were received without much comment.

Gradually Winston's condition improved – but there were bad days and good days. Statements were issued to the press; it was known that he was seeing quite a number of people, and reading papers, and generally keeping in touch with affairs. But, despite the silence from Fleet Street, inevitably there were rumours and speculation, and reports circulated in the foreign press that Winston was completely paralysed. This particular lie would be laid when he was photographed on the doorstep at Chartwell on 24 July before leaving to drive to Chequers, where he stayed for the next week or two, to give the staff at Chartwell a much-needed rest. Nearly a month later the *Daily Mirror* gave great prominence to a report in the American press that Winston had had a stroke, and that although he had made an astonishing recovery, it was thought unlikely that he would be able to resume active leadership.

A most welcome visitor at Chequers on 6 August was Lady Violet Bonham Carter: she had a talk with Clementine before she saw Winston, and in her diary she recorded: 'Clemmie said that she felt sure he ought to retire in the autumn & begged me not to urge him to stay on if he asked my advice . . . His mind was quite clear but tired more easily. He shld go now rather than wait & peter out (these were not her words, only their sense). She was wonderfully brave, sensible & unemotional.'[18] Lady Violet noticed Winston was 'walking with a stick but not too badly, his face quite normal – pink & no distortion . . . He said "I'm better – but you know I've been quite paralysed. I cldn't walk. I had no pain – it was like a heavy weight all down one side. Now I can move my fingers – & my toes even." '[19] At luncheon he spoke 'thickly sometimes', and she perceived 'a very slight crookedness in his mouth'.

During the meal and the afternoon the two old friends ranged over many people and (mostly political) subjects: Lady Violet remarked that at moments 'he became suddenly & unreasonably angry – like a violent child'. After luncheon they sat in the garden until it was time, at four o'clock, for his massage. On the way out, Winston took her to see 'his fish – 4 large tanks of them – illuminated – which he takes with him wherever he goes – to & fro from Chartwell where he returns next week . . . Again I realised his eternal childhood. His passion for his toys.'[20]

After her visit, Violet wrote to Clementine:

> I must tell you darling, what intense admiration I felt for your courage, wisdom & dispassionate judgment, when your emotions must have been so deeply involved –
>
> I cannot imagine anything more difficult & agonising than these last weeks must have been for you – with all their conflicts of hope, fear, anxiety & decisions for the future.[21]

Various milestones marked Winston's progress: on 2 August he visited the Queen at Royal Lodge, Windsor; and on 18 August – eight weeks after his stroke – he presided at his first Cabinet meeting since his illness: it went very well, and he grappled vigorously with a long agenda, but it tired him very much, and the next day, on his return to Chartwell, he looked dazed and grey.

From the very earliest days of his illness, the great and anguishing question, 'Can he go on?', hung over us all. He brooded on it all the time. Nearly all of us who saw him constantly could not believe that his recovery would be complete enough for him to wrestle again with public life: we were a loving but defeatist band, I fear.

In the early days of his affliction, Winston's goal was to hang on until the autumn, when Anthony Eden would be sufficiently recovered to take over the succession. Then there was a further horizon – the Queen was due to go on a prolonged tour of the Commonwealth, leaving England towards the end of November, and returning in May 1954; he would hold the fort till then. But as his strength increased, the 'holding on' theme was more and more superseded by the idea of 'carrying on'. During these long weeks Winston suffered untold frustration; he felt power had been snatched from him just when he might have made moves of vital and lasting importance in the field of international relationships. Stalin's death in early March 1953 had led to changes in the chemistry of Soviet policy. Churchill had warned the world about the 'Iron Curtain': now he felt his last task was to assist the 'thaw' in international relations. He felt quite simply, and without conceit, that he could play a unique part in this process.

These were agonizing weeks for Clementine. She saw, all too clearly, that Winston meant to battle on – she did not doubt the need for his presence in the counsels of the world, or that he had a unique gift to offer – she simply could not bear to contemplate the prospect of the slow diminution of his powers, or the possibility of a public breakdown. She bore up wonderfully well throughout this time, although she herself had a fall and cracked some ribs. She scarcely left Winston for more than a few hours at a time, and she gave him unwavering love, comfort and companionship: but he knew her views without asking – as she knew the depths of the dilemma in which Winston found himself.

If Churchill was to return to his job as Prime Minister, various deadlines had to be met. The first one had been the Cabinet meeting on 18 August, where all had gone well; but the real test would be the Conservative Party Conference in early October at Margate, when he would have to speak for at least forty minutes to a large meeting, and endure the keen scrutiny of the press. Meanwhile, he started to lead a more normal life: he diverted himself with a little croquet; he saw more people, and read an increasing number of official papers. While Christopher and I were away in Scotland, my mother wrote from Chequers on 5 September about his progress:

> I am sad about Papa; because in spite of the brave show he makes, he gets very easily tired & then he gets depressed – He does too much work & has not yet learnt how & when to stop. It just tails off drearily & he won't go to bed. He is making progress, but now it is imperceptible. If no setback occurs the

improvement can continue for 2 years . . . I expect you have seen from the newspapers the tremendous 'va-et-vient' of ministers. Papa enjoys it very much. Incidentally they are even more tired than he is by the sitting over the dinner table till after midnight! . . .

A month before, the Queen had invited Winston and Clementine to be her guests at Doncaster Races for the St Leger, and then to travel with her and Prince Philip on the Royal Train to Scotland and to spend the weekend at Balmoral. Winston had accepted this delightful invitation; but, as the date drew near, his doctors and Clementine thought the expedition would be too tiring – and indeed might jeopardize his recovery. Winston (speaking on the telephone from Chequers) and Clementine (from No. 10) had had 'words' on the subject! Later that same evening Winston telephoned her again to apologize. The next day – 3 September – she wrote to him:

My darling Winston,

It was sweet of you to ring me up last night & to say loving & forgiving words to me –

I would like to persuade you to give up Doncaster & Balmoral.

First Doncaster. You will be watched by loving but anxious & curious crowds – It would be rather an effort to keep up steady walking – It may be a longish way to the Paddock & there will be much standing about. Altho' you sit in the Queen's Presence in intimate Court Circles – If you sat in public when she was standing it would be noticed.

Then Balmoral –

You are improving steadily though slowly, but I fear you are not up to a night in the train and so on yet. And you don't want to have a set back before the Margate Speech; but rather you must husband your strength for that important event, & for Parliament.

I will be with you this afternoon –

Your devoted
Clemmie

For the record: Winston and Clementine did go to the races! The day of the St Leger, Saturday, 12 September, was their forty-fifth wedding anniversary, and everyone from the Queen and her family downwards made much of them – and Winston had a wonderful welcome of his own from the crowd. The two days spent at Balmoral were most agreeable, and when they returned home Winston was – although a little tired – refreshed and encouraged.

Max Beaverbrook had offered the loan of La Capponcina at Cap d'Ail, and Winston, Christopher and I flew out there on 17 September. Clementine did not come with us: Winston was now so much better she felt she could entrust him to us, and she and Nellie went off for a few days sightseeing and theatre-going in Stratford. From La Capponcina I reported truthfully, if rather sadly:

> Papa is in good health – but alas, low spirits – which Chimp and I are unable to remedy. He feels his energy and stamina to be on an ebb tide – He is struggling to make up his mind what to do. I'm sure you know the form – you have been witnessing it all these months . . .
>
> He thinks much of you & wonders what you are doing . . .
>
> Chimp & I are having a lovely time – bathing – reading – cards & we love being with Papa – Only we yearn to be able to do more than be the mere witnesses (however loving) of his sadness.[22]

Winston wrote several long letters to Clementine, all in his own hand – the writing is surprisingly clear; it was evident how much he missed her, and how unsettled he was still in his own mind about the future. On 21 September he wrote:

> My darling One,
>
> The days pass quickly & quietly. I have hardly been outside the garden, & so far have not had the energy to paint in the sunlight hours . . . I do not think I have made much progress, tho as usual I eat, drink and sleep well. I think a great deal about you & feel how much I love you. The kittens [Christopher and myself] are vy kind to me, but evidently they do not think much of my prospects. I have done the daily work and kept check of the gloomy tangle of the world, and I have dictated about 2000 words of a possible speech for Margate in order to try & see how I can let it off when it is finished to a select audience. I still ponder on the future and don't want to decide unless I am convinced.

This letter, which went on to describe in a quite lively way thc various guests he had entertained at the Villa, ended: 'Ever your loving & as yet unconquered W. [with a very debonair pig]'.

Clementine, who was at Chequers, replied on the 24 September:

> My darling Winston,
>
> . . . Your letter reached me only this morning . . . I wish my dearest that you did not feel so sad and melancholy – I feel at our age it takes a little time to become acclimatized to the soft relaxing air of the Riviera – It would probably be good if you could be there a month – But that's not possible –
>
> It will be lovely to welcome you back & I am making all arrangements for us to spend the week-end . . . at Chartwell; as I know that is what you would like . . .
>
> All my Love Darling
Your devoted
Clemmie

When Christopher and I arrived home a few days ahead of my father, we brought my mother another letter from him (dated 25 September), in which he wrote that he had got going with his painting and certainly felt

> the necessary vigour & strength to be as bad as I used to be . . .
>
> I do hope my darling, you have found the interlude restful and pleasant. I must admit I have had a good many brown hours. However the moment of action will soon come now . . .
>
> I wish you were here for I can't help feeling lonely.

Clementine was deeply aware of how difficult the decison – to go or to stay – was for Winston: at the end of a long letter on 26 September she wrote: 'I think much about you Darling & your problem, so difficult to resolve. I have faith that you will know best.'

Winston returned home at the end of the month, and on 10 October he addressed the Conservative Party Conference. It was a great ordeal, and he knew – and Clementine knew – just how much depended upon his performance, which would be judged and commented upon, not only by the large audience in the hall, but by a host of unseen witnesses and interested commentators. We were all tensed up for this formidable test. It was an unqualified success: *The Times* wrote of a 'triumphant return to public life'; doctors, closest colleagues, loving relations – all of us had been confounded. As we looked back over the harassments and anxieties of the past months, it seemed miraculous.

There was no more talk, for the present, of resignation or retirement, as Winston once more took up the normal load of his work as Prime Minister. But Clementine, deeply thankful though she was at the return of his health and strength, knew that he was working, if not living, on borrowed time, and that at any moment he might be struck down again. She had had an exhausting summer, and in response to a warm invitation, she went to Paris for a week in October to stay with Lord and Lady Ismay (he being, at that time, Secretary-General to NATO): she greatly enjoyed this short 'breather'. Winston wrote to her three times during this short absence – in his 'own paw'. There was a new kitten at No. 10 (called Jock, after its donor) which was 'behaving admirably & with its customary punctilio! Rufus [the poodle] is becoming gradually reconciled'.[23] He reported quiet on the domestic front (which must have been a great relief to Clementine). There was also the pleasant and gratifying news that Winston had been designated as that year's winner of the Nobel Prize for Literature. But despite public and private preoccupations, his loneliness without Clementine emerges in every letter: 'I was lonely last night but Pitblado [Joint Principal Private Secretary] dined. I am reading "The Dynasts" [by Thomas Hardy] & getting into it.'[24] And three days later, just before leaving London for a brief visit to Chartwell, he wrote: 'I am writing in the Cabinet room, & the little cat is holding the notepaper down for me. I miss you vy much. One night I had dinner in bed as I did not want anyone but you for company. I do hope you are enjoying yourself and finding the days interesting.'[25]

The postponed Bermuda Conference was rearranged to take place in the first week in December. Lord Moran accompanied Winston, and Christopher was also of the party in his official capacity now as the Prime Minister's

Parliamentary Private Secretary*. The new date for the Conference unfortunately coincided with the date of the presentation of the Nobel Prizes, which takes place every year in Stockholm in early December, and King Gustav Adolf and Queen Louise had invited Winston and Clementine to be their guests at the royal palace for the occasion. When a Nobel Prize winner is unable to receive the prize personally, it is customary for the Ambassador of the recipient's country to receive it on his or her behalf; it was therefore unusual, and a signal honour, that Clementine was invited to accept the Prize for Literature on his behalf. I also was invited as a companion to my mother. And so it was that we spent four unforgettable days in Stockholm.

The social occasions and festivities which form the annual programme surrounding the actual ceremony of presentation of these world-famous prizes are most beautifully and imaginatively organized, and the press and population of Stockholm take a keen interest in all the events and personalities involved. Although the Swedes were disappointed that Winston could not himself come to Stockholm, the reason for his absence was completely understood, and Clementine arrived on the evening of Tuesday, 8 December, to be received as his most welcome representative. But if she came to Stockholm as a substitute, she was quickly accepted and acclaimed in her own right. Her first engagement was a party for the press given at the British Embassy by the Ambassador, Mr (later Sir) Roger Stevens. Here she made a great hit: the combination of her natural manner, her friendliness and her looks (which stunned them) made a good start to the visit. The Swedish press thereafter paid detailed attention to all she said, did and wore. It was delightful staying at the Palace, as nothing could have exceeded the King's and Queen's hospitable kindness: King Gustav Adolf and Clementine got on extremely well together; the King was a great savant and very interested in a number of subjects. Queen Louise was also most charming, and there was an immediate bond in that she was a sister of 'Dickie' Mountbatten.

The presentation of the Nobel Prizes took place in the great concert hall in Stockholm, and afterwards, at a banquet for over 900 people, Clementine read Winston's speech of thanks for this exceptional honour. After the banquet there was a ball, and during the evening nearly a thousand students entered the hall, wearing their white caps and carrying banners. The student choir sang folk-songs, and then, as the dancing was in progress, suddenly the band broke into the time-honoured strains of 'My Darling Clementine', the words being taken up by the students: Clementine was quite overcome by the charming tribute. When my mother and I arrived back in England on Saturday, 12 December – a day after the return of the Bermuda party – there was a great reunion; we had a golden account to give of those last few days, and Winston was thrilled and gratified by the wonderful reception his 'darling Clementine' had received.

* Christopher Soames had entered politics in 1950 as Conservative MP for Bedford, a seat he was to hold until 1966.

But for all the high days and holidays, and for all the perks and privileges that go with high official position and general public esteem, Clementine was finding the going hard and heavy. Her heart had never been in this second term of office, and although she would always put on a good show in public, her morale was often low in private. She found the constant entertaining during the week in London, often followed by weekend parties at Chequers, a great strain. Christopher and I had a room at No. 10, and nearly always went to Chequers if there were official guests, in order to help, and so I was often the distressed witness of my mother's moments of weariness and depression. But although her spirits were often desperately low, I never ceased to be amazed at the way in which she could snap out of the doldrums in order to do whatever had to be done. I remember just before one large evening reception at No. 10 her being in an extremely agitated state, to such an extent that I begged her to allow us to say she was not well, and for her to go to bed: she agreed, and I went to tell my father. The first guests were just about to arrive when suddenly my mother appeared, looking beautiful, not a hair out of place, dripping with diamonds and decorations – and, taking command of the whole situation, carried on through the evening without one betraying sign that half an hour before the party she had been in a state of nervous, and near-hysterical, collapse.

When she was driving herself hard, Clementine became difficult to live with and to work for. Towards the secretaries and the servants, to whom she was normally polite and considerate (and most understanding of any personal problems), she would become, when under strain herself, super-perfectionist and cuttingly cold. To her family she would be generally 'tricky' and demanding. Nervously restless, she would fuss endlessly over what seemed to others to be trifling matters; above all, when she was 'down' the present held no enjoyment, and the future loomed with problems great and small. It was at these times also that she would be most difficult and sometimes quite aggressive with Winston – producing objections to any proposals he made, and making it quite clear to him that everything was a great burden.

It is idle to pretend that Winston himself was not, at times, maddening. He could, on occasion, behave like a spoilt and naughty child – though a loyal and affectionate staff often contrived for these usually brief outbursts to pass unnoticed. Most of Winston's colleagues were indulgent towards him, but Clementine used to become enraged if such incidents happened in front of other people, when she would feel mortified both for him and for herself. She would (usually) contain her feelings at the time, but they would burst forth as soon as they were in private: she was absolutely unafraid of him, and she used to berate him thoroughly. But for all that he sometimes deserved her scoldings, those closest to my parents felt there were periods now when she really harried him too much – without regard to the load he himself was bearing, and finding heavier all the time.

Strangely enough, although there were sometimes explosive scenes, Winston endured a good deal with remarkable patience. He knew that he had imposed this extra mileage on her against both her own wishes and her judgement of what was right for him, and he saw how exhausting she was finding her life. Clementine herself was often conscious that she had been too abrasive, and she suffered remorse and unhappiness when she felt she had been unkind or unreasonable; both were always eager to make up any quarrel. This touching 'olive twig' survives among Winston's papers; it is not dated, but belongs to the 1954 period – it was probably pushed under Clementine's door:

> Darling
>
> Fondest love. I am so sorry I was awkward at dinner.
>
> My heart was full of nothing but love, but my thoughts were wayward.
>
> Your ever devoted
>
> W

Looking back, it is not difficult to trace the causes of Clementine's change of demeanour and attitude during this last period at Downing Street. The most continuing reason was to be found in her own state of health and strength; it must be remembered that when Winston returned to office at the end of October 1951, she had only just finished her convalescence after a major operation. In obedience to the instructions of her doctors, who were adamant that she must go right away from the strains and worries of home life, Clementine had been for a fortnight to Montecatini in early July 1952, and later she had joined Sarah and Antony in Capri. Later on, that September, she and Winston went to La Capponcina at Cap d'Ail, lent to them by Max Beaverbrook. Before their visit, on 17 August 1952, Clementine wrote to Max:

> It is indeed kind and thoughtful of you to have lent us your lovely home & garden. We look forward to the sunshine, the peace & the rest.
>
> I am much better, but my doctor was vexed that I came home after a month as he wanted me to stay longer than that . . . Now I can resume my 'rest cure' in wonderful surroundings. I hope Winston will paint. He says, 'There are a 1000 pictures to be painted from Max's garden!'

The autumn found her much restored, and she started taking up public engagements again, but, on her doctors' advice, she now limited her commitments. For the early part of 1953 Clementine's health was fairly good, but the extra strain of Coronation entertaining and then Winston's grave illness had put her under further pressure, and during the summer she started having trouble in her right arm and shoulder from what was eventually diagnosed as neuritis: this painful and extremely wearing affliction was to dog her for a long time. During March and April 1954 the condition became acute, and she was in severe pain; several doctors and treatments were tried, and she had to wear a surgical collar for some weeks. Apart from its being disagreeable to wear, she hated being seen in it, which made her shun company which might otherwise have cheered her. A fellow sufferer was Max Beaverbrook, and they had a lively correspondence on symptoms, doctors and remedies. He sent her some flowers one day with a little note saying: 'From one victim of neuritis who deserves it, to another victim who does not deserve it. Max.'[26] He could be irresistible.

In a bid to conquer the neuritis, in May Clementine took a three-week cure at Aix-les-Bains, Grace Hamblin (who had rheumatic trouble) going with her as a companion and fellow patient. The weather was very cold, and the cure rigorous. During her absence much was afoot on the political scene, and Winston kept her regularly informed. In particular a tremendous row blew up in the House of Commons about a motion to raise the salaries of Members of Parliament: Sir John Mellor, a Conservative, resigned the Government Whip on 2 June in protest against these proposals. On the Chartwell front one of the black swans had flown off, to Winston's great concern.* On 8 June Clementine, who had been following home news in the foreign press, wrote to him:

> My Darling,
>
> The Monde is exercised about your relations with the Tory Party – It recounts in detail the pros & cons of the increased payment of Members of Parliament – And it feels that you are above all these petty considerations & are meditating on the great problems which affect humanity as a whole. It concludes the article by saying that in one week you have had the bad luck to lose Sir John Mellor and a black swan; & that you are much affected by the loss of the young bird. (I thought this would make you laugh.)
>
> Cure over to-day. Hooray! Must now recover from Cure which has been quite something! Longing to see you Saturday.
>
> Tender Love
> Clemmie

The treatment at Aix did her some good, but the trouble still recurred from time to time. One way and another, Clementine was tussling with indifferent health throughout Winston's last tenure of office and, in addition, the plain fact was that she was getting older – and feeling it.

During these years, too, there were causes of strain for Clementine apart from those relating to her health. In 1953 Diana suffered a severe nervous breakdown, which continued to affect her over a considerable period of time. Both Winston and Clementine grieved deeply for her, seeing with dismay the great unhappiness this illness caused Diana, and the difficulties and anguish it brought into her life. As is often the case in such illnesses, those who are nearest and dearest are sometimes the least able to help. The relationship between Clementine and Diana had for many years been a fragile construction, and now there were moments when it came near breaking-point; yet also, there were times when Diana was most tender and

*The swan was eventually located in the Netherlands.

loving towards her mother, and the rift of incompatibility and mutual misunderstanding was bridged. But Clementine suffered not only the temporary pain from harsh verbal wounds; she must also have gone through the heart-searchings that all parents feel as they see a child in the grip of a bewildering illness, the root causes of which may well go back to adolescence and childhood. Although Winston found psychological troubles and their explanation beyond his ken, yet he had memories of his 'Black Dog', and he nearly always had a calming and soothing influence on Diana in her moments of distress; and she found in him tenderness and rock-like stability.

For the greater part of these years Sarah was in America, where she had plenty of work both in the theatre and on television. Clementine missed her very much, for she and Sarah were close to each other. Diana and Sarah were always devoted friends, and Sarah was often able to be a helpful and mediating influence whenever she happened to be in England. Sadly for her, from about 1953 her own marriage began to run into serious difficulties, and for a good deal of the time while she was working in America, Antony was in London. After periods of temporary separation, they finally parted in 1955. Although Clementine had never really liked Antony, and had had forebodings regarding the marriage from the beginning, she naturally grieved deeply for this further unhappiness in Sarah's life. Sarah's own letters home are full of references to long and understanding letters from her mother, which were a lifeline of comfort and support to her through the professionally hectic and emotionally arid months during which she was touring in America and wrestling with painful personal problems.

Finally, there was a sad personal loss for Clementine to bear: in July 1954 her sister Nellie was diagnosed as suffering from inoperable cancer. In August and September she was at intervals in St George's Hospital having massive doses of ray treatment, which made her feel dreadfully ill and weak. Clementine had her driven down to Chartwell at the weekends, and cherished her tenderly. Throughout that autumn and winter Nellie was constantly either at Chequers or Chartwell whenever she felt strong enough; but although she made a brave appearance at Winston's birthday party in November, at Christmas she was not up to joining the big family gathering at Chequers. In the New Year it was clear she had not much longer to live, and Clementine made it possible for her to be nursed at home. She died aged sixty-six on 1 February 1955.

The two sisters had totally different temperaments and characters, and the circumstances and course of their lives were in sharp contrast. Nellie's life from the time of her marriage to Bertram Romilly was full of difficulties: he always suffered from fragile health as the result of the head wound he had sustained in the First World War, and they were very badly off financially. Nellie loved her two sons, Giles and Esmond ('the Lambs'), with a blind devotion, standing loyally by them through a series of adolescent peccadilloes and their espousal of communist political views. Bertram died in 1940, and throughout the war she had had a sad and lonely time, with

Giles a prisoner-of-war in grim Colditz, and Esmond, a navigating officer in the RAF, lost over the North Sea in November 1941. But despite all her trials and tribulations she remained gay, gallant and undaunted. Of the two sisters, Nellie had a much happier, easier-going temperament than Clementine, and with her winning charm and gregarious nature she made a host of friends in all walks of life. The two sisters were devoted to each other, but from time to time there would be mild tiffs, and Nellie used sometimes to earn a good scolding from Clementine for her perennial propensity for gambling; but there was never a real rift. When she died, the very great number of letters which Clementine received were not only a true tribute to Nellie, but also showed that people realized the gap which would now be left in Clementine's life.

* * *

In August 1954 Clementine had been for a fortnight's holiday to France with Rhoda Birley*, based at Ste Maxime-sur-Mer, Beauvallon, on the Riviera. Winston was ensconced at Chartwell with a cold. On 4 August, the day after their arrival, Clementine telegraphed: 'How is the cold Darling? Don't look at the new moon through glass. Love. Clemmie' – receiving the reply (5 August): 'Cold better. No moon visible. Hope all well. Love. Winston.'

Clementine and Rhoda had a very nice time, and after bathing and beach life in the morning, she told Winston (9 August), 'Rhoda & I are hunting (in our rattle-trap little car) for a villa for you for next year!' Rhoda was a friend of Jacques Cousteau, the famous underwater explorer and commander of the oceanographic research vessel *Calypso*, and Clementine's letters were full of their plans to go and visit him at Toulon: 'I can see Rhoda means to dive too. I don't think I shall – You wear a lung on your back & can go down 300 feet without a line and stay 2 hours,' she wrote on 9 August. This news much perturbed Winston, who telegraphed on 11 August: 'Your letter of 9th arrived. Please no diving. Love. W.' Clementine tried to reassure him in a letter on 13 August: '[W]e shall not go deeper than 10 meters [*sic*] as it is very difficult to swim down deep without being weighted . . . We shall see octopuses, but I shall not encourage any familiarity.' Unfortunately, the correspondence does not recount any details of their visit to Cousteau and the *Calypso*. But Clementine returned home from this holiday in good spirits and better health.

* * *

The highlight in Winston and Clementine's personal lives in 1954 was his eightieth birthday, interest in which began to be evinced several months

* Widow of the portrait painter Sir Oswald Birley (d. 1952). Both were great friends of Winston and Clementine, and Rhoda herself a gifted painter.

beforehand. Private people and public organizations alike conspired to make it perhaps one of the most widely celebrated birthdays on record.* Both Houses of Parliament decided to make Winston a joint and all-party presentation on this occasion, in the form of a portrait of himself by Graham Sutherland. The House of Commons was to give him a handsomely bound book recording the gift, and containing the signatures of those who had joined to make this generous and imaginative gesture (in which only a very few had declined to participate). Sittings for the portrait started in mid-August at Chartwell, and both Winston and Clementine took at once both to Graham Sutherland and his charming wife. On 1 September my mother wrote to me (Christopher and I were in Scotland): 'Mr Graham Sutherland is a "Wow". He really is a most attractive man & one can hardly believe that the savage cruel designs which he exhibits, come from his brush. Papa has given him 3 sittings & no one has seen the beginnings of the portrait except Papa & he is much struck by the power of his drawing.' There were several more sessions at Chartwell, and in mid-November the Sutherlands spent a weekend at Chequers for the final sitting.

On 13 November, Winston visited Harrow (for the fifteenth year in succession) to take part in 'Songs'; a new verse had been added in his honour to 'Forty Years On', and the boys presented him with a cigar cabinet.

During the weeks immediately preceding and following his actual birthday, individuals and organizations continued to shower him with wonderful presents. Thousands of people both in Britain and overseas sent contributions to a 'Winston Churchill Eightieth Birthday Presentation Fund', and on the day itself Lord Moynihan, on behalf of the Fund, presented Winston with an 'interim' cheque for £150,000. When the Fund was closed a year later the total, subscribed by over 300,000 people, came to £259,000. Winston made this amazing present into a Trust Fund from which he was able during the ensuing years to make handsome contributions to various causes, and in May 1958 he gave £25,000 from this Trust towards the founding of Churchill College, Cambridge.

When, on his actual birthday, Tuesday, 30 November, Winston went to Buckingham Palace for his usual Tuesday evening audience with the Queen, she gave him four beautiful silver wine coasters, engraved with the ciphers of herself and the Duke of Edinburgh and the other members of her family who had joined to give this handsome present. During the day presents and greetings from all over the world poured into Downing Street – the telegrams and cards numbered over 23,000; it seemed that the love and regard in which Winston was held by all sorts and conditions of men and women had welled up and spilled over. Both Winston and Clementine were deeply moved, and quite taken aback by these demonstrations.

On the morning of 30 November the moving and, in parliamentary history, unique ceremony took place, when both Houses of Parliament assembled in Westminster Hall to salute Winston Churchill, and to give

* At that time the 80th and 100th birthdays of Queen Elizabeth the Queen Mother lay many years ahead.

him their presents – jointly the Sutherland portrait and (from the House of Commons) the commemorative book. The State Opening of Parliament by the Queen had taken place earlier in the morning, and afterwards members of both Houses and their spouses proceeded to Westminster Hall for the presentation and addresses. The speeches, made by the Speaker; Lord Salisbury; the 'Father of the House', Mr David Grenfell (who presented the book on behalf of the Commons); and Mr Attlee, were all felicitous and shot through with shafts of humour in true parliamentary vein, transcending all barriers of political partisanship. Winston, who made his speech after the portrait had been unveiled, remarked of it with wryness: 'The portrait is a remarkable example of modern art. It certainly combines force and candour.'

But the picture itself had already become for him a source of deeply wounded feelings. Although he had seen some of the early sketches, he did not see the work again until it was finished, about a fortnight before his birthday – and he then took an instant loathing to it. He felt he had been betrayed by the artist, whom he had liked, and with whom he had felt at ease, and he found in the portrait causes for mortal affront. Clementine, who at first sight had thought it remarkable, also came to be repelled by the work – she thought it was a cruel and gross travesty of Winston, showing all the ravages of time and revealing nothing of the warmth and humanity of his nature. They were both deeply disturbed by it, but when she realized the violence of Winston's reaction – to the extent that he even considered refusing to accept the picture – Clementine threw her influence into calming his wounded sensibilities, and tried to make him concentrate on the feelings which the commissioning of this portrait represented, to which indeed he was very much alive, and to which he paid warm and heartfelt tribute in his speech. But all this made a great and emotional upset behind the scenes in the days prior to the presentation.

All shadows, however, were banished on the day, by the warmth of the ceremony, by the tributes from political friends and foes alike, and by the acclamation of the great assembly as, at the end of the proceedings, Winston and Clementine (who was looking beautiful and radiant) walked slowly down the centre aisle. The thunderous applause seemed to envelop them both as they passed out of the great historic hall into the November sunshine, where more cheers awaited them from the crowds in Parliament Square.

Our own family celebration had to wait until the evening of the following day, when Clementine gave an after-dinner party of about 170 people for all the family, and for close friends and colleagues. It was at once cosy and splendid; with champagne, and hugs all round, and some sentimental tears. Round the intricate pink and white edifice of the birthday cake was a banner, upon which were inscribed the Shakespearian words:

> Take him all in all he is a man.
> We shall not look upon his like again.

A postscript to these wonderful days is provided in my diary for the ensuing weekend, which my parents spent at Chartwell:

> Mama has collapsed with fatigue and a streaming cold; and we all feel flat and stale and tired – WSC is fresh as a daisy, and enjoying mulling over his presents and letters![27]

CHAPTER TWENTY-SEVEN

Does the Road Wind Up-hill All the Way?

FROM THE MOMENT OF WINSTON'S TAKING OFFICE ONCE MORE AS PRIME Minister in the autumn of 1951, there lurked in the background of many people's minds the question: 'How long can he last?' In those early days he had told his own intimate circle that he thought he would stay in office only for a year, but by the end of that year imminent resignation was far from his thoughts. During his serious illness in the summer of 1953 the idea of his retirement from office was discussed – and in some quarters advocated. But then there had been his vigorous return to public life, dating from his speech at the Conservative Party Conference in October.

From then on, however, until his resignation some nineteen months later, there not unnaturally continued to be speculation from time to time, either openly in the House of Commons and in the press, or privately in social and political circles. Winston took all this in his stride, and Clementine realized that as long as he continued in office there was bound to be comment and criticism, which neither surprised nor upset her. But she felt only contempt and disgust (as did many people) for the snide comments on the level of an exceptionally unchivalrous cartoon in *Punch* (3 February 1954), where the effects of a stroke on Winston's face were clearly indicated. Although she was ever of the same mind about his 'second innings', which was certainly known by those in their immediate circle, Clementine was intensely loyal, and gave no encouragement to the voicing of similar views by others.

Winston knew that, despite his amazing recovery, the stroke he had sustained had left its effects. In a letter to Clementine (who was at Aix-les-Bains) on 25 May 1954, he wrote in reference to an important speech he had shortly to make: 'This is a toil which lies ahead of me and I do not conceal from you that original composition is a greater burden than it used to be, while I dislike having my speeches made for me by others as much as I ever did.' But periods of worrying about his capacities and feeling a lack of confidence alternated with a stubborn resolve to continue in office. When talking about the question of his possible retirement, Winston was wont to remark cheerfully that he meant to carry on until 'either things become much better, or I become much worse'.

That Clementine was able to take a realistic and unemotional attitude when dealing with the difficult question of Winston's resignation was demonstrated in an incident that took place in the middle of July 1954, shortly after Winston had returned from a visit to the United States and Canada. There was unrest in the Cabinet, which demonstrated itself in various ways, and came to involve Clementine when Harold Macmillan (then Minister of Housing) asked to see her privately at No. 10. He told her that he came on behalf of himself and several close colleagues in the Government, to say that they thought Winston should retire in favour of Anthony Eden. Clementine listened attentively, and said she would of course convey this information to Winston. Immediately Mr Macmillan left her, Clementine sent for Jock Colville and told him what had happened. This was during the course of the morning, and she and Winston were to lunch alone that day: Clementine begged Jock to join them, as she felt his presence would be helpful. Jock noticed that Clementine seemed very ill at ease at the beginning of luncheon, which was unusual, as she was never afraid of Winston and could normally broach thorny questions without demur. But she must have been fully aware of the importance of Harold Macmillan's action, knowing that he would never have taken such a step but for real and valid reasons, and that the existence of such feelings in the heart of the Cabinet was, of course, of utmost significance.

Presently however, during the course of luncheon, she took the plunge, and told Winston what had happened. She did so admirably – with calm detachment – merely giving the facts of the interview, and refraining from any comment. Winston took it rather well, and with complete calm; he merely asked Jock if he would go and telephone Mr Macmillan's office to say that the Prime Minister would be grateful if the Minister of Housing would 'step round' in the course of the afternoon.

During the ensuing interview, it appears that Winston was most urbane, but told Harold Macmillan that he thought he would soldier on, as there were several matters on hand that he would like to 'see through' – but of course, if the Tory Party wanted to get rid of him, they were at perfect liberty to do so.[1] Winston knew very well not only that he commanded the loyalty of the Tory Party as a whole, but also that his prestige transcended party barriers, and that neither his own party, nor the country as a whole, would look kindly upon a 'palace revolt' of discontented, impatient colleagues.

One of the reasons for the feelings of frustration and irritation (despite their undoubted admiration and affection for him) among his colleagues was the fact that, during the last twelve months before he finally resigned, Winston was continually changing his mind about whether or when he would go. His own health and energy, combined with events, made it a difficult as well as, for him personally, a painful decision. Churchill had a genuine and deep desire to see accomplished, or at least set on their way, a few major objectives. First among these was the rebuilding of that close and special relationship between the British Prime Minister and the President of the United States which had been so vital in the war, and which he thought held as great a significance now, in this time of uneasy peace.

Churchill had been the first to point to the silent, sinister encroachment of Russian power over Eastern and Central Europe; but after Stalin's death in 1953 he looked for any opportunity to make a fruitful contact with the new regime, despite the unhelpful orientation of American policy at that time.

Finally, when in February 1954 details of the American tests of the hydrogen bomb became known, Churchill at once grasped the immense gulf in power that now yawned between the atomic and the hydrogen bombs, and its dread import for mankind. He sought another opportunity to visit President Eisenhower, specifically to discuss these matters, and a fruitful meeting took place in June – it would be his last as Prime Minister. It was due largely to Churchill's realistic appraisal of the changed nuclear situation that the implications of the hydrogen bomb were faced and thought through in a paper on global strategy by the British Chiefs of Staff, which had an influence on American thought and policy.

These objectives which Churchill hoped so earnestly to achieve before laying down the burden of power and responsibility were not unworthy; nor was he wrong in feeling that (if time and strength were granted him) he could make a unique contribution. But also there was, of course, the reluctance of an old man to relinquish power: he cherished no illusions – there could never be a 'next time'. His long, varied and splendid public life would be over, and he did not relish the prospect of brooding over the cooling embers.

The uncertainty of Winston's intentions was not only an inconvenience to his colleagues in the Government: Clementine too found it perplexing in terms of their domestic arrangements. The tenants at No. 28 Hyde Park Gate left early in 1954 at the conclusion of their lease, and, unwilling to let the house again for a long period, Clementine managed to find a tenant for the summer months only; thereafter the house was not let again.

During the summer of 1954, Winston continued to be in a state of indecision. On 29 July, on the eve of the parliamentary recess, I wrote in my diary: 'WSC is a marvel & a mystery. And none of us really know what his intentions are – Perhaps he doesn't himself!' A month later the whole question must have been in the balance, for in a letter to my mother on Sunday, 29 August from Scotland, where Christopher and I were staying, I wrote:

> Papa rang up Chimp, & told him all had gone well on Friday [there had been a Cabinet meeting]. We are so glad. I'm sure indecision was the worst of all possible climates to live in for everyone – & liable to breed unrest.
>
> I'm afraid Papa's decision to stay must in some ways be a blow to you – I only hope that you are feeling better from yr neuritis & more rested, and that the winter ahead will not be too burdensome.

There was no more wavering for quite a little time after this, and indeed in the middle of October Winston made several important changes in Cabinet appointments; but everyone knew that the final decision could not be far off now, and only the right moment had to be found. We knew, as we gathered at Chequers for Christmas, that this would be the last one there. It seems appropriate that the record number staying in the house party was reached that year, with the twenty-two signatures in the Visitors' Book showing a complete muster of children and grandchildren.

In the New Year Churchill accepted that his hope of a summit meeting with the Russians would not be realized in a near foreseeable future. The present Parliament was in its fourth year, and he recognized that he had not the strength in him to fight another General Election as Prime Minister. He now faced up to the fact that he must fix a definite date to resign: the first week in April seemed to be the most suitable, but up to the last possible minute the day itself remained uncertain. On 1 March he made a magisterial speech in the House of Commons on the hydrogen bomb. It was a month before his final resignation, and it was his last great pronouncement as Prime Minister.

During the weekend of 18 March, Christopher and I were at Chequers; the date – 5 April – had been fixed, but was still not public knowledge. My father was in low, sad spirits that weekend, and bitter at moments. Clementine's neuritis had returned a little while before, and she had retreated to Chequers for a fortnight to rest and to try to throw it off. Although she was intensely relieved that the decision had at last been taken, she understood with deep compassion how dark a moment this was for Winston. In my diary for Saturday, 19 March, I wrote:

> From now till the 'day' it will be hard going. He minds so much. It is sad to watch –
>
> Mama too feels it – she said 'It's the first death – & for him, a death in life'.

* * *

For nearly a month from 25 March there was a major strike in the newspaper industry, and during that period London national newspapers ceased publication. But in the provinces the *Manchester Guardian* continued to appear and provided a running commentary on the events leading up to Winston's resignation. The strike meant, from a private and personal point of view, that we were spared the full pressure of press comment and speculation which would otherwise have been inevitable in these last few weeks, and which would have made life for Winston, his close circle and his government colleagues, that much more difficult. As it was, the sequence of events as finally arranged took place with a measured dignity which did credit to everyone.

Replying on 23 March to two 'loving letters' from Violet Bonham Carter, Clementine filled her in confidentially on the plans which were now in place, continuing:

I write from here [Chequers] because the neuritis has come back with vengeance & I must nurse myself up for my 70th Birthday when my dear Winston is giving a party for me (to which I hope you and Bongie are coming) & also for the Queen's Dinner Party.

I have been promised a dope for both occasions, so I shall be as gay as a lark . . .

I think what you write is true that (politically) there will be 'nothing left remarkable beneath the visiting moon'.*

Your loving
Clemmie[2]

The weekend of 25–27 March was their last at Chequers. On Saturday, the 26th, Winston visited Woodford to open the Sir James Hawkey Memorial Hall, where he gave no indication as to his immediate intentions. Clementine was unwell and remained quietly at Chequers. The rest of the weekend was spent very quietly, with just a few members of the family staying. Clementine's rest in the country had done her good, and they returned to London on the Monday for the 'last lap'.

On Tuesday, 29 March, Winston answered Prime Minister's Questions in the House of Commons, and later had his usual Tuesday evening audience with the Queen. Clementine had a busy week, making personal preparations, but 'business as usual' was very much the order of the day for her and for us all. On her seventieth birthday, Friday, 1 April, there was a party in the evening at No. 10 – a mixture of family and friends, including the Attlees, the Macmillans and the Morans. Winston and Clementine both stayed in London for that last weekend; Winston went down to Chartwell on the Sunday for the day, but Clementine remained at No. 10, garnering her strength.

On Monday, 4 April, the Queen and the Duke of Edinburgh came to dinner at No. 10. The occasion had been arranged some time before, and it had then been envisaged as a small party, but when it was realized that the dinner would, in fact, take place on the very eve of Winston's resignation, it was felt that this was no longer appropriate, and the party was expanded to about fifty. Arranged somewhat hurriedly, the party represented an inspired jumble of Winston's several lives. The political world was represented by the Speaker, the Leader of the House of Lords (Lord Salisbury) and other close colleagues in the Cabinet, including, of course, his successor, Anthony Eden; the Leader of the Opposition, Clement Attlee, and the Liberal Leader, Lord Thurso (Archie Sinclair) – all these, among others, with their wives. Military glories were remembered by the Alexanders and Monty. Other connections echoed in the presence of Anne Chamberlain (a graceful touch), and Anne (Nancy), Duchess of Westminster, the widow of Winston's great old friend Bendor (who had died in 1953). The faithful 'in the wings' friends – Brendan and 'The Prof' – were of course included; and Winston's two

* William Shakespeare, *Antony and Cleopatra*, Act IV, sc. xv.

Principal Private Secretaries, David Pitblado and Jock Colville (with their wives), who had borne loyally and discreetly the heat and burden of many days. And lastly came his closest family – Diana and Duncan, Randolph and June, Christopher and myself; but, sadly, no Sarah, who was in America.

Despite the absence of London newspapers, there had been in the last few days a general sense of expectancy – and there was nearly always a crowd watching the comings and goings in Downing Street. Of course, everyone at the dinner knew or guessed its significance. It was splendidly done. Clementine sailed through it all, looking beautiful and calm. Winston, in proposing the Queen's health, made a touching, graceful and debonair speech, reminding the company that it was a toast 'which I used to enjoy drinking during the days when I was a cavalry subaltern in the reign of Your Majesty's great-great-grandmother, Queen Victoria'; and he spoke of 'the wise and kindly way of life of which Your Majesty is the young and gleaming champion'. When he sat down, the Queen rose and with infinite charm proposed the health of 'My Prime Minister'.

The next morning, Tuesday, 5 April, Winston presided over his last Cabinet, and in the afternoon he went to Buckingham Palace and tendered his resignation to the Queen.

The following day there was a tea party of about a hundred for all the staff at No. 10 – the Private Secretaries, the telephonists, the messengers, the drivers – so that he and Clementine could thank them and take leave of them all. Winston then left for Chartwell; Clementine remained behind in London as she had many things to arrange. He was cheered away by his staff, who lined the long corridor that leads from the Cabinet Room to the front door.

When he arrived in the gloaming at Chartwell there was a small crowd at the gates to welcome him back. Christopher and I both dined with him; he was in quite good form, and messages were flowing in. So all had been accomplished with fitness – and, as Winston had said to a press reporter on the steps of Chartwell, 'It's always nice to come home.'

* * *

The weekend following Winston's resignation was Easter. Fortunately no one was staying, for Clementine was quite exhausted and the wicked neuritis had returned; so she spent a good deal of time in her bedroom. However, on Easter Sunday the Sandys family came down to Chartwell in force for luncheon, as well as three or four old friends and colleagues. The combination of family and friends was very welcome, and Winston was in buoyant form. Two days later he and Clementine flew to Sicily, with Jock Colville and 'The Prof' for company. They were there a fortnight, but for Clementine the holiday was a failure: she had little respite from pain, and in addition the weather was cold and grey. Although Chartwell and Hyde Park Gate were both at this moment uninhabitable because of alterations and decorations, she was only too glad to get back to England earlier than planned, and they both stayed for a short time at the Hyde Park Hotel.

While they had been away the date of the General Election – 26 May – had been announced. Winston came back eager to help the Government in some way; but no suitable suggestion was made. It was quite natural that the new Prime Minister should wish to stand on his own feet, and not seem to be leaning on Winston's prestige; but the implicit rebuff was hurtful. Winston took it rather more philosophically than the rest of the family – who seethed. Anyway, he had his own seat to fight, and he addressed a few meetings outside his own constituency – one of them in Bedford, where he spoke for Christopher.

Churchill was once again elected at Woodford by a large majority. Throughout the country the election campaign was remarkable for its lack of excitement, and the Conservatives were returned with an increased majority. In the new Parliament, Churchill took the place he had habitually occupied in the thirties – the first seat below the gangway on the Government side.

With the election over, Winston and Clementine had to settle down to a new tempo of life. All their family and closest friends were deeply concerned for Winston in this period, when the active and fascinating part of his public life had been sliced off for ever, and there remained none of the stimulation that the prospect of a return to the fray had previously always provided. Everybody rallied round to keep him company and to try to divert him. The gallant buoyancy of the days immediately following his resignation did not last, and was succeeded by long periods of depression, boredom and lethargy. On 11 June he had a 'spasm' – a mild one, but it made us all anxious.

For both Winston and Clementine the summer of 1955 was a dark valley: the strain of the uncertainty and the anxiety of the last months of Winston's tenure of office had left Clementine wrung out, and with the removal of the stimulus which the multitude of her obligations had provided, she gave way to low spirits. Sarah and I worried about her. In a letter to her mother in Sicily, Sarah had written: 'Don't be sad – I cannot believe that all the sparkle and gaiety & care that you have given cannot in some little measure be given back –'[3] And at the end of July I wrote in my diary: 'I am oppressed by the sadness of old age . . . CSC I know suddenly feels old & done for. And him – He just seems slower & sleepier every day.'[4]

Although Clementine had longed for Winston to resign, and was particularly glad personally, in view of her own ill-health, to be relieved of entertaining and official obligations – in fact, the problem of adjustment to a more leisured way of life was, ironically, much more acute for her than for him. For when his spirits revived – which they did – and when he was in reasonable health, life for him still had pleasures and many occupations. He now took up the completion of the four volumes of his *History of the English-Speaking Peoples*, started before the war, and laid aside ever since. Painting kept him occupied for many happy hours, and the resources of Chartwell never failed. His racing activities continued to prosper, and afforded him many pleasurable outings; there were cards, reading, and the affection and company of family and friends. And everywhere he went he was showered with loving, admiring attentions, which greatly touched and gratified him. But for Clementine the problem was more difficult. First, her poor state of health, which was to persist for a long period, coloured her whole outlook on life. Second, she had for so long lived a life of dedication to Winston and her public duty that her capacity to enjoy herself seemed to have withered – like a muscle that atrophies through lack of use.

She had no hobbies such as gardening – that great solace and refuge for countless Englishwomen was for her more a matter of administration than an absorbing or satisfying occupation; her own active involvement stopped after dead-heading roses and irises. Nor did tapestry, knitting or embroidery – those other excellent forms of occupational therapy – appeal to her. Although she had driven a car in the past, during the years of their official life Clementine had had a chauffeur, and she had lost the habit of driving. Now, when she tried to grapple once more with gears, and the increased volume and buzz of traffic, she found herself unequal to it; and so (to the barely concealed relief of her nearest and dearest) she abandoned driving. In these financially more palmy days she could always have the services of a chauffeur, but to drive oneself gives independence and flexibility in plans.

Social life as such, depending on charming fringe-friends, had never held anything for her, and her highly developed critical faculty combined with the puritanical streak in her character made her often at odds in the gilded temples (or villas) of Mammon. Clementine's splendid independence of mind, which characterized her views and actions so strongly throughout her life, and which saved her from being a 'yes' woman to a dominating personality, did not extend into the territory of her own life and pleasures. If she had been more egotistical, more pleasure-loving or more personally ambitious, she might have been less at a loss when the calls of duty ceased to be so imperious and consuming. But if she had been any or all of these things, the history of her relationship with Winston, and her influence over him, would certainly have been different – and so, to some extent, might have been the history of our times.

* * *

After Winston's resignation Clementine was determined to deal seriously with the neuritis which had been plaguing her since the end of 1953. In May 1955 there is a record of a series of appointments with a new doctor; then in early June she fell and broke her left wrist, and had to wear a plaster for many weeks. The pain from this injury and from neuritis in her right arm was so continuous and wearing that her doctors prescribed injections of pethidine, which she learnt to administer to herself every few hours.

Quite apart from her own health, this period was a sad and worrying one for Clementine on the family front: Winston had a 'spasm' in June; Diana was nervously ill again; and Sarah and Antony were in the throes of parting for good: there seemed to be no respite for her from pain or anxiety.

Although she was loath to leave Winston at this time, she decided to go in early August to St Moritz, where there was a centre for treatment which had been recommended to her. She went there for a month, taking with her Heather Wood, her secretary; they stayed at Suvretta House. Later in her visit I was able to join her.

In her own personal life, this was the first gleam of sunshine for a long time: the bracing mountain air, a spell of good weather, and encouraging results even after a few days of treatment, all combined to raise her spirits. Winston's care and anxiety for her is reflected in his constant flow of letters, nearly all of them written in his 'own paw', which always pleased her. 'Do write to me & don't give up hope,' is how he ended a letter on 8 August, written shortly after her departure. He thought about her constantly, and I reported to her on 9 August: 'Darling Papa really does seem in distinctly better form these last five days. He was awfully worried & low about you, & I think knowing you had safely arrived & having had a letter – he feels much better about everything.'

Clementine found one or two people she knew staying in or near St Moritz, and she made the acquaintance of a courtly and erudite elderly widower, Mr Lewis Einstein, who was also staying there. He was agreeable company, and we all trundled about in his large and comfortable car viewing panoramic sights. Nor did Clementine want for reading material, for Winston kept her supplied with his revised and polished chapters for *The English-Speaking Peoples*, which were engrossing his whole attention just now. He liked to know what she thought of them, and on receiving her comments on the early chapters he wrote, on 17 August: 'I am so glad you like it, and what you say about it is a great encouragement to me . . . I am delighted at what you say about helping a lot of people to read history.'

When Clementine returned to Chartwell early in September, she was in better spirits. Later in the month she and Winston went to La Capponcina (once more lent to them by the generous Max), taking Christopher and me with them. This visit proved a great success all round. Winston wrote and painted away, while my mother had us both for company. Christopher was a great enlivener of Winston, making him laugh; playing cards with him; and, at times, teasing him affectionately. Winston and Clementine were lent a comfortable yacht, which was anchored at Monte Carlo, and in which we could all go for expeditions, which Clementine particularly enjoyed. She stayed in France for a whole month; Winston remained on until the middle of November.

Alas, the improvement in her health was not of long duration. As the autumn advanced the pain returned, and with it bad nights and despair of spirit. The pethidine brought periodic relief, but cured nothing; and in the sleepless hours, dark thoughts and cares multiplied. A Roman Catholic friend sent her a rosary, which she tried to learn to use. Her distress was obvious, and yet one did not seem to be able to help.

Sarah, who with Diana had gone out to France to be with their father, wrote on 9 November:

> I wish all was better for you – When you hold your rosary at night – do not wish for sad things – hold it for comfort – or pray we can all yet be happy – I am powerless & incapable in front of your despair – but I thank you for sharing it with me – it makes me feel closer to you. I love you very much
>
> Your ever loving child. Sarah

Sarah came very close to her mother in these dark days. She herself was going through the mill of personal unhappiness with the final break-up of her marriage; she stayed quite a lot at Hyde Park Gate that winter, and was able to be with her mother more than usual. Diana was walking a narrow tightrope of nervous and physical health, and, being close to them both, Sarah was able to play a helpful role between Diana and her mother.

Christmas, which Winston and Clementine spent at Chartwell (as they were to do for the next six years), brought a more cheerful interlude. Sarah had to go back for a while to America, but Randolph and June came down, bringing with them Arabella, now aged six, who stayed with us and our brood at the Farm. Both the 'big' house and the 'small' combined continually in entertaining each other; the Sandys family descended on Boxing Day, and a day or two later there was a lovely Christmas party for all the family and estate children.

Clementine had borne up over Christmas, but the New Year found her again at a low ebb, and early in January 1956 she went into University College Hospital for a complete overhaul. It was expected that she would be there for no longer than three or four days, but unluckily she caught a streptococcal infection, fending off pneumonia only with the aid of antibiotics. Winston meanwhile, before she became so ill, had left as planned for the South of France, where he stayed with his friend and literary agent, Emery Reves, at Roquebrune. Clementine was glad for him to be out there, and knew he was happy and well looked after. As usual, Winston wrote to her every two or three days, and was in daily touch with her doctors. At one point he very nearly returned home to be with her, but we persuaded him that she really was improving, and that he was better in the sunshine.

Clementine was in hospital for three weeks, and even when she was allowed home again she was still so lacking in energy that it was over a week before she wrote her first letter to Winston, on 29 January: 'My darling, I am ashamed that this is my first letter to you. But I have been so sick & weary that I have not been able to hold a pen or read a book. But now I am really better.' 'Sick and weary' she may have been, but the state of the world had not passed her by, and a sure sign of her improved health is shown in the spirited postscript to this letter; commenting on the grave crisis in Cyprus, she wrote: 'I am shocked that the Government mean (perhaps) to "jam" Athens Radio. I can't imagine anything more ineffective. And all the time we have been scolding Russia for "jamming" us – It's lucky the Government have got a long term before them or I think Gaitskell [Leader of the Labour Opposition] would get them out.'

During Clementine's time in hospital a plan for her convalescence was devised: it was arranged that she and Sylvia Henley (who had herself been far from well) should go by sea to Ceylon. This was an ideal plan for Clementine, as it would combine a long sea voyage, a warm climate and wonderful sightseeing, all of which things she enjoyed; and in addition she would have the bracing company of Sylvia, to whom she was devoted. Winston thoroughly approved, and he came home from France so as to have some time with her, returning to Roquebrune after her departure. From there he wrote her frequent accounts of his days, which were pleasantly divided between agreeable company, painting, writing and house-hunting (of which latter activity more will be recounted later).

Clementine was a much less good correspondent, and although she had written or cabled from various stops on the outward journey, there was evidently a long gap, for Winston wrote rather reproachfully on 17 March (a month after her departure from home): 'My darling One – No news from Ceylon by letter or telegram! I do hope this is because you and Sylvia are enjoying yourselves.' They were; and in fact letters were on the way: one with a description of a visit they had made to one of the great 'buried' cities of Ceylon, which had thrilled and interested them. The travellers returned home from their eight-week journey on 13 April; as Clementine wrote to Sarah on 1 May, 'We set out two old crocks & have returned quite set up again. I still tire very easily but, Thank God, the neuritis has faded away & both my arms feel equal.'

With the improvement in her health, her spirits and her capacity to enjoy things also revived. Just over a month after she returned from her long journey, Clementine went for a week's jaunt to Paris, staying with her St Moritz friend, Mr Einstein. They went to four plays and did some serious sightseeing; she saw the Ismays; and, with her host, she lunched at the British Embassy – a packed week, which she thoroughly enjoyed. During that summer life was quite busy at Chartwell, with a good many visitors coming and going, including the President and Mrs Truman, who came to luncheon towards the end of June. In early July Winston and Clementine gave a garden party for a large contingent of Winston's constituents – an occasion which became an annual event for several succeeding years.

During August, Clementine again returned to St Moritz. Despite the previous winter's relapse, she felt the cure there had done her great good. This time she stayed at the Palace Hotel, to be more in the centre of things and nearer her treatments. Unluckily, a day or two before she left home Clementine had a bad fall, and arrived in St Moritz hobbling and covered with bruises; fortunately she had her maid with her (as she generally did now when she travelled). Most opportunely her crony, Mr Einstein, was once more taking the mountain air, and the trundles in the big car were more than ever a welcome diversion.

Meanwhile, the Suez crisis was brewing. In June 1956 Egypt declared that the Suez Canal Company's concession would not be extended, and ordered British troops to leave the Suez base. Subsequently Colonel Nasser, the newly elected President of Egypt, nationalized the Canal, and the following month (August) he boycotted the London Conference on Suez and rejected the American proposals for the future international use of the Canal. Clementine followed all this as best she could from the newspapers and the radio. Winston was kept informally in touch with the situation but, as he explained to her on 3 August: 'As I am well informed, I cannot in an unprotected letter tell any secrets.' My mother described to me in a letter on 9 August how she and some English acquaintances in the hotel had been 'breathless & ears cocked round the Wireless trying to hear Anthony Eden. We gathered most of what he said (altho' we were told that Russia was jamming). Everybody was rather disappointed – I was loyal and said it was excellent –'

Clementine came home to Chartwell towards the end of August and we all rejoiced to see her at last in very much better health. It had been such a long, hard grind, and now she really did seem to have made strides towards recovery. She and Winston remained at Chartwell together until the middle of September, when he went to Roquebrune, Clementine staying on only a short time at Chartwell before going to London, where she indulged in a positive orgy of theatre-going. All Winston's letters from France spoke of his joy and relief at her recovery. On 24 September he wrote: 'My darling One, It is such a pleasure to receive your letters. The handwriting is so strong and you can dash them off with a vigour wh shows your troubles and their consequences are now steadfastly relegated to the background.'

* * *

Soon after Winston's resignation, Anthony Montague Browne, who had been in the Private Office at No. 10 since 1952, came to look after Winston's affairs. Anthony had been a pilot in the RAF during the war, and had been awarded the DFC. After the war he had entered the Foreign Service, and in due course became one of the Prime Minister's Private Secretaries. He had, during his time there, gained the esteem and confidence of both Winston and Clementine, and when Winston resigned Anthony was seconded by the Foreign Office to work for him.* Having ceased to be Prime Minister, Winston remained until the day of his death a world figure, and it was essential that his affairs, public and private, should be watched over and handled by someone trained in the tradition and ways of the public service. It would be impossible to over-rate the service which Anthony rendered Winston in the last ten years of his life. Winston confided all his affairs to him, and Clementine too depended much on him for help and advice in many matters. Giving up his prospects of a promising career in the Diplomatic Service, Anthony remained to help Winston long after there was any glamour or excitement in the job. In those last years of my father's slow, sad decline he never failed in his loyal and devoted service and

* Anthony's Foreign Office ranking was maintained, and he continued to receive his salary, which WSC refunded to the Foreign Office.

companionship to him, and became a trusted friend of all of us in the family. It was indeed fitting that on the last cold, grey morning he would be with us all.

* * *

For several years, Winston (in those moments when he contemplated life in retirement) had toyed with the idea of buying a villa in the South of France. Clementine had viewed this project with dismay from the start, but the idea of a house of his own on the Riviera had taken deep root in Winston's imagination. He had always loved that sunshine region, and the thought of being able to spend months on end there, installed in his own surroundings, painting and writing and sitting in the sun, glowed in his mind. With his resignation the scheme became a matter for active and immediate consideration. Because Winston had so set his heart upon a house in France, Clementine really tried to consider the proposition with an open mind – but the more she thought about it, the larger loomed the difficulties: first, to find a really nice property; then the enormous cost of the purchase and subsequent alterations; and finally (and for her most worrying of all) how to staff and run it. Many of her objections could be (and were) swept aside – but two remained immovable: her dread of having a third house to run, and the fear that although Winston was now much better off than at any other time in his life, yet this project might well over-stretch him financially; and she dreaded the prospect of a return to their economic cliff-hanging of former years. In her view, the only real advantage she saw in Winston having his own house was that it would remove the necessity of accepting hospitality from very rich and, in her opinion, not always very suitable people. This last consideration carried real weight with her.

Winston pooh-poohed Clementine's prudent and practical objections with an airiness which greatly irritated her, and the 'villa' discussions could become explosive. The family was somewhat divided in their views and sympathies over this thorny question. We could see both sides of it, but one sometimes had the feeling that Clementine made too much of the difficulties, and it was easy to sympathize with Winston's wish for a holiday home in the sun.

But he was not to be easily put off, and from 1955 for two years he was seldom in the South of France without making enquiries, and sending out reconnaissance parties to view prospective villas. The regional newspapers daily reported details of his search, and, of course, once it was known that 'Monsieur Churchill' was looking for a residence, villas appeared from behind every cactus – at double the asking price! Winston did not conceal from Clementine that he was house-hunting, but when she was with him he wisely refrained from sending her on hot and tiring expeditions. He just hoped something lovely would 'turn up' and that she would be charmed by it. When she was in England, he would send her accounts of 'Hunt-the-Villa', always reassuring her that nothing precipitate would be done. These activities caused poor Clementine much anxiety, which was well understood by Sarah who, while staying with her father in France in November 1955, wrote to her mother: '[W]e have been going to see villas (mostly monstrous!) with vague plans of buying – don't worry – he would never without you – I don't think he even wants to really – but he does love the sun so.'[5]

Notwithstanding the elusive nature of the 'Dream Villa', Winston, after his resignation, spent long periods of the year abroad. He could have gone to hotels; and he could now afford the very best; but to people of his generation the charm and seclusion of 'private life' in private houses, with private service, were very powerful. Moreover, to anyone of Winston's eminence, a private domain is the only protection from prying eyes, whether they are admiring, malicious or merely curious. Winston was blessed in his friends, and he found now, as he had found in the past, hospitable havens where he could be undisturbed.

Since 1949 Winston and Clementine had often stayed at La Capponcina: sometimes Max Beaverbrook would be there himself, at other times he would lend them the house with its charming housekeeper, excellent chef and smiling staff. Winston loved staying there, and Clementine, for her part, disliked being at La Capponcina less than anywhere else on the Riviera. Indeed, she had some very happy visits there, especially when any of us children, or Anthony Montague Browne and his wife Nonie, or Jock and Meg Colville, of all of whom she was very fond, were staying. She had no qualms about Winston accepting princely hospitality from such an old friend, and she and Max had long since called a truce in their warfare. In these last years Max showed all his winning charm to her and, when challenged, Clementine had to admit that she had become quite fond of him. Winston was of course greatly gratified to see the end of hostilities, for not only was Max Beaverbrook a friend of fifty years' standing, but, as others slipped away, he became the survivor of them all.

In 1956 Winston went to stay for the first time with Emery Reves at his home near Roquebrune. La Pausa is a large house (built in the 1930s by Bendor Westminster for his paramour Coco Chanel), standing high on the shoulder of the mountainside above Cap Martin, commanding sensational views towards Menton on one side and Monte Carlo on the other. Emery Reves, by origin a Hungarian Jew from Budapest, had come into Winston's life before the Second World War, when he handled the foreign rights of many of his newspaper articles. In 1940 he became a British subject; during the war most of his family in Europe were liquidated, and he himself was badly wounded in a London air raid. After the war Reves handled the American rights in Churchill's memoirs, and himself bought the world rights in these, as well as in *A History of the English-Speaking Peoples*, making thereby a great deal of money both for his client and for himself. He lived mainly in the United States, but also owned this house on the Riviera, which he filled with his remarkable collection of chiefly Impressionist pictures. It was here that, in the fifties, he hospitably invited Winston and Clementine, and members of their family, on many occasions and for long visits. Emery's companion was Wendy Russell, a beautiful and charming American model a good deal younger than himself, whom he would marry in 1964.

In January 1956, when Winston paid his first visit to La Pausa, Clementine was ill in hospital, going very shortly afterwards to Ceylon to convalesce. Winston therefore went without her to Roquebrune, accompanied by Anthony Montague Browne, and joined later by Diana. His letters to Clementine were full of the beauties and comforts of the house, and of the care, kindness and thoughtfulness of his hosts. During this visit he was seriously searching for a furnished villa to rent, and Emery and Wendy accompanied him on several inspections. But it was due to their hospitable blandishments and to the fact that he was then, and on successive visits, so comfortable and happy in their house, that he abandoned the search for a villa of his own.

Clementine paid her first visit to La Pausa later that year, joining Winston there on 1 June after her whirlwind week in Paris: she could stay only five days as she had to get back to England, where she had engagements. During the next few years Winston was to make some eleven visits to La Pausa, staying often for many weeks at a time, but Clementine was there with him on only four occasions, and (except in the early spring of 1958, when Winston was seriously ill) she made her visits as short as was compatible with politeness. Emery and Wendy repeatedly invited her, and to their welcoming motions Winston added his pleading. But the fact was that Clementine did not enjoy herself there: she found life at La Pausa claustrophobic. Not having a car, it always meant hiring one specially; and to take a walk outside the villa grounds was difficult, as the mountainside was very steep. Winston worked all morning in bed, and after luncheon painted or rested; in the evenings he liked to play cards for a while: as always, he loved her being there – but he was not companionable. Emery and Wendy were full of care for Winston, and solicitous for all their guests' pleasure, but Clementine found she was really the prisoner of kindness and of other people's plans. And although it was mostly smiles and pleasantness on the surface, there was very little she and her hosts had in common. Seeing therefore that Winston wanted for nothing at La Pausa, and was surrounded by care for his every need, she preferred to stay in England or to make some other plan for herself.

It must be remembered also that Clementine had a long and abiding dislike of the Riviera and all it stood for. When, after 1959, Winston took to staying at the Hôtel de Paris in Monte Carlo, she usually went out to keep him company for at least part of his visits; but it really was a sort of penance. In a letter to Sarah from there in January 1960, she burst out: 'God – the Riviera is a ghastly place. I expect it's all right if you keep a flower shop or if you're a waiter!'[6] And in the autumn of that year, staying again in Monte Carlo, she wrote to me: 'I am suffocated with luxury & ennui as you feared.'[7]

Presently, however, a new feature of sunshine holiday life developed which was to prove pleasurable to both Winston and Clementine. During Winston's first visit to La Pausa he had made the acquaintance of Aristotle

Onassis, the Greek millionaire shipowner. He wrote to Clementine, who was in hospital, on 17 January 1956: 'Randolph brought Onassis (the one with the big yacht) to dinner last night. He made a good impression upon me. He is a vy able and masterful man & told me a lot about Whales. He kissed my hand!' A little while later the villa party were all invited to dine aboard the *Christina* with Onassis and his wife Tina. Winston told Clementine that Ari had offered to lend them the yacht, an offer Winston did not accept, though he told Clementine in his letter of 8 February that it was 'the most beautiful structure I have seen afloat'. The following February, when both Winston and Clementine were at La Pausa, they lunched on board the *Christina*, and Clementine also had the opportunity of meeting this compelling and most unusual man, and the lovely and charming Tina. These were the beginnings of a friendship which was to bring much enjoyment and pleasure to Winston and Clementine, for at the end of September 1958 Ari Onassis invited them both for a ten-day cruise in the Mediterranean. They had been staying at La Capponcina, where they had celebrated their Golden Wedding: both enjoyed the cruise, and it made the perfect epilogue to a golden summer.

Sadly, the Churchills' new friendship with Ari and Tina Onassis was the cause of a deep rift with Emery and Wendy, and although Winston would stay again at La Pausa (his last visit there was in August 1959), things were never the same thereafter.

In the next two years Winston and Clementine were the guests of Ari and Tina on four more journeys, including a three-week cruise in Greek waters in the summer of 1959. And in March 1960 they embarked at Tangier for nearly a month's journey which took them to the West Indies. Clementine, who had visited many of these islands before the war with Lord Moyne in the *Rosaura*, particularly enjoyed this cruise. She was able to revisit Nelson's Dockyard in Antigua, in raising funds for the restoration of which she had been most active a few years before. She had always loved sea journeys, and now here again were holidays that she and Winston could share. Ari and Tina were the epitome of thoughtfulness and kindness, inviting as well as Winston and Clementine people who they knew would be a pleasure to them. 'Cruising' became a new feature of Winston's life, and in these luxurious circumstances, and with such delightful hosts, it is not surprising that he found the journeys highly agreeable. Moreover, as the years slid by and Winston became less mobile, this was the ideal holiday for him.

They were both very sad when Ari and Tina parted and, shortly afterwards, in 1961, were divorced. Although Tina was a much younger woman, her charming manners, her gaiety and her kindness always made life on board the *Christina* enjoyable for everybody. Clementine said she missed her very much after her departure. They were, however, to meet each other again in a different context quite soon, for in 1961 Tina married Winston's cousin, Sunny Blandford*. But after 1960 Clementine did not go with Winston for any of his three remaining cruises in *Christina*.

* The Marquess of Blandford; he succeeded his father as 11th Duke of Marlborough in 1972.

* * *

Reading old diaries and letters reminds one strongly how much one lives life from day to day. It is only on looking back over a span of years as a whole that patterns emerge in such things as holidays, habits and health. During the nearly ten years of life that were left to Winston after his resignation, he weathered numerous minor incidents of ill-health, as one would expect in someone of his age, which varied in degree from the odd bronchial cold, annoying 'tickles' and such-like passing indispositions to the more worrying 'dizzy spells', and the occasional 'spasm', which were significant and always exacted a toll, however slight or imperceptible. Hanging over him was the possibility of another major stroke, which might kill or, worse, disable him. In this last decade of his life Winston also survived another serious bout of pneumonia, and two bad falls. As a result of the first fall he broke a bone in his neck in November 1960; in the second, two years later, aged eighty-eight, he broke his hip in the Hôtel de Paris at Monte Carlo, and had to be flown back to London, where he was in hospital for nearly eight weeks.

Even after his retirement from public life, the press throughout the world kept a constant watch on Winston's general health, and the smallest indisposition (if noticed) aroused their interest, and also their suspicion that there might be 'more to it' than was given out in any statement. The solicitude of the public for his well-being continued, letters and enquiries from all over the world literally pouring in whenever he suffered some publicized illness or accident. But this solicitude was somewhat overpowering to deal with, and Clementine and the rest of the family were therefore at considerable pains not to draw attention to Winston's minor indispositions, which so often passed off in a day or two. In the case of rumours circulating, or during any of his graver illnesses, statements or regular bulletins were always made to the press. It was at times like these that Anthony Montague Browne's efficiency, judgement and calm were especially invaluable.

Looking back over those years one can see that Winston had reasonably good health for about two and a half years after his resignation. Clementine, and all of us, were always watchful and anxious – sometimes perhaps too anxious. But what one sees now to have been a passing and maybe trivial episode could have been the start of a grave, or even of the final, illness. In retrospect the two features of Winston's pattern of health seem to have been (I speak, of course, as a layman) bouts of melancholy or lethargy, and increasing deafness. After the few sad months which succeeded his resignation his interest in life and all his 'toys' and occupations revived, but in these latter years he tended from time to time to sink into periods of silence and gloom, from which he would often emerge with startling suddenness – rather like a sinking fire which suddenly burns up bright and clear.

By 1955 Winston was becoming increasingly deaf, and like many people he put up an obstinate resistance to the idea of wearing a hearing-aid. Clementine sometimes scolded him quite fiercely about his reluctance to do so, pointing out that it was needed not only for his own sake but for other people's. He gradually came to realize his need for a device; but it was a necessity he loathed, and he often had to be prodded into wearing it.

In the middle of January 1958 Winston went to stay at La Pausa. Clementine planned to join him there for a week later on in February. The latter end of the previous year had not been good for her healthwise: a minor foot operation had necessitated a week in hospital, and in December she had been laid low with two recurring attacks of influenza, which had a debilitating effect. Now, in the New Year, disturbing news came from America about Sarah. In August 1957, Antony Beauchamp had taken his own life by an overdose of sleeping pills. He and Sarah had been separated for nearly two years, but his death came as a tremendous shock to her; and suicide leaves such a cruel legacy of unanswered, often unanswerable questions. During the winter Sarah went back to America, where she had plenty of work on offer, and in the New Year she was living alone in a house on Malibu Beach near Los Angeles, while she was working on a television programme.

On 14 January 1958 it was reported in the British press that Sarah had been arrested and charged with being drunk; there were, of course, distressing photographs and accounts of the episode. In the subsequent court case, she pleaded guilty and was fined. Clementine was deeply distressed for Sarah, and also humiliated by the blazing publicity which inevitably attended the affair. On this occasion, as when there would be similar incidents, Winston minded very much, and grieved for her, but he always took it less hard than did Clementine. As he grew older he seemed to acquire a degree of insulation from sad or unpleasant news about those he loved, which is perhaps one of the kinder gifts time bestows. But, of course, there were no shut doors: when Sarah arrived back from America in early February Winston was staying at La Pausa, and she went straight there to be with him. Ten days later her mother flew out to join them. Sarah was in France with them both for two or three weeks, writing to her mother after she left: 'I was so grateful for your sweetness and reticence with me – . . . Thank you both – you and Papa for being so gentle and understanding –'[8]

In mid-February, Winston became ill with a chest cold which developed rapidly into pneumonia. By coincidence, Clementine had arrived for an already planned visit the same day as the first bulletin saying Winston was 'indisposed' was issued on 18 February: but the one week's visit she had envisaged turned into a month. Lord Moran flew out the next day; Winston was really quite ill, but he responded well to the treatment, and was after about ten days in a fragile, but convalescent, state. He was due to go to America in the third week of April, and up to that point had been determined to stick to his plans; but on 15 March my mother wrote to me:

> Papa, for the first time shews hesitation about going to America –
>
> I think Alas – he feels definitely weaker since his illness – He certainly made a marvellous recovery, but without the mass use of antibiotics he would have sunk and faded away . . . If Papa does go to America . . . would you go

> with him in my place? I don't feel strong enough, I am ashamed to say; but I think a member of his family should go . . . Of course – I hope he won't go – If he does not make one or two speeches & television appearances, the visit will be a flop as regards the American People – who . . . want to see and hear him. Then if he lets himself be persuaded to make public appearances it will half kill him. Monty, the dear creature has just arrived & thinks Papa is crazy to contemplate the idea.

Monty's visit had been most opportune, and a comfort to them both. A few days later, on 18 March 1958, my mother wrote to me: 'Monty returns to Versailles [his HQ as Deputy Supreme Commander, Europe] this morning. His visit has been a great tonic.' One day he and Clementine visited Menton: 'We went into a beautiful old church & burnt 2 candles for Papa.' In his illness Winston felt more than ever that he wanted Clementine there. She knew this, and although these weeks must have seemed very long, there was no question of her not staying with him: 'I shall stay & return with Papa, even if he lingers on after recovery,' she told me in the same letter.

At about this time Winston had a relapse, which proved rather worrying. On top of the recurrence of the pneumonia, he was also suffering from jaundice. There were more bulletins, and his return home, which had been planned for the end of the month, was postponed. Lord Moran, who had left him in the excellent care of a local British doctor, now returned to take stock of the new situation. After a few days, Winston again improved, and this time did not look back. But the illness and relapse had left him in a weak state.

He was finally well enough to return home on 3 April, two days after Clementine's seventy-third birthday, and in time to spend Easter at Chartwell. He had made a remarkable recovery, but good sense prevailed, and about a week later an announcement was made of the cancellation of his intended visit to the United States. It is hardly surprising that this long and tiring illness should have affected Winston's state of health, and it was soon after his return from France that it was thought he needed more continual care than could be given him by a valet who had other duties in the household as well. A full-time nurse, Roy Howells, was therefore engaged, who looked after him in every way with devotion and kindness until his death.

One realizes in retrospect that it was from the time of this illness that Winston started to decline generally. But the twilight was long, and as after a glorious day the beams of the sun linger on, lighting parts of the slowly darkening landscape – so it was to be with Winston's last years.

* * *

This year of 1958 which had started so badly had, however, a happy and glorious finale. On 12 September Winston and Clementine celebrated their Golden Wedding. They were staying with Max Beaverbrook at La Capponcina and spent the day quietly, but very happily. Messages of affectionate congratulation poured in: and one of Winston's horses, Welsh

Abbot, had enough sense of occasion to win a race at Doncaster. Of the family, only Randolph and Arabella were able to be there; but (aided by Arabella, who recited reams of a classical poem, 'The Garland of Meleager', for the edification of her grandparents) Randolph gave them on behalf of all their children a token of our present, which was a border of golden roses for Chartwell. Since the summer months are not good planting months, Randolph had conceived the wonderful idea of asking a number of famous artists and friends each to paint one of the roses that would be in the border. Paintings, engravings and drawings in great variety were contributed by (among others) Augustus John, Ivon Hitchens, Duncan Grant, Vanessa Bell, Cecil Beaton, John Nash, Oliver Messel, Adrian Daintry, André Dunoyer de Segonzac and Sir Matthew Smith. Gifted friends and relations also added their offerings, making a dazzling and unique 'catalogue' of roses. The pictures had all been sumptuously bound in a large album, and the script and the illuminations were exquisitely done by Denzil Reeves of Colchester. Later that year the border was duly planted. It runs like an avenue through the centre of the kitchen garden, circling the sundial where the Bali dove lies buried. It was a romantic idea, and soon looked lovely in reality; and I believe it gave our parents pleasure.*

I minded dreadfully not being able to be with my parents on this glorious day, but a combination of children, domesticities and politics held me at home. I wrote to my mother:

> [W]e thought & spoke about you & Papa so much over the Glorious Golden 12th and had such loving telegraphic exchanges; . . . What moving and touching tributes! I felt so proud & so gratified that your great achievements, & all your long years together should receive the praise due. You see now what you both together mean to hundreds & thousands of people.[9]

Sarah, writing from London, echoed the same thought: 'I wonder if you know how many people share in the joy & pride of today? And what another aspect of radiance you both have shed upon daily life by the way you have led your private lives.'[10]

One final gleam of glory in this year came in November, when General de Gaulle (as head of the French Government) invested Winston with the *Croix de la Libération*.† It was a signal honour, and a gesture which moved Winston and Clementine deeply. They stayed at the British Embassy with Sir Gladwyn and Lady Jebb, and on 6 November Clementine watched the ceremony of the Investiture by the General in the garden of the Hôtel Matignon. Whatever had been the differences between these two men over the years, only the strong link which bound them – a fierce love of liberty and of France – was remembered on this day.

* Now, over forty years on, carefully replenished with newer varieties, the border is a favourite feature at Chartwell, delighting the crowds of visitors for many weeks in the summer and early autumn.

† This is the highest award designated for those who served with the Free French Forces or in the Resistance. Only two Englishmen have received it: King George VI and Winston Churchill.

CHAPTER TWENTY-EIGHT

Twilight and Evening Bell

CLEMENTINE'S HEALTH HAD BEEN BETTER FOR ABOUT A YEAR AFTER SHE HAD finally thrown off the accursed neuritis. But from the summer of 1958 she started to suffer from shingles, which affected her face, eyelid and eye. Shingles is a painful, lowering and persistent complaint, and although it developed slowly and fitfully, it all but ruined a delightful holiday in Morocco in August 1958. She had been invited by Bryce and Margaret Nairn to stay with them in Tangier; this was a friendship which brought Clementine much happiness and pleasure over the years. After their earlier meetings at Marrakech and Hendaye they had next met again in Madeira, where Winston and Clementine had spent the New Year of 1950, and where the Nairns were then *en poste*. Now, Bryce Nairn was Consul-General at Tangier, and Clementine went out to spend three weeks with them at the British Legation. Apart from the shingles, which was a harassment, she had a most enjoyable time, and flew back to France just in time to join Winston for their Golden Wedding celebrations.

In the New Year of 1959, they went back to their old haunt, Marrakech, staying at the Mamounia Hotel. Since their last visit there eight years earlier, Morocco had become independent, and Clementine noticed changes: the hotel was in a rather dilapidated state, and there seemed to be hardly any tourists. The Nairns came up from Tangier, and the Churchill party had also imported its own fun, as my mother reported to me on 28 January:

> The first ten days were enlivened by Jock, Meg and Biddy Monckton* – When Papa heard that Biddy was coming without Walter, he was rather sulky . . . But soon he took to her like a house on fire & kissed her tenderly on departure. As for Meg, she & Papa flirted outrageously & almost romped . . . When they all went away poor Papa fell into the doldrums – He is better now & has started a picture from the terrace outside his bed-room.

Most of the party had been ill part of the time with one minor affliction or

* Second wife of 1st Viscount Monckton.

another, but, my mother continued:

> Thank God Papa is blooming in his health. His memory fails a little more day by day & he is getting deafer. But he is well. I have learnt to play poker & enjoy it very much.

Their visit to Marrakech lasted a month, and on 19 February they boarded the *Christina* at Safi, joining the hospitable Ari Onassis for a cruise in the Atlantic which took them to the Canary Islands.

A great event this year in Winston's 'life as an artist' was the one-man show of his paintings mounted by the Royal Academy. In early March he and Clementine together went to a private view of 'his' exhibition. The following day Winston flew off to stay at La Pausa, and Clementine, writing on 13 March, told him about the public enthusiasm aroused: 'Long before you get this letter you will have heard that yesterday, Thursday, when your pictures were shewn to the General Public 3,210 people visited the Exhibition – The crush was so great that last night a third room was allocated & the pictures were all re-hung. The Academy officials are wildly excited & say this is a record for a "One Man" Show.'

In May of this year, Winston's health was sufficiently stable for him to undertake a visit to the United States. He had not been well – he had had circulatory troubles, and both Lord Moran and Clementine were anxious lest the strain of the visit should be too much for him: but he was determined to go. Clementine had not planned to make this journey, nor would Winston allow Lord Moran to accompany him, as he hated to appear an invalid. So he and Anthony Montague Browne flew to Washington, where he was warmly received by President Eisenhower, and stayed with him at the White House. During the course of his visit, Winston made a speech at a dinner. From Washington he flew to New York, and stayed a day or two with Bernie Baruch. After this stimulating but tiring week in the United States he returned home safe, if only fairly sound, to England: he had 'got away with it' again.

Clementine was plagued by shingles for a year, but at last it abated, although leaving her with a drooping eyelid. Fortunately this could be put right: and at the end of August 1959 she was operated on by Sir Benjamin Rycroft, who completely cured the blemish. It was while she was recovering from the operation at the Queen Victoria Hospital, East Grinstead, that Sir Benjamin took her into a large ward and told her that a great number of the patients there could have their sight restored if only there were enough corneal tissue available. Clementine was so struck by this that she forthwith willed her eyes to Moorfields Eye Hospital for therapeutic purposes.

With Winston's resignation in 1955, the load of public obligations had been greatly reduced for Clementine, and in the succeeding eighteen months, in consequence of her poor health, she undertook only the minimum of engagements. But as her health improved she once more started to accept invitations in connection with various causes which either appealed to her personally or with which she already had some link through Winston. Her diary in the succeeding year or two shows a sprinkling of plaque unveilings, hospital ward or department openings, and other similar functions – and, of course, she was ever-faithful to Winston's constituency. But in these latter years she avoided engagements far from London, as she found the travelling too long and tiring.

Winston and Clementine were included in nearly all great State occasions, such as royal weddings, and State banquets for visiting potentates. They both went regularly to the Garter Service at Windsor every June until 1960, after which Winston found the long procession and service too exhausting. And a feature of their winter programme which Clementine enjoyed as much as he did was their annual visit to Harrow for 'Songs'.

In these later years Clementine took part in two major public appeals. In the course of World Refugee Year she made a radio appeal at Christmastime 1959, which brought in £48,000. The following year Winston gave one of his pictures to be auctioned at Sotheby's in aid of refugees (it fetched £7,400), and during the summer one of the Chartwell garden openings was in aid of the same cause.

Two years later Clementine lent her name to, and followed with interest, the appeal for the buildings of New Hall, Cambridge, the University's third college for women. Ever since her Berkhamsted days, when Miss Harris had fired her with a desire to go to university herself, Clementine had kept a strong sense of the necessity for the opening up, and levelling up, of opportunities for women. Now, because of her age and health, she felt unable to do more than lend her name. But the Chairman of the Council of New Hall assured her that her support and backing would be a tremendous help. A year or two previously Clementine had used her influence to ensure that Churchill College, Cambridge, would open its doors to women. At one of the early meetings of the Trustees, Winston had said that 'My wife thinks women should be admitted to Churchill . . . and I think so too.'[1] Churchill became the first college in Oxford or Cambridge to receive women students on the same terms as men, and living in the same college.

But, as throughout all her married life, Clementine's first priority was to run her own home. Her standards of perfection never altered; nor did her attention to minute detail. Her right hand in all her domestic arrangements was, as ever, Grace Hamblin. Good servants were not easy to find. There were the true stalwarts like Mrs Landemare, who had been in our lives for over twenty years; Vincent, the Head Gardener, who remained and stayed on after the National Trust took over Chartwell; and Bullock, the chauffeur, who was with them until Winston's death. But these lynch-pins apart, other members of the household tended not to stay for very long. They nearly all disliked the shuttling to and fro between Chartwell and Hyde Park Gate, which must indeed have been tiring and restless. Although 'time off' was meticulously planned (and Clementine's diary is full of little notes as to whose day off it was), the entertaining in both houses, although no longer on an official scale, was almost continuous, and there were also the secretaries and nurses (two now) to be fed. Of course, there were quite long periods when Winston was abroad, but it was rare for both houses to be shut, as Clementine spent so much more time at home than Winston.

Until 1962 (in the summer of which year Winston broke his hip), Christmas and the New Year were always spent at Chartwell, as were most of the Easter and Whitsun recesses. Although now, especially during the first dark dreary months of the year, Winston used to go for longer and more frequent visits abroad, when he was in England weekends at Chartwell were sacred; and, as time went on, 'the weekend' became longer and longer, Friday-to-Monday stretching into Thursday-to-Tuesday. The summer saw them almost continuously at Chartwell; September was a favourite 'cruising' or 'abroad' month. October, November and December were again 'weekends at Chartwell' months.

The reconciliation between Clementine and Chartwell, which had its beginnings soon after the end of the war, blossomed in these last ten years of their life there into a period when she took increasing pleasure and satisfaction in this place which had always represented an earthly paradise to Winston, yet up to now had afforded her mainly worry and fatigue.

The relief from gnawing financial care was a major factor in this change in her attitude, and the knowledge that Chartwell would one day, under the National Trust, be open in perpetuity to the public, gave her an incentive to take a longer perspective than she had ever had before. In the gardens she now planned with an eye to the future. In the past she had so often tried to dissuade Winston from undertaking large-scale works – but now the boot was on the other foot! Clementine conceived and carried out several major improvements, while Winston made demurring noises – although he nearly always admired and approved the final result. The most successful (and the most expensive) of all her projects was the wholesale removal of the ratty old greenhouses, dilapidated frames and potting-sheds from the site they had hitherto occupied, and the erection of a splendid range of glasshouses in a much better place, where the buildings were concealed behind a wall. The space made vacant by this operation was transformed into a wide and sunny terrace, which looks out across the kitchen garden and away over the Weald of Kent.

Clementine had always found the gardens too spread-out and exhausting to cope with; now that she was older, this was even more the case. The parts of the garden she liked the best were the croquet lawn, where she spent many happy hours, and the grey-walled rose garden, which was not too far from the house. About 1949, she moved her bedroom from up aloft in the Tower Room (to where she had ascended in the first wave of post-war alterations) down onto the ground floor, converting the blue sitting-room into a delightful bedroom which gave out on the pink terrace. Here Clementine made a 'trough' garden, where she loved to sit with her guests. In the spring the deep wooden troughs and tubs were filled with white tulips, forget-me-nots and polyanthus, while in the summer they spilled over with single white geraniums, cherry pie, fuchsias and verbena.*

* These charming planting schemes are faithfully reproduced every year by the National Trust.

The frequent openings of Chartwell gardens to the public made her intensely 'garden proud', and Vincent and his gardeners found it a considerable challenge to produce the standard they all aimed at two, three or four times in the summer. Clementine liked to be there on the open days, and took a keen and competitive interest in the numbers of people who came (often as many as six thousand), and the amount of money raised for various good causes.

In these later years she happily spent more and more time at Chartwell, particularly in the spring and summer, even when Winston was away. The fact that Christopher and I and our children lived at the bottom of the garden undoubtedly made a difference to her life in the country: our plans were carefully concerted, and hardly a day went by without some contact. And if Clementine was away and Winston alone at Chartwell, she confided him to our care, and we dined or lunched with him, or he with us, almost daily.

By the time Christopher and I had four children Chartwell Farm was bursting at the seams, and so we had to look for a larger house. It was a great wrench for us all when we moved away in 1957 to Hamsell Manor at Eridge Green near Tunbridge Wells. It was about forty minutes' drive from Chartwell: but although 'dropping in' was no longer possible, the to-and-fro between Chartwell and Hamsell was frequent. My parents would often come over to luncheon or dinner, bringing their guests with them, and vice versa. During the considerable time the alterations and decorations were in progress at Hamsell, my mother frequently visited the scene of operations, picnicking with us amid the dusty chaos. Every detail fascinated her, and she had many practical suggestions and ideas. She always thought of lovely presents for house or garden: camellias; white iron Victorian garden seats; and one year (having noticed the howling winds on our Sussex hilltop), a revolving summer-house to bask in. Whenever my mother came to stay with us, she entered into our life, greatly enjoying meeting our friends in the neighbourhood.

Clementine and Christopher now had a most affectionate relationship; and she relied upon his judgement in many things. She followed his political career with keen interest, thoroughly appreciating his vicissitudes first as a junior minister, and then as a member of the Cabinet. She had a sensitive understanding, and when Christopher's father – with whom he had had a difficult relationship – died in 1962, she felt instinctively that the purely conventional expressions of sympathy would not strike the right note. In a letter full of comprehension Clementine wrote:

My dearest Christopher,

You are much in my thoughts.

I know that it has been a deprivation that you never knew your Father well – Winston had the same experience with his Father, and like you, minded very much.

Although he did not shew it, I feel your Papa must have been proud of what you have made of your life and of what you are –

It's not the same thing I know, but you have a sure abiding place in Winston's heart and mine

Your loving Mama in law
Clemmie[2]

Both Winston and Clementine enjoyed their role as grandparents, and counted the steadily rising tally with great satisfaction. By 1959, with the birth of our Rupert, the total of ten was complete: two Churchills, three Sandys and five Soameses. In 1962 Edwina Dixon (Sandys) produced their first great-grandchild, and Winston and Clementine were delighted with their new, positively patriarchal standing. The arrival of a new baby in the family always thrilled Clementine, and she would hasten to see it as soon as possible. She was always troubled that modern mothers were, in her view, 'hounded' out of bed in a most barbaric fashion – in her time, three weeks in bed without a toe to the floor was the normal practice; and although she conceded that medical practice in this field had progressed since her day, she still thought we all did too much too soon.

Four of our children were born while we were living at Chartwell Farm, and both my parents were usually in close attendance to encourage before, and to rejoice afterwards. My mother always insisted I should telephone her the moment the baby showed any sign of starting. She would come down the hill at once, always disappearing discreetly at exactly the right moment, taking Christopher away with her to be distracted and fed. (This was before fathers were expected to assist in the process of childbirth.) When Jeremy was born in 1952, I see from my diary that I telephoned my mother at five in the morning and that 'She arrived like lightning looking too soignée & pretty for words.'[3]

Clementine was a much-loved grandmother, with an aura of glamour. In the summertime grandchildren usually came to stay; and as babies, our 'country bumpkins' often had a London season, staying with their grandparents at Hyde Park Gate. The various nannies liked Clementine very much, partly because she was so considerate for their comfort and happiness, and partly because she never interfered. Until the farms were sold in 1957 there was a Christmas party at Chartwell for all the children on the estate and as many of her own grandchildren as could be mustered. Thereafter she organized a mass family outing to a pantomime in London; the highlight of these annual treats came in 1958, when we all went in force (including Winston) to see Sarah as Peter Pan.

As well as giving parties, Clementine also loved to be invited to them, and from the grandchildren's earliest years attended their birthday parties. As they grew older and outdoor sports took over from nursery games or conjurors, she was a keen spectator at their home-made 'Sports Galas' or cricket matches (being always in demand afterwards to present the prizes). As the children's tastes became more sophisticated, it diverted her very much to be included in their teenage parties, and when Edwina Sandys came out in 1957 her grandmother gave a dance for her at Claridge's. As well as taking interest in all the minutest details of the arrangements, Clementine enjoyed herself enormously on the night.

* * *

One of the penalties of longevity is that some of the dearest and most familiar faces depart, leaving gaps which cannot be filled. Winston and Clementine had especially strong bonds of affection and friendship with those who had known, and borne with them, the storms and struggles of the thirties. Among these was Venetia Montagu (Clementine's cousin, and a friend since childhood), who died of cancer in 1948. A decade later 'The Prof' and Brendan Bracken – two friends of over thirty years' standing – departed: 'The Prof' died in 1957, aged seventy-one; and Brendan thirteen months later, aged fifty-seven, of agonizing cancer of the throat. It is true that Clementine had for long mistrusted this flaming-haired friend of Winston's, but in these latter years no one had appreciated his loyal friendship more than she, or been more grateful for his wise and understanding advice in our family's affairs.

Among survivors – soldiering on – was Sylvia Henley (Venetia's elder sister): ramrod straight, defying time's statistics, and proving the truest of the true to both Winston and Clementine. There was Violet Bonham Carter, with whom they shared so many memories of past political triumphs and disasters; and Horatia Seymour – pale and frail-looking, but still the ghost of a great beauty – was another friend from their long past.

Clementine had never been an easy person to get to know well, and for long periods the need for total discretion had added an outer defensive wall for would-be friends to breach in addition to those already constructed by her own innate reticence and shyness. But in the years after the Second World War Clementine made new friendships, and revived some old ones, which were to give her much happiness and let fresh air into her life, which might otherwise have been too inward-looking and claustrophobic. Among the older generation was an Ogilvy cousin, Lady Helen Nutting, a beautiful, intelligent and original woman, who came a great deal to Chartwell in these later years.

A consideration of Clementine's closest friends reveals that they all had at least one thing in common – Winston liked them too. Among latter-day friendships which they shared were those with two most unalike people: the beautiful and gifted Rhoda Birley and the irrepressible Monty. He was devoted to Clementine, and accepted with good grace some not undeserved scoldings from her from time to time for various 'Montyish' lapses. He was to prove in sadder years an assiduous and faithful friend to them both.

But her friendships were not confined to those from long ago, or of an older generation. She very much liked the company of younger people – they cheered her up, and gave her hope. Among these, the Colvilles, the Montague Brownes, the Deakins and the Rowans*, all spring at once to

* Leslie and Judy Rowan. He had been WSC's assistant and later Principal Private Secretary, 1941–5.

mind. And in her charming young secretaries she found delightful companions for expeditions of all sorts. She greatly admired the way younger people grappled with their lives, and she always loved to know details of their housekeeping arrangements, and of children, boyfriends or husbands. Through her own children she also made some friends she greatly enjoyed, and who would bring glimpses of quite differing worlds and generations.

For a woman in her seventies with fragile health and long exhausting years behind her, Clementine's life in this last decade which remained to Winston and her together was still a full and demanding one. There were many pleasurable and interesting aspects to her life, and no lack of true friends to give her companionship and affection. But Clementine's capacity to relish all these blessings was directly linked to her own state of health and spirits, the latter in turn being closely bound up with her constant preoccupation with Winston's well-being: her life still revolved entirely around him. And, as in earlier days, she could always brace herself to meet her obligations, so often leaving only the lees of her energy for her own friendships, interests and pleasures.

* * *

In ceasing to be Prime Minister, Winston still remained what he had been for half a century – a Member of the House of Commons; and for forty years the same constituency had faithfully returned him to Westminster. He cared greatly for his role as a Member of Parliament, although after 1955 he was a less constant attender at the House's deliberations, largely owing to the long periods he spent abroad. He made very few speeches after he ceased to be Prime Minister, and the first time he spoke in the House after his resignation was on 30 November 1959, when he thanked Members for their congratulations and wishes on his eighty-fifth birthday. But although he may have been a silent presence in Parliament, he followed events very closely; and Anthony Eden, and later Harold Macmillan, always kept him well informed. When Anthony Eden resigned through ill-health early in 1957, the Queen sent, among others, for Churchill, and sought his advice on the succession to the Prime Ministership.

He rarely now made a speech in his constituency, but he always acquainted his electors of his views on crucial issues in the form of letters to the Constituency Chairman which were released to the press. Any Parliamentary Questions which needed to be raised on behalf of Churchill's constituents were looked after by the Conservative Member for the neighbouring seat of Walthamstow East, Mr John Harvey, a former Chairman of the Woodford Conservative Association – this was true and loyal neighbourliness. Constituency correspondence certainly did not suffer, as it was dealt with with extreme care by his own secretariat and by his Agent, Colonel Barlow-Wheeler, a most able and dedicated man. But the main task of visiting Woodford and of attending functions fell (as it had always done) on Clementine. She also, through her close and friendly contacts with the officers of the constituency party, kept a vigilant eye on its affairs.

There were from time to time rumblings of discontent among the younger elements in the local party, who felt the constituency was not properly represented, and that Winston's constituents did not get their fair share of attention. But these murmurings were subdued, if not entirely silenced, by the steadfast loyalty and affection the senior party members felt for him and Clementine. Most people in Woodford were undoubtedly proud to have him as their Member of Parliament.

As the time drew near once more for a General Election there was speculation as to whether Churchill intended to contest his seat again. At a meeting in London in early January 1959 of the Executive of the Woodford Conservative Association he made a vigorous address, but let fall no hint of his intentions. On 20 April of the same year, however, at the end of a 22-minute speech to a crowded meeting in Woodford, he told the audience that he would be ready once again to offer himself as their candidate,[4] this announcement was received with tumultuous applause. And so in October 1959 Winston and Clementine (aged nearly eighty-five and seventy-four respectively) fought yet another election: it was their fifteenth campaign together.

Clementine for some time now had held the view that Winston was too old to go on in politics, and that while the House of Commons would continue to accept him with proud and tolerant affection, sooner or later Woodford would become restive, despite the loyalty of the local party leadership. Winston knew what she thought; but she did not engage in any strong private campaign of persuasion: she knew his public life was the last stimulus left to him, and that he hoped to die – as he had for so long lived – the Right Honourable Member for Woodford. As in former years the decision was his, and she loyally, albeit somewhat wearily, braced herself for this last fight.

Winston made two speeches in his own constituency; the only occasion on which he spoke outside Woodford was at a meeting for John Harvey at Walthamstow. Two days before polling day, which was on 8 October, Winston made his usual open-car tour of the division, Clementine following in another open car. They were received with loyal affection and enthusiasm. Nationally the Conservatives were returned with an overall majority of 100 seats; Winston himself, although winning by a majority of more than fourteen thousand over his one Labour opponent, did less well than in 1955. It may well be that the feelings of discontent among some of his own supporters were reflected in this result.

* * *

After she had recovered from shingles in 1959, Clementine's health was quite good for a while, and 1960 was to be a busy year for her. She was particularly active in the constituency, trying as always to compensate for Winston's absence. Together with him, or alone, she also attended quite a number of 'gala' occasions, such as the functions surrounding the State Visit in April of General and Mme de Gaulle, who during the course of the few crowded days they were in England paid a private visit to Winston and Clementine at Hyde Park Gate – a gesture which touched them greatly. Later that same month they attended Princess Margaret's wedding in Westminster Abbey; and yet one time more they were at Windsor for the Garter Service in June.

In mid-November 1960 Winston had a fall at Hyde Park Gate, breaking a small bone high up in his neck; he made an astonishingly good recovery, but inevitably the accident left its mark, and he was unable to go to Edwina Sandys' wedding to Piers Dixon a few days before Christmas. Nevertheless, he made his reappearance in the House of Commons towards the end of January 1961. In April he visited the United States – it was to be for the last time – during the course of a month-long voyage in the *Christina*. It was a purely private visit, and while in New York he saw again his old friend Bernie Baruch.

In the winter of 1960, when Winston had had his fall, there was for Clementine all the worry and strain of the succeeding weeks, and by Christmastime and the New Year, which they spent at Chartwell, she was feeling exhausted; but she fended off her mounting fatigue with a course of injections. Knowing that Winston was going for a spring cruise, Clementine acquiesced when her doctors insisted that she should go into hospital for a complete rest and thorough check-up: she was admitted to St Mary's, Paddington on 1 March, where Winston visited her a few days before he left to join the *Christina*.

Clementine underwent rigorous medical tests and, beyond the normal changes of age, was found to have no organic disease: her trouble was nervous fatigue, depression and a state of anxiety – complementary conditions – brought on by the long strain of a lifetime. Also, the constant worry of Winston's declining health had increased the strain upon her; and the organization of their domestic life was still a perpetual harassment at a time when she herself was less able to cope.

After over three dreary weeks in hospital, she came down to spend Easter at Hamsell. When she left us, she went on to stay in Brighton at a hotel; for the doctors were most anxious that she should have as long a break as possible. The seaside always suited her, and, as Horatia Seymour was then living at nearby Hove, she had pleasant company. When Winston arrived home, he found Clementine much restored, and her 'batteries recharged' for the next lap.

While Winston's health and mental alertness were slowly declining, there were still bright moments, and sources of enjoyment. Although he always showed pleasure when any of his close family were around, old friends often stimulated him more than 'nearest and dearest' (perhaps both made more effort). Backgammon and bezique (especially the latter) were daily pleasures: Clementine, Sylvia Henley and Anthony Montague Browne were the most regular players, and meticulous count was kept of the 'debts'. After years of official papers, novels (mostly historical), old and new, now absorbed him by the hour. All his outings and public appearances were carefully prepared and stage-managed so that, impeccably attired, he gave a debonair impression. He was always attended by Anthony, or at the Other Club* dinners by Christopher, who also, with a kind fellow member, Rolf Dudley-Williams, escorted him on both sides during his visits to the House.

And so the months went slowly on. But it is noticeable from her engagement diary that from 1961 Clementine was always at home when Winston was in England, making a few brief visits away only when he was comfortably and safely installed at the Hôtel de Paris in Monte Carlo, where he now habitually stayed, or with the kind and protective Ari Onassis in *Christina*.

A fall in the Hôtel de Paris the following summer of 1962, in which Winston broke his hip, necessitating two operations and a prolonged recovery period, marked another definite stage in his slow decline. His mobility was considerably affected, and while he was in the Middlesex Hospital Clementine set about making alterations at Hyde Park Gate which would enable him to lead his life without negotiating stairs. The Office, which had hitherto occupied the main reception room of No. 27, was turned into a lovely bedroom for him: it was quiet, with bay windows looking out over the garden. Through in No. 28 a lift was installed from the ground floor to the dining-room on the lower ground floor, so that Winston could still come down to meals, and also have easy access to the garden, where he so loved to sit. Although he was nearly two months in hospital, it was a great rush to get all the work done. On his return home Winston was much pleased with the new arrangements, and particularly liked his new bedroom.

After this last accident it was a whole year before he went to Chartwell again. As Winston climbed slowly yet one more time up the steep slope of recovery, Clementine was hardly ever absent from his side. During the whole of the summer of 1962 she did not stir from London. It was a long vigil. Violet Bonham Carter wrote to her on 27 October 1962, with sympathy and true perception:

> My Darling Clemmie, I must write you a line (not to be answered) to say how moved I am, whenever I see you, by your amazing courage under this long strain, & by your gaiety & tenderness to beloved W. It is as though you alone could reach him with comfort & amusement. Your 'private line' with him has remained intact. Most people can be brave in short spasms – but the steadfast endurance of the 'long haul' is attained by few. You have had so many years of – sometimes intermittent, sometimes continuous – anxiety & strain with never a let-up – & now W needs you & claims more from you than ever before. I am filled with admiration & emotion when I see you together.[5]

Lady Violet understood also that if Clementine was to last the course, she

* The Other Club, founded in May 1911 by WSC and F. E. Smith, met fortnightly to dine at the Savoy Hotel while Parliament was in session. (It still does; and it now has women members.)

must have periods of let-up, and in this same warm and understanding letter she told Clementine that she would brush up her bezique, so that she could come and play with Winston from time to time. She was as good as her word, and with Sylvia Henley and Sybil Cholmondeley* formed an unofficial 'committee' who would in turn come to dine and play cards with Winston, while Clementine had a 'night off', going to the theatre or just dining out quietly with friends.

Christmas 1962, like the two more that remained to Winston, was spent in London. Clementine contrived to make it a true feast, and she wrote in gallant vein to Sarah, who was in Spain:

> Yesterday, Christmas Day we had a Patriarchal luncheon – Four generations, Papa & me, – Diana, Edwina & Piers – & Mark (5 weeks old). Papa was photographed by Fritz [the Swiss chef] holding Mark in his arms – We had poor Lord Moran whose wife is still very ill though better . . . Last week I took all Mary's children and Arabella to the Circus – It was great fun & we were much feted by Coco the Clown.[6]

Now not only was Winston physically less strong, but mentally he had become more lethargic. Gradually the silences became longer, and he was content to sit gazing into the fire, finding faces in the quivering glow; or, in the two summers that were left to him, he would lie on his 'wheelbarrow' chair contemplating from the lawn the view of the valley he had loved for so long. If he were not in good form, meals could be conducted in almost total silence. He rarely initiated a subject, and his deafness was an added and most daunting barrier to communication. But he hated to be alone – and indeed he hardly ever was. Sometimes after a long silence he would put out a loving hand, or say apologetically: 'I'm sorry I'm not very amusing today' – which wrung one's heart. He still loved to play bezique, but the games were now more drawn out, and he sometimes became confused with the score. This natural but infinitely sad decline was slow and uneven. There would be bright clear spells, and then dull and rather hazy days. Life held little for him now: he said to Diana, 'My life is over, but not yet ended.'[7]

Faithful as were old friends, and loyal and assiduous as was his entourage, the main burden of companionship fell, of course, on Clementine. Her devotion never failed, but sometimes her spirits flagged. Although she was glad for a slowing of the day-to-day pace of life, she was ten years younger than Winston, and her mind, accustomed to the brisk mental barter and exchange of political affairs, at times fretted (albeit unconsciously) at the dwindling flow of thought and conversation. The process of increasingly silent meals, the effort of communication and even the over-warm temperature of the house itself all contributed to a claustrophobic atmosphere which over a long period was chafing to her sensitive and highly strung

* The Marchioness of Cholmondeley, sister of Philip Sassoon. She and her husband were lifelong friends of Winston and Clementine.

nervous constitution. We in the family, and her close friends, became increasingly aware of this – but there was no easy solution.

In family terms the years 1962 and 1963 brought a full quota of joy and tragedy. In April 1962, after so much unhappiness and loneliness, Sarah married Henry Audley (the 23rd Baron Audley), a charming, sensitive and gifted man, with somewhat fragile health; he was nearly fifty (almost the same age as Sarah), and they had met each other early in 1962 in Marbella in Spain, where she was staying, and where he had a house. Telling her mother that she was going to be married, Sarah wrote, on 22 March 1962: 'I never ever believed I would ever find anyone ever again who could make me take heart and believe that happiness & love were yet ahead of me.' Shortly after this they both came back to England, and Henry was presented to the family. Clementine liked him very much and Winston, although now rather too old really to 'take in' new personalities, rejoiced for Sarah in her new-found happiness – as did we all. They were married in Gibraltar on 26 April, and Diana flew out to be with them. There is a series of happy letters from both Sarah and Henry to Clementine during their Moroccan honeymoon, and later from Marbella, where they planned to live for most of the time.

Their happiness, however, was to be of short duration: in early July 1963, while they were staying in Granada, Henry was suddenly struck down by a massive coronary and died. They had been married less than fifteen months, and Sarah's new-found happiness and the promise of security in which she had hardly dared to believe were brutally shattered. Diana, who had been the witness of their joy, now flew out to sustain Sarah and be with her when Henry was buried in the British Cemetery at Malaga.

During the summer of 1963, Clementine began once more showing signs that the heat and burden of the long day were indeed too much. Nervous tension and fatigue slowly built up, and she had real cause for anxiety on Winston's account when in mid-August he had a vascular stoppage just above his left ankle. Circulatory troubles always rang a warning bell, and he had to go to bed, which made him low and sad. It was school holiday time, and I was away with my family in France. I received worrying accounts from Chartwell, both from Grace Hamblin and from Monty, who had been to stay with my parents and gave me a detailed report on them both. Winston was evidently improving physically, and his circulation was restored. 'But, as you know,' Monty wrote on 1 September 1963,

> he can't now read a book or a paper; he just lies all day in bed doing nothing.*
>
> This has been a great strain on Clemmie, and she finally collapsed under it all and took to her bed the day after you and the children looked in on your way to France. She really is worn out. Winston dislikes being left alone all day

* It must be remembered that Winston's state varied very much, and this was a particularly bad period following on the circulatory trouble in his foot; he was by no means continuously bedridden.

> with his nurses, and dislikes having meals alone; Clemmie found it a strain having to talk loudly to make him hear.
>
> Since I have been here I have been with him all day, trying to interest him in things and showing him photographs of us two in the war. He is now definitely on the mend. He will recover. My view is that Clemmie is now the problem; she is worn out and needs rest.

Subsequently Clementine was examined by several doctors; they could find nothing specific, but all agreed that she was exhausted in mind and body.

I was greatly concerned, but could be of little help until the school holidays were over. But on 20 September my mother came to stay for what was meant to be a nice long restful visit with agreeable local distractions. Quite soon I realized, to my dismay, that she was really ill: our local doctor confirmed that her state of nervous exhaustion and mental anxiety were such as to be quite beyond ordinary sympathetic care – she needed expert attention. So her visit to us at Hamsell was unhappily curtailed and, heavily sedated, she was driven up to the Westminster Hospital, where she remained, receiving proper treatment, for over three weeks.

It was while Clementine was in hospital, on the night of 19–20 October, that Diana took a massive overdose of sleeping tablets, and died.

After she had parted from Duncan some time in 1956–7, Diana had settled in a charming house in Chester Row, SW1, where she lived with her daughters, Edwina (until her marriage in 1960) and Celia. The misery her bouts of nervous ill-health had caused her had been accentuated by the break-up of her marriage. Diana and Duncan had been married nearly twenty-five years when they were divorced in 1960: they had three children, and had known great happiness together in earlier days. The ending of their marriage was a grievous and deeply felt experience for Diana. In April 1962 Duncan married Marie-Claire, Viscountess Hudson. Around that time Diana made it officially known through her solicitor that she wished in future to be known as 'Mrs Diana Churchill'.

Undoubtedly the lack of understanding between herself and her mother was a cause of deep insecurity and unhappiness to her, but in this last period of her life Diana and her mother came to be, if not close to each other, at least on warmer and more understanding terms. Clementine felt deeply sympathetic to her in the break-up of her marriage, and Diana and her children were often at Chartwell, either for the day, or at weekends, and invariably at Christmastime. Diana also went on several occasions to the South of France or on cruises on the *Christina* to keep her father company. Now, during the distressing and perplexing illness which Clementine suffered in September and October 1963, Sarah was in Spain, and Diana and I had kept closely in touch about our mother's condition and the treatment she was undergoing. Diana was a source of wise counsel and compassionate understanding: she knew, all too well, the labyrinthine miseries of nervous and spiritual fatigue and despair.

From about the middle of 1962 Diana had started to work in a voluntary capacity for the Samaritans, the organization which gives instant and round-the-clock help to anyone in despair or on the brink of suicide: the work had gripped her enthusiasm, and she devoted many hours a week to it. Another source of satisfaction to her was Edwina's marriage and, later, the birth of her first grandchild: indeed, during the summer of 1963 she seemed in better spirits, and in a calmer and more robust state of mind, than she had been for a long time.

Diana had a great gift for friendship, and there were in particular three or four people to whom she could turn in moments of depression or worry. In the family, her closest relationship was with Sarah: together they had faced much, and each would always rush to the other's rescue in moments of crisis. As ill-chance would have it, at the time she took her life, all her main 'props and stays' were away from London, and Celia also was staying away in the country over that weekend. Edwina had seen her mother on the afternoon of Saturday, 19 October, and had found her in good spirits; Diana was due to lunch with her mother at the hospital the following day, and to dine with her father on that same Sunday evening. And so the dreadful news, when it broke upon us all, was not only totally unexpected but seemingly inexplicable.

Both Winston and Clementine, although through different causes, were spared the extreme shock and grief that such a sudden and tragic event must normally bring. Despite her illness, it was absolutely necessary for my mother to know about Diana without delay, for she always listened to the news on the radio, and usually read the newspapers. Owing to the nature of her illness, she was under fairly heavy sedation, and so what I had to break to her was mercifully filtered and softened. The lethargy of extreme old age dulls many sensibilities, and my father took in only slowly the news I had to tell him: but then he withdrew into a great and distant silence. It was a merciful dispensation that both my parents were spared the chill formalities of the inquest.

Clementine was enough recovered to be able to return home the day before Diana's funeral, but neither she nor Winston was well enough to go to the service. However, both were present at the crowded and moving Memorial Service which was held the following week in St Stephen's, Walbrook, the church in the City of London the crypt of which houses the headquarters of the Samaritans. Diana's ashes lie near her parents' grave at Bladon.

* * *

From the time of the 1959 General Election everybody concerned with the matter, and who knew Winston's true state of health, realized that he could not fight another campaign. He had said himself that if the Labour Party was victorious in 1959 he would fight again next time, but that if the Conservatives won, he would give Woodford a by-election. However, as the months following the Conservative victory went by, no further mention was made of this. In private, Clementine put forward her view that, even if Winston meant to stay on as Member of Parliament, he should let his local Association know that he did not intend to fight the seat again, so that they could look for a suitable successor. She often discussed this delicate problem with Christopher, who completely shared her view; and he also quite often raised the subject with Winston, who usually resisted discussion of the question, and turned an obstinately deaf ear to these reasonable but, to him, distasteful arguments. The Woodford Executive was too loyal to make an issue of the question themselves, but of course they hoped that a decision would be forthcoming sooner rather than later.

Towards the end of 1961 Clementine began to put pressure on the Woodford Executive Council to take the initiative, and themselves to raise with Winston the question of his continuing as their Member: but the Council was very loath to make the first move. Their Chairman at this time was Mrs Doris Moss, a most able and charming woman, devoted to Winston and Clementine, and greatly liked by both of them. She was fully aware of the delicacy and difficulty of the situation, and she had said to Winston privately, and in public on many occasions, that while she was Chairman, the Council would never ask for his resignation. However, both she and her colleagues continued to hope that he would make the first move, or that Clementine would raise the subject with him afresh. But, writing to Mrs Moss in early February 1962, Clementine told her that it was extremely unlikely that they could expect the first solution, and that she, for her part, was not prepared to be the go-between in this matter.

In all the record of their long life together, this is the first example I can find of Clementine lacking the will and the nerve to raise and argue a matter with Winston which she deemed to be of vital importance. Her realistic appraisal of the situation in the constituency was not new: as long ago as 1949 she had expressed herself strongly to him on his position as Party Leader and his relationship with his constituency in a 'House-post' letter:

> You often tease me and call me 'pink' but believe me I feel it very much. I do not mind if you resign the Leadership when things are good, but I can't bear you to be accepted murmuringly and uneasily – In my humble way I have tried to help – the political lunches here, visits to Woodford, attending to your Constituency correspondence – But now & then I have felt chilled & discouraged by the creeping knowledge that you do only just as much as will keep you in Power. But that much is not enough in these hard anxious times.[8]

Now, thirteen years on, her perceptiveness was still as keen, and her pride in him and for him made her view with dismay the prospect of a slow, and perhaps humiliating, end to his parliamentary career. Moreover, there was the constantly lurking fear that Winston might fail or collapse in public.

Clementine had never before shirked an issue: but now she knew that in urging him to resign his seat she would be counselling him to cut finally the last link between himself and the ebb and flow of public life which had for so long been the throbbing impulse of his own. The question of his final retirement had been discussed between them from time to time, but always inconclusively – and usually unhappily. Enfeebled by age and the illnesses he had so marvellously survived, and although now only an infrequent attender at the House of Commons, Winston still drew satisfaction from being a Member of Parliament; and visitors in the public galleries always noticed his presence in the House and liked to see him in his accustomed place.

Now, Clementine could not bring herself to urge Winston to sever this main artery of his public life. They were both older, and more and more dependent on each other's company as friends departed and life bustled on. Whenever she discussed the infinitely painful subject with me, she always said that she could not bear for Winston to be estranged from her, and she dreaded to initiate a suggestion she knew he would regard as cruel – and, indeed, disloyal.

For a while, therefore, an impasse having been reached, the situation remained unchanged. Clementine continued to hold the fort in the constituency; often Anthony Montague Browne would go down with her to help, and quite often she invited visiting speakers to address meetings, at which she was nearly always present herself. But although she soldiered on, she was sensitive about the situation, and it made her ill-at-ease. In 1962 the summer garden party at Chartwell to which Winston and Clementine annually invited a large party of constituents from Woodford had to be cancelled when Winston broke his hip; and naturally enough, during the weeks of his illness and slow convalescence, the whole matter remained in abeyance.

But early in 1963 another General Election was at most eighteen months off, and Mrs Moss and her colleagues on the Divisional Executive realized that the issue of Winston's resignation must be grasped and resolved; and this time she raised the matter with Winston herself. Clementine invited her, with her husband, to luncheon at Hyde Park Gate on 10 April; at the end of the meal, by previous arrangement, Clementine and Mr Moss withdrew, leaving Mrs Moss alone with Winston. After talking to him in general terms about the situation, she left with him (it was Clementine's suggestion) a carefully reasoned memorandum, setting out the whole problem and its urgency. But no decision was arrived at on this occasion.[9]

The next day Winston left London to spend Easter at Monte Carlo, taking Anthony and Nonie Montague Browne with him for company. Clementine remained in England, coming down to us at Hamsell for the Easter weekend. The whole Woodford problem was troubling her greatly, and she talked to both of us about it. On Easter Day, Christopher wrote a long letter to Winston putting forward once more, with necessary but stringent candour, all the arguments which had already over the past months been advanced, and urging him to accept the fact that, however unpalatable, the time had come when he really must give Woodford a decision – and that there could be only one.

In a later exchange of affectionate letters between Winston and Clementine, the painful question was not mentioned, but it is clear from her letter to him on 19 April that she knew he was still turning over in his mind the possibility of actually contesting his seat at the next election. 'I hope, Darling,' she wrote,

> you are thinking carefully about the letter Christopher wrote to you – He read it to me before he despatched it & I agree with all he says.
>
> I don't see how you can stand next year without campaigning & fighting for your seat. And it would be kind to let your Executive Council know now, before they become too restive.

Winston flew home on 25 April, and the following day Mr and Mrs Moss again came to luncheon. Clementine had been struck down by a virus, and was in bed, but Anthony Montague Browne, who was present, gave her a full account of what had passed: it had been painful and difficult – but Mrs Moss had been admirable in the handling of the situation, although there had been no positive response from Winston whatsoever. But if he had not wished to commit himself then, only a few more days passed before he brought himself to take the – for him – infinitely depressing decision. On 1 May Clementine wrote to Christopher:

> In tomorrow's newspapers you will see Winston's letter to his Chairman telling her that he will not stand again at the next General Election.
>
> I feel your excellently reasoned letter was a contributory cause to it being written at last!
>
> I am much relieved as the situation at Woodford was becoming increasingly uneasy.

The anguish of this long-drawn-out dilemma may well have contributed to Clementine's illness later in the summer of 1963.

This sad and difficult episode, however, had a dignified and seemly epilogue in the following year. The Government and leaders of the Opposition parties wished to mark in some special way Churchill's departure from the House of Commons, and they decided to do so by moving a vote of thanks and appreciation of his services just before the House rose for the summer recess. This had been, until early in the century, the manner in which the House paid tribute to distinguished and victorious senior officers in the services, and to Speakers and servants of the House. The custom had fallen into disuse, and it was proposed to revive it for Winston Churchill. It was felt by everyone concerned that it would be too testing for Winston to reply to the Vote of Thanks in the House, especially with the emotion such an occasion would obviously engender. It was proposed, therefore, that the Resolution should be moved in his absence, and that a special committee from the House should then call at Hyde Park Gate to hand him a copy of the Resolution; Winston would then thank them, and hand them a written speech of thanks which at a convenient moment would be read out to the House.

Clementine was in entire agreement with the course proposed, and she incorporated the suggestion in a short memorandum which she gave to Winston on 18 April 1964, asking him if such a procedure would be 'agreeable' to him. She sent a copy of the memorandum to Christopher with a covering note:

> Dearest Christopher,
>
> Yesterday I gave Winston this little note. At first he seemed to pretend not to understand it, but later on said he thought it would be very suitable. You will notice that I have carefully left out any suggestion that he could possibly be in the House when the Vote of Thanks is proposed. I did not show him the wording of the Resolution.
>
> Later on in the afternoon he seemed very sad and depressed.
>
> Yours affectionately,
> Clemmie[10]

Christopher, who in all this was the chief link, had sent a copy of the proposed Resolution to Clementine. It read as follows:

> That this House offers to the Right Hon. Gentleman the Member for Woodford, its grateful thanks in deep appreciation of his outstanding service to the nation and expresses its sincere regret that the Right Hon. Gentleman, after a Parliamentary career so long and so distinguished, will not seek to be re-elected to this House.

This somewhat banal effusion touched off an immediate and fiery reaction in Clementine, who, having done some historical research, wrote to Christopher on 18 April:

> This is what Mr. Speaker said to the Duke of Wellington when he was thanking him for his services:
>
> 'My Lord,
>
> Since last I had the honour of addressing You from this place, a series of eventful years has elapsed; but none without some mark and note of your rising glory.
>
> The Military Triumphs which your valour has achieved upon the banks of the Douro and the Tagus, of the Ebro and the Garonne, have called forth the spontaneous shouts of admiring nations. Those Triumphs it is needless on this day to recount. Their Names have been written by your conquering sword in the Annals of Europe, and We shall hand them down with exultation to our children's children.'
>
> As a tribute compare this with the mangy resolution you enclosed with your letter.
>
> Yours affect[ly]
> Clementine S.C.

The essence of this broadside was tactfully communicated by Christopher to the right quarters, and presently a redrafted Resolution was sent to Clementine, who signified that she liked the new version very much better.

Winston Churchill was present for the last time in the House of Commons on 28 July 1964. He had been a Member almost continuously for over half a century. The following day, in the afternoon, Sir Alec Douglas-Home, the Prime Minister, accompanied by the Opposition Party Leaders Mr Harold Wilson and Mr Jo Grimond, and the Leader of the House, Mr Selwyn Lloyd, together with the two 'elders' of the House, Sir Thomas Moore and Mr Emmanuel Shinwell, called at Hyde Park Gate to present to Winston the Resolution which had been carried earlier in the day in the House – *nemine contradicente*. The words of the Resolution were:

> That this House desires to take this opportunity of marking the forthcoming retirement of the right honourable Gentleman the Member for Woodford by putting on record its unbounded admiration and gratitude for his services to Parliament, to the nation and to the world; remembers, above all, his inspiration of the British people when they stood alone, and his leadership until victory was won; and offers its grateful thanks to the right honourable Gentleman for these outstanding services to this House and to the nation.[11]

This private but historic ceremony took place in the dining-room at Hyde Park Gate. Clementine had asked some of us children to be present also to witness this rather sad, muted occasion. But it made a dignified end to a long, proud chapter.

CHAPTER TWENTY-NINE

Port after Stormy Seas

A MARKED FEATURE OF THESE LATTER YEARS OF WINSTON AND CLEMENTINE'S long life together was the decrease in the number of letters between them. The decline in Winston's health after he suffered from pneumonia in 1958, accentuated by his neck injury in 1961 and breaking his hip in the following year, combined with Clementine's long-drawn-out ordeal with neuritis and her subsequent bout of shingles, served to make letter-writing for them both a considerable effort, and sapped the mutual capacity for constant communication which had been such a strong characteristic of their relationship for more than fifty years. Short notes from Winston which survive are full of affection and concern; occasionally they are somewhat muddled, and the writing is wandery, but the message shines through. Clementine's bold hand is still fluent, but she too became a meagre letter-writer.

Three examples suffice:

Undated [October 1962, from London]

Darling,

I hope you are going on well & that we may come together again tomorrow. I have found it quite lonely & will rejoice to see us joined together in gaiety and love. Dearest one I place myself at your disposal & intend to take a walk in the park hand in hand

With many kisses
Ever loving
W

[dated by CSC 8 April 1963]

My darling One,

This is only to give you my fondest love and kisses <u>a hundred times repeated</u>. I am a pretty dull & paltry scribbler; but my stick as I write carries my heart along with it.

Yours ever & always,
W

My darling one,

This is only to give you my fondest love and kisses a hundred times repeated.

I am a pretty dull & paltry scribbler; but my wish is to write carries my heart along with it.

Yours ever & always,

W

October 1962

Darling,

I hope you are going on well & that we may come-together again tomorrow.

I have found it quite lonely & will rejoice to see us joined together in gaiety and love.

Dearest one I place myself at your disposal & intend to take a walk in the park hand in hand with many kisses

Ever loving

W

July the 4th 1963

WESTERHAM 3344

CHARTWELL.
WESTERHAM,
KENT.

My Darling,

The Time has seemed long with-out you –

I shall be on the door-step to wel-come you Home.

Your devoted
Clemmie

Letters between Clementine and Winston in the 1960s.

And, on Winston's return from France:

July the 4th, 1963
Chartwell

My Darling,
The Time has seemed long without you –
I shall be on the door-step to welcome you Home.

Your devoted
Clemmie

The tragic deaths of Henry Audley and Diana in 1963, and Clementine's illness, combined to make that winter a quiet one. The manner of Diana's death left for her, as for all of us, a legacy of painful thoughts and questions. After Clementine came out of hospital, she needed a little time to readjust to life once more; but she seemed now much more able to accept the slow tempo which Winston's infirmities imposed on his own life, and upon those who lived with him. Moreover, once his impending retirement from the House had been settled and announced, a whole weight of worry and fret had been lifted from her; she continued to keep in touch with the constituency, but she was no longer haunted by embarrassment, or by a sense of fatiguing obligation.

Although after 1963 Winston did not go abroad again, he was not totally inactive: he faithfully attended the Other Club dinners, where old and younger friends and former colleagues all joined to make him feel the glow of their affection and regard. During the months left to him as the Member for Woodford, he went on several occasions to the House of Commons; indeed, towards the end of February 1964 he caused surprise and pleasure when he attended the House one night, staying to vote in more than one division.[1]

Winston and Clementine went down to Chartwell for the first time that year for Easter; two days later, on 1 April, Clementine's seventy-ninth birthday, they lunched with us at Hamsell, bringing with them Sylvia Henley and Monty and one of his godsons, Nigel Hamilton, who were staying with them. In addition to the quiverful of Soameses, we had invited 'Cousin Moppet' and Horatia Seymour (both of whom lived nearby). It was a happy feast: and I am so glad we did not know it was the last time my father would be at Hamsell.

In early June they returned to London, and there on 9 June they received news of Max Beaverbrook's death: Winston was greatly saddened and depressed. Since Clementine had 'laid up her sword', her relationship with Max had been mellow; his kindness in these last years had touched her heart. It must also have moved her to have received this 'accolade' from him, less than a year before his death, written in his own hand on the occasion of Winston's eighty-ninth birthday the previous November: 'My dear Clemmie, What a burden you have borne over so many years – and with what charm & dignity. How much the Nation & the World owes to you for all your labours. And on this 89th birthday I send you this message. Many many intimates sending Winston messages of love & devotion will think of you. Max.'

Winston and Clementine stayed in London until the end of July, when, after he had made his last visit to the House, and received their Resolution of Thanks, they both returned to Chartwell. There they remained tranquilly for nearly three months, enjoying visits from friends and relations; and Clementine made one or two brief expeditions to London. In mid-October they packed up, and went back to winter in London. And Winston left – for the last time – the place which for over forty years had given him happiness, occupation and peace of mind.

Throughout 1964 Clementine's health was in general much better, although she did have some trouble with her eyes, and she too began to find the need of a hearing aid – a necessity she disliked as much as did Winston. But apart from these tedious but minor troubles, she seemed to be on a much more even keel, notwithstanding the fact that, other than two week-ends spent with us at Hamsell (one in February and one in July), and three days in Paris in October, she had no holiday, and was constantly at home. Undoubtedly she could have done with more 'time off'; but she felt Winston needed her more than ever, as slowly but perceptibly his health and spirits declined. Clementine, like all of us, was conscious that the sands were gently but inexorably running out.

However, she had distractions, and friends rallied round either to divert Winston or to take her out; she loved the theatre, and exhibitions were as ever a great pleasure and interest. As usual, during Wimbledon tennis fort-night she was a regular spectator, the Committee always giving her tickets in the Royal Box. She used often to take a grandchild with her, and these Wimbledon outings were regarded by them as beautiful treats.

During this year Clementine undertook one or two public engagements. On 13 May she attended the opening of Howard House, a home for retired nurses; the previous year she had signed a letter with Lord Moran appealing for support for this home, and it was a cause she cared about – she always realized that nursing was a hard and exhausting career.

Churchill College, Cambridge, founded by Winston in 1958 (and to which he had subscribed a great part of his Eightieth Birthday Fund), was now a living, working entity, and the formal opening by the Duke of Edinburgh took place on 5 June 1964. Winston was not well enough to attend such a long and, of necessity, tiring event; but Clementine repre-sented him, and all the family was invited. It was, of course, a thrilling day for all of us to see this brainchild of his old age so gloriously launched on its useful and significant career.

In the latter part of the year she visited Paris, staying at the British Embassy to unveil a bust of Winston by Oscar Nemon, presented by Mr Alan Spears and his wife. Clementine wrote to Sarah, who was in Rome, a lively account:

> I have just got back from spending 3 days in Paris – I went to unveil a bust (by Nemon) of Papa which is placed in the British Embassy. It is extremely good. I lunched with General de Gaulle which I enjoyed very much. He was very mellow. I stayed at the British Embassy with Edwina's parents-in-law, Sir Pierson & Lady Dixon. Edwina was there & we did sightseeing together.[2]

On the family front the year had been variable, but fortunately less tragic and dramatic than 1963. Although Randolph had a lung operation and was seriously ill during March, he made a good recovery; while he was in hospital his mother visited him frequently. And in early August, on the eve of our departure for a family holiday abroad, Christopher had a riding accident and injured his pelvis, and had to spend ten weeks in hospital. My mother at once offered practical help, by having Nannie and the younger children to stay at Chartwell, so that I could spend my time with Christopher.

But the most important family event, and one which gave true joy and pleasure to both Clementine and Winston, was the marriage of Randolph's son, Winston, to Minnie d'Erlanger. Although his grandfather was unable to go to the ceremony or the reception, Winston and Minnie and her parents came afterwards to No. 28 to greet and be greeted by him.

Winston's ninetieth birthday brought an avalanche of good wishes. The Queen sent him flowers, and many people gathered in the street outside No. 28 on the day before, when, carefully dressed and looking benevolent, Winston appeared at the open window of the drawing-room so that the press could take some birthday photographs. The little crowd of well-wishers cheered and clapped and sang 'Happy Birthday to You'. Before luncheon on the actual day Clementine arranged for all his secretarial, nursing and domestic staff to gather in his bedroom to drink his health in champagne. She took this opportunity to thank them for their loyal and devoted care. Her present to him was a small golden heart enclosing the engraved figures '90'. It was to hang on his watch-chain, and joined the golden heart with its central ruby 'drop of blood' which had been her engagement present to him fifty-seven years before. During the afternoon the Prime Minister called to bring Winston good wishes from the Cabinet.

On the evening before, the BBC showed *Ninety Years On*, a birthday tribute devised by Terence Rattigan and Noël Coward: it was perfect, full of Winston's old favourites – songs from the Boer War and the Edwardian music halls to 'Run, Rabbit, Run'.

On the night of the 30th itself there was the usual hallowed family dinner party: Randolph, Sarah, myself and Christopher; Winston and Minnie, and Arabella; Julian Sandys, Edwina and Piers Dixon, and Celia Kennedy (Sandys); and Cousin Sylvia. The only guests not members of the family were Jock and Meg Colville, and Anthony and Nonie Montague Browne. Monty had been invited but was himself ill in hospital. The house glowed with candlelight and flowers, and we were united yet one more time in drinking first Winston's health and then Clementine's. But this birthday evening had for us all a poignant quality – he was so fragile now, and often so remote. And although he beamed at us as we all gathered round him, and one felt he was glad to have us there, in our hearts we knew the end could not be far off.

* * *

On Monday, 11 January 1965 my four elder children and I were due to go up to London to see my parents, and to dine and go to the theatre with my mother. During the morning my mother telephoned me at Hamsell to say that my father was far from well, and seemed to have had another 'spasm'. Lord Moran had been to see him on the previous day, and was bringing Lord Brain, the neurologist, to examine him. She insisted that our plans should not change, so the children and I went to No. 28 that evening as arranged. My mother told me that in the doctors' opinion Winston had suffered another stroke, but that it was not possible at this stage to judge its gravity. I went in to see my father, who was in bed. He was silent, and I do not think he recognized me. We all went to the theatre, but our minds were full of anxious thoughts.

During the next day I visited my parents. My father's eyes were open, but he seemed very remote and said nothing. He was still eating a little, but was having to be fed. Christopher came to see him during the afternoon, and was alone with him. Desperately trying to kindle a spark, he said: 'Wouldn't you like a glass of champagne?' Winston looked at him vaguely. 'I'm so bored with it all,' he said. These were, I think the last coherent words he spoke to any member of the family.

During that afternoon I had seen Lord Moran. He seemed gloomy. 'Is this it?' I asked. 'I'm afraid so,' he replied.

Over the next days the effects of the stroke became gradually more pronounced, with paralysis affecting his left arm and hand; at this stage there was still an element of doubt as to whether Winston might yet once more recover; but it was an anguishing prospect, for Lord Moran warned us that he would be gravely impaired, and would not even regain the level of existence of the last sad months. The question had at once to be faced as to whether a bulletin should be issued; it was decided that nothing should be announced at the present time, but that should his state change decisively, or if there were to be rumours and specific press enquiries – then a statement should be made at once. Meanwhile, only those members of the family and friends who were closest and could be easily contacted were told.

For the next two days there was not much change in his condition. He was semi-conscious at times, but even when he slipped into unconsciousness, the strength of his grasp, as one sat holding his hand, was warm and strong. On Thursday, 14 January, he seemed a little worse: although I had been coming to the house every day, I now moved in, so that I could be close at hand if needed. To pass the time and to get some fresh air and exercise, my mother and I went for walks, and, on one bitter afternoon, we paced the long galleries of the Victoria and Albert Museum in a somewhat distracted way. As these days went by without rumours, we marvelled at our luck; we all felt that there was everything in favour of keeping the secret for as long as possible, for it seemed perfectly clear that whatever the course of my father's illness might be, it was likely to be a long-drawn-out ordeal, and that nothing would be the same again once the news was known to the public.

Friday, 15 January was a cold, grey day, but my mother and I went for a brisk walk in Hyde Park during the morning; as we were returning home and were approaching the top of Hyde Park Gate, we saw Anthony Montague Browne coming quickly towards us. The same thought flashed instantly through both our minds – he was coming to tell us that Winston had died. But Anthony was only hastening to warn us that the news was out, that there had been several enquiries from the press and that an interim bulletin had been issued. We hurried on home, and a few minutes before we got into the house, the first cameramen and reporters were arriving. By the middle of that afternoon the street was crowded and No. 28 was in a state of siege. During the afternoon a bulletin was put out signed by Lord Moran and Lord Brain. It said: 'After a cold, Sir Winston has developed a circulatory weakness and there has been a cerebral thrombosis.'

From now until, at Clementine's personal request some five days later, the press withdrew to the end of the street, we were completely hemmed in, not only by reporters, cameramen and television crews, but by numbers of the general public, calling to deliver notes and flowers, or just standing quietly in the chilly street: it was quite difficult getting in and out of the house. Reporters were, without exception, kind and courteous, and once it became clear that a steady flow of informed, official bulletins would be issued, we were not stopped and questioned. An account was published daily of the comings and goings at No. 28; people were asked – through the medium of the press – to refrain from ringing up for news and to wait for the bulletins, as the two telephone lines in the house would otherwise have been completely drowned by the flood of enquiries.

My mother said she would like a priest to come, and I telephoned Philip Hayllar, the vicar at Eridge, whom she had often met on her visits to Hamsell, and whom she liked very much. He slipped into the house, unremarked, on the Friday evening. My mother and I knelt on either side of my father's bed while Philip Hayllar said some simple prayers and blessed him. Winston was now unconscious, and slipping gradually into deeper sleep. We knew in our hearts that he was dying.

During that evening, the Prime Minister called. My mother, although quite calm, was exhausted, and had gone upstairs to rest, so I received Mr Wilson on her behalf. Throughout his illness, Buckingham Palace and Downing Street kept in close touch with Winston's condition through Anthony Montague Browne. The following week a delegation bearing a personal message from the Pope called, and Clementine with several members of the family received them. Messages poured in from all over the world. Anthony and the secretaries were on duty almost round the clock. The nursing team, headed by Mr Howells, was increased to four.

Sarah arrived back from Rome in the early hours of Saturday, 16 January, and came straight to the house to see her father; every day children, grandchildren and closest friends called, and through all their visitations Winston slept tranquilly. His room became the heart and centre of the house; we all slipped quietly in and out, to sit just a little while more with him. It was such a peaceful scene – the gently lighted room full of flowers; the quiet form in the big bed, with the beautiful hands spread on the quilt. Our children came up to say farewell to their so dearly loved grandfather – they were rigid and apprehensive until they had been into that peaceful room, where their grandmother was sitting quietly by their grandfather's side holding his hand, and with the marmalade cat he so much loved curled up at his feet. Afterwards I found Jeremy (aged twelve) in tears in the passage, and murmured some banal words of comfort: 'Oh, I know that!' he said almost roughly. 'But it's just seeing *them* together, for the last time.'

As the days of their long journey together were thus drawing to their gentle end, Clementine seemed to withdraw into a remote inner quietness. Although Anthony and I dealt with as many problems and queries as possible, all major points were of course addressed to her. She saw a great number of the messages, letters and telegrams that poured in every day, and dealt with many of them personally. She looked immaculate at all times, and the house continued to be impeccably run: there had never been any room in her life for slipshod ways.

Almost every day my mother and I went out for a drive or a walk, or visited in a somewhat robot-like trance the National Gallery and the Wallace Collection. It was, however, quite an ordeal, running the gauntlet of the crowd at our door, and in public places one could feel people's loving but curious sympathy reaching one in waves: so, on the whole, we marched about the parks in the grey chill days – killing time, while time killed him. Clementine never liked to be out for long, for although Winston was sinking slowly, there were one or two crisis moments when we thought the end was coming.

Public concern throughout the world mounted through all the days Winston lay dying. On Sunday, 17 January prayers were offered for him in churches of all denominations throughout the country. The Morning Service was broadcast from St Mary's, Westerham, and my mother and I listened, sitting in her bedroom. It seemed quite natural that together we knelt to join in the prayers. Meanwhile the state of siege in which we were living was becoming a great trial: the reporters and cameramen were keeping a round-the-clock vigil. Nearly every time someone called, or Lord Moran gave out the bulletins, or made a visit (sometimes twice or three times in the twenty-four hours), the crowd pressed closer round the front door and the television arc-lights were turned on. Clementine's bedroom was on the first floor, and directly over the front door; and on the ground floor the drawing-room windows gave directly on to the street. Although the crowd was as quiet as possible, its presence penetrated even inside the house. The strain was considerable – and particularly so for Clementine. She was also very conscious of the great inconvenience inevitably caused to other residents in the street (which is a cul-de-sac) by the constant presence of so many people, clustered on the pavements and often spreading over the roadway as well. On Tuesday, 19 January, after consulting with Anthony, the following request was issued on her behalf:

> Lady Churchill expresses her true thanks to members of the press who have shown great kindness and restraint. However, she would be most grateful if the press, radio, T.V. and film representatives could withdraw from Hyde Park Gate. The numbers have grown to such an extent that the cameras' floodlights and inevitable disturbance have become a severe strain, apart from obstructing the street. In future, medical bulletins will be issued over the telephone at the news agencies and not read aloud at the front door.

This most courteous request met with an immediate response – within about half an hour of the statement being issued all reporters and cameramen had withdrawn, and from now on they posted themselves at the top of the street. Their instant co-operation and understanding were greatly appreciated by Clementine (and, indeed, by all of us living and working in the house), and rendered less arduous the four days of the long vigil that were left.

During all this time Clementine's demeanour was calm and quiet. I saw her break down only once – and only briefly. One day she said to me: 'I don't know where all my tears have gone.' Members of the family kept her company at meals, and Sylvia Henley came often to lend her unfailing strength. Apart from our walks, my mother went out rarely; but one day she lunched quietly with Christopher and me at our flat. During the day she would go and sit with Winston, either alone or with one of us, sitting for long periods holding his hand. The nurses told me that, even after Winston had sunk into a deeper level of unconsciousness, they were convinced he was aware of her presence; and sometimes Clementine found it quite hard to withdraw her hand from his clasp. Just before she went to bed she would go to say 'goodnight', and to sit with him for a little while. She nearly always woke during the night, and at one or two in the morning she would come downstairs to see how he was.

While she was resting on the evening of Friday, 22 January, Randolph came round with the happy news that Minnie, young Winston's wife, had given birth to a son. Not wishing to disturb his mother, Randolph left a note for her – adding at the end: 'in the midst of death, we are in life.'

The bulletins about Winston's condition continued to be published at least twice a day, although there was often nothing new to report. On the evening of 20 January the bulletin noted: 'The weakness of Sir Winston's circulation is more marked.' Two days later: 'Sir Winston has had a restful day, but there has been some deterioration in his condition.' On

23 January: 'The deterioration in Sir Winston's condition is more marked.'

I woke up at half-past five on the morning of Sunday, 24 January and went downstairs. I met Nurse Huddleston (who was on duty with Mr Howells) coming up to wake me – my father's condition had deteriorated, and she told me she did not think 'he could go on very long'. We decided, however, not to wake my mother for a little while. About half-past six I was just going up to fetch her when she appeared. Lord Moran was sent for, and I started telephoning the other members of the family and Anthony Montague Browne. My mother and I sat one on each side of my father. About a quarter-past seven Randolph and young Winston arrived and, soon after, Sarah and Celia. They all joined Clementine in Winston's room: Lord Moran and Anthony were also there.

Clementine did not move from Winston's side all this time, but sat by the bed holding his right hand. Once she looked up at me: 'Go and tell Mrs Douglas we shall be . . .' – she broke off to count those around her – '. . . eight for breakfast.' She then turned again to Winston, and concentrated her whole thought upon him. When I returned from carrying her message I sat at the foot of the bed. Clementine was, throughout, on Winston's right, and young Winston at his left: the others had gone into the drawing-room for a while. Sitting exactly opposite my father, I suddenly noticed a change in him – his breathing seemed to alter. Through the open door I signalled to one of the nurses, who observed the change and instantly ran along the passage to fetch the others. We gathered around him – Sarah, Celia and I kneeling at the foot of the bed, Randolph standing near the pillow by his son, Lord Moran on his right by my mother. I was dimly aware that at the back of the room the two nurses and Anthony had fallen to their knees. There was absolute silence – Winston gave two or three long, long sighs – nobody moved or spoke. Presently Clementine looked up at Lord Moran: 'Has he gone?' she asked. He nodded. Presently, one by one we got up and silently left the room. I went back and remained with my mother for some little time. Then we both kissed his hand and then his brow and left him.

At half-past eight that morning the news was given to the press: 'Shortly after 8.0 this morning, Sunday, January the 24th, Sir Winston Churchill died at his London home. Moran.'

Some of the family went quickly away; those of us remaining had breakfast (rather chokily) with Clementine. We listened in as the news the world had been waiting for was broadcast. Clementine then went upstairs to dress, and when she came down again she went to the (now) candlelit room. Death had smoothed all wrinkles away – there was no sign of feebleness or age: the alabaster-like presence distilled around it a majestic sense of peace and finality. During that day my mother and I spent long periods sitting there, in silent contemplation and recollection. Otherwise Clementine remained most of the time in her room. She was perfectly calm, and addressed her mind to those matters which had to be discussed or decided. Anthony, who had through all these long days been a tower of strength, continued now to take an immense burden off Clementine, and dealt with the enquiries and messages which once again were at flood-tide.

My mother had asked me to telephone Violet Asquith* and to invite her to come and take her leave if she so wished. Violet had visited Winston soon after he had become ill: 'Clemmie came down', she had written in her diary on 14 January, 'very brave but looked worn out. She asked if I wld like to come & see him & I went into his bedroom. He was propped up on pillows – breathing quite easily & quietly & his beloved face looked at peace.'[3] Now Violet came to say a last goodbye. I took her to my father's room and left her.

It was hard indeed to tear ourselves away from that great and peaceful presence, but presently Clementine too took her very last farewell. Later, she sent me down to see the undertakers, to assure myself that all was seemly and to instruct them to close the coffin.

* * *

Seven years before, the Queen had told the then Prime Minister, Harold Macmillan, that in the event of Churchill's death she wished him to be accorded a Lying-in-State, followed by a State Funeral – a signal and rare honour to be granted to a commoner, and the first since the Duke of Wellington. The Queen's wish was in due course transmitted to Winston, who was deeply touched and gratified by this gesture from the Sovereign whose family he had so long served, and for whose person he had such a great regard and deep affection. A myth has taken root that Winston took a detailed interest in the planning of his own funeral, down to the choice of hymns: this is not true. Although I myself once heard him express the hope that there would be 'plenty of bands', the matter did not greatly occupy his thoughts: he was more than content to leave these arrangements to others.

From the moment of Winston's quiet end, 'the plan' took over. There had for some time been a discreet liaison between the Earl Marshal's office and Anthony Montague Browne, and he in his turn had consulted Clementine on various points or communicated to her such information as was timely and necessary without being distressingly cold-blooded. At every step her feelings, and those of the family, were taken into consideration.

On Monday, 25 January, the Houses of Parliament paid their tributes to Winston. Clementine did not go to the Commons, but various members of the family went, and Sarah and I watched and listened from the Speaker's Gallery, while political friend and foe alike recalled what Winston Churchill had been to the House and to his country. His old place – the seat below the gangway on the government side – had been left symbolically empty. Many of the speakers made especial reference to Clementine, and to her constant and steadfast role at Winston's side.

Winston lay in his room at No. 28 Hyde Park Gate from the Sunday morning on which he died until the evening of the Tuesday. During those

* Lady Violet Bonham Carter had been made a Life Peeress in 1964 and had taken the title of Baroness Asquith of Yarnbury.

two days his room was still the centre of the house, and Clementine and all of us would go and sit there very often. On the evening of Tuesday, 26 January the closest members of the family gathered at No. 28, and at nine o'clock the Lord Chamberlain, Lord Cobbold, arrived to escort the coffin to Westminster Hall where Winston was to lie in state until the State Funeral on the following Saturday. From the moment his body left his home, he, and all of us, became integral parts of a great pageant of state, upon which were fixed the interest and emotions not only of his own countrymen, but of people throughout the world. The coffin, covered by the Union Flag, was driven in a motor hearse, and we all followed in other cars. Although it was a dark and cold night, there was a crowd outside Westminster Hall. At the door the Bearer Party found by the Grenadier Guards bore the coffin into the vast gloom of the historic hall. The small family group followed, led by Clementine (in black and veiled) and Randolph. When the coffin had been placed on the catafalque, the Archbishop of Canterbury said some prayers and gave a benediction. Then the first four officers from among those from all three services who were to find the guard during the days and nights of the Lying-in-State, took up their positions by the catafalque. And guarded thus – we left him.

During the next three days Westminster Hall was open twenty-three out of the twenty-four hours so that people could come and pay their respects. In the bitter cold weather the queue was often more than a mile long, and people had to wait for up to four hours to reach the hall. When the doors were finally closed to the public early on the morning of Saturday, 30 January, the day of the funeral, over three hundred thousand people had passed by the catafalque.

For Clementine these days were spent mostly in seclusion at home; but every day she went to spend some time at the Lying-in-State, usually taking me with her. Members of the family were allowed in by a side-door, and we were able to stand at the side of the hall without disrupting the flow of people. It was an amazing, an unforgettable sight. The only gleam of light in that vast place seemed to come from the great cross at the head of the catafalque, and from the candles round the flag-covered coffin, on which reposed the Insignia of the Order of the Garter. The guarding officers stood like statues, heads bowed, hands clasped on the hilts of their naked swords. The change of guard, which took place every twenty minutes, was performed without a spoken command and in almost total silence. Carpeting had been laid, and so the stream of people who passed on both sides of the catafalque without interruption were equally noiseless – as one stood and watched, it had the mesmeric effect of a river flowing past.

Although all the arrangements were made by the Earl Marshal (the Duke of Norfolk) and his officials, there were, of course, many points on which Clementine was consulted. One of her earliest expressed wishes was that there should be no flowers sent, other than those from the family. The Cathedral authorities sought her views as to what hymns should be incorporated in the Order of Service; she in her turn consulted with some of us, and between us we chose hymns that had been Winston's lifelong favourites: 'Fight the good fight'; Bunyan's 'He who would valiant be'; 'O God, our help in ages past'; and the great Battle Hymn of the Republic – 'Mine eyes have seen the glory of the coming of the Lord', the words of which Winston knew from start to finish.

For a long time it had been Winston's wish that he should be buried at Chartwell; then, a few years before he died, after a visit to Blenheim, when he had visited Bladon churchyard (just outside the park walls), where his parents and brother Jack already lay, he changed his mind and told Clementine that he would like to be buried there with them; and these wishes were expressed in his last will. Accordingly it was arranged that following the funeral service at St Paul's his coffin would be borne by river to Waterloo, and from the station by train to Long Hanborough, near Bladon.

The flood of letters and telegrams from all over the world was overwhelming. Anthony helped Clementine to draft answers to Heads of State and Government and official bodies. Now, and during many weeks to come, she herself wrote many letters; and three or four secretaries grappled with many hundreds more.

During the days before the funeral few people other than members of the family came to see her, out of consideration and delicacy. There was one exception, however: the faithful friend and valiant colleague of Winston's, Sir Robert Menzies, Australia's Prime Minister, called on the morning of Friday, 29 January; his was a warm and consoling presence, and Clementine was glad to see him. Immediately after the funeral he made a moving broadcast from the crypt of St Paul's, and at the end he referred to Clementine: 'A great and gracious lady in her own right. Could I today send her your love, and mine?'

Although Saturday, 30 January was bitterly cold, it was a fine day, and sometimes a gleam of pale sunshine broke through the leaden grey sky. An official car took my mother and me to Westminster Hall, where the procession assembled; here we met the rest of the family. We all stood by the catafalque as the coffin was lifted down and carried by the Bearer Party found by the Grenadier Guards to the waiting gun carriage, drawn by naval ratings. Immediately behind the gun carriage walked Randolph and Christopher, followed by other male members of the family and Anthony Montague Browne. Clementine and the rest of us rode in five of the Queen's carriages. In the first carriage were Clementine, with Sarah and myself sitting opposite her. We had been given rugs for our knees, and each of us was handed a small hot-water bottle for our hands. This was typical of the detailed thoughtfulness which marked all the arrangements. The processional journey to St Paul's was to take an hour, and we would indeed have been very cold but for this kind and practical provision. As Big Ben struck a quarter to ten, the funeral procession led by the Duke of Norfolk set out. We were very silent in the carriage; it swayed gently and creaked a little, like a small boat at sea. The music of the bands and the throb of the drums reached us only faintly. From St James's Park came the crash of the ninety-gun salute – one for each year of Winston's life. Near the Cenotaph we noticed the crowding banners and faces deeply marked by emotion of the French Resistance groups, come to take leave of one whose voice alone had given them heart and strength. The way was lined by great, silent crowds – all sorts and conditions of men, women and children. Through the large uncurtained windows of the carriage one could see clearly the expressions on people's faces. I was deeply struck by how noble – often how beautiful – is the human countenance in recollection, gravity or sorrow.

Up the Strand, up Fleet Street, past a City church where the priest and his white-robed choir crowded out onto the church steps, golden cross held aloft in hope and blessing, and – at last – into the forecourt of St Paul's. The Bearer Party once more shouldered their heavy burden, and moved slowly and painfully up the great stone steps. Randolph, who all the day was to care tenderly for his mother, gave her his arm, and together, followed by the rest of us, they followed the coffin up into the shining vastness of St Paul's.

There were, we were afterwards told, about three thousand people in the cathedral. But my impression, as we entered, was almost of solitude. The organization had been so perfect, so timed to the second, that even the official ushers were still as stone. The vast congregation had already risen to its feet. Ahead of us, swaying gently, went the shoulder-high coffin, and there seemed to stretch before us a limitless vista of pale blue carpet. And somewhere, beneath the shimmering cupola, the piercingly beautiful voices of the choir sang the words that dignify and lend hope to the humblest as to the greatest of funerals: 'I am the Resurrection and the Life'.

The Queen had broken all custom in attending the funeral; and had laid aside her usual precedence. She and her family, already in their places, awaited the arrival of her greatest commoner. The coffin was laid on a catafalque in the chancel, and we took our seats on the right. Opposite to us were the Heads of State and Government. Among them, General de Gaulle stood out, distinguished by his great height and the pallor of his gaunt countenance. The service was inexpressibly beautiful, but in a grandly simple way: the matchless words of the liturgy and of the 'old' Bible alternating with the hymns that are the family heritage of the Church of England. There was indeed 'Nothing . . . here for tears'* – only for honourable pride and loving, grateful recollection. Ever since his death we, his family, had realized that he belonged as much to others as he belonged to us – perhaps more – and that we were only a small part of the laying-to-rest of Winston Churchill.

After the service, the coffin was embarked at Tower Pier in the Port of London Authority launch *Havengore*. Massed pipe bands wailed him aboard, and there was a seventeen-gun salute. As the launch drew away from the pier the band played 'Rule Britannia'. As *Havengore* turned upstream and headed for the Festival Pier, the cranes on the shore opposite slowly dipped their

* John Milton, 'Samson Agonistes'.

giant giraffe-necks. Overhead, Lightnings of Fighter Command flashed by. The river was lead and pewter, and on the deck of the launch the coffin was still guarded by Grenadiers.

The nation now handed Winston Churchill back to his family. At Waterloo a new Bearer Party took over, found by the regiment with which Winston had first served – the 4th Hussars (now the Queen's Royal Irish Hussars). During the nearly two-hour journey to Long Hanborough, Winston lay, once more, 'In State' – in the rattling luggage-van – where, even here, soldiers of his old regiment kept watch around him.

One felt one had touched the outer limits of emotion – yet, in some ways, what those of us in the train saw now was most moving of all: the platforms of the stations through which the train slowly rumbled were crowded with people, waiting to take their leave of him. A Thames lock-house keeper, all alone at his post, came to attention and saluted. The winter fields had little groups of people – families with their children and dogs; a farmer, taking his cap off; children on shaggy ponies – all waiting in the chill of a winter's afternoon, to watch Winston Churchill's last journey home.

From the earliest announcements of the arrangements for the funeral, stress had been laid on the strictly private nature of the burial service at Bladon. To the official statements Clementine added her own personal request that there should be no press or TV coverage for the last part of the journey and the interment: and her wish was respected in the most remarkable way. Bladon village was sealed off from outsiders by the police, but there is a limit to what can be done by official restraint in our country; also, there were press representatives from all over the world in London, sent expressly by their papers to obtain the fullest possible story and detailed picture coverage, and restraint and respect for privacy on such occasions are better understood by the British press than by others. All I know is that if there were spying eyes and lenses or uninvited persons in the churchyard, they were not noticed by any of us. Mostly local people lined the sides of the road from Long Hanborough station to the entrance to Bladon village, and those whose houses directly overlooked the churchyard behaved with kind and touching discretion.

The coffin, still borne by soldiers of his regiment, was met at the lych-gate of St Martin's church by the Rector, the Revd J. E. James, who led the way to the graveside. Apart from members of the family, the only others with us at this last act were Anthony Montague Browne, Jock Colville, Leslie Rowan, Lord Moran and Grace Hamblin – and last, but not least, the man who had, at the Queen's command, directed the whole of this unique happening: the Duke of Norfolk. He stood, a dignified and watchful figure, at the head of the grave, presiding over the last details of this sad, but infinitely proud day.

After the committal and the lowering of the coffin into the grave, Clementine first of all, and then all of us one by one, filed past and bade our last farewell. The whole ceremony took only a little while. Before we left, two wreaths were placed on the grave: Clementine's red roses, carnations and tulips, bearing a card, 'To my Darling Winston. Clemmie'; and a wreath of exquisite spring flowers from the Queen, with a card in her own hand, 'From the Nation and Commonwealth. In grateful remembrance. Elizabeth R.'

Twilight was falling as we made our way back in silence to the cars, and to the train, which carried us back to London. Clementine had not faltered throughout this long day – one would not have expected her to. Now, on the homeward journey, she talked to some of those who had come on this last lap. Bernard Norfolk came and sat by her, and she thanked him for all the arrangements, so perfectly carried out in every detail. When he got up to leave her, he suddenly bent over and kissed her hand.

We were very tired when we got back to No. 28. It had been a long, long day. My mother and I and Grace Hamblin had an early dinner together, and after Grace left us we watched for a while part of the replay of the funeral on television. Then my mother got up to go to bed, and I busied myself switching out the lamps. As she reached the door she paused, and turning round said, 'You know, Mary, it wasn't a funeral – it was a Triumph.'

CHAPTER THIRTY

Afterwards

A DAY OR TWO AFTER THE FUNERAL, MY MOTHER CAME DOWN TO US AT Hamsell for a fortnight. She had not sustained the trauma of shock, but she bore the burden of the fatigue accumulated over many years, as well as the strain of the prolonged emotion of these last weeks. Clementine did not collapse; her mood was one of proud and loving recollection; and only slowly did the great void that now gaped in her life assume its inescapable dimension. As well as resting a good deal, she continued to answer great numbers of the letters she had received. One was from the Prime Minister, Harold Wilson; his consideration touched her particularly in the matter of Anthony Montague Browne, who he suggested should remain on the Foreign Office strength, but stay on with her for a few months more to help and advise on the many matters which would arise immediately after Winston's death. She was also conscious of the great back-up of sympathy and support she had received from official quarters in many ways, and she wrote to Mr Wilson: 'May I express to you my true and heartfelt thanks for the wonderful and sustaining way in which many Government Departments have helped me in recent days? I am thinking not only of the State Funeral, but of many other matters. I know that much of this is due to your personal sympathy.'[1]

In these first weeks after Winston's death, Clementine took two major decisions. She was quite clear that, although she had a right to live at Chartwell for her life, she wished to hand it over straight away to the National Trust. Secondly, after some consideration, she decided to sell both the houses in Hyde Park Gate. She had been left comfortably off, but her practical good sense did not desert her; even the original No. 28 would be too big for just one person – expensive, and a 'business' to run. Soon, she would look for a flat. Most of these plans were made by her during her visit to Hamsell. After she left us to return to the empty shell that now was No. 28, she wrote me a letter I have always treasured:

February the 15th 1965

My darling Mary,

I shall never forget your strong loving help to me during this last month. I have felt comforted and supported by you.

Hamsell has been my refuge and resting place this last fortnight & I feel revived

Your devoted
Mama

Kind friends and relations stretched out helping hands at this moment. Ronnie and Marietta Tree* invited Clementine to Heron Bay, their romantic and lovely house on the seashore in Barbados, and the Duke of Marlborough sent a warm invitation to visit him and his family at his house Woodstock near Montego Bay in Jamaica. In both these invitations Christopher and I were included. It was an ideal plan, and would provide the necessary change and break before Clementine started to remake her life.

A few days before we left my mother visited Chartwell, taking Grace Hamblin and me with her, to meet representatives of the National Trust and to make preliminary plans for the takeover. It was, naturally, a rather muted occasion. Neither my mother nor I had been to Chartwell since she and my father had left there for the last time. The sensitive understanding of both Robin Fedden and Ivan Hills of the National Trust helped her, but she was already calmly fixed in her mind about her intentions, and she at once addressed herself to the main theme of her thoughts concerning Chartwell's future. By the terms of the original sale of Chartwell in 1946, only the house and grounds were designated for the National Trust; there was no commitment about furnishings, and the house could have been handed over an empty shell. But from the beginning both Clementine and Winston felt that the house should be 'so garnished and furnished as to be of interest to the public'.[2] This intention Clementine now reaffirmed to Mr Fedden and Mr Hills, telling them at the same time that she wished the house to be seen as it was in its heyday during the twenties and thirties, rather than in its latter-day practical, but topsy-turvy, arrangement. My mother delegated to Grace Hamblin and me the task of working closely with the National Trust in carrying out all the details which the takeover and re-adaptation of the house would involve.

Shortly before she left for the Caribbean, Clementine received a warm and understanding letter from Randolph, who was spending the winter months in Marrakech, working on the official biography of his father. He wrote:

I have been thinking so much about you, dearest Mama, and am very glad to hear that you are going to the Barbados and to Jamaica. I know what a terrible time you have had in the last ten or fifteen years and trust that you have realised my understanding of this. When you have had a good holiday

* Ronnie and Nancy Tree had been divorced, and he had married Marietta Fitzgerald in 1947.

and rest you must try to create a new life for yourself. To begin with you will probably feel a vast void.[3]

We left Southampton in the *Queen Mary* on 24 February, bound for New York, on our way to Barbados, where we spent ten days with the Trees. Clementine was very tired, and this was a perfect and peaceful interlude. From there we flew to Jamaica, where Clementine stayed with Bert Marlborough's eldest daughter, Sarah Russell, while Christopher and I stayed with her father: the houses were quite close, and we made daily plans together. But through these weeks, despite great kindness and solicitude, blue skies and sunshine, Clementine, although ever pleasant and appreciative, seemed to be living in a world of her own, remote from us all. She arrived home towards the end of March only somewhat rested and refreshed by her month-long holiday. And of course, once home, decisions and problems awaited her. She began at once looking for a flat, and starting life anew – and alone.

Soon after Winston's death Anthony Eden* had written Clementine a most moving letter from on board the *France*, in which he and Clarissa were going to America. Now, a few days after her return from Jamaica, Clementine replied: it is only recently that I have discovered this letter, which moves me deeply even after the lapse of so many years:

My dearest Anthony,

I returned to find you had sent me your book with such a loving inscription & I also must thank you for your affectionate and understanding letter written as you were crossing the Atlantic.

As the days pass I think more & more of Winston's past life – of its triumphs & tragedies & then of his failing years & death –

He passed away so softly & looked young & beautiful when he was dead.

I do hope dear Anthony you are better – I am distressed at these recurring bouts of fever which must be weakening –

I send you and Clarissa a great deal of love

Your very affectionate
Clemmie[4]

Anthony and Clarissa in the coming years would prove affectionate and faithful friends to Clementine, sending her flowers and messages on birthdays, and visiting her when they were in England; they spent several winter months in their house in the West Indies, on account of Anthony's fragile health. Clementine greatly enjoyed a weekend she spent with them at Broadchalke, near Salisbury, in the summer of 1967.

* * *

* Created 1st Earl of Avon in 1961.

It was during the voyage to New York that my mother told Christopher and me that the Graham Sutherland portrait of my father was no longer in existence. We were both flabbergasted. We knew the picture had been sent to Chartwell not long after the presentation, and had been stored away there, but the information Clementine now imparted to us was a total surprise. She told us that Winston's deep dislike of the picture, and his bitter resentment at the manner in which he had been portrayed, had weighed more and more on his mind during the months that followed its presentation in November 1954, to such an extent that Clementine told us she had promised him that 'it would never see the light of day'. As far as I know, she consulted no one about her intention, and it was entirely on her own initiative that some time in 1955 or 1956 she gave instructions for the picture to be destroyed. I do not believe she ever specifically told Winston the steps she had taken: but her original pledge had calmed him, and had reassured him that the actual portrait which he so heartily loathed would not be seen by generations yet unborn. (But of course photographs had been taken of the picture.)

As time went by there were spasmodic but persistent enquiries about the portrait, and requests for it to be exhibited on permanent loan or for specific exhibitions. These enquiries were at first easily parried, both Winston's and Clementine's dislike for the picture soon becoming well known. The very small number of people who came to know of its fate were all of the same mind – namely, that whatever the enquiries, or the speculation, the fact of its destruction should not be revealed in Clementine's lifetime. It was therefore inevitable that from time to time a few of us were forced to be less than candid when questioned on the subject. Clementine herself simply refused to discuss the matter at all, except with those of us who knew the story; and even with us, it was not a topic frequently touched upon.

Within a month of Clementine's death in December 1977 her Executors (of whom I was one) issued a statement telling the bare facts of what had befallen the portrait. I had written a personal letter to Mr Sutherland, making sure he had received it before the statement was released. When the storm of controversy blew up – as we had anticipated it would – those of us who had known the facts for so long were glad we had consistently advised my mother not to allow the story to be revealed in her lifetime.

Although a small number of people (among them the artist himself) probably guessed that the portrait had been destroyed, the news came as a complete surprise to the public. To generalize – the artistic world had the vapours; some of the press were governessy; and there was some valid criticism. But as far as one can judge from the letters published in the correspondence columns of the newspapers, and from those received by members of our family, a poll on the subject would have shown an overwhelming endorsement of Clementine's right to dispose of a picture of her husband which they had both regarded as offensive.

There is, however, one point about which some doubt has been expressed, and which should be cleared up. The idea circulated that it had always been the intention of the Members' Parliamentary Committee that, in time, the picture should be returned, and hung somewhere in the Palace of Westminster in perpetuity. It was claimed by some, therefore, that Clementine did not have the right to cause it to be destroyed, since it was not in fact Winston's outright property. On investigation, it seems the idea that the picture should eventually revert to the House of Commons did form part of the Committee's thinking at an early stage; but this condition was certainly not formulated in any document that I know of; nor was it mentioned in any of the speeches at the presentation of the picture. There was no question in the minds of those in close contact with Winston and Clementine, at this time or later, that the portrait was anything other than an outright gift. Moreover, there exists a document drawn up and signed on behalf of the Members' Parliamentary Committee assigning the 'sole and exclusive copyright' of the portrait to Winston Churchill.[5]

It is right that the destruction of the Sutherland portrait should be put in its proper context. It was not an isolated case of 'destructivitis' on Clementine's part: fiercely protective of Winston's honour, she had all along been the ruthless guardian of his 'image', in all senses of the word.

Although Clementine greatly admired Sickert's work, and although he was a friend from her girlhood days, this had not deterred her from putting her foot through a sketch he had done of Winston some time around 1927 which she disliked. And seventeen years later, in September 1944, when she and Winston were staying with the Roosevelts at Hyde Park after the Second Quebec Conference, she had written home: 'I paid a visit to the President's Museum and managed to hook out of it a horrible caricature* of your Father by Paul Maze. I boldly told the President I did not like it and he said, "Nor do I". So I said, "May it come out?" and he said, "Yes" and so now it is destroyed.'[6]

Clementine's strong dislike of Graham Sutherland's portrait was not an instantaneous reaction. She had quickly 'taken to' both Mr and Mrs Sutherland, when they had come to Chequers for sittings, and she was predisposed to view this 'state' portrait of Winston favourably. She first saw the finished picture at the end of October 1954 when lunching with Sir Kenneth and Lady Clark, who were old friends, at Saltwood Castle, near Hythe; Mr Sutherland had deposited the picture with the Clarks, hoping that they would show it to Clementine. I was present too, and Sir Kenneth† later reminded me that my mother's first reaction had not been unfavourable. She studied the picture for some time, and then spoke of it with approval, and praised its truthfulness. I certainly remember that her antipathy was not immediate; but it worked up pretty quickly. I think in her heart of hearts the picture probably rather stunned her, and when she realized how much Winston hated it, his views coloured her own. But, as her whole life shows, she was perfectly capable of taking an independent view,

* In fact the picture was not a 'caricature', although she regarded it as such: it was, as far as I can remember, a charcoal sketch.
† Created Baron Clark, OM, CH (a Life Peer) in 1969.

and so I think her early approval of the picture was not very wholehearted; and when she saw how deeply Winston felt, she herself came to hate the picture too.

Lord Clark told me that he never thought Winston would be pleased with it; he wrote to me: 'He had come to accept an image of himself, which had become an international symbol; & this Graham had completely disregarded.'[7]

Whatever her views, however, Clementine still had a friendly contact with Mr Sutherland. Early in December 1954, shortly after the presentation of the portrait, he sent her a sketch of a hand he had made during the sittings. In her reply Clementine wrote: 'I am touched that you should have given me the sketch of Winston's hand. Thank you very much indeed.'[8] That was all; she gave no hint of her own or Winston's feelings.

In October 1957 Clementine wrote to thank Max Beaverbrook for inviting them both to Nassau. He was a great admirer of Sutherland's paintings, and, from her letter, it looks as if they had discussed his work, and Winston's portrait in particular, for she went on:

> I visited the Tate Gallery last week & contemplated the two brilliant and ferocious portraits of Madame Rubinstein – I must say I should hesitate to have my face lifted by her! I'm afraid that if I have any say in the matter, Winston's portrait by Mr Sutherland will never see the light of day. This gift which was meant as the expression of the affection and devotion of the House of Commons caused him great pain & it all but ruined his 80th birthday – It wounded him deeply that this brilliant . . . painter with whom he had made friends while sitting for him should see him as a gross & cruel monster.[9]

The portrait was probably already destroyed when she wrote this letter, but it is a complete and personal record of how Winston had felt, and of how she herself had come to feel. It suffices to add that Clementine never regretted what she had done, and indeed only a few months before her death she confirmed that she had not changed her mind on this subject by one iota.

* * *

On 1 April 1965, a few days after her return from the Caribbean, Clementine celebrated her eightieth birthday. Randolph organized a splendid luncheon party for her at the Café Royal, attended by an almost complete muster of children and grandchildren. It was like a launching into that new life which Randolph had so rightly said she must create for herself.

Within a short time she found herself a flat at No. 7 Prince's Gate, SW7. It was only about half a mile from Hyde Park Gate, so she would not be moving out of her neighbourhood. The flat was on the sixth floor of a handsome apartment block: it was large and sunny, all the main rooms looked southwards over gardens, and there was a wide and pleasing panorama of buildings and rooftops. Clementine began at once on the necessary alterations and redecorations. But although she had herself, without prompting, decided not to continue living at Hyde Park Gate, the wrench of leaving it was, understandably, severe. She moved into her new home early in September, and the Hyde Park Gate houses were sold at auction late in October.

Happily for her, Clementine did not start her new life bereft of all domestic continuity. From Hyde Park Gate came with her Nonie Chapman, her Private Secretary, and Lily Douglas, her cook. Nonie had been Clementine's secretary since April 1964; she was to be with her now until her death. Nonie veritably 'structured' her life. Apart from dealing with Clementine's considerable volume of correspondence, she organized the household and drove my mother about. The companionship Nonie gave Clementine, and the unfailing and tender devotion she showed towards her, contributed more than it is possible to evaluate to the large measure of happiness and serenity Clementine was to know in these last twelve years of her life.

Mrs Douglas had come as cook to Hyde Park Gate in 1964, and now she followed Clementine into her new flat and new life, in which she became a great feature, not only as an excellent cook, but because she was an interesting and highly intelligent person. She was devoted to Clementine, and shouldered the responsibility of watchfulness, especially at weekends, when Nonie was away. There was a succession of mainly sweet and kind Filipino maids, but continuity and security were assured by the presence of Nonie Chapman and Mrs Douglas. It was a great grief to my mother (and indeed to us all) when dear 'Mrs D' died in March 1976 of cancer.

This year of 1965 saw Clementine involved in many things flowing from the aftermath of Winston's death. On 5 April, she and Randolph lunched with the Queen at Windsor. In a private ceremony, Randolph handed back to the Queen in the customary fashion his father's Insignia of the Order of the Garter.* In the afternoon, Clementine, Randolph and several other members of the family attended the beautiful and moving service held over Winston's Banner, which always takes place in St George's Chapel, Windsor Castle, following the death of a Garter Knight. A duplicate Garter Standard now hangs from the great oak beam in his study at Chartwell.

Later in the year, after the twenty-fifth Battle of Britain Service in Westminster Abbey on 19 September, Clementine and more than forty members of our family were present to watch the Queen unveil a memorial stone laid between the Great West Door of the Abbey and the Tomb of the Unknown Warrior. It bears the inscription: 'Remember Winston Churchill.' Every year thereafter, on the anniversaries of Winston's birthday and of his death, Clementine would lay flowers on this memorial: and since my mother's death I have continued to do the same.

Within a month of Winston's death, Field Marshal Alexander launched an appeal to form the Winston Churchill Memorial Trust, the purpose of which would be to fund travelling fellowships for British men and women to travel abroad in pursuit of their own projects. (Similar Trusts were

* Subsequently graciously loaned by the Queen to Chartwell.

established in Australia and New Zealand.) The idea had been discussed with Winston and Clementine, and now it had her full blessing and support. Nearly £3 million was raised, and 1966 saw the first batch of over sixty Churchill Fellows. During the succeeding years Clementine kept a lively interest in this 'living' memorial to Winston.*

In May 1965 an event occurred which caused her much pleasure and gratification. On the recommendation of the Prime Minister, Mr Harold Wilson, the Queen conferred a Life Peerage on Clementine. She assumed the title of Baroness Spencer-Churchill of Chartwell, and her introduction into the House of Lords took place on 15 June; her sponsors were Lord Normanbrook† and Lord Ismay. We were all so proud and overjoyed for her, and felt this would be a new outlet and interest. Clementine herself was full of enthusiasm. She took her seat on the cross benches, which caused astonishment in some quarters, but certainly not among those who knew her well. In 1965 she attended the House thirteen times, voting for the Abolition of the Death Penalty Bill on 20 July. The following year seven attendances for her are recorded; but she did not vote again, and she never made a speech. It was a great disappointment to her that after 1966 her deafness was such that she could not hear the debates, which of course rendered her attendance pointless.

During the last year of Winston's life, Clementine had been very loath to leave him for any length of time. Now, we all felt it was important for her health and spirits that she should come away as often as possible from London and her endless preoccupations with house-moving and other matters. She made a number of weekend visits to Hamsell, and started the (for us joyful) ritual of spending Christmas with us, which she did for eight consecutive years. Randolph had a beautiful house and garden – Stour, at East Bergholt, Suffolk – and she would make day trips here to see him, which she very much enjoyed. Young Winston and Minnie, with their enchanting and growing family, lived near Haywards Heath in Sussex, and their home too was a frequent port of call. Staying with other than relations Clementine found something of an effort; but during these next few years she made several visits to old friends. And in August 1967 she went to Scotland, staying first with Captain and Lady Victoria Wemyss at Wemyss, and going on to Airlie Castle to stay with her cousins, Lord and Lady Airlie. This was a pleasurable and nostalgic journey into the long-distant past.

In London her social life quickly revived: old friends invited her out, and she started entertaining in her new home. Her diaries are sprinkled with details of small luncheon parties, in addition to the numerous occasions when her family came to see her. Evening engagements she already found tiring, and they were few and far between; and after 1966 the only people other than her immediate family who dined with her were Jock and Meg

* Now, over thirty years on, the Trust's capital stands at £23 million, and an average of 100 Churchill Fellowships are awarded annually.

† As Sir Norman Brook he had been Secretary to the Cabinet during WSC's 1951–5 prime ministership, and he and his wife were great friends of the Churchills.

Colville, Sylvia Henley and Grace Hamblin. Sylvia was a frequent luncheon and dinner guest, and her devoted friendship supplied a continuity of companionship right to the end of Clementine's road. They played a running tournament of backgammon, with the small stakes meticulously recorded and paid up. Sylvia's always keen and active mind was a stimulus, and she often accompanied Clementine to exhibitions and on expeditions. And my mother would always see, when visiting us at Hamsell, the other close surviving friend of her youth, Horatia Seymour: she was greatly saddened by Horatia's death in November 1966.

Clementine also carried out a considerable number of public, or quasi-public engagements. From now to the end of her life, if she was in London and well, she always made a point of attending the annual Battle of Britain Service at Westminster Abbey; and Harrow 'Songs', so long a feature of her life with Winston, now became a regular fixture in her winter programme.

Churchill College, Cambridge, was a continuing source of pride and interest to her. After his father's death, Randolph had suggested to his mother that the College would be the most appropriate place for Winston's treasured collection of Napoleonic books, most of them beautifully bound: Clementine agreed and presented the College with some 150 books in early December 1965. Some years later she gave the Bevin Library at Churchill the copies of General de Gaulle's speeches, inscribed by him. But her greatest benefaction (with the agreement of the Churchill Archives Settlement) was to arrange the deposition in the College of the eighteenth-century tortoiseshell casket enclosing 630 original manuscript letters from John Churchill, 1st Duke of Marlborough, to Antonie Heinsius, the Grand Pensionary of Holland. This priceless collection of letters had been the gift of the Netherlands Government in 1945, Queen Wilhelmina having flown over to England to present them to Winston herself as a thank-offering for his services in the war to her country and to the world. Clementine greatly enjoyed visiting Churchill College, which she did from time to time after Winston's death. She was always received with the utmost warmth and kindness by the Master, Fellows, students and staff.

It was most unfortunate that, just when her health and spirits were beginning to revive after Winston's death, Clementine should have had an accident. In mid-October 1965 she was walking alone in Hyde Park, on a path which runs alongside a football pitch. A high ball in its descent struck her on the side of the head, throwing her down and breaking her right shoulder. Although the injury was not severe in itself, it shook her badly, and she endured several weeks of pain. The boys who had been playing were deeply concerned and apologetic and brought her round a lovely bunch of flowers. One of her younger grandsons, hearing about his grandmother's misfortune, remarked that he was so very sorry about Grandmama, but he didn't actually know she played football!

The highlight of 1966 for Clementine was the opening of Chartwell to the public towards the end of June. She was thrilled by this event, and in the year it had taken to prepare the house and gardens for a continual flow-through of people, and to remake the rooms as once they were, she had taken a keen supervisory interest. Colours, chintzes and upholsteries had been as faithfully copied as possible, and the original furniture had been put back as before. Clementine loaned a number of Winston's paintings to hang in the house and studio; many of these she ultimately left to Chartwell.

Earlier, in May, prior to the public opening, the National Trust gave a large luncheon party at Chartwell, in honour of the original donors and of Clementine. All her family and some of her closest friends were invited. It was a lovely occasion, and to the eyes of Randolph, Sarah and me, the Chartwell of our youth had miraculously been revived before us. During the years to come Clementine loved to visit Chartwell several times in the spring and summer. Everything interested her, from the flower arrangements to the lavatories. She kept a keen eye on the garden where, under the aegis of the celebrated garden designer Mr Lanning Roper, much of her own original plantings and taste is perpetuated. Vincent, her own former head gardener, was still there, and she enjoyed at first walking, and then later being pushed round the gardens with him as guide.

The first Administrator of Chartwell was Grace Hamblin, and Clementine wanted to know the details of the successes and also the teething pains of the first season or two; the admission figures were regularly telephoned to her. When Grace retired in 1973, her successor Jean Broome continued to keep Clementine in touch. I truly believe that my mother had more real pleasure and satisfaction from Chartwell in these last twelve years of her life than in all the forty she lived there up to my father's death.

* * *

In May 1966, Lord Moran published a book about Winston Churchill, whose doctor he had been for twenty-five years. The work – entitled *Winston Churchill: The Struggle for Survival 1940–65* – had been serialized the previous month in the *Sunday Times*. At once a storm of controversy arose, based mainly on the ethical tenet of confidentiality propounded in the Hippocratic Oath. When, in addition to the waiving of this traditional and basic rule of confidence between doctor and patient, it became known that his widow and closest family were in total opposition to the publication of Lord Moran's book, the fires of public argument were further fanned.

Clementine had learnt in 1964 that Lord Moran intended to publish a book based on his relationship with Winston: she was much upset, and she raised the matter *viva voce* with Lord Moran himself. Also, a letter to him, marked 'Private and Confidential' and dated 30 July 1964, put the matter clearly:

> I am seriously disturbed by our conversation yesterday. I think that you should have told me of your intentions: it was only through the matter having been mentioned in the Press that I found out you intended to write a book about Winston, and I subsequently raised the matter with Jock.

> I had always supposed that the relationship between a doctor and his patient was one of complete confidence.
>
> Had you been writing your own biography, with passing references to Winston, it would have been understandable, though I would have hoped that you would tell us what you intended to say. But I do not see how you can justify your present course. An impartial observer would, I think, consider that your career had been a successful one and had not been damaged by your association with Winston, and I think he has not shown himself ungrateful to you.
>
> . . . I do urge you to reconsider your intentions.

To this letter Clementine received no reply. In the months that followed she asked on more than one occasion that she might be allowed to see the proofs of the book. All her requests were ignored.

When Clementine had learnt of Lord Moran's intention to write his book, she had not raised the subject with Winston. At that time (some seven months before his death), his health had deteriorated to a point where it would not have been feasible to discuss such a matter with him in any detail, nor did she want to arouse anxieties, and so disturb his peace of mind in this last phase of his life.

In January 1966, a year after Winston's death, Lord Moran wrote blandly to Clementine asking her permission to use a photograph of the Orpen portrait of Winston (painted in 1916) in his book; on 21 January Clementine replied:

> With regard to the Orpen picture, I would rather that you had not asked my permission to publish this painting in your book, for I regret very much that you should write about Winston. You will no doubt remember that I asked if I might see the proofs more than a year ago, but you ignored my request. I am very sorry about this, but I think you should know my feelings.

This letter, at least, received a reply: on 24 January 1966, Lord Moran wrote Clementine a long, diffuse letter covering various points.

About his not having sent Clementine the proofs, he pleaded that most writers are averse to having incomplete versions of their work examined; and he expressed the hope that when she read the book in its final form she would not disapprove of it.

As to the doctor–patient relationship, Lord Moran based his argument on the approval – nay, encouragement – he had received from the eminent historian Professor G. M. Trevelyan, who had taken the view that everything would eventually become known about such an historical figure as Winston Churchill. Lord Moran also averred that Field Marshal Smuts had approved his intention, and that Brendan Bracken, only a few days before his death, had pressed him to write the book.

In addition he explained to Clementine how he had sought to do justice to Winston and to portray the physical odds against which he battled, particularly after his major stroke in 1953.

Finally, he returned to his request that he might be allowed to use the Orpen picture for the book. There is a short note attached to the correspondence, dated 25 January 1966, stating that Clementine had telephoned Lord Moran to tell him that as the portrait now belonged to Randolph, the matter was out of her hands. The portrait is not included among the illustrations in the book.

Of course, Charles Moran's care of Winston and his frequent absences from London to accompany him on his journeyings resulted in serious interruption of his own private practice. But much resentment was felt in our family circle that it became quite widely accepted that Lord Moran had been justified in writing this book because he had received no remuneration for his services to Churchill. As is made clear in Richard Lovell's biography, *Churchill's Doctor: A Biography of Lord Moran*, Sir Edward Bridges (the Cabinet Secretary) had asked to see him in early May 1945 and had told him that Churchill was concerned about the matter, and that a stipend could come from public funds: Lord Moran had not liked the idea, but had been touched by his suggestion. Churchill shortly after this broached the subject *viva voce* with him, urging him to consent to some public remuneration: 'You must – this is a public matter. It doesn't come out of my pocket.'[10] Charles Moran continued to demur, and the matter lapsed.

Since 1942 Winston had in fact been paying, under a seven-year deed of covenant, an annuity of £500 (free of income tax) to Lady Moran. This deed was renewed from time to time until his death. In 1949 he executed two further deeds of covenant for £300 net per annum in favour of John and Geoffrey Wilson, Lord Moran's two sons; again, these deeds were renewed up to Churchill's death.[11] I have before me as I write the charming and grateful letters from John and Geoffrey to my father in 1956, when the covenants were renewed for the first time.

Of course, it is not possible truly to evaluate the care of a dedicated doctor in terms of money, be he or she an eminent consultant or one's local GP. All the same, I think it should be on record that my father was neither ungrateful for, nor unmindful of, Charles Moran's devoted service to him, and that he tried to express his gratitude in a practical form from his own resources.

The publication of the serial and the book caused Clementine real distress, although she did not intervene herself in the long-drawn-out argument which raged in the correspondence columns, and in the press generally. But Randolph was a doughty champion; and from Winston's own close entourage Jock Colville and Anthony Montague Browne also took up the cudgels.

Randolph's letter to *The Times* (26 April 1966) dealt specifically with the point Lord Moran had made (*The Times*, 25 April) in relating how he had sought the advice of Professor G. M. Trevelyan (who had, incidentally, died in 1962), who had told him: 'This is history, you ought to get it on paper.' Randolph commented: 'Professor Trevelyan did not say: "You ought to publish it"; certainly not within 15 months of his patient's death, and against the wishes of his family.'

In a further letter (*The Times*, 3 June 1966), Lord Moran took up some points which had been raised in the numerous letters published in the intervening weeks. He stated that the book had Winston's knowledge and approval, and that Brendan Bracken, at Winston's request, had arranged a meeting for Lord Moran with his lawyer, Mr Anthony Moir, about a certain business aspect of the book. The following day (*The Times*, 4 June) Mr Moir, Winston's lawyer for twenty-three years, replied that

> Lord Moran sought my advice on a literary matter in 1958 but according to my own recollection, supported by the note which I made immediately after the interview, which was confirmed by a letter to him on the following day, no mention was made of a book to be solely based upon a part of Sir Winston's life and no reference was made to any approval having been given by him to any book.

As to Lord Moran's declaration in the same letter that he had had Winston's permission and approval to write the book, Anthony Montague Browne wrote (*The Times*, 6 June):

> He claims that he obtained Sir Winston's approval for the writing of his book. Why did Sir Winston not inform Lady Churchill, from whom he had no secrets, of his blessing on this somewhat unusual medical procedure? Why did Lord Moran not inform her himself?
>
> I served Sir Winston for the last 12 years of his life. He was meticulous in keeping his Private Office informed of decisions taken in their absence: he warned none of us of this one.
>
> . . . Might it not have been prudent of Lord Moran to have confirmed his remarkable release from the Hippocratic Oath to just one member of Sir Winston's family, just one of his advisers or staff, by reference to Sir Winston during his lifetime?

My mother said to me of Lord Moran's book: 'It shows Winston in a completely false light.' She bitterly resented, as did all of the family, the wholesale quoting of Lord Moran's conversations with Winston and others close to him, certainly without Winston's permission, and in a number of cases without the consent of other persons quoted. Several of them complained at this breach of confidentiality in relation to essentially private conversations: among these were Sir Russell Brain (the eminent neurologist called in by Lord Moran); Sir William Haley, Editor of *The Times*; and Sir Norman Brook, the Secretary to the Cabinet.

The book's account of Winston's conversations, procured largely during the discharge of Lord Moran's medical duties, tends to suggest that Winston had at times a nervous preoccupation with his health, and gives the impression that he was almost perpetually ill. But eminent people, as well as more ordinary folk, chiefly seek interviews with their medical advisers when they are concerned about their state of health, and not in order to discuss with them such things as war strategy and politics.

Clementine was also deeply sceptical about Lord Moran's claim that he had Winston's approval for writing the book. With Anthony, she wondered greatly that Winston had never mentioned this to her or to others close to him; and she was amazed that, in the circumstances, Lord Moran had not done so himself.

But ultimately the real and unchanging objection to Lord Moran's book held by Clementine and her family, and shared by many others in all walks of life, including many doctors, was founded on the simple proposition that – prince or pauper, prime minister or ploughman – a person should be able to repose complete trust in the inviolable confidence of his priest, his lawyer and his doctor.

* * *

In these latter years, and especially since Winston's death, the relationship between Randolph and his mother had become less strained. He made efforts to please her; and she, in her turn, sought, and often accepted, his suggestions and advice. After the writing of his father's official biography had been entrusted to him in 1960, Randolph had bent his energy and his brilliant, tempestuous mind to the accomplishment of this great work – the task he cared for above everything else in his life. But, tragically, his health began to fail, and after a lung operation in 1964, although he seemingly made a good recovery, he cannot have felt that time was with him. Nevertheless, he battled on, and the first two volumes of *Winston S. Churchill* were published in 1966 and 1967 respectively.

On 6 June 1968, Randolph died suddenly at Stour; he was fifty-seven. Although his health had been visibly deteriorating for some months, his death came as a shock. As always in sorrow, Clementine had little to say. Now, of the five children she had borne, only Sarah and I survived. At Randolph's funeral at East Bergholt, and later at the crowded Memorial Service in St Margaret's, Westminster, she remained a mute, dignified figure.

But despite the penalty that time inexorably exacts from those who live on, in terms of family and friends; despite occasional gusts of controversy, or hurtful slaps at Winston's departed greatness; in the face of increasing deafness and failing sight, and the toll which two falls (in which she broke first one and then the other hip) took of her strength – looking back over Clementine's last decade, one sees what a long and benign sunset it was.

Since Winston's death, Clementine truly lived in the glow of what so many people throughout the world had come to feel for him. The awe and admiration of what he had been as a man, and of his long life's extraordinary story; gratitude beyond the grave for what he had meant to the history of our country, and indeed to that of civilization; love, for this very lovable and human human being – these feelings now found expression in touching solicitude for the woman he had loved so dearly, and who, by her single-minded love for him, her loyalty, and her tenacious, fierce integrity, had made also for herself a special place in people's affection and esteem.

Even now in her old age many demands were made upon her: to visit this; to open that; to launch a ship; to approve a bust or picture of Winston (tricky ground here). Her considerable correspondence made her address her mind to many different things. Reading the newspapers (wearing white cotton gloves to protect her hands from the grubby newsprint) and listening to the radio news were sacred rituals. The radio gradually had to be turned to shouting pitch, and presently Nonie Chapman had to read her the papers: but she was astonishingly well-informed. The correspondence columns of *The Times* particularly interested her; and every now and then she would quite spontaneously fire off a letter herself – always signed in the old-fashioned way: 'I am, Sir, your obedient servant . . .'

Nor did she react only through the post to events or matters which roused her interest, her indignation or her pity. In April 1970, the German Ambassador to Guatemala was kidnapped and shot: a book was opened at the German Embassy for people to sign as a mark of their sympathy, and the then German Ambassador to London told me how the Embassy staff were as astonished as they were moved to recognize Lady Spencer-Churchill in the rather frail old woman who, helped by her young secretary, slowly climbed the Chancery steps to record, in the only way she could, her revulsion and disapproval of such a barbarous act.

Through all these last years, Clementine maintained her contacts with societies with which she had long connections. The YWCA kept her informed of any aspects of their work which they knew would particularly interest her; and it was not until 1966 that her Special Fund to help hostels with their improvements was finally exhausted. She was to remain their Honorary Vice-President for life.

In the same way, she kept in close touch with Winston's former constituency. She would be their Conservative Association's President until the end of her life. Although she no longer attended meetings or functions in Woodford, every year the Chairman of the Association would write her a long and detailed letter putting her in the picture as to the affairs of the Association, and in her reply she always picked up various points which were of greatest interest to her.

Clementine did not take on many new obligations – but two are of note. In 1966 she became one of the Trustees of Lord Attlee's Memorial Foundation; and in 1968 she actually took part in a film made to raise funds for the Appeal. As late as 1972 she accepted the Presidency of the National Benevolent Fund for the Aged. Throughout her last years she continued to try to make whatever contribution it was still in her power to do.

Her family in all its several generations was to be an abiding source of interest and stimulation to her. She loved attending their weddings, christenings and confirmations. We were all ever welcome at Prince's Gate, where her old standards of hospitality and delicious food were still maintained. The company of her grandchildren she found revitalizing, and their jobs, examinations, friends and ambitions were all interesting to her. My daughter Emma, nearly seventeen, had dined with her in January 1966, after which my mother wrote to me: 'She seems to love her school. She told me she was learning German. I think she should be learning Russian & in turn her children Chinese! so as to keep up with the trend.'[12]

* * *

In the autumn of 1968, my husband was sent as Ambassador to Paris. His appointment had come as a complete surprise to everyone, including ourselves. He had lost his seat in Bedford at the General Election of 1966, and in early 1968 he was busy looking for a new parliamentary seat when Harold Wilson, the Prime Minister, and George Brown, the Foreign Secretary, offered him this challenging job, with the special mission of helping to bring about Britain's entry into the Common Market. When my mother learnt from Christopher of this new and surprising turn in our lives, she was not only delighted for us, but from the first absolutely certain that Christopher was right in accepting this new assignment, although it removed him for the time being from the domestic political scene.

Excited though I was by the prospect of this new adventure in our lives, and of Christopher's job in which I would have a share, I was concerned that it would of necessity make me less immediately available for my mother. She had, in fact, nine more years to live, but at the time of our going to Paris she was already eighty-three; Sarah was often in America, and I knew that my mother to a large extent depended on me. But never at any moment did she allow this thought to cloud the natural excitement and anticipation she shared with us. And on the wet September morning when we set out from Victoria on the *Golden Arrow*, there on the chilly platform to bid us Godspeed were my mother and Sarah.

During the next four years a regular feature of Clementine's life, and ours, were her visits to us in Paris. Always staying for a week, and eschewing the speed and easy comfort of the air, she invariably travelled by the overnight train ferry. We felt at the time that she really did enjoy herself, and now, reading her 'thank-you' letters again, I have wept happy and grateful tears.

The British Embassy in Paris is a glorious mid-eighteenth-century house, which had later become the home of Pauline Borghese, Napoleon's sister. It must be among the most beautiful and historic houses in Europe; and if one has a large family, this glittering, romantic house has room for all, and responds to all the demands of both diplomatic and family festivities. At Christmas we would try to assemble all our own children, Christopher's mother (Hope Dynevor), his sister (Dido Cairns) and her family, and my mother. Clementine, although getting older and deafer, fitted in perfectly, and enjoyed the hurly-burly of the different generations. On New Year's Day

1971, she wrote to me from London: 'Darling, this is the third Christmas I have spent with you in Paris. All equally delightful. It is something to look forward to all the year.'

When she came to stay at other times, usually in the spring, I tried to choose a relatively quiet patch, so that I could 'concentrate' on her. My mother had known and loved Paris all her life; now we revisited many famous and favourite sights. After she broke her hip in 1969, walking long distances was out of the question, but, luckily, she had no false inhibitions about using a wheelchair for expeditions; and so, with a *gardien* (or more usually with me) pushing her, we bowled through kilometres of galleries and exhibitions. We became quite ambitious, and with my Nannie to help as 'pusher', we managed parks and gardens as well – once narrowly avoiding calamity, when we nearly capsized my mother over an obstinate tree root in the gardens of Le Petit Hameau at Versailles.

Apart from the expeditions, Clementine took great interest in all the political aspects of Christopher's job, about which he would talk to her at length; and she enjoyed meeting her old friends and ours, and other Embassy guests. And, needless to say, she wanted to know every detail of my now rather glossy housekeeping arrangements.

On 9 November 1970, General de Gaulle died very suddenly at his home, Colombey-les-Deux-Églises. Clementine had just returned home from a visit to us. Always an admirer of and a believer in this great man, she was much saddened. Ever since the war she had kept up a spasmodic, friendly but formal correspondence with both the General and Mme de Gaulle.

On the first anniversary of Winston's death, Clementine had received a letter from the General, then President of France, which moved her deeply:

> Voici venir le triste et émouvant anniversaire. Laissez-moi vous dire, qu'en portant en cette occasion ma pensée sur la grande mémoire de Sir Winston Churchill, je ressens, mieux que jamais la dimension de sa personalité, l'étendue de son oeuvre, enfin la force et la qualité des liens qui m'attachaient à lui et qui, à travers nous deux, unissaient l'Angleterre et la France.[13]

Twice Clementine had sent him messages at moments of political crisis: once in 1953, at the time of his *Rassemblement du Peuple Français*, and again during the weeks of late April and early May 1969 when, after the referendum which resulted in his narrow defeat, he resigned on 28 April, as he had vowed he would. To this last message of hers, the General replied: 'Il n'y a pas de message qui m'ait touché plus que le vôtre. En le recevant de votre part, il m'a semblé qu'il me venait, au même temps, au nom du grand et cher Winston Churchill. Je vous en remercie de tout coeur.'[14]

Now, in November 1970, within twenty-four hours of her return home she learnt of the General's sudden death. In her thank-you letter to me, dated 10 November, she wrote: 'I have just heard of the death of General de Gaulle; & although it is true that at times he was no friend to England, it saddens me very much – It is the end of an epoch.' I wrote to her a week or two later, describing the solemn Requiem which had been held in Notre Dame, and to which Britain had sent a splendid representation headed by the Prince of Wales, the Prime Minister (Edward Heath) and three former Prime Ministers. I described to my mother how I had been to Notre Dame, although Ambassadors' wives had not been asked, as a generality, and also how Clarissa Avon (Eden) and I had both been specially invited to President Pompidou's reception later that same day. 'So I felt Papa's memory had been honoured as well as the General's – As you say a whole epoch seems to have ended with his death –'[15]

* * *

A faithful and constant friend in all these last years was Edward Heath. Clementine followed his fortunes – and misfortunes – with an affectionate, almost maternal eye. He, for his part, was never failing in charming attentions. He would sometimes give a luncheon party for her, at Albany, and later at No. 10 Downing Street; and he lunched with her from time to time. One year he accompanied her to Harrow 'Songs', and in the summer of 1972 he took her to see the Tutankhamen Exhibition.

One very agreeable occasion was in November 1971, when Ted Heath, then Prime Minister, invited Clementine and a small family party to luncheon at Chequers, to inspect in its new surroundings a picture by Winston which Clementine had given to Chequers. During his Prime Ministership, Heath had presided over some major changes in the decoration and arrangement of the house. Needless to say, Clementine was much interested to see these alterations, and she found them a great improvement on the rather gloomy decor she had known in her day. She was also able to inspect the progress in growth of the avenue of beech trees which Winston and she had presented to Chequers after his resignation in 1955, in commemoration of their time there.

* * *

Although, as the years went on, Clementine found staying away for holidays an effort, and she did not like flying or making complicated journeys to get to a chosen destination, in 1970 and 1971 she found a very good solution: taking Nonie Chapman with her, she embarked on 'cruise' life. In September 1970 she went on the *Andes*, a ship of the Royal Mail Lines, for a three-week cruise in the Mediterranean. It was a great success. She made some agreeable shipboard acquaintances, and towards the end of the cruise she took part in a fancy dress competition, appearing as 'A Lady in Black'. To the warmly expressed delight of the other competitors, and Clementine's great surprise and unconcealed pleasure, she was awarded first prize. Her trophy was an enamelled powder case, which she often used thereafter.

The following year she repeated her holiday plan, and set out on another enjoyable Mediterranean cruise. But in 1972 she contented herself with a visit to us in Paris. In November of that year she had her second fall, which resulted in a broken hip: this really did slow her down very considerably, and thereafter no cruises were undertaken. But they remained a pleasant memory.

There were some very special highlights in these last twelve years of Clementine's life. In December 1969 she unveiled a statue of Winston in the House of Commons. Executed in bronze by Oscar Nemon, the statue stands on the left of the Churchill Arch, through which Members of Parliament pass as they enter or leave the Chamber. When the Chamber of the House of Commons was rebuilt after being destroyed by bombing in the Second World War, Winston had suggested that the archway leading into the Chamber should be left with all its battle scars, to remind future generations of the perils through which free democratic institutions had passed in those terrible years: it had been thereafter known as the Churchill Arch.

In 1970, the Lord Mayor of Westminster launched a public appeal for funds to commission and erect a statue of Winston Churchill in Parliament Square. The statue, chosen by the Royal Fine Art Commission from a number submitted, was by Ivor Roberts-Jones. It combines a sombre likeness with considerable allegorical allusion. Some months before the unveiling, the Queen's Private Secretary had written to Clementine saying that the Queen would so much like her actually to unveil the statue: Clementine was greatly touched. The occasion itself, on 1 November 1973, was moving and memorable. It was a fine, grey day, and a great concourse of people gathered in Parliament Square. As well as the Queen, Queen Elizabeth the Queen Mother attended the ceremony, as did the Prime Minister, four former Prime Ministers and a great number of notabilities. Of our family, more than twenty were present, representing four generations. There were some very nice touches to the proceedings: the Band of the Royal Marines played some of Winston's favourite tunes – from the Edwardian music halls and Gilbert and Sullivan, as well as the Harrow song, 'Forty Years On'. The bells of St Margaret's, Westminster, rang out, and among the bell-ringers were two who had rung the peal at Winston and Clementine's wedding sixty-five years before.

At the end of her speech, the Queen said: 'For more than fifty eventful years, Lady Churchill was his deeply loved companion and I think it would be right, therefore, for her to unveil the statue of her husband.' And, turning to Clementine, the Queen handed her the silken cord which would release the Union Flag shrouding the statue. So it was that, in the presence of his revered Sovereign, Winston Churchill's statue was unveiled by his beloved 'Clemmie'.

The year 1974 was the centenary of Winston's birth. It was chiefly marked by the launching of an appeal for a Churchill Centenary Trust, from which both the Winston Churchill Memorial Trust (Churchill Fellows) and Churchill College, Cambridge, were to benefit. The Appeal was launched at a large party attended by, among others, the three party leaders, at the Banqueting House, Whitehall. It coincided with Clementine's eighty-ninth birthday. After the reception at Banqueting House, her family whisked her away and gave her a birthday luncheon at Claridge's.

In May, Clementine was present when the Queen Mother opened a large 'Churchill' Exhibition at Somerset House. In the lovely rooms of that eighteenth-century 'palace' was displayed every kind of Churchilliana, and the long pageant of his whole life was vividly recorded. Clementine was thrilled by the exhibition, and made several visits to it.

The culmination of Winston's centenary year for the family came on 30 November – the hundredth anniversary of his birth. A large number of members of his family, in their several generations, and some of his closest friends, attended a commemoration service at St Martin's, Bladon, arranged by Sunny Marlborough, who had succeeded to the dukedom on the death of his father in 1972. The church service was, for us all, deeply moving, and afterwards her grandson Winston pushed his grandmother to her husband's grave in the churchyard, where she laid some flowers and sat for a little while in contemplation. She said to Winston: 'I hope I shall not be long now.'

Clementine had not, since her youth, been a great churchgoer; but in these later years, she 'took' to it again. When she was settled at Prince's Gate, she made a regular habit of walking every Sunday to nearby Holy Trinity, Brompton. As her deafness increased, she would sit right at the back of the church, slipping out, with the kindly help of the verger, before the sermon. After her first hip accident in 1969, when she could no longer go out unaccompanied, her expeditions to Holy Trinity ceased. But whenever she stayed with her family, we shepherded her carefully to and from church. Later, both at Christmas and at Easter, if she was in London, the Vicar from Holy Trinity would bring Holy Communion to her.

Clementine never talked much about her inner feelings, but I am sure she thought about her death, and was quite prepared to receive her summons, whenever it might come. She would often talk to me in a brisk, practical way about various arrangements, and her wishes for her possessions: I (coward that I am) would shy away from such 'tomb talk'. Clementine thought much about Winston: one evening in the year before her death, when I had been dining with her, she said: 'I do miss Papa so – really more than I did just after he died.'

* * *

In the summer of 1975, Clementine received yet another sign of appreciation of herself as an individual personality. During the July Congregation of Bristol University, of which Winston himself had for more than thirty years been Chancellor, Clementine was awarded the honorary degree of Doctor of Laws. She was greatly gratified, but she did not feel strong enough herself to go to Bristol to receive this honour, and she deputed me to receive it on her behalf.

In these last years of her life I came to realize fully just what a special place my mother had carved for herself in the minds and hearts of all sorts of men and women. After her several accidents, two of which resulted in her becoming gravely ill, and which were widely reported, I helped to answer the letters and messages which poured in from the general public. I saw then how her devotion as a wife, her steadfast adherence to her ideals and high standards, and her tenacious struggle with all life had confronted her with, had been recognized, and that people seemed to have found in her an example and an inspiration.

* * *

To celebrate Clementine's ninetieth birthday in April 1975, her grand-daughter, Celia, and her husband Dennis Walters, lent their house, and the family gave a party in her honour. To this party came a great gathering of children, grandchildren and great-grandchildren, as well as cousins and her closest friends. She enjoyed it very much.

But now the rhythm of her daily life was perceptibly slowing. Her increasing deafness made it difficult for her to hear, or join in, general conversation; and although she had many people to luncheon, she usually only had them one or two at a time. But in this summer she made three expeditions to Chartwell, and gave a tea party at Prince's Gate for all the Chartwell guides and helpers.

After Christopher and I returned from Paris, we were in England only a few months before in January 1973 we went to Brussels, where he took up his post as one of the first two British Commissioners to the European Economic Community. The tour of Commissioner lasts three or four years, but since I was not needed 'in the shop', as I had been when Christopher was Ambassador, and thanks to his understanding and generosity, I was able to come home to England a good deal, and to stay for longer periods than when we had been in Paris. I was very glad of this, because now my mother's health did from time to time cause us all anxiety: one felt that slowly the clock was winding down. Although we always came home to England for Christmas with our children, we had sold Hamsell, and now lived in a much smaller house, Castle Mill House, near Odiham, in Hampshire, where, sadly for us, we could no longer have my mother to stay: so she used to make day trips to see us. But Winston and Minnie made three out of the four remaining Christmases of Clementine's life happy and joyful with them.

For some little time, those of us who had the knowledge and responsibility for my mother's financial situation had become increasingly anxious about it. When she went to live at Prince's Gate, she could afford her way of life; but inflation, and, as time went on, the need for more and more nursing, wrought ravages in her personal economic situation. After considerable discussion and consultation, she decided to sell several pictures, including two by Winston. When her intentions became public knowledge, there was a loud and, for the greater part, sympathetic outcry in the press. Most touching letters began to arrive at Prince's Gate, with offerings from pensioners, retired officers and many other people. Nonie Chapman was kept very busy expressing Clementine's deep thanks, and returning these infinitely generous contributions. Inevitably politics and public argument combined to make the whole thing embarrassing. In 1972 the Heath Government had enacted legislation providing pensions for the widows of Prime Ministers, Cabinet ministers and Members of Parliament, but it was not retroactive, and so Clementine had not benefited from it. There were suggestions, both in the press and in Parliament, that special provision should be made for her by Parliament, and a kindly disposed individual wrote to *The Times* with the idea of launching an appeal to alleviate her circumstances. Clementine was, predictably, horrified, and immediately issued a statement to the effect that while she was deeply touched by the motives which prompted these suggestions, she would greatly deplore any legislation or appeal being initiated on her behalf. Although she was rather sad to have to part with some treasured possessions, she was always eminently practical, and did not make 'heavy weather' of it. When, several weeks later, the pictures were sold by Christie's, they realized an astonishing sum. Clementine was stupefied – and delighted – by the result. For some time she had felt real anxiety about her future, but now the worries were lifted from her brow. As to Winston's pictures – he had left them all to her unconditionally, and it would have given him great satisfaction to know that 'my little daubs', as he called them, would help one day to secure for Clementine comfort and dignity to the end of her days.

* * *

Clementine now led a real 'old lady's life'. She rested more, and had good days and less-good days: but her twilight was a happy one, lit by a golden haze of love and esteem. The serenity so long attributed to her had come at last, along with a stoical acceptance of those dreary companions of old age – deafness and loss of sight.

The disciplines of a long lifetime remained: her house was well-run; the food was delicious – she and the cook conferred most days. Her appearance was as lovely and as immaculate as ever: even when dining alone she always changed into one of her ravishing housecoats.

Being read aloud to, or listening to her 'talking book', formed an important part of her daily routine. When she was well, and consequently her concentration good, she enjoyed biographies, memoirs and histories. It was a treat for her when Martin Gilbert came to Prince's Gate to read her some chapters of Winston's official biography, the writing of which he had taken up when it fell from Randolph's hand; and she managed to read the greater part of the volumes published up to her death. When she was feeling tired, she fell back on light romances and historical novels, the works of Georgette Heyer and Barbara Cartland alternating agreeably with the more serious works already mentioned. And, of course – every day – she read the newspapers.

The progress of her own biography greatly interested my mother, and she read all the earlier chapters, including some of the First World War period, and supplied useful corrections of fact. When I found old letters or papers which I thought would divert her, I always sent them to her. Her memory became patchy – but the remembered episodes and personalities were clearly lit.

Most afternoons she would set forth, often with Sylvia Henley, in her red Rover driven by Nonie Chapman, with the wheelchair in the boot, for a drive. As Clementine's sight failed, exhibitions and galleries lost their point; but the public parks remained to the last a pleasure. She became quite an expert upon them – comparing their states of maintenance and planting. Bushey Park was a favourite, as were Isabella's Plantation in Richmond Park at azalea time, and (as she ceremoniously called it) 'The' Regent's Park, particularly when the roses were out.

But Clementine's most continuing pleasure right to the end of her days was to see her family and friends. Perhaps these years of her widowhood were the first period of her life when she had ever had the leisure to cultivate her friends. And she now discovered how many she had, and how true and constant they were, as she picked up the threads of old – and newer – acquaintance.

Clementine suffered many of the almost inevitable inconveniences of great old age. Apart from her publicized illnesses, there were bouts of pneumonia (they always made us very anxious) and odd high fevers; and less serious aches and pains, which nevertheless are a burden towards the end of a long day. And it made one's heart ache after any illness to see how painfully slow and wearying was the climb back uphill to the plateau of her now fragile normal health.

But among her blessings – and she did count them – was that she always had company, and did not therefore suffer what must be one of the worst horrors of old age – loneliness. And although she could not any longer pass the hours by watching television, or by reading, yet there was for her, in addition to her family and friends, the boon of Nonie Chapman by day, and a faithful team of nurses at the weekends and by night. Those long, weary, dark hours, when soul and body seem to be at the lowest ebb, were made lighter for her, not only by the skill, but above all by the companionship of her young nurses. I believe they kept her in touch with the throb of life; and that they came to have a real affection and regard for her was testified to in the letters I received from them after my mother's death, and by their presence, almost without exception, at both her funeral and Memorial Service.

Clementine's interest in her children, grandchildren and great-grandchildren continued undimmed. Unless prevented by illness, she would never fail to attend a gifted granddaughter's – Edwina Sandys' – highly successful exhibitions of painting and sculpture. Sarah, to acting, writing, reciting poetry and singing songs, had now added the wielding of brush and pen, and her mother always visited the galleries where her work was being shown. And she could often be counted upon as an early visitor to a family cocktail party, leaving only when the crowd and the noise really became too overpowering. When our Charlotte, married at nineteen to Richard Hambro, gave birth to a daughter – Clementine the second – her great-grandmother, very frail now, attended her christening in a freezing church on a freezing January day in 1977. Later that year her fourteenth great-grandchild was born: Julian and Elisabeth Sandys' youngest son, Roderick.

Nor was it only for the family that Clementine, now over ninety, would get herself 'on parade'. When the French President and Mme Giscard d'Estaing came to London in June 1976 on a State Visit, at the President's request time was carved out of his programme of official engagements for him and his wife to call upon Clementine. It was a graceful gesture, and Clementine was much complimented and pleased, summoning Winston and Minnie, and Christopher and me, to be with her when she received the President and Mme Giscard d'Estaing. The time allowed for the visit was five to ten minutes: the visitors stayed nearly half an hour, the President and Clementine sitting conversing agreeably on the sofa. She presented him with the volumes of Winston's life; and when a few months later another volume was published, she remembered to send it to him.

Other enjoyable, if less ceremonious occasions were when, in the winters of 1975 and 1976, Harrow boys came to Prince's Gate to sing for Clementine the songs Winston had taught her to love, but which she could no longer go to Harrow to hear for herself. She greatly enjoyed the boys' visits, and made sure there was a large tea, with plenty of rich chocolate cake.

One of Winston and Clementine's oldest friends, Lord De L'Isle, having tea with her early in 1976, asked if she had ever met Mrs Margaret Thatcher: Clementine replied that she had not, but would very much like to do so. Shortly afterwards, Lord De L'Isle brought Mrs Thatcher to see her. They talked about the suffragette days, and how Clementine, unlike Winston, had been strongly in favour of 'Votes for Women'. My mother told me about their meeting later; she had been much pleased and interested to meet Mrs Thatcher, who would soon make history by becoming Britain's first woman Prime Minister.

But there were sad occasions when, with a great sense that she stood there for Winston too, Clementine made dignified *actes de présence*. On 1 April 1976 – her own ninety-first birthday – she attended the military funeral of Field Marshal Lord Montgomery. She had not been in touch with Monty for some time – his health had been such that he had lived the last years of his life in seclusion; but the war had forged strong links between Winston and Clementine and this great, but bizarre, man, and he had been a faithful friend in the sadder years. It was quite a long expedition for her to go to St George's Chapel, Windsor, and a very long wait. She could not hear the splendid hymns, nor see (although she was seated in honour, quite close by) the touch of panache in the placing of his famous 'beret' on the great Field Marshal's coffin.

Nearly a year later, Clementine mourned the death of a man who had been a part of the tapestry of her life with Winston for over forty years – Anthony Eden (the Earl of Avon). After the tragic debacle of the end of his career through illness, his wife Clarissa, Winston's niece, had cherished him through more than a decade of fragile health. Now he was among the last of Winston's political and wartime comrades and friends to depart. And in the previous summer, the charming, cultivated, sympathetic Ronnie Tree had died. It is one of the heavier penalties of great longevity that the muster of contemporaries dwindles away, leaving an emptier scene.

Clementine attended the Battle of Britain Service at Westminster Abbey for the last time in September 1976. Her presence at this annual commemoration had become a looked-for feature of this service which had such a particular significance for people. Wearing her medals, Clementine was gently shepherded from her car and wheeled down the long aisle of the thronged Abbey. The previous year, Sister A. M. Johnson (one of Clementine's nurses in this last period of her life) had accompanied her to the Abbey, and she wrote to me:

> Last Sunday I went with Lady Churchill and your sister to the Battle of Britain Memorial Service. I always go to it anyway, – but this year it was particularly poignant . . . It was very moving indeed to see the rapturous welcome of the crowd when Lady Churchill came out of her car, – & – the gasp of admiration as the entire congregation within the nave rose instantly to their feet as she entered the Abbey.[16]

* * *

Towards the end of January 1977, my husband had a major coronary bypass operation. It was a success, but he suffered long weeks of pain and weakness. My mother was full of tender enquiries, sending flowers and 'baby' bottles of champagne – which cheered us both! And Christopher was well enough for us to give a small luncheon party at our flat to celebrate Clementine's ninety-second birthday on 1 April. The family included, of course, dear Cousin Sylvia, herself ninety-five, and still ramrod straight. Also there to drink my mother's health were Nonie Chapman, Grace Hamblin and Meg Colville. It was a loving 'demonstration', which I believe she enjoyed. But even small, intimate parties were now really too much for her.

Life went on for her like a gentle stream – but it flowed much more slowly now. Hearing, even with an aid, was a great effort and, as my father had done, she would withdraw sometimes into great and distant silences. But on 'good' days there were still pleasures and interests. It was Jubilee Year, and Clementine was glad she had lived to see so happy and glorious the young Queen whose first Prime Minister Winston had been. In May, Prince Charles opened an exhibition of Winston's pictures at Knoedler's Gallery in aid of the Jubilee Appeal. Clementine was much disappointed that she was not well enough to go herself to the opening, but a few days later she had a private viewing 'after hours', and although she could not see the pictures in detail, we described each one to her: mostly lent by the family, she remembered nearly all of them. They had been beautifully lit, and she received the impression of how jewel-like they looked, hung against the grey velvet walls of the gallery.

In mid-June, Clementine made an expedition to Chartwell. It was looking lovely: attendance figures were up, and the takings from the souvenir shop were booming. It all gave her great satisfaction; she enjoyed her luncheon in the restaurant, and returned home laden with flowers. A few weeks later, my mother drove down to lunch with us at Castle Mill, and this made another happy day.

On 26 July Clementine was suddenly taken ill, and rushed to King Edward VII Hospital, where she was quickly operated on for an abdominal stoppage. Amazingly, she weathered the operation, which was a success and revealed nothing sinister – but it was a severe ordeal for her to suffer. Her own nurses, and Pat and Matilda (her dear and devoted Filipino maids), scarcely left her: they gave her confidence, and knew all her little ways. Among the loveliest of the flowers she received was a bouquet from the Queen Mother, with a personally written note of sympathy and encouragement. Clementine improved enough for her to be brought home, with gentlest care, about a fortnight after the operation.

Then followed weeks of desolating weakness and fatigue. However, slowly, she began to pick up: she started eating better, and enjoying company more; gradually she resumed a gentle routine of a guest or two to luncheon; and she started her afternoon drives again. But we all realized this latest illness had taken a heavy toll. However, my mother was well enough for me to feel I could go to America in November to visit my daughter Charlotte and her husband (who were living in New York) and to see the 'new' Clementine.

Ever an early Christmas shopper, Clementine started arranging all her presents with Nonie, giving her usual detailed thought to what everyone would like best; and Nonie was busy as usual with discreet enquiries. But, for the first time this year, Clementine did not send any Christmas cards. She always chose a reproduction of one of Winston's pictures, and her Christmas card list for 1976 had shown a hundred names. Writing in all of them herself, she used to do batches of them on her 'good days' throughout the autumn months. In the last year or two the message and signature were quavery, and sometimes ran off the card – but: 'Keep right on to the end of the road . . .'

On 30 November, the anniversary of Winston's birthday, Winston and Minnie invited Clementine to luncheon with them at the Berkeley. It was the first time she had been out to a meal since her operation in the summer, and she enjoyed herself very much. Winston and Minnie had again asked her to stay with them for Christmas, and they talked about plans: she was looking forward to it.

I liked it best of all now when I lunched or dined alone with my mother.

Sometimes we did not speak very much – but that did not matter. Even as a middle-aged woman I rejoiced, and felt sustained to be received as a beloved child. I saw my mother for the last time when I lunched alone with her on 6 December: she was in rather good form, and she gave me her usual open-arm loving welcome. Although drawn and frail now, she still looked beautiful. Above all, I dreaded for her another test of pain or wearying weakness: but death was to come kindly.

On Monday, 12 December, she was feeling pretty well: Nonie Chapman always lunched with Clementine on Mondays and, the day being fine, they decided they would go out for a drive afterwards. Her coat, gloves and scarf were ready in the front hall. At luncheon, which she was quite enjoying, Nonie suddenly noticed a change in her breathing: she called the maids, and between them they got her to her bedroom. There, in a few minutes more, Clementine died peacefully and without pain.

* * *

Messages and letters – official and public, personal and private – came flooding in for Clementine's family from all over the world. The tributes and appreciations in the press, and on radio and television, were moving and remarkable. We had felt her death would evoke a response from many people, but the reality of the expressions of love and esteem for her far exceeded any expectation we could have had.

Her funeral service, attended by her family and close friends, was held at Holy Trinity, Brompton, where she herself had worshipped during these last years. Later she was cremated according to her wishes, and on 16 December her closest family were at Bladon when her ashes were laid in Winston's grave.

The Service of Thanksgiving for Clementine Spencer-Churchill's life was held in Westminster Abbey on 24 January 1978, the thirteenth anniversary of the death of Winston Churchill. The Queen and the Royal Family sent representatives; the Prime Minister, the Speaker, the Leaders of the Conservative and Liberal Parties, and many Members of both Houses of Parliament were there. Sixty countries were represented by their Ambassadors, their High Commissioners or other representatives. People from many aspects of civic and official life, and organizations for which Clementine had worked or with which she had had connections, were also represented. Winston's old regiment, the Queen's Royal Irish Hussars (4th Hussars) were there, helping to show people to their places, as were officers and men from the submarine HMS *Churchill*. Applications for seats had poured in – and there was room for all. And, apart from those who had received tickets, there were a great number of people – they filled the nave – who had come simply and anonymously, to join with us all, to remember, and to give thanks.

Sarah and I placed a wreath of white flowers for both our parents on the marble stone near the Great West Door upon which are engraved the words:

'Remember Winston Churchill', before our family made its way to our places in the Lantern near the High Altar. Then the long, gleaming procession of clergy, with the Dean and Chapter, the Archbishop of Canterbury and the Apostolic Delegate, moved up the aisle, and the service began. It was moving and splendid; sad, and above all, triumphantly thankful. We prayed the prayer of St Ignatius Loyola, which asks for 'a steadfast heart which no unworthy affection may drag downwards . . . an unconquered heart which no tribulation can wear out . . . an upright heart which no unworthy purpose may tempt aside.'

It reminded me so much of her.

The former Dean of Westminster, Dr Eric Abbott (a friend of Clementine's), read from St Paul's Epistle to the Corinthians of the love that 'beareth all things . . . hopeth all things, endureth all things. Love never faileth.' And that reminded me of her too.

And young Winston read an extract from the sermon preached by old Bishop Welldon nearly seventy years before in the Church of St Margaret's, Westminster nearby, at the marriage of Mr Winston Churchill to Miss Clementine Hozier:

> The sun shines upon your union today; the happy faces of your friends surround you; good wishes are lavished upon you; many prayers ascend to Heaven on your behalf. Will you suffer me to remind you how much you may be to the other in the coming days? There must be in the statesman's life many times when he depends upon the love, the insight, the penetrating sympathy and devotion of his wife. The influence which the wives of our statesmen have exercised for good upon their husband's lives is an unwritten chapter of English history, too sacred to be written in full . . . In the sunny hours of life, in the sombre hours, if they, too, come, may you recall the feelings and the resolves that were yours when you knelt today before the altar of God. So it is with deep feelings that I commend you to the blessings of the Eternal and Supreme. May your lives prove a blessing each to the other and both to the world, and may you pass in the Divine mercy from strength to strength and from joy to joy.

As we came out into the gusty winter world, the bare branches of the trees were blowing about against a leaden sky. Away across Parliament Square, Winston Churchill's statue was trudging forward, doggedly, into whatever future may be, and somewhere, far above us, the bells of the Abbey were ringing out – full peal.

Notes and Sources

The correspondence between Clementine Churchill (CSC) and her husband Winston Churchill (WSC) is housed at the Churchill Archives Centre, Churchill College, Cambridge, in three separate collections: the Baroness Spencer-Churchill Papers (CSCT), denoted here by the symbol *; the Chartwell Papers, WSC papers up to his resignation in July 1945 (CHAR), denoted here by the symbol ø; and the Churchill Papers, WSC papers post-resignation July 1945 (CHUR), denoted here by the symbol □.

Letters dated in the text are not noted again here. In all letters quoted the original spelling has been retained so far as possible and '[*sic*]' inserted where appropriate. Some punctuation has been amended slightly for easier comprehension.

For WSC, *The World Crisis*, of which there have been several editions, a chapter reference is noted as well as the page reference in the first edition.

RSC is an abbreviation for Randolph Spencer Churchill.

Chapter 1: Forebears and Early Childhood

1 Quinn, *Cavalier of Christ*, p. 18.
2 The author is indebted for many of these details concerning Lady Airlie to the memoirs of her daughter-in-law, Mabell, Countess of Airlie, *Thatched with Gold*.
3 Blunt, 'Secret Memoirs', Fitzwilliam Museum, Cambridge, MS 31–1975, vol. XIV, p. 125; see also Soames, ed., *Speaking for Themselves*, introduction, p. 3.
4 Blunt, 'Secret Memoirs', vol. XIV, p. 125.
5 Much of this family background is derived from the author's correspondence with the late Lady Helen Nutting (Blanche Hozier's niece) and from correspondence shown her by the present (13th) Earl of Airlie.
6 Mary S. Lovell, *The Mitford Girls*, p. 25.
7 Lees-Milne, *Caves of Ice*, p. 155.
8 Blunt, 'Secret Memoirs', MS 32-1975, vol. XV, p. 36.
9 This diary was shown to the author by the late Peregrine Spencer-Churchill.
10 Pearson, *Citadel of the Heart*, p. 117.
11 The picture is now owned by the author.
12 Blunt, Diary, Fitzwilliam Museum, Cambridge, MS 16–1975, Part III,

23 August 1916; see also Longford, *Pilgrimage of Passion*, p. 386.
13 CSC wrote *c.*1930 some 'sketch' chapters about her childhood entitled '*My* Early Life', from which account this quotation and much subsequent information have been taken. Unfortunately the account ceases in about 1904.
14 Letter from CSC to Giles Romilly, 2 January 1957.
15 Earl of Airlie to Blanche, Countess of Airlie, 29 December 1890.
16 Census of Scotland, Edinburgh 685/I, vol. 7, schedule 30, MS, New Register House, Edinburgh.

Chapter 2: Dieppe and Afterwards

1 Pakenham, *60 Miles from England*, pp. 179–80.
2 This story and many other details of their Dieppe life have been taken from the papers of Nellie (Mrs Bertram Romilly).
3 Baron, *Sickert*; also Pakenham, *Pigtails and Pernod*.
4 From CSC, '*My* Early Life'.
5 Cynthia Asquith, *Remember and Be Glad*, p. 87.
6 Sir Alan Lascelles to the author, 26 November 1964.

Chapter 3: To Thine Own Self Be True

1 Airlie, *Thatched with Gold*, p. 125.
2 RSC, *Winston S. Churchill*, vol. II, p. 252.
3 Letter from Lady Blanche Hozier to her brother, the Hon. Lyulph Stanley, 22 October 1906.
4 WSC to CSC, 16 April 1908*.
5 CSC to Nellie Hozier, undated, but probably 13–15 August 1908*.
6 Airlie, *Thatched with Gold*, p. 125.
7 Blunt, Diary, 14 August 1908, MS 395–1975.
8 WSC to CSC, undated, written at Blenheim, probably 13 August 1908*.
9 CSC to WSC, undated, written at Blenheim, probably 13 August 1908*.
10 Airlie, *Thatched with Gold*, p. 125.
11 Ibid.
12 CSC to WSC, undated letter written from Batsford Park, August 1908*.
13 WSC to CSC, undated letter but must have been written immediately after their engagement*.

Chapter 4: Early Days Together

1 Blunt, *My Diaries*, vol. II, p. 222; quoted by Longford, *Pilgrimage of Passion*, p. 386.
2 Ibid.
3 Marsh, *A Number of People*, p. 154.
4 Bonham Carter, *Winston Churchill As I Knew Him*, p. 219.
5 Webb, *Our Partnership,* p. 416, entry for 16 October 1908.
6 Cynthia Asquith, *Remember and Be Glad*, p. 88.

7 See Pearson, *Citadel of the Heart*, pp. 109–11.
8 Gilbert, *Winston S. Churchill, Companion Vol. II, Part I*, p. 672.
9 Lady Randolph to WSC, 21 November 1907, ibid., p. 705.
10 Lady Randolph to WSC, 5 December 1907, ibid., p. 718.
11 WSC to CSC, 28 April 1909 [WSC writes 1910]*.
12 CSC to WSC, 4 November [1909]*.
13 WSC to CSC, 12 September 1909*.
14 CSC to WSC, 12 September 1909*.
15 CSC to WSC, 11 September 1909*.
16 CSC to WSC, 16 September 1909*.

Chapter 5: At the Home Office

1 CSC to WSC, 18 September 1909*.
2 *The Times*, 9 December 1909.
3 At a meeting in Dundee; see *The Times*, 4 February 1909.
4 *The Times*, 15 November 1909, and CSC's personal account.
5 Bonham Carter, *Lantern Slides*, p. 162.
6 Ibid., pp. 162–3.
7 Quoted by Roy Jenkins, introduction to Bonham Carter, *Lantern Slides*, p. xxvi.
8 Violet Asquith to Venetia Stanley, 20 November 1908, Bonham Carter, *Lantern Slides*, p. 171.
9 WSC to CSC, 10 November 1909*.
10 WSC to CSC, 19 December 1910*.
11 Quoted in Moran, *The Struggle for Survival*, p. 167.
12 Ibid.
13 WSC to CSC, 11 July 1911*.
14 WSC to CSC, 15 March 1925*.
15 CSC to WSC, 21 September 1910*.
16 CSC to WSC, 19 December 1910*.
17 Blunt, *My Diaries*, vol. II, pp. 279–80, entry for 5 September 1909.
18 Esher, *Journals and Letters*, entry for 1 December 1909.
19 CSC to WSC, dated 'Saturday evening' [3 June 1911]*.
20 WSC to CSC, 29 June 1911*.
21 Ibid.

Chapter 6: Those that Go Down to the Sea in Ships

1 WSC, *The World Crisis 1911–1914*, ch. 5, p. 119.
2 Neville Lytton to Nellie Hozier, 4 September 1911.
3 CSC to WSC, dated 'Tuesday evening' [16 January 1912]*.
4 CSC to WSC, 7 February [1912]*.
5 Sir Edward Grey to CSC, 18 February 1912.
6 *Belfast Weekly News*, 15 February 1912.
7 Riddell, *More Pages from my Diary*, p. 37.
8 Ibid.
9 *The Times*, Special Correspondent, 8 February 1912.

10 CSC to WSC, 25 March [1912]*.
11 H. H. Asquith to WSC, 1 April 1912, quoted in Gilbert, *Winston S. Churchill, Companion Vol. II, Part III*, p. 1483.
12 Margot Asquith to CSC, 2 April [1912]*.
13 Bonham Carter, *Winston Churchill As I Knew Him*, p. 265.
14 Violet Asquith to Venetia Stanley, 26 May 1912, Violet Bonham Carter Papers, Bodleian Library, Oxford [as yet uncatalogued]; quoted in Bonham Carter, *Lantern Slides*, p. 316.
15 CSC to WSC, dated 'Sunday' [14 July 1912]*.
16 CSC to WSC, dated 'Monday Evening' [22 July 1912]*.
17 CSC to WSC, wrongly dated 'Wednesday the 23rd' [24 July 1912]*.
18 CSC to WSC, 23 July [1912]*.
19 CSC to WSC, 24 July [1912]*.
20 CSC to WSC, dated 'Tuesday Evening' [30 July 1912]*.
21 *The Times*, 14 and 17 August 1912.
22 *The Times*, 13 September 1912.
23 CSC to WSC, 4 November [1913]*.
24 WSC to CSC, 1 February 1913*.

Chapter 7: The Sands Run Out

1 CSC to WSC, 31 January [1913]*.
2 CSC to WSC, 6 April [1913]*.
3 Margot Asquith to WSC, 10 February [1910]ø.
4 Margot Asquith to CSC, 2 April [1912]*.
5 H. H. Asquith to Venetia Stanley, 27 April 1915 [ex Montagu Papers].
6 H. H. Asquith to Venetia Stanley, 30 March 1914; as quoted in Brock and Brock (eds), *H. H. Asquith: Letters to Venetia Stanley,* p. 62.
7 H. H. Asquith to Venetia Stanley, 10 January 1915.
8 H. H. Asquith to Venetia Stanley, 26 January 1915.
9 H. H. Asquith to Venetia Stanley, 1 February 1915.
10 H. H. Asquith to Venetia Stanley, 7 March 1915; quoted in Brock, p. 465.
11 Neville Lytton to Nellie Hozier, 8 February 1911.
12 WSC to CSC, 23 July [1913]*.
13 WSC to CSC, 23 April 1914*.
14 WSC to CSC, 27 April 1914*.
15 Violet Asquith to Hugh Godley, 19 May [1913], quoted in Bonham Carter, *Lantern Slides*, p. 383.
16 Diary of Margot Asquith, Bodleian Library, Oxford, Ms.Eng.d.3210, fos 97–113.
17 CSC to WSC, undated, but headed 'In the train to London, 7 p.m.' [20 October 1913]*.
18 CSC to WSC, 23 October 1913*.
19 This whole incident has been pieced together from CSC's own recollections and from the accounts given to the author by Margaret, Countess of Birkenhead (then Mrs F. E. Smith) and the 10th Duke of Marlborough (Sunny's son, Blandford).

20 WSC to CSC, 3 November 1913*.
21 CSC to WSC, dated 'Wednesday the 22nd' [April 1914]*.
22 CSC to WSC, 24 October [1913]*.
23 Quoted in RSC, *Winston S. Churchill*, vol. II, p. 703.
24 CSC to WSC, 24 April [1914]*.
25 CSC to WSC, 4 June [1914]*.
26 CSC to WSC, 5 June [1914]*.
27 CSC to WSC, 6 July [1914]*.
28 WSC to CSC, 13 July 1914*.
29 WSC, *The World Crisis 1911–1914*, ch. 8, p. 190.
30 Ibid., ch. 9, p. 197.
31 CSC to WSC, undated*.
32 WSC to CSC, 28 July 1914*.
33 CSC to WSC, 29 July [1914]*.
34 CSC to WSC, 31 July [1914]*.

Chapter 8: All Over by Christmas

1 CSC to WSC, undated, but almost certainly just before 26 September 1914*. See Gilbert, *Winston S. Churchill*, vol. III, p. 92.
2 WSC to CSC, 26 September 1914*.
3 CSC to WSC, 9 August [1914]*.
4 CSC to WSC, 12 August [1914]*.
5 CSC to WSC, 14 August [1914]*.
6 CSC to WSC, 14 August [1914] (2nd letter)*.
7 H. H. Asquith to Venetia Stanley, 5 October 1914, quoted in Brock and Brock (eds), *H. H. Asquith: Letters to Venetia Stanley*, p. 263.
8 Bonham Carter, *Winston Churchill As I Knew Him*, p. 332.
9 Figures taken from British, French and German official casualty returns; see WSC, *The World Crisis 1916–1918*, Part I, appendix I.

Chapter 9: The Dardanelles

1 WSC, *The World Crisis 1915*, ch. 2, p. 44.
2 Hankey, *The Supreme Command*, vol. I, pp. 265–6.
3 WSC, *The World Crisis 1911–1914*, ch. 4, p. 73.
4 Bonham Carter, *Winston Churchill As I Knew Him*, p. 362, from which also the account of the review is taken.
5 H. H. Asquith to Venetia Stanley, 26 February 1915, quoted in Brock and Brock (eds), *H. H. Asquith: Letters to Venetia Stanley*, p. 450.
6 WSC, *The World Crisis 1915*, ch. 12, p. 252.
7 Gilbert, *Winston S. Churchill*, vol. III, p. 431.
8 WSC, *The World Crisis 1915*, ch. 18, p. 353.
9 Ibid., p. 358.
10 Bonham Carter, *Winston Churchill As I Knew Him*, p. 393.
11 H. H. Asquith to Venetia Stanley, 14 May 1915; quoted in Brock, p. 596; also by Gilbert, *Winston S. Churchill*, vol. III, p. 446.

12 H. H. Asquith to WSC, 21 May 1915, quoted ibid., p. 464.
13 H. H. Asquith to Venetia Stanley, 22 May 1915, quoted in Gilbert, ibid., p. 466.
14 WSC, *The World Crisis 1915*, ch. 18, p. 371.
15 Gilbert, *Winston S. Churchill*, vol. III, p. 473.
16 See Taylor, *English History 1914–1945*, p. 31 n. 3.
17 CSC to H. H. Asquith, 20 May 1915, Bodleian Library, Oxford, Ms Asquith 27, fos 172–5.
18 Maurice Bonham Carter to Violet Asquith, 22 May 1915 (continuation of letter of 21 May), quoted in Bonham Carter, *Champion Redoubtable*, p. 57.
19 Diary of Margot Asquith, Bodleian Library, Oxford, Ms.Eng.d.3212, fos 48–50.
20 H. H. Asquith to Venetia Stanley, 20 May 1915, quoted in Gilbert, *Winston S. Churchill*, vol. III, p. 459.
21 Edwin Montagu to Venetia Stanley, 26 May 1915; quoted ibid., p. 472.
22 Edwin Montagu to CSC, 26 May 1915; quoted ibid.
23 Diary of Margot Asquith, Ms.Eng.d.3212, fos 111–16.
24 H. H. Asquith to Sylvia Henley, 9 June 1915, Bodleian Library, Oxford, Ms.Eng.lett.c.542/1, fos 89–92.
25 Thurso Papers, Churchill College, Cambridge, THSO I 1/2.
26 WSC, *Painting as a Pastime*, p. 16.
27 Ibid., p. 18.
28 Ashmead-Bartlett, *The Uncensored Dardanelles*, p. 121. Diary entry, 10 June 1915.
29 H. H. Asquith to Sylvia Henley, 19 August 1915, Ms.Eng.lett.c.542/2, fos 357–60.

Chapter 10: The Place of Honour

1 Beaverbrook, *Politicians and the War*, vol. II, p. 74.
2 Bonham Carter, *Winston Churchill As I Knew Him*, p. 429.
3 WSC to CSC, 18 November 1915*.
4 CSC to WSC, 19 November [1915]*.
5 CSC to WSC, 21 November [1915]*.
6 WSC, 'With the Grenadiers', in *Thoughts and Adventures*, p. 101.
7 CSC to WSC, 22 November [1915]*.
8 WSC to CSC, 25 November 1915*.
9 WSC to CSC, 27 November 1915*.
10 Ibid.
11 WSC to CSC, 1 December [1915] (third letter, marked 'Later still')*.
12 CSC to WSC, 28 November [1915]*.
13 CSC to WSC, 19 November [1915]*.
14 CSC to WSC, 25 November [1915]*.
15 CSC to WSC, 28 November [1915]*.
16 WSC to CSC, 1 December 1915*.
17 Ibid. (second letter, marked 'Later')*.
18 Ibid. (third letter)*.
19 Ibid.

20 WSC to CSC, 3 December 1915*.
21 Ibid.
22 WSC to CSC, 1 December 1915 (third letter)*.
23 WSC to CSC, 23 November 1915*.
24 WSC to CSC, 25 November 1915*.
25 WSC to CSC, 27 November 1915*.
26 WSC to CSC, 4 December 1915*.
27 WSC to CSC, 8 December 1915*.
28 WSC to CSC, 12 December 1915*.
29 WSC to CSC, 4 December 1915*.
30 Ibid.
31 Ibid.
32 Ibid.
33 CSC to WSC, 6 December [1915]*.
34 CSC to WSC, 28 November [1915]*.
35 CSC to WSC, 6 December [1915]*.
36 Ibid.
37 WSC to CSC, 8 December 1915*.
38 WSC to CSC, 8 December 1915*.
39 CSC to WSC, 9 December [1915]*.
40 Ibid.
41 WSC to CSC, 10 December 1915*.
42 CSC to WSC, 12 December [1915]*.
43 WSC to CSC, 12 December 1915*.
44 WSC to CSC, 18 December 1915, from which letter the ensuing quotes and account of the 'Brigade' drama is drawn in general*.
45 *Hansard, Parliamentary Debates: Commons*, vol. 78, col. 2218.
46 WSC to CSC, 15 December 1915 (second letter)*.
47 Ibid.
48 CSC to WSC, 17 December [1915] (first letter)*.
49 WSC to CSC, 18 December 1915*.
50 Ibid.
51 Ibid.
52 Ibid., 'Later'.
53 Ibid.
54 Ibid.
55 WSC to CSC, 20 December 1915*.
56 Ibid.
57 Ibid.
58 Ibid.

Chapter 11: Waiting in Silence

1 WSC to CSC, 2 January 1916*.
2 WSC to CSC, 1 January 1916*.
3 CSC to WSC, New Year's Day, 1916ø.
4 WSC to CSC, 3 January 1916*.

5 WSC to CSC, 5 January 1916*.
6 CSC to WSC, 7 January 1916ø.
7 CSC to WSC, 9 January 1916ø.
8 WSC to CSC, 13 January 1916*.
9 CSC to WSC, 16 January 1916ø.
10 WSC to CSC, 6 January 1916*.
11 WSC to CSC, 10 January 1916*.
12 WSC to CSC, 7 January 1916*.
13 CSC to WSC, 11 January [1916]ø.
14 WSC to CSC, 16 January 1916*.
15 WSC to CSC, 7 January 1916*.
16 WSC to CSC, 10 January 1916*.
17 WSC to CSC, 17 January 1916*.
18 WSC to CSC, 19 January 1916*.
19 WSC to CSC, 7 January 1916*.
20 WSC to CSC, 19 January 1916*.
21 CSC to WSC, undated, marked 'Later' [16 January 1916]ø.
22 CSC to WSC, 18 January [1916]ø.
23 CSC to WSC, undated [16 January 1916]ø.
24 CSC to WSC, 20 January [1916]ø.
25 WSC to CSC, 22 January 1916*.
26 WSC to CSC, 24 January 1916*.
27 CSC to WSC, 21 January [1916]ø.
28 Ibid.
29 CSC to WSC, 24 January [1916]ø.
30 CSC to WSC, 5 January 1916ø.
31 WSC to CSC, 26 January 1916*.
32 WSC to CSC, 20 January 1916*.
33 CSC to WSC, 24 January [1916]ø.
34 WSC to CSC, 27 January 1916*.
35 CSC to WSC, 30 January [1916]ø.
36 CSC to WSC, 27 January 1916ø.
37 WSC to CSC, 27 January 1916*.
38 Ibid.
39 CSC to WSC, 31 January [1916]ø.
40 CSC to WSC, 2 February [1916]ø.
41 CSC to WSC, 31 January [1916]ø.
42 WSC to CSC, 31 January 1916*.
43 WSC to CSC, undated [1 February 1916]*.
44 CSC to WSC, 4 February 1916ø.
45 Ibid.
46 Ibid.
47 WSC to CSC, 4 February 1916*.
48 CSC to WSC, 9 February [1916]ø.
49 WSC to CSC, 10 February 1916*.
50 CSC to WSC, 23 February [1916]ø.
51 WSC to CSC, 14 February 1916*.
52 WSC to CSC, 20 February 1916*.

53 WSC to CSC, 22 February 1916*.
54 CSC to WSC, 16 February 1916ø.
55 CSC to WSC, 25 February [1916]ø.
56 CSC to WSC, 12 February [1916] (letter begun on 11 February and continued at 5 a.m. the next day)ø.
57 CSC to WSC, 21 February [1916]ø.
58 CSC to WSC, 25 February [1916]ø.
59 CSC to WSC, 27 February 1916ø.

Chapter 12: Day Must Dawn

1 Bonham Carter, *Winston Churchill As I Knew Him*, p. 449. The author verified this story with CSC.
2 Gilbert, *Winston S. Churchill*, vol. III, p. 722.
3 CSC to WSC, 14 March [1916]ø.
4 Ibid.
5 Ibid.
6 CSC to WSC, 13 [14] March [1916], 7 p.m.ø (wrongly dated by CSC).
7 CSC to WSC, 15 March [1916]ø.
8 WSC to CSC, 13 March 1916*.
9 WSC to CSC, 19 March 1916*.
10 WSC to Archibald Sinclair, 22 March 1916, Thurso Papers, THSO I 1/4.
11 WSC to CSC, 21 March 1916*.
12 CSC to WSC, 22 March [1916]ø.
13 CSC to WSC, 24 March [1916]ø.
14 Ibid.
15 CSC to WSC, 22 March [1916]ø.
16 CSC to WSC, 24 March [1916] (second letter)ø.
17 WSC to CSC, [22 March 1916]* (WSC writes 22.1.16 but CSC was sure it was March).
18 CSC to WSC, 25 March [1916]ø.
19 Ibid.
20 WSC to CSC, 23 March 1916*.
21 WSC to CSC, 28 March 1916*.
22 CSC to WSC, 1 April 1916ø.
23 CSC to WSC, 4 April 1916ø.
24 CSC to WSC, 6 April [1916]ø.
25 WSC to CSC, 10 April 1916*.
26 CSC to WSC, Monday 27 [March 1916]ø.
27 WSC to CSC, 7 April 1916*.
28 CSC to WSC, 12 April [1916]ø.
29 Ibid.
30 CSC to WSC, 14 April [1916]ø.
31 Ibid.
32 WSC to CSC, 13 April 1916*.
33 *Dundee Advertiser*, 27 July 1917.
34 CSC to WSC, 16 February 1916ø.

35 RSC, *Twenty-One Years*, p. 15.
36 Jean, Lady Hamilton Diaries (microfilm), Liddell Hart Centre for Military Archives, King's College, London, ref. Hamilton 20/1/3; see also Celia Lee, *Jean, Lady Hamilton 1861–1941*, pp. 198–9.
37 CSC to WSC, 23 July 1918ø.
38 CSC to WSC, 15 August 1918ø.
39 WSC to CSC, 17 August 1918*.
40 CSC to WSC, 15 August 1918ø.
41 CSC to WSC, 29 October 1918*.
42 WSC to CSC, 12 September 1918ø.

Chapter 13: Bleak Morning

1 Gilbert, *Winston S. Churchill*, vol. IV, p. 166.
2 Pelling, *Winston Churchill*, p. 242.
3 Gilbert, *Winston S. Churchill*, vol. IV, p. 173.
4 RSC, *Twenty-One Years*, p. 19.
5 CSC to WSC, 9 March 1919*.
6 Ibid.
7 CSC, in personal recollections to the author.
8 CSC to WSC, 31 March 1920ø.
9 RSC, *Twenty-One Years*, p. 12.
10 WSC to CSC, 27 January 1921*.
11 CSC to WSC, 13 February 1921ø.
12 Ibid.
13 Ibid.
14 WSC to CSC, 14 February 1921*.
15 WSC to CSC, 27 January 1921*.
16 WSC to CSC, 6 February 1921*.
17 CSC to WSC, 21 February 1921ø.
18 WSC to CSC, 14 February 1921*.
19 WSC to CSC, 21 February 1921*.
20 WSC to Lord Curzon, 29 June 1921.
21 Sarah Churchill to CSC, undated, but obviously August 1921.
22 Told to the author by WSC.
23 Margot Asquith, *Autobiography*, vol. I, p. 23.
24 WSC to CSC, 19 September 1921*.
25 CSC to WSC, 22 September 1921ø.
26 Ibid.
27 Ibid.
28 WSC to CSC, 29 December 1921*.
29 Ibid.
30 WSC to CSC, 4 January 1922*.
31 CSC to WSC, 4 January 1922ø.

Chapter 14: The Twenties

1 CSC to WSC, 3 February 1922ø.
2 CSC to WSC, 17 August 1922ø.
3 CSC to WSC, 4 August 1922ø.
4 WSC to CSC, 10 August 1922*.
5 CSC to WSC, 18 February 1921ø.
6 Quoted in Gilbert, *Winston S. Churchill*, vol. IV, p. 878.
7 *Dundee Courier*, 8 November 1922.
8 WSC, 'Election Memories', in *Thoughts and Adventures*, p. 213.
9 Recounted by Henry Pelling in *Winston Churchill*, p. 292.
10 CSC to WSC, 29 January [1916]ø.
11 CSC to WSC, 11 March 1925ø.
12 CSC to WSC, 20 March 1926*.
13 CSC to WSC, 25 March [1926]*.
14 WSC to CSC, 28 March 1926*.
15 CSC to WSC, 21 April 1924*.
16 WSC to CSC, 22 March 1925*.
17 WSC to CSC, 3 [August] 1929*. (The letter is clearly dated, in WSC's hand, '3/9/29', but this must have been a mistake.)
18 WSC to CSC, 8 August 1929*.
19 WSC to CSC, 19 September 1929*.
20 CSC to WSC, 8 February 1921ø.

Chapter 15: Chartwell

1 CSC to WSC, 11 July 1921ø.
2 CSC to WSC, dated 'Wednesday' [16–18 July 1921]ø.
3 Sarah Churchill, *A Thread in the Tapestry*, p. 22.
4 WSC to CSC, 27 January 1923*.
5 CSC to WSC, 29 January [1923]*.
6 CSC to WSC, 'Thursday' [1 February 1923]*.
7 WSC to CSC, 2 September 1923*.
8 CSC to WSC, dated 'Thursday the 26th' [March 1925]ø.
9 CSC to WSC, Memorandum [April 1926]*.
10 CSC to Margery Street, 14 July 1933.
11 WSC to Sir John Rothenstein, quoted in Rothenstein, *Time's Thievish Progress*, p. 129.
12 WSC to CSC, 25 September 1928*.
13 CSC to Margery Street, 14 January 1934.

Chapter 16: Herself and Others

1 CSC to WSC, 31 March 1920ø.
2 Letter to the author from CSC's solicitor, John Turing.
3 RSC, *Twenty-One Years*, p. 25.
4 Diana Churchill to CSC, 28 [November 1923].

5 Diana Churchill to WSC, 4 December 1923.
6 RSC to CSC, 21 May 1926.
7 Diana Churchill to WSC, 6 October 1926.
8 Diana Churchill to CSC, 19 November 1926.
9 RSC, *Twenty-One Years*, p. 86.
10 Ibid.
11 WSC to CSC, 8 April 1928*.
12 CSC to WSC, 13 April 1928*.
13 RSC, *Twenty-One Years*, p. 104.
14 CSC to RSC, 21 December 1930.
15 CSC to WSC, 17 February [1931]*.
16 Ibid.
17 CSC to WSC, 22 February 1935*.
18 WSC to CSC, 2 March 1935*.
19 Ibid.
20 Ibid.
21 CSC to Margery Street, 9 June 1933.
22 Ibid.
23 CSC to Margery Street, 18 September 1934.
24 CSC to Margery Street, 21 August 1935.
25 CSC to Margery Street, 7 December 1935.
26 CSC to WSC, 20 December 1935*.
27 CSC to Margery Street, 30 December 1935.
28 CSC to WSC, 31 August 1929*.
29 CSC to WSC, 1 January 1936*.
30 CSC to WSC, 11 January 1936*.
31 CSC to WSC, 23 January 1937*.
32 CSC to WSC, 29 January 1937*.
33 Bonham Carter, *Winston Churchill As I Knew Him*. p, 218.
34 Rhodes James, *Churchill: A Study in Failure*, p. 292.
35 Boyle, *Poor Dear Brendan*, pp. 15–18.
36 CSC to RSC, 18 January [1931].
37 CSC to WSC, 10 September [1926]ø.
38 Winston S. Churchill, *His Father's Son: The Life of Randolph Churchill*, p. 44.
39 Told by John S. Churchill to John Pearson; see Pearson, *Citadel of the Heart*, p. 213.
40 Conversation between the author and the late Peregrine Spencer-Churchill.
41 CSC to WSC, 29 January 1927, from Lou Sueil in the South of France*.
42 CSC to WSC, dated 'Friday the 30th' [March 1928]*.
43 WSC to CSC, 4 April [1928]*.
44 WSC to CSC, 15 March 1925*.

Chapter 17: Holiday Time

1 CSC to RSC, 12 January 1932.
2 Ibid.
3 CSC to WSC, 22 December 1934*.

4 WSC to CSC, 23 January 1935*.
5 WSC to CSC, 1 January 1935*.
6 WSC to CSC, 2 March 1935*.
7 CSC to WSC, 17 January 1935*.
8 Maryott Whyte to CSC, 16 January [1935].
9 Sarah Churchill to CSC, 22 January 1935.
10 Sarah Churchill to CSC, 17 February 1935.
11 CSC to WSC, 20 January 1935*.
12 Ibid.
13 CSC to WSC, 22 February 1935*.
14 CSC to WSC, 26 February 1935*.
15 Ibid.
16 WSC to CSC, 30 January 1935*.
17 CSC to WSC, 26 March 1935 (continuation of letter begun on 18 March)*.
18 Information kindly supplied by Lord Zuckerman, President of the Zoological Society of London, from the Zoo's archives.
19 CSC to the author.
20 Ibid.
21 CSC to WSC, 1 April 1935 (further continuation of letter begun on 18 March)*.
22 WSC to CSC, 9 March 1935*.
23 CSC to WSC, 2 April 1935*.
24 WSC to CSC, 10 March 1935*.
25 WSC to CSC, 5 April 1935*.
26 WSC to CSC, 13 April 1935*.
27 WSC to CSC, 11 April 1935*.
28 WSC to CSC, 10 March 1935*.
29 WSC to CSC, 5 April 1935*.
30 WSC to CSC, 13 April 1935*.
31 CSC to Margery Street, 21 August 1935.

Chapter 18: The Thirties

1 CSC to WSC, 7 January 1936*.
2 From the official report of the Conference.
3 See Broad, *Winston Churchill, The Years of Preparation*, pp. 352–3.
4 Rhodes James (ed.), *'Chips': The Diary of Sir Henry (Chips) Channon*, entry for 26 February 1937.
5 Winant, *A Letter from Grosvenor Square*, p. 98.
6 WSC to CSC, 2 February 1937*.
7 CSC to WSC, 19 January 1939ø.
8 CSC to Margery Street, 6 April 1939.
9 CSC to WSC, 13 December [1938]ø.
10 CSC to WSC, 24 December 1938ø.
11 CSC to WSC, 1 January 1939ø.
12 CSC to WSC, 19 January 1939ø.
13 Paul Maze, diary, quoted in Gilbert, *Winston Churchill, Companion Vol. V*, pt 3, p. 1592.

Chapter 19: Year of Destiny

1 WSC, *The Second World War*, vol. I, p. 363.
2 Ibid., pp. 363–4.
3 Ibid., p. 365.
4 Ibid.
5 CSC to Nellie Romilly, 20 September 1939.
6 WSC, *The Second World War*, vol. I, p. 375.
7 Cooper, *Trumpets from the Steep*, p. 37.
8 WSC, *The Second World War*, vol. I, p. 601.
9 *Hansard, Parliamentary Debates: Commons*, vol. 360, col. 1502.
10 WSC, *The Second World War*, vol. II, p. 38.
11 Ibid., p. 100.
12 Ibid., p. 87.
13 *Hansard, Parliamentary Debates: Commons*, vol. 361, col. 796.
14 The account of this incident was given to the author by CSC; the facts were confirmed and amplified by the late Hon. Mrs Averell Harriman (Pamela Churchill), who was present at the luncheon.
15 Buckle (ed.), *Self-Portrait with Friends: The Selected Diaries of Cecil Beaton 1926–1974*, pp. 73–4.
16 Vickers, *Cecil Beaton*, pp. 245. The photograph was published in *Picture Post*, 23 November 1940.
17 Beaton, unpublished diary, by courtesy of Hugo Vickers.
18 Ibid.
19 Colville, *The Fringes of Power*, p. 277.
20 Ibid., p. 379.

Chapter 20: Taking up the Load

1 Harrisson, *Living through the Blitz*, p. 214.
2 WSC, *The Second World War*, vol. III, p. 345.
3 Ibid., p. 421.
4 CSC to Lord Beaverbrook, 10 May 194[1]. Dated 1942, but this is clearly an error.
5 CSC to the author, 15 September 1941.
6 WSC, *The Second World War*, vol. III, p. 540.
7 WSC to CSC, undated but headed 'At Sea but now White House' [21 December 1941]*.
8 WSC, *The Second World War*, vol. IV, p. 81.
9 CSC to WSC, 11 April 1942*.
10 WSC, *The Second World War*, vol. IV, p. 344.
11 Ibid., p. 428.
12 Ibid., p. 541.
13 Nicolson, *Letters and Diaries 1939–45*, pp. 257–60.
14 Ibid.
15 Ibid.
16 Author's diary, 6 November 1942.
17 From account of the visit in Lash, *Eleanor and Franklin*, pp. 661–3.

18 CSC to Franklin D. Roosevelt, 1 November 1942.
19 Diary of Eleanor Roosevelt, quoted in Lash, *Eleanor and Franklin*, p. 663.
20 Lash, *Eleanor and Franklin*, p. 664.
21 Ibid.
22 WSC to Eleanor Roosevelt, quoted in Lash, *Eleanor and Frankin*, p. 668.

Chapter 21: All in the Day's Work

1 Statement issued from 10 Downing Street, 28 October 1943, quoted in *The Times* the following day.

Chapter 22: Conference and Crisis

1 RSC to CSC, 4 July 1944.
2 CSC to WSC, 14 January 1943*.
3 John Martin to Private Office, undated [*c.*19–22 January 1943]*.
4 Public Record Office, PREM 4/72/1.
5 Ibid.
6 Public Record Office, Cabinet Papers, CAB 120/77.
7 Author's diary, 21 February 1943.
8 CSC to author, 7 March 1943.
9 CSC to author, 9 March 1943.
10 CSC to WSC, 13 May 1943*.
11 CSC to WSC, 20 May 1943*.
12 WSC to CSC, 29 May 1943*.
13 CSC to WSC, 30 May 1943*.
14 Suckley, *Closest Companion*, p. 238.
15 Author's diary, 12 September 1943.
16 Public Record Office, CAB 120/120.
17 WSC, *The Second World War*, vol. V, p. 339.
18 CSC to WSC, 6 December 1943*.
19 Colville, *Footprints in Time*, p. 134.
20 Ibid., pp. 134–5.
21 Public Record Office, CAB 120/120.
22 CSC to the family, 21 December 1943.
23 Ibid.
24 Colville, *Fringes of Power*, p. 459.
25 CSC to the family, 10 January 1944.
26 CSC to the family, 12 January 1944.
27 Cooper, *Trumpets from the Steep*, p. 179; letter to Conrad Russell dated 12 January, from Marrakech.
28 WSC, *The Second World War*, vol. V, p. 401.
29 Cooper, *Trumpets from the Steep*, p. 181.
30 Ibid., p. 182.

Chapter 23: The Last Lap

1 WSC to CSC, 'In the air over Algeria', undated but certainly 11/12 August 1944*.
2 Reminiscence from Mrs Jean Hale of Barbados to the author, 3 January 1965.
3 CSC to author, 31 August 1944.
4 Author's diary, 1 September 1944.
5 Sherwood, *The White House Papers of Harry L. Hopkins*, vol. II, p. 806.
6 CSC to the family, 8 September 1944.
7 CSC to the family, 14 September 1944.
8 Ibid.
9 Ibid.
10 Sarah Churchill to CSC, 20 September 1944.
11 WSC, *The Second World War*, vol. VI, p. 348.
12 CSC to author, 21 February 1945.
13 Author to CSC, 18 February 1945.
14 CSC to author, 23 March 1945.
15 CSC to author, 27 March 1945.
16 Ibid.
17 CSC to Sarah Churchill, 29 March 1945.
18 CSC to WSC, 30 March 1945*.
19 Field Marshal Montgomery to CSC, 8 April 1945.
20 WSC to CSC, 2 April 1945*.
21 CSC to WSC, 5 April 1945*.
22 WSC to CSC, 6 April 1945*.
23 CSC, *My Visit to Russia*, p. 17. Much of the account of her journey is taken from this short book, written on her return.
24 This was recounted to the author by Mrs Whitfield, of Warminster, Wilts., to whom Miss Johnson told the story. CSC confirmed that it was true.
25 CSC, *My Visit to Russia*, p. 26.
26 CSC to Averell Harriman, 13 April 1945, Harriman Papers, Library of Congress, Box 178.
27 CSC, *My Visit to Russia*, pp. 28–30.
28 Ibid., p. 34.
29 Ibid., p. 36.
30 CSC to WSC, 24 April 1945*.
31 CSC, *My Visit to Russia*, pp. 37–8.
32 Ibid., p. 48.
33 WSC to CSC, 2 May 1945*.
34 CSC to WSC, 5 May 1945*.
35 CSC, *My Visit to Russia*, pp. 56–7.
36 WSC, *The Second World War*, vol. VI, p. 477.
37 CSC, *My Visit to Russia*, p. 59.

Chapter 24: The Two Impostors

1 CSC to author, 28 January 1945.
2 The author was obliged to the late Lord Duncan-Sandys for this reminiscence.
3 CSC to author, 21 June 1945.
4 WSC, *The Second World War*, vol. VI, p. 549.
5 Author to CSC, 24 July 1945.
6 WSC, *The Second World War*, vol. VI, p. 583.
7 Author's diary, 26 July 1945.
8 Ibid.
9 Facts and figures concerning the election are taken from McCallum and Readman, *The British General Election of 1945*, ch. 2.
10 Author's diary, 26 July 1945.
11 Author to CSC, 6 August 1945.
12 CSC to Mrs Pleydell-Bouverie, 9 August 1945.
13 Author to CSC, 18 August 1945.
14 *Hansard, Parliamentary Debates: Commons*, vol. 413, col. 97.
15 WSC to CSC, 8 September 1945*.
16 WSC to CSC, 13 August 1947*.
17 Author to CSC, 21 October [1945].

Chapter 25: Swords into Ploughshares

1 Author to CSC, 20 January 1946.
2 Author to CSC, 12 March 1946.
3 Author's diary, 26 March 1946.
4 Author's diary, 1 April 1946.
5 Ibid.
6 CSC to Horatia Seymour, November 1947.
7 CSC to Lady Desborough, 22 April 1948, Grenfell Papers, Hertfordshire Archives, D/ERv C503/8.
8 CSC to WSC, 26 December 1947□.
9 CSC to WSC, 11 August 1947□.
10 CSC to author, 16 February 1947.
11 Sarah Churchill to CSC, 29 August 1949.
12 CSC to Antony Beauchamp, 20 December 1949.
13 CSC to Sarah Churchill, 20 December 1949.

Chapter 26: A Woman's Work Is Never Done

1 *The Midwife*, 21 December 1942.
2 Lady Proctor to CSC, 23 March 1947.
3 YWCA Hostel Wardens' Conference, 28 January–1 February 1949, *Official Report*.
4 CSC speech at YWCA Hostel Wardens' Conference, 1949.
5 Mrs Eva K. Mackenzie of Glasgow to author, 25 November 1964.
6 CSC to author, 31 August 1952.

7 CSC to author, 26 August 1952.
8 CSC to Ronnie Tree, 28 May 1951.
9 CSC to WSC, dated 'Monday Evening' [29 October 1951]□.
10 Author's diary, 31 December 1952.
11 HRH The Princess Royal to CSC, 26 May 1953.
12 W. H. Cartmell to CSC, 27 April 1953.
13 Pelling, *Winston Churchill*, p. 605; see also Macmillan, *Tides of Fortune*, p. 516.
14 Colville, *Fringes of Power*, p. 669.
15 Told to the author by the late Sir John Colville.
16 *The Times*, 29 June 1953.
17 RSC to CSC, 30 June [1953].
18 Bonham Carter, *Daring to Hope*, p. 127.
19 Ibid.
20 Ibid., p. 129.
21 Lady Violet Bonham Carter to CSC, 13 August 1953.
22 Author to CSC, 22 September [1953].
23 WSC to CSC, 13 October 1953*.
24 Ibid.
25 WSC to CSC, 16 October 1953*.
26 Lord Beaverbrook to CSC, 22 March [1954].
27 Author's diary, 3–5 December 1954.

Chapter 27: Does the Road Wind Up-hill All the Way?

1 The author was indebted to the late Sir John Colville for details of this incident.
2 CSC to Lady Violet Bonham Carter, 23 March 1955, Bonham Carter Papers.
3 Sarah Churchill to CSC, dated 'May' [1955].
4 Author's diary, 26 July 1955.
5 Sarah Churchill to CSC, 9 November 1955.
6 CSC to Sarah Churchill, 14 January 1960.
7 CSC to author, 7 October 1960.
8 Sarah Churchill to CSC, 12 March 1958.
9 Author to CSC, 19 September 1958.
10 Sarah Churchill to CSC, 12 September 1958.

Chapter 28: Twilight and Evening Bell

1 Told to the author by the late Sir John Colville.
2 CSC to Christopher Soames, 6 July 1962.
3 Author's diary, 25 May 1952.
4 *The Times*, 21 April 1959.
5 Violet Bonham Carter to CSC, 27 October 1962.
6 CSC to Sarah Churchill, 26 December 1962.
7 Gilbert, *In Search of Churchill*, p. 316.
8 CSC to WSC, 5 March 1949□.

9 The author was much indebted to the late Mrs Doris Moss, OBE, for many details concerning this period.
10 CSC to Christopher Soames, 19 April 1964.
11 *Hansard, Parliamentary Debates: Commons*, vol. 699, col. 1237.

Chapter 29: Port after Stormy Seas

1 *The Times*, 20 February 1964.
2 CSC to Sarah Churchill, 24 October 1964.
3 Bonham Carter, *Daring to Hope*, p. 298.

Chapter 30: Afterwards

1 CSC to Harold Wilson, 9 February 1965.
2 WSC to Charles Nicoll of Messrs Nicoll, Manisty, Few & Co., 11 April 1946.
3 RSC to CSC, 14 February 1965.
4 CSC to the Earl of Avon, 28 March 1965, Avon Papers, University of Birmingham, AP19/3/6A.
5 Dated 30 November 1954.
6 CSC to the family, 18 September 1944.
7 Lord Clark to author, 10 August 1978. The author was most grateful to the late Lord Clark for other details.
8 CSC to Graham Sutherland, OM, 4 December 1954.
9 CSC to Lord Beaverbrook, 13 October 1957.
10 Quoted in Richard Lovell, *Churchill's Doctor*, p. 265.
11 Letter from Anthony F. Moir of Fladgate & Co., London SW1, to Lady Spencer-Churchill, dated 15 June 1966.
12 CSC to the author, 24 January 1966.
13 President de Gaulle to CSC, January 1966.
14 General de Gaulle to CSC, 19 May 1969.
15 Author to CSC, 21 November [1970].
16 Miss A. M. Johnson, later Matron of the Star and Garter Home, Richmond, to the author, 28 September 1975.

Select Bibliography

Manuscript Sources

Family Papers

Baroness Spencer-Churchill Papers (Churchill College, Cambridge)
Chartwell Papers (Churchill College, Cambridge)
Churchill Papers (Churchill College, Cambridge)
Churchill family papers
Romilly Papers
Mary Soames, personal papers and diaries
Diaries of Blanche, Countess of Airlie

Other Papers

H. H. Asquith Papers (Bodleian Library, Oxford)
Margot Asquith Papers (Bodleian Library, Oxford)
Avon Papers (University of Birmingham)
Beaverbrook Papers (House of Lords Record Office)
Wilfrid Scawen Blunt Papers (Fitzwilliam Museum, Cambridge)
Violet Bonham Carter Papers (Bodleian Library, Oxford)
Cabinet Papers and Prime Minister's Office Papers (Public Record Office, London)
Grenfell Papers (Hertfordshire Archives)
Jean, Lady Hamilton Diaries (Liddell Hart Centre for Military Archives, King's College, London)
Moran Papers (Wellcome Library for the History and Understanding of Medicine, London)
Thurso Papers (Churchill College, Cambridge)

Printed Sources

Airlie, Mabell, Countess of, *Thatched with Gold: The Memoirs of Mabell, Countess of Airlie*, ed. Jennifer Ellis, London, 1962
Airlie, 12th Earl of, *The Bonnie House of Airlie*, ed. Katherine Thomasson, privately printed, Brechin, 1963
Arnold-Baker, Charles, *The Companion to British History*, Tunbridge Wells, 1996
Ashmead-Bartlett, E., *The Uncensored Dardanelles*, London, 1928

Asquith, Lady Cynthia, *Diaries 1915–1918*, London, 1968
—— *Remember and Be Glad*, London, 1952
Asquith, Margot, *Autobiography*, 2 vols, London, 1920, 1922
Baron, Wendy, *Sickert*, London, 1973
Beaverbrook, Lord, *Politicians and the War*, 2 vols, London, 1928; repr. 1966
Bennett, Daphne, *Margot: A Life of the Countess of Oxford and Asquith*, London, 1984
Best, Geoffrey, *Churchill: A Study in Greatness*, London, 2001
Birkenhead, 2nd Earl of, *F. E.*, London, 1959
Blanche, Jacques-Emile, *Cahiers d'un artiste*, 6 vols, Paris, 1915–20, vol. I
Blunt, Wilfrid Scawen, *My Diaries*, 2 vols, London, 1912, 1920
Bonham Carter, Lady Violet, *The Letters and Diaries of Violet Bonham Carter*: vol. I, *Lantern Slides, 1904-1914*, ed. Mark Bonham Carter and Mark Pottle, London, 1996; vol. II, *Champion Redoubtable, 1914–1945*, ed. Mark Pottle, London, 1998; vol. III, *Daring to Hope, 1946–69*, ed. Mark Pottle, London, 2000
—— *Winston Churchill As I Knew Him*, London, 1965
Boyle, Andrew, *Poor Dear Brendan*, London, 1974
Broad, Charles Lewis, *Winston Churchill: The Years of Preparation*, London, 1963
Brock, Michael and Brock, Eleanor (eds), *H. H. Asquith: Letters to Venetia Stanley*, Oxford, 1982
Buckle, Richard (ed.), *Self-Portrait with Friends: The Selected Diaries of Cecil Beaton*, London, 1979
Butler, D. Edgeworth and Freeman, J., *British Political Facts 1900–1960*, London, 1963
Cambray, P. G. and Briggs, G. G. B., *Red Cross and St John War Organisation 1939–47*, London, 1949
Churchill, Clementine S., *My Visit to Russia*, London, 1945
Churchill, John Spencer, *A Crowded Canvas*, London, 1961
Churchill, Randolph S., *Twenty-One Years*, London, 1965
—— *Winston S. Churchill*, vols I and II, London, 1966, 1967; also *Companion Volumes* I and II, London 1967, 1969 (for later volumes see under Gilbert, Martin)
Churchill, Sarah, *A Thread in the Tapestry*, London, 1967
Churchill, Winston S., *Painting as a Pastime*, London, 1948
—— *The Second World War*, 6 vols, London, 1948–54
—— *Thoughts and Adventures*, London, 1932
—— *The World Crisis*, 6 vols, London 1923–31
Churchill, Winston S. (b. 1940), *His Father's Son: The Life of Randolph S. Churchill*, London, 1996
Colville, John, *Footprints in Time*, London, 1976
—— *The Fringes of Power: Downing Street Diaries 1939–1955*, London, 1985
Cooper, Lady Diana, *Trumpets from the Steep*, London, 1960
Coward, Noël, *Diaries*, ed. Graham Payn and Sheridan Morley, London, 1982
Cross, Colin, *The Liberals in Power 1905–1914*, London, 1963
Esher, Lord [2nd Viscount], *Journals and Letters*, London, 1934
Gibb, D. E. W., *Lloyd's of London: A Study in Individualism*, London, 1957
Gilbert, Martin, *In Search of Churchill*, London, 1994
—— *Winston S. Churchill*, vols III–VIII, London, 1971–88; also the *Companion Volumes* III–V, London, 1972–82 and their continuation, *The Churchill War Papers*, 2 vols, London 1993, 1994

Hankey, Lord, *The Supreme Command*, 2 vols, London, 1961
Hansard: Parliamentary Debates: Commons, various dates
Harrisson, Tom, *Living Through the Blitz*, London, 1976
Jenkins, Roy, *Asquith*, London, 1964
—— *Churchill*, London, 2001
Keesing's Contemporary Archives, various dates
Langworth, Richard M., *A Connoisseur's Guide to the Books of Sir Winston Churchill*, London and Washington DC, 1998.
Lash, Joseph P., *Eleanor and Franklin*, London, 1972
Lavery, Sir John, *The Life of a Painter*, London, 1940
Lee, Celia, *Jean, Lady Hamilton 1861–1941: A Soldier's Wife*, London, 2001
Lee, John, *General Sir Ian Hamilton 1853–1947: A Soldier's Life*, London, 2000
Lees-Milne, James, *Diaries: Caves of Ice, 1946–1947*, London, 1983; *A Mingled Measure, 1953–1972*, London, 1994; *Deep Romantic Chasm, 1979–1981*, London, 2000
Leslie, Anita, *Jennie: The Life of Lady Randolph Churchill*, London, 1969
Lloyd George, David, *War Memoirs*, 6 vols, London, 1933–6
Longford, Elizabeth, *A Pilgrimage of Passion*, London, 1979
Longmate, Norman, *How We Lived Then*, London, 1971
Lovell, Mary S., *The Mitford Girls*, London, 2001
Lovell, Richard, *Churchill's Doctor: A Biography of Lord Moran*, London, 1992
McCallum, R. B. and Readman, A., *The British General Election of 1945*, London, 1947
Maclean, Fitzroy, *Eastern Approaches*, London, 1956
Macmillan, Harold, *Tides of Fortune*, London, 1969
Magnus, Sir Philip, *King Edward VII*, London, 1964
Marsh, Sir Edward, *A Number of People*, London, 1939
Martin, Sir John, *Downing Street: The War Years*, London, 1991
Minney, L. J., *No. 10 Downing Street*, London, 1963
Montague Browne, Sir Anthony, *Long Sunset*, London, 1995
Moran, Lord, *Winston Churchill: The Struggle for Survival 1940–65*, London, 1966
Nicolson, Harold, *Letters and Diaries 1939–45*, London, 1967
O'Brien, Terence, *Civil Defence* (Official History of the Second World War series), London, 1955
Oxford Companion to the Second World War, Oxford, 1995
Pakenham, Simona, *Pigtails and Pernod*, London, 1962
—— *60 Miles from England: The English at Dieppe 1814–1914*, London, 1967
Pankhurst, E. Sylvia, *The Suffragette Movement*, London, 1931
Panter-Downes, Mollie, *London War Notes 1939–45*, London, 1972
Pawle, Gerald, *The War and Colonel Warden*, London, 1963
Pearson, John, *Citadel of the Heart: Winston and the Churchill Dynasty*, London, 1991
Pelling, Henry, *Winston Churchill*, London, 1974
Quinn, James, SJ, *Cavalier of Christ: John Ogilvie, Priest and Martyr*, Glasgow, 1976
Rhodes James, Robert, *Churchill: A Study in Failure 1900–1939*, London, 1970
—— 'Churchill the Outsider', series of four articles in *The Observer*, 4–25 Sept. 1966
Rhodes James, Robert (ed.), *'Chips': The Diary of Sir Henry (Chips) Channon*, London, 1967

Riddell, Lord, *More Pages from My Diary 1908–1914*, London, 1934
Rothenstein, John, *Time's Thievish Progress*, London, 1970
Sherwood, Robert B., *The White House Papers of Harry L. Hopkins,* 2 vols, London, 1948, 1949
Soames, Mary (ed.), *Speaking for Themselves: The Personal Letters of Winston and Clementine Churchill*, London, 1998
—— *Winston Churchill: His Life as a Painter*, London, 1990
Steen, Marguerite, *Pier Glass: More Autobiography*, London, 1968
Suckley, Margaret, *Closest Companion: The Unknown Story of the Intimate Friendship between Franklin Roosevelt and Margaret Suckley*, ed. Geoffrey C. Ward, Boston and New York, 1995
Taylor, A. J. P., *English History 1914–1945*, London, 1965
The Times Diary and Index of the War, London, 1921
Tree, Ronald, *When the Moon was High*, London, 1975
Tunney, Christopher, *Biographical Dictionary of World War II*, London, 1972
Vickers, Hugo, *Cecil Beaton: The Authorized Biography*, London, 1985
Webb, Beatrice, *Our Partnership*, London, 1948
Welcome, John, *The Sporting Empress: The Story of Elizabeth of Austria and Bay Middleton*, London, 1975
Wharton, Edith, *A Backward Glance*, New York and London, 1934
Wheeler-Bennett, Sir John (ed. and intr. by), *Action This Day: Working with Churchill*, various contributors, London, 1968
Wilson, W., *The House of Airlie*, 2 vols, London, 1924
Winant, John G., *A Letter from Grosvenor Square*, London, 1947
Woods, Frederick, *Bibliography of the Works of Sir Winston Churchill*, rev. edn, London, 1969
Young, Kenneth, *Churchill and Beaverbrook*, London, 1966

Newspaper Sources

Principally *The Times*, but most national papers consulted at various dates; also, among others, *Belfast Weekly News*; *Dundee Advertiser*; *Dundee Courier*; *Woodford Guardian*; *Woodford Times*, for specific reports.

Index

The following abbreviations have been used for members of the family: CSC (Clementine Hozier/Spencer-Churchill); DC (Diana Churchill/Bailey/Sandys); MC (Mary Churchill/Soames); RSC (Randolph Spencer Churchill; SC (Sarah Churchill/Oliver/Beauchamp/Audley); WSC (Winston Spencer Churchill). Other abbreviations are: C (Churchill); H (Hozier); HH (Henry Hozier); WWI/WWII (First and Second World Wars). Maiden names, knighthoods, peerages and service ranks acquired later than their mention in the text appear in brackets. The courtesy title 'Hon.' has been omitted throughout.

With some important exceptions, the correspondence quoted between CSC and close members of the family is not entered separately but is incorporated into relevant subject-headings.

© ROBIN MATTHEWS

Mary Soames, born in 1922, is the youngest and only surviving child of Winston and Clementine Churchill. During World War II she served in mixed antiaircraft batteries in England and northwestern Europe and accompanied her father as an aide on several wartime overseas journeys. In 1947 she married Captain Christopher Soames, later Lord Soames, the politician and diplomat, a vice president of the European Commission and the last governor of Southern Rhodesia; he died in 1987. The Soameses have five children. Mary Soames is the editor of *Winston and Clementine: The Personal Letters of the Churchills* and the author of *A Churchill Family Album, The Profligate Duke,* and *Winston Churchill: His Life as a Painter*.

WINSTON AND CLEMENTINE: THE PERSONAL LETTERS OF THE CHURCHILLS

A LOS ANGELES TIMES BEST BOOK OF THE YEAR

"Beautifully edited . . . a sweeping yet accessible view of British politics in the twentieth century." —*Washington Post Book World*

A fascinating read of more than eight hundred intimate letters of the Churchills, edited by their daughter. ISBN 0-618-08251-4

WINSTON CHURCHILL CLASSICS

THE SECOND WORLD WAR

Churchill's six-volume history of World War II

"A major document of our time." —*Time*

VOLUME I: THE GATHERING STORM The step-by-step decline into war, with Churchill becoming prime minister as "the tocsin was about to sound." ISBN 0-395-41055-X

VOLUME II: THEIR FINEST HOUR The eight uneasy, dangerous months from May to December 1940, as Britain stands isolated and Germany follows its path to war. ISBN 0-395-41056-8

VOLUME III: THE GRAND ALLIANCE The momentous year 1941: the *Bismarck* is sunk, Hitler marches on Russia, Japan strikes Pearl Harbor—and the Grand Alliance is formed. ISBN 0-395-41057-6

VOLUME IV: THE HINGE OF FATE From uninterrupted defeat to almost unbroken success: a year when Rommel is gradually thrown back in North America, and in the Pacific the tide turns. ISBN 0-395-41058-4

VOLUME V: CLOSING THE RING The drive to victory between June 1943 and July 1944, as the Allies consolidate their achievements, with enormous difficulty and great divergence of opinion. ISBN 0-395-41059-2

VOLUME VI: TRIUMPH AND TRAGEDY From the Anglo-American landings in Normandy in June 1944 to the unconditional surrender of Japan in September 1945. ISBN 0-395-41060-6

ALSO AVAILABLE AS A SIX-VOLUME BOXED SET. ISBN 0-395-41685-X

MEMOIRS OF THE SECOND WORLD WAR

The quintessence of the war as seen by its greatest player, in a one-volume abridged edition that captures all the drama of the original volumes. ISBN 0-395-59968-7